TOM COPE

The Wild Flora *of* Kew Gardens

A Cumulative Checklist from 1759

Kew Publishing
Royal Botanic Gardens, Kew

First published in 2009 by
Royal Botanic Gardens, Kew
Richmond, Surrey, TW9 3AB, UK

www.kew.org

ISBN 978 1 84246 401 4

British Library Cataloguing in Publication Data
A catalogue record for this book is available from the British Library.

Production editor: Ruth Linklater
Typesetting and page layout: Margaret Newman
Publishing, Design & Photography
Royal Botanic Gardens, Kew

Cover design: Lyn Davies

Printed and bound in the United Kingdom by Cambrian Printers Ltd

For information or to purchase all Kew titles please visit
www.kewbooks.com or email publishing@kew.org

All proceeds go to support Kew's work in saving the world's plants for life

The paper used in this book contains wood from well-managed forests, certified in accordance with the strict environmental, social and economic standards of the Forest Stewardship Council (FSC).

Contents

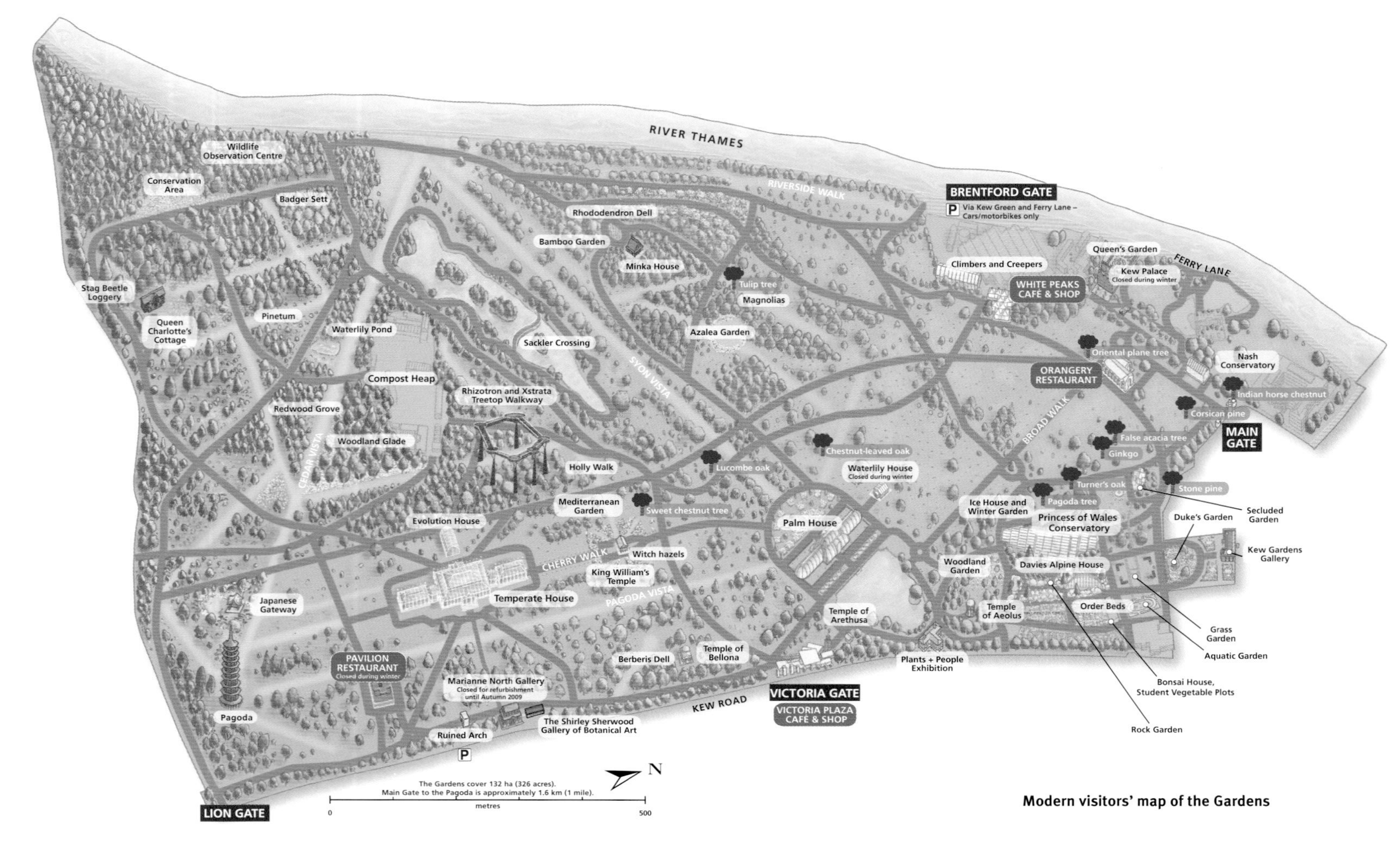

Modern visitors' map of the Gardens

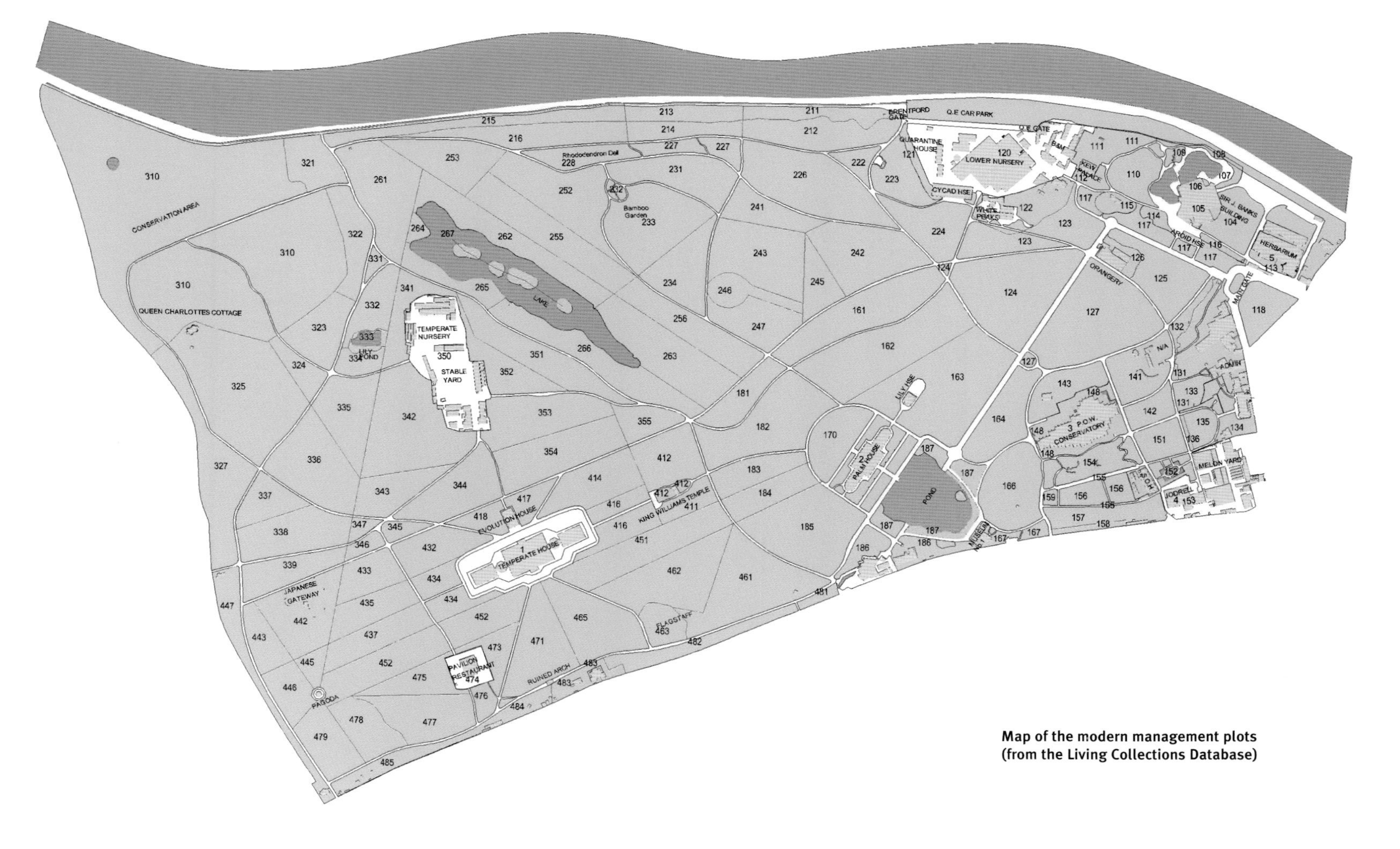

Map of the modern management plots (from the Living Collections Database)

Foreword

Our 250th Anniversary Year provides an eminently suitable time in which to celebrate every aspect of the rich legacy left to us at Kew by history. This publication detailing all the wild and naturalised plants which have been found at Kew over the last 230 years, both celebrates that legacy and becomes an important element in the inheritance we will leave for future generations. It is the culmination of an interest in Kew's flora sustained over 35 years while Tom worked as a grass taxonomist in the Herbarium. It all began with informal surveying and recording then gathered momentum during the 1990s with a full plot by plot survey which took over three years to complete since it was done in Tom's own time, early in the mornings, at lunchtimes and after work. Formal recognition came in 2000 when surveying was absorbed into Tom's work. Since then he has also completed a survey of the wild plants of the Royal Mid-Surrey Golf Club since its estate and flora are contiguous to Kew's. While so much green space in lowland Britain has been irreversibly modified by so-called improvements to pasture and crop production areas involving fertilisers and herbicides, Kew's flora has escaped and remains relatively diverse. Plants that were common when Tom and I were children have become rare, but some of these persist at Kew, whether by purely good fortune or overtly careful habitat management. There have, of course, also been not a few new introductions of species 'alien' to the British flora, some of which have become invasive. These facts make the Royal Botanic Gardens and its environs much more interesting botanically than might be expected of this West London acid grassland habitat on Bagshot Sands & Gravels. Thus, I am delighted to see the results of such a long and devoted interest published and all the more since it will provide an invaluable aid in conserving our wildflowers for the future.

Dr Nigel Taylor, Curator, RBG Kew

INTRODUCTION

After several decades of *ad hoc* observations, a new survey of the wildlife of Kew Gardens was inaugurated by the newly-formed Wildlife Recording Group in 1998. The WRG was not only interested in those species currently found wild within the Gardens, but also those that had been recorded in the past, and to this end I began to accumulate historical records of the wild flora dating back as far as the herbarium and literature would take me. It was my original intention to catalogue just those species that have been seen growing wild in the Gardens but it soon became evident that there is a vast grey area between the truly wild species and those that had — or might have — escaped from cultivation. As the scope of my historical research expanded it became clear that since almost the whole of the known British flora had been in cultivation from the very beginnings of the Botanic Garden there must be doubt about the provenance of any species found in the wild. I have, therefore, elected to consider every British species that occurs, or has occurred, within the Gardens whatever its original source, although species from Ireland and the Channel Isles not known as wild plants in Great Britain have been excluded.

The catalogue includes natives, archaeophytes (naturalised before 1500AD), neophytes (naturalised since 1500AD) and casuals currently on the British list, and any other exotics that have escaped from cultivation. (For the purposes of this catalogue natives and archaeophytes are both deemed to be native.) It covers all plants that have ever been recorded growing in a wild state within the Gardens or their periphery, and all natives cultivated in formal beds or other plantings, either currently or in the past, for which there is reliable documentation.

BOTANICAL RECORDING AT KEW

The earliest known wild-collected specimens from Kew are those in the herbarium of Bishop Samuel Goodenough (1743–1827) which are provisionally dated at between 1779 and 1805. It is not always possible to determine exactly where or when these plants were found (and they may not all have been Goodenough's own collections since he moved to Rochester in 1802), but the oldest surviving species that can be found among his specimens seems to be *Leersia oryzoides*. This plant occurs on the Royal Mid-Surrey Golf Course, an area that in Goodenough's time was very much a part of the Royal Estate, and although the pond in which it now grows is relatively new, it surely cannot be far from the site of the original find. Goodenough found *Chenopodium vulvaria* by Kew Bridge in 1779 and this is the earliest specimen we have with a known date of collection. The earliest record I can trace of a species that came from the Gardens as we know them today is that of *Arabis glabra* found on the boundary wall by Borrer in 1805, but both the *Arabis* and the *Chenopodium* are now extinct in the area. In that same year — 1805 — Goodenough recorded *Trifolium glomeratum* on Kew Green where it can still be found today. The next record did not appear until 1833 when an unknown collector found *Juncus bufonius* in 'Kew'. Thereafter, observations slowly gained momentum until ultimately George Nicholson undertook the first full botanical survey of the Gardens (see below). Prominent among the early collectors was John Gilbert Baker (1834–1920) who also identified many of Nicholson's plants in the fledgling Kew Herbarium.

The first full vascular plant survey of the Gardens was that mentioned above undertaken in the years 1873 and 1874 by George Nicholson (1847–1908). Nicholson joined the staff in 1873 and was curator of the living collections from 1886 until 1901. The results of his survey were published in 1875 (see below, p. 21 [para. 2], for details), but there are no supporting exsiccatae in the Kew herbarium. He continued to keep records of the wildlife after its publication and a revised list, which also included non-vascular cryptogams, animals and fungi, was published in 1906. Nicholson himself was too ill to complete this part of the project and it was left to his colleagues to do so on his behalf. The second list was edited from his notes initially by Henry Harold Welch Pearson (1870–1916), until he left the country in 1903, and then by Sir William Turner Thistleton-Dyer (1843–1928), who had just retired as Director of the Gardens.

New records of plants, animals and fungi have intermittently been noted since 1906, and many were published in an occasional series of supplements in the *Bulletin of Miscellaneous Information* and its successor, *Kew Bulletin*. The first 17 of these supplements appeared in quick succession between 1906 and 1922 and then resumed, after a long gap, with a further two just before the Second World War. It is not clear who had assumed editorial responsibility for the early supplements (presumably it was the editor of the *Bulletin*) but after the War the task was taken on by Herbert Kenneth Airy Shaw (1902–1985). After his death he was succeeded by Bernard Verdcourt until ill-health finally forced him to give up in 2008. Since 2001, information collected on the wildlife of Kew has also been entered into a Wildlife Database.

According to current taxonomy, Nicholson recorded 394 vascular plant taxa in his 1875 checklist. He concentrated on natives and naturalised exotics but deliberately excluded all native species known only from cultivation with the exception of *Sonchus palustris, Lysimachia vulgaris* and *Typha angustifolia*. These three occurred in a 'half-wild condition' and he expected the last two, at least, to hold their own (as indeed they have; the *Sonchus* has not survived). He listed several exotics that were naturalised, but for some reason excluded *Luzula nivea* except in an observation in his Introduction, and he deliberately omitted a large number of casuals which were mostly confined to the neighbourhood of the Herbaceous Ground (today's Order Beds). The 1906 list added a further 36 taxa although seven from his first list were inexplicably omitted. Not all of Nicholson's records were, however, the first for Kew Gardens and their immediate vicinity. Table 1 shows the history of botanical recording in terms of new records and includes accepted subspecies, varieties and hybrids.

Table 1. History of plant recording in Kew Gardens before and after the new survey began; first records only and excluding species known only from cultivation.

Era	New taxa recorded
Pre-Nicholson (1779–1874)	98
Nicholson's first survey (1875)	354
Between Nicholson's lists (1876–1905)	64
Nicholson's revised list (1906)	25
Post-Nicholson, before New Survey	306
Sub-total	847
New Survey	185
Total	1032

The changing wild flora of Kew Gardens

The new survey, now well advanced, has revealed a significant decline in the diversity of Kew's wild flora. Of the 430 taxa originally recorded by Nicholson only 307 (71%) are still present. Of the taxa in this catalogue that were noted before the new survey began 43% have not yet been refound, and some of these were first recorded only as recently as 1990 when the grassland area between the Herbarium and the Banks Building was restored. To some extent, the loss has been balanced by new discoveries made during the course of the new survey, but these number only 185. Of Nicholson's 430 taxa, 92% were native; of the total taxa recorded in this catalogue only 65% are native; and of the taxa added in the new survey only 36% are native. Of all the native species once recorded from the Gardens 35% have been lost. However, these losses must be considered in a wider context: outside of the Gardens and their immediate vicinity, loss of native flora has been almost total.

There can be little doubt that modern horticultural practices are unsympathetic to the survival of wild plants, especially in what is essentially a garden, but the expansion of building works has taken a further toll. When Nicholson prepared his first list no extensions had yet been added to the Herbarium (the first was opened in 1877, three years after the survey), and the building itself was apparently surrounded by open woodland with cowslips and primroses. There are now four wings to the Herbarium (with a fifth under construction), a staff car park and a lawn (this itself badly damaged by further building works in the late 1990s and again in the early 2000s before restoration in 2009). Much of the remainder of the once extensive grounds of Kew Palace are now occupied by the Sir Joseph Banks Building and its two ponds. The grassland between the Herbarium and the Banks Building, once very rich in wild plants, is now partially buried beneath a temporary car park with the remainder derelict. (This area has been known to staff of the Herbarium for many years as 'The Paddock' but the name is unofficial, no-one knows where it came from and it does not appear on any map; it is, however, useful as there appears to be no other name for the area and it is used throughout this Catalogue.) The gravel paths that Nicholson knew are now covered in asphalt (except in the Conservation Area). On a more positive note, the digging of gravel to provide the foundations of new glasshouses has resulted in the creation of two new wetland areas: the Lake was dug partly to supply gravel for the Temperate House plinth (but replaced open heathland containing gorse and broom); and the Brick Pit (or Gravel Pit) in the Conservation Area was largely the result of gravel extraction for the foundations of the Alpine House (which has recently been demolished to make way for an extension to the Jodrell Laboratory). The Palm House Pond, reputed to have been based on an existing ox-bow lake of the River Thames, is but a shadow of its former self; at one time it reached almost to the Holly Walk and comprised 6 acres of water with a 3-acre island accessed by an ornate bridge.

The natural grasslands have suffered most, however, from their conversion — through the progression from hand-cutting with scythes to closer and more frequent cutting with ride-on power mowers — to ornamental, close-cropped lawns (although the 'Ancient Meadow' near the Isleworth Gate is still in good condition; a patch of acidic grassland within the Conservation Area is notable for its waxcap fungi; and a small area of grass to the north of the Rhododendron Dell is remarkable for its distinctive physiognomy). In the 18th century the meadows were grazed by sheep and the pastures by cattle; the meadows were also scythed and swept every two weeks (more frequently in summer) and fed only by the sheep-droppings. The removal of hay began to decline with the end of hand-cutting in the mid- or late 19th century when horse-drawn mowers were first introduced. Indeed, many areas of grassland, particularly in the southern two-thirds of the Gardens, provided both hay and grazing, and horses were a feature of the landscape until the death of the last one in 1961. Many

The recently cut 'Ancient Meadow' near Isleworth Gate. The rhododendrons to the left have since been removed.

The Mycology Building, now demolished, was the only known site of wild clary (*Salvia verbenaca*) in the Gardens, visible in the foreground.

modern rotary and cylinder mowers tend to leave their clippings in place and the resultant mulch enriches the soil, suffocating the smaller herbs and leading inevitably to a decline in floristic diversity. Coarse grasses, clovers, docks and thistles benefit from the increased fertility to the exclusion of less aggressive, and often more desirable, species. In parts of the Gardens suffering from heavy public usage, the worn or damaged areas of grass must be quickly repaired by turfing or reseeding with commercial seed-mixes instead of allowing the natural communities to regenerate over a longer period from the seedbank in the soil and this can introduce not only new species into the Gardens but also different genotypes of existing species. The light, sandy soil of Kew is prone to compaction by heavy machinery and this too contributes to the changing floristic profile. Because of falling staff numbers and constantly rising costs there has been an inevitable increase in the use of herbicides and imported topsoil which also effect the natural flora. The past decline in the wild flora of Kew was paralleled by a decline in its fauna, particularly amongst invertebrates, although birds also suffered, but despite that Kew is still an important feeding and breeding station linking Richmond Park and rural Surrey beyond with more urban areas, especially the Wetland Centre at Barnes. Recent surveys of invertebrates have demonstrated an astonishing richness in various groups especially, among others, soldier-flies, butterflies, dragonflies, damselflies and spiders.

Maiden pink, mistletoe and adder's-tongue, as wild plants, have probably gone from the Gardens for ever (although the last of these is cultivated in the Woodland Garden; attempts to reintroduce mistletoe have mostly failed but the latest reintroduction programme is looking more promising). Goldilocks buttercup was thought to have been lost, but was refound in another area after passing un-noticed for twenty-five years. Wild clary, a relative newcomer in the 1930s, barely survived in grass outside the Mycology Building in an area scheduled in 2006 for redevelopment, but the plants were taken into cultivation and its seeds have been banked. Seedlings have since been planted in other areas and a single individual, possibly resulting from seed in soil moved from the original site during building operations, appeared in 2008 at the opposite end of the Herbarium. Subterranean clover, once described by Nicholson as 'nearly as common as White Clover', is now found in only a handful of localities, all of them vulnerable. The same is true of meadow saxifrage which was once so abundant that 'the grass was coloured white by its flowers'. Common centaury, rough chervil, broad-leaved helleborine, green-winged orchid, petty whin, hoary plantain, clustered bellflower, greater burnet-saxifrage and both native species of dodder, to name but a few, have not been seen for many years and have probably also been lost.

An equally worrying problem, shared with other gardens, and indeed much of the rest of the country, is the increase of certain alien species. Perfoliate Alexanders, least duckweed, California brome, Canadian fleabane and the bramble 'Himalayan Giant' have increased to plague proportions

Wild clary (*Salvia verbenaca*) in front of the partially demolished Mycology Building.

Cut-grass (*Leersia oryzoides*) with pickerelweed (*Pontederia cordata*) in a pond covered in duckweed (*Lemna* spp.).

in recent years, while giant hogweed and Japanese knotweed are quietly waiting in the wings. Floating pennywort has already shown up in the Moat but, thankfully, New Zealand pygmyweed has not yet made an appearance.

At times, least duckweed is so abundant that it forms an unbroken carpet over the surface of the water. The lower pond outside the Banks Building has been covered for several years, summer and winter, and nothing grows in the water beneath. Waterfowl and ornamental fish have always been popular with the visiting public but their excrement, as well as uneaten food offered by the public, have contributed to eutrophication of the water of the Palm House Pond and the Lake leading to a decline in the submerged aquatic vegetation. The Lake itself has recently been drained for restoration and although its marginal flora was lifted and later replanted the submerged aquatics were lost. In the 1840s Canadian waterweed was said to be in all the ponds in Kew and was a particular problem in the Palm House Pond, but it can no longer be found anywhere.

On the more positive side, since the National Heritage Act of 1983 there has been a requirement for Kew to safeguard its biodiversity alongside its collections, landscapes and buildings, a move subsequently reinforced by Kew's award of World Heritage Site status. Pieces of grassland in many of the lawns are left uncut for certain periods (six months, one year and two years) and this has had a significant beneficial effect on invertebrate populations; it is vital for the health of the Gardens that this practice should continue. Low-intervention management of part of the Queen's Cottage Grounds (the Conservation Area) has maintained a high level of biological diversity (although perfoliate Alexanders and the bramble 'Himalayan Giant' are a serious problem) and plans are in hand to make further upgrades: an insect loggery has already been installed, the pond is due for cleaning and repuddling and a new wildlife pond has recently been created nearby. Periodic forays are made into the Conservation Area to pull out perfoliate Alexanders and it does appear to be declining. Botanically, this area is consistently the most diverse at Kew; the highly disturbed lawns around the Herbarium and in the Paddock have held a higher cumulative species-count, but high counts here are sporadic and the number of species recorded has steadily declined in recent years. However, plants are amazingly resilient and given the right conditions many species that have been lost may well begin to regenerate naturally.

The Paddock immediately after its restoration in 1990 following construction of the Banks Building (visible in the background).

Bluebells carpeting the Queen's Cottage Grounds in spring.

Bluebells in front of Queen Charlotte's Cottage in spring.

Cultivation of the British flora at Kew

The Botanic Garden at Kew was founded in 1759 by Princess Augusta, wife of Frederick, Prince of Wales and mother of George III. Prince Frederick acquired the Kew Estate in 1727, the same year that his parents acceeded to the British throne as King George II & Queen Caroline and in which he was installed as Prince of Wales. Frederick married Augusta in 1736 but in 1751, at the age of 44, he died (allegedly after being struck on the head, or chest, while playing cricket at Cliveden, but the story may be apocryphal; he probably died of pneumonia). Throughout the 1750s and 1760s Augusta continued to create Frederick's dream garden and the walled 9-acre Botanic Garden was formally inaugurated in 1759 with its first curator, William Aiton, being appointed in 1761.

Princess Augusta died in 1772 and the Kew Estate was inherited by her son and daughter-in-law, King George III & Queen Charlotte. Aiton was unable to cope with the full responsibility of the job after Augusta's death so Joseph Banks (1743–1820) was appointed as overall manager, a post he held for nearly 50 years. In 1831 the walls around the Botanic Garden were demolished to allow its expansion. In 1841 William Jackson Hooker (1785–1865) was appointed Kew's first official Director and he immediately asked Queen Victoria for 17 acres to be added to the Botanic Garden. The Queen responded by giving him an impressive 46 acres, and the rest of the Pleasure Ground was handed over during the next four years. By 1845 the Pleasure Ground, now including the whole Botanic Garden, comprised 178 acres. Between 1845 and 1851 Hooker was also in control of the 350-acre Deer Park to the south of the Gardens, but this was lost as the result of a reorganisation of Government departments. Kew was handed 13.5 acres of the Deer Park in order to straighten the boundaries between the two properties and part of this parcel of land has been incorporated in the Conservation Area; the rest comprises the Ancient Meadow adjacent to Isleworth Gate. There was always some doubt about who was responsible for the management of this awkward parcel of land and as a consequence very little was ever done to develop it. Today it remains botanically one of the most valuable parts of the Gardens and the northern end, immediately south of Isleworth Gate, comprises almost unadulterated grassland. Ultimately, with the incorporation of the Palace Grounds, the land around Queen Charlotte's Cottage, and various parcels of land still owned by the House of Hanover, the Gardens reached their present size of 300 acres.

In 1789 Aiton published his *Hortus Kewensis*, a catalogue of the plants that were being cultivated in the Botanic Garden and among them were 1043 British natives and archaeophytes, a significant proportion (c. 95%) of the known British flora at the time. Aiton died in 1793 and was succeeded by his son, William Townsend Aiton. Between 1810 and 1813 the younger Aiton published a revised catalogue (*Hortus Kewensis* ed. 2); the number of British taxa in cultivation had now grown to 1201, rising to 1222 on publication of the *Epitome* in 1814. The geographical origin of each species was given in the lists but there was no indication of the origin of the actual specimens in cultivation. Thus, the entry for *Draba aizoides* (see below) reads 'native of the Alps of Europe' in the first edition, but 'native of Wales' in the second. There is nothing to suggest that the material in cultivation at the time of the second edition was not the same European stock as that growing in 1789.

A number of species brought into cultivation at the beginning of the Botanic Garden were recorded by Aiton as exotic, but were subsequently discovered to be British natives. Among them were *Draba aizoides*, noted by Aiton as a European species in 1789 but found wild in Britain in 1805, *Lychnis alpina* (found in 1811), *Erica ciliaris* (found in 1829), *Monesis uniflora* (found in 1793) and several others. Between them, the Aitons reported a further 109 species in cultivation that were thought to be British natives but are now believed to be introductions; among these were *Asarum europaeum, Aristolochia clematitis, Epimedium alpinum, Roemeria hybrida, Holosteum umbellatum, Coronopus didymus* and *Cyclamen hederifolium*; a few of these were corrected in the second edition of *Hortus Kewensis* and a few others in the *Epitome*.

Aiton's *Hortus Kewensis*, though well-known, was not the first such catalogue of Kew's cultivated flora. In 1768 John Hill published a list, also entitled *Hortus Kewensis*, but it provided almost nothing on the provenance of any of the plants. They are listed simply by binomial and it has been largely a matter of guesswork to interpret the names (and doubtless numerous errors will eventually come to light). Nevertheless, Hill reported 876 natives and archaeophytes in cultivation. Between them, the various versions and editions of *Hortus Kewensis* recorded 1256 British natives in cultivation between 1768 and 1814.

Bluebells, perfoliate Alexanders and ramsons in the Queen's Cottage Grounds in spring.

Of all the taxa listed in this new catalogue 42% are British natives and archaeophytes known only from cultivation. Of those taxa observed in the wild, only 13% have not knowingly been cultivated at some time since 1759.

The proportion of British species in cultivation has fluctuated over the years and is probably lower now than it has ever been; according to the Living Collections Database there are currently 613 native and archaeophyte taxa in cultivation, just 70% of the number recorded by Hill. The areas around the Order Beds (formerly the Herbaceous Gound) have been the source of many records of plants that have escaped from these beds and many of them have subsequently spread to other parts of the Gardens; some of them would have been known British natives when brought into cultivation. *Myosurus minimus*, for example, undoubtedly originated in the Gardens as an escape from cultivation since there have never been records of it in the wild from the area, but other species which could well be genuinely native in the Gardens may also have originated as escapes. Some species in cultivation may even have been sourced from wild plants within the Gardens. There is now no way of being sure. Nicholson was quite careful about distinguishing native from escape and any native plants recorded from the Gardens before 1875 but not listed by Nicholson are very likely to have been escapes (or from beyond Nicholson's boundaries of the Gardens).

The boundaries of Kew Gardens

The recording areas employed by Nicholson in his original lists were abbreviated as shown in Table 2. Their boundaries are apparent on a map prepared by the Office of Works and published in 1885 (Map 1).

The Herbarium and Palace grounds were bounded in parts by a hedge, a wooden fence and a ditch; the boundary ran more or less eastwards from Brentford Gate almost to the Broad Walk, then more or less northwestwards parallel to the Broad Walk and finally northeastwards to run along the present boundary of the public Gardens north of the Nash Conservatory (the former Aroid House) towards Ferry Lane. The Botanic Gardens proper, in the northern part of the Gardens, were bounded by a wire fence, erected in 1843, that ran from the middle of the southern boundary of the Palace grounds in an arc more or less southwards towards King William's Temple and then eastwards to the Unicorn Gate. The Queen's Cottage Grounds were bounded by an open rustic trellis — which in 1845 replaced a tall wooden fence — that ran in an arc from the Isleworth Gate to a point midway between the cottage itself and the Oxenhouse Gate. The Arboretum comprised the remaining area of the southern part of the Gardens. The Strip ran along the river frontage from Brentford Gate to the sunken fence (ha-ha) that in 1845 replaced the existing wall and fence that separated Kew Gardens from the Old Deer Park.

As far as other literature and herbarium records are concerned, the main difficulty has been that of judging when to include and when to exclude an observation. From the table on p. 11 it is clear that Nicholson did not stray beyond the accepted boundaries of the Gardens except along the 'Strip.' Other recorders were often less precise; 'towpath between Kew and Richmond' could equally apply to the stretch alongside the Old Deer Park south towards Richmond as it could to Nicholson's 'Strip' between the Old Deer Park and Brentford Gate. In order that the historical list be comparable with the new survey some minor extensions to the boundaries of the Gardens have perforce been made, thus embracing several species not recorded by Nicholson. Any records from the Mortlake side of Kew Bridge have been ignored, but the whole of the towpath between the Old Deer Park and the Bridge has been included. For practical purposes the riverside limit of the Gardens has been taken as the Surrey/Middlesex boundary along the River Thames except for a small diversion to remain mid-stream on the Surrey side of Lot's and Brentford (or Kew) Aits; the siting of this boundary would account for the record of the Common Seal to be found in Supplement 34. All records from Kew Green (including St Anne's Churchyard) have been allowed except when it was clear that they came from the eastern side of Kew Road. Westerley Ware was the land between Kew Green and the river; it stretched from Kew Ferry (about where the Queen Elizabeth Gate now is) to just beyond Kew Bridge, so part of it would be outside the current recording area. The northern boundary of Kew Gardens strictly lies along the fence beside Birdcage Walk which runs across the Green from the Director's House to Ferry Lane. The eastern boundary of the Gardens follows the wall along Kew Road; and the southern boundary is marked by the sunken fence along the northern edge of the Old Deer Park.

Map. 1. Map of 1885 showing the boundaries of Nicholson's recording areas (1906 revision).

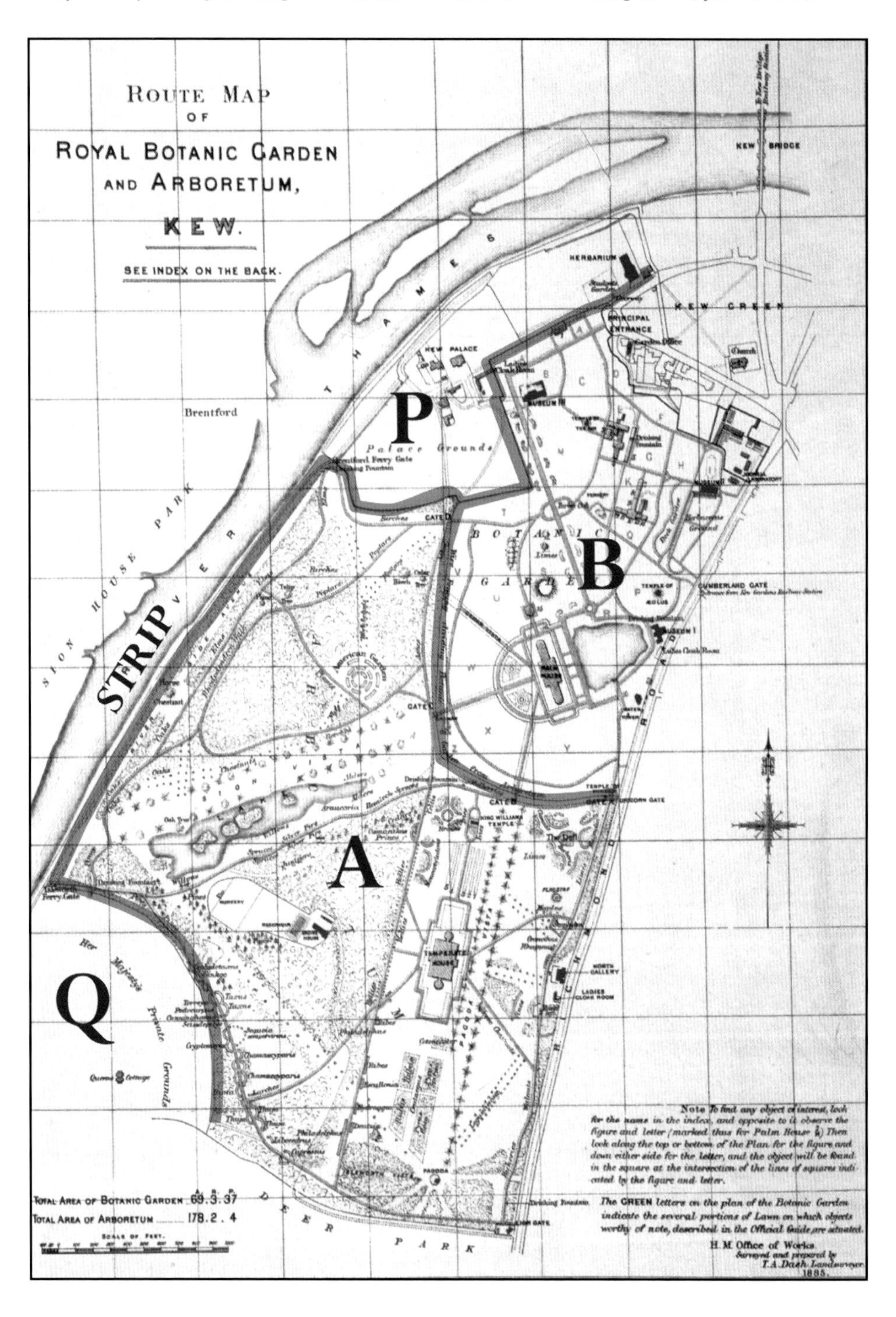

Table 2. Nicholson's subdivisions of Kew Gardens.

1875 survey	Subdivision	1906 revision
B	Botanic Garden proper	B
P	Pleasure Grounds / Arboretum	A
Pal	Grounds about the Herbarium and Kew Palace	P
Q	Queen's Cottage Grounds	Q
Strip	Towpath from Brentford Gate to the Old Deer Park	Strip

For the purposes of listing observations and exsiccatae, the following convention has been adopted:

Kew Gardens, to denote plants known to have been collected or seen within the boundaries of the Gardens as defined by Nicholson, including the towpath between the southern end of Brentford Gate car park and the northern edge of the Old Deer Park;

Kew Environs, to denote plants collected or seen outside of — but in close proximity to — the Gardens, especially from the towpath north of Brentford Gate (but not beyond Kew Bridge), Kew Green, Kew Road (as far as the track to the south of Lion Gate) and the Old Deer Park; also included here are plants of uncertain provenance that may or may not have come from 'Kew Gardens.'

Early localities

Holly Walk, formerly 'Love Lane', which follows the old boundary between the Kew and Richmond Estates.

One of the problems with attempting to rediscover plants in their original haunts is that of knowing exactly where those haunts were. The Gardens have changed out of all recognition since the amalgamation of the two royal estates in 1802. This process began much earlier with an Act of Parliament in 1785 to close Love Lane, a public right-of-way bounded by high walls that ran between the two estates to link Richmond Green with the ferry at Kew and which more or less followed the line of today's 'Holly Walk' (it continues beyond the Deer Park as 'Kew Foot Road' and, on the far side of the A316, as 'Parkshot'). Even since Nicholson's time there have been many significant changes: the removal of internal boundary fences, walls, hedges and ditches, the demolition of the ferneries and T-range, the building of the Princess of Wales Conservatory, Cycad House and two successive Alpine Houses, the Victoria Gate development, a new art gallery and the addition of 4 new wings to the Herbarium are just a few. Some locations have simply changed their names and are not difficult to resolve, but others have been lost and their position can sometimes only be guessed at. Many buildings and topographical features have come and gone over the last two and a half centuries, and the history of the development of the modern Kew landscape is a fascinating study all of its own. Very little of the estate remains in anything approaching a natural condition and the entire landscape has been remodelled more than once. It is impossible even to summarise the history of the Kew landscape in this book and the reader is referred to Ray Desmond's monumental *The History of the Royal Botanic Gardens Kew* (1995, 2007).

The occasion of the confirmation of the rediscovery of cut-grass in the Old Deer Park in 2003. l to r: Margaret Leigh, conservation officer of the Royal Mid-Surrey Golf Club, the author and Madeline Holloway, the ecological surveyor who made the find.

The more significant problem locations are these:

AMERICAN GARDEN. The 1:2500 Ordnance Survey maps of 1893 and 1894 place the American Garden behind the Palm House, on the site of the present Rose Garden (zone 170), where Nesfield intended it to go in his plans of 1848. In a map of the Pleasure Ground prepared by J. D. Hooker in about 1886 the American Garden is indicated on the site now occupied by the Azalea Garden (zone 246). Nicholson was clear that it was in the Botanic Garden near the Palm House and not in the Pleasure Ground.

CANAL BEDS. These are the rectangular beds — immediately south of the Temperate House and now partly grassed over — that the Hookers originally planted with examples of Rosaceae.

DOUGLAS SPAR. Another name for the Flagpole between Victoria Gate and the Marianne North Gallery (zone 463). The Flagpole itself, in an advanced state of decay, was finally removed in 2007.

FILTER BEDS. These were adjacent to the Engine House in the Stable Yard (zone 350).

HOLLOW WALK. This is the current Rhododendron Dell, but when it was constructed by 'Capability' Brown in 1773 it was planted with laurels.

KITCHEN GARDEN. The original walled Kitchen Garden occupied about 10 or 12 acres. It comprised two main sections, known as Methold's Ground (to the west, from near the Main Gate to the present Administration Block) and Home Ground (to the east), separated by a paddock that belonged to the King of Hanover but which was later (1851) incorporated (and is now the site of the Rock Garden and the Duke's Garden; Cambridge Cottage stands at its northern end). In 1899 part of Home Ground became known as the Melon Yard, a name it retains to this day; the remainder became known as the Herbaceous Ground and is currently occupied by the Order Beds and new student allotments. Part of the original boundary wall of the paddock still stands between the Order Beds and the Rock Garden. Methold's House, attached to the Kitchen Garden, is currently occupied by the Director.

MERLIN'S CAVE. The location of Merlin's Cave is not known for certain and was discussed at some length by Ray Desmond (see above). It was designed by William Kent and built in 1735 at the southwest corner of the Duck Pond in Richmond Gardens, but was demolished in 1766 by 'Capability' Brown. It was sited about 300 yards southwest of the 'Hermitage,' another folly also designed by Kent and built in 1731 (its location is probably in or near the Beech Circle north of the Azalea Garden). The Hermitage initially escaped Brown's landscaping of the gardens but, after languishing for many years in a ruinous state, it had finally gone by the mid-nineteenth century. To add to the confusion, the last remnants of the Hermitage were often mistaken for Merlin's Cave. In 1865 an Ordnance Survey map identified a number of stones a few hundred yards northwest of 'Mossy Hill' (Moss or

Mosque Hill, the present site of the Japanese Gateway) as Merlin's Cave, but these were in fact the remains of another building, called 'Stone House' (a stone potting shed built for the royal children). The 1894 O.S. map corrected that error, but put the Cave on the central of the three largest islands in the Lake (this was created in 1856 although not filled until 1861). According to W. J. Bean, *The Royal Botanic Gardens, Kew* (1908), the Cave stood about 50 yards from the southern corner of the Lake in a northeasterly direction, but there is a map in the same book placing it elsewhere, at the southernmost edge of the Lake. It is a mystery that Nicholson should have cited Merlin's Cave on several occasions as though it were an extant locality since it had been demolished some 80 years before he was born. This suggests that he was drawing on earlier records of the Kew flora, but these have not been traced.

MOAT. The ditch bordering the Gardens along the River Thames was constructed in 1767, and subsequently repaired in 1792 and 1810. Nicholson initially called it the Moat but later, and somewhat confusingly, referred to it as the 'Ha-ha.' The boundary between the Gardens and the Old Deer Park is a true ha-ha, being a ditch containing a fence. The Moat is currently a brick- and concrete-lined trench, and the term 'moat' seems more appropriate.

ORCHID HOUSE. The original Orchid House was once a part of the T-Range which was demolished in 1983 to make way for the Princess of Wales Conservatory.

PRINCESS'S GATE. This was an entrance to the Kew Palace Lawn, across the boundary ditch, from the Pleasure Ground. Old Ordnance Survey maps show a break in the ditch due south of the Dutch House (Kew Palace) with a path, known as Princess's Walk, coming up to it from the Rhododendron Dell (according to the Ordnance Survey map of 1894, that is; Kew's own map, issued in the 1960s and 1970s, incorrectly shows Princess's Walk as a northern continuation of the Holly Walk). The path entered the lawn from the Pleasure Ground near Gate D (one of four gates which linked the Pleasure Ground with the Botanic Garden) and this entrance was probably the Princess's Gate. Its position today more or less coincides with the small triangular outlying segment at the southwest corner of zone 124.

RAILWAY GATE. Properly called 'Queen's Gate' this was a proposed new entrance to the Gardens that never quite materialised. With the coming of the London & South Western Railway to Kew in 1869 Joseph Hooker realised that this was an opportunity to link the new 'Kew Gardens' railway station directly with the Temperate House. A local landowner, J.G.D. Engleheart, who was at that time constructing new housing in and around Kew, agreed to support Hooker's plan and The Avenue was laid out as the link. A section of the boundary wall of the Gardens was demolished, the Temperate House Lodge was built as the gatehouse in readiness and the new gates were installed. Shortly afterwards the railway company decided to relocate the proposed station six hundred yards to the north thereby destroying the alignment. The gates, which had never been utilised, were taken down but by public demand the wall was never rebuilt; instead iron railings were installed on a low brick parapet allowing a view of the Temperate House from Kew Road. The gate itself was to have been wide enough to admit carriages, and once it was operating Lion Gate to the south would have been closed (pedestrian access on Kew Road was at that time also available via the now defunct Unicorn Gate; the small Cumberland Gate was enlarged by Engleheart, at his own expense, and was opposite the newly created Kew Gardens Road which then served as the approach to the Gardens from the resited railway station; this gate, too, is now closed and visitors are directed to the newly developed Victoria Gate which connects to the railway station via Lichfield Road. The sections of the old Queen's Gate were brought out of storage in 1889 and installed as the Victoria Gate).

RICHMOND GARDENS. So far, this locality has only been found on a single sheet collected by or for Bishop Goodenough. Richmond Gardens, in his day, comprised the entire riverside estate between Kew Ferry and Richmond Palace. The Gardens were amalgamated with the Kew Estate in 1802 but in 1851 the Department of Woods and Forests, responsible for the running of the Gardens, was split and that part of the Estate known as the Old Deer Park was retained by this department; the modern Botanic Gardens were handed over to the Office of Public Buildings and Works.

SEVEN SISTERS LAWN. To the northwest of the Palm House (probably in zone 162) was a collection of elms known as the Seven Sisters. They were reputed to have been planted by the daughters of George III but the last of them was lost to a storm in March 1916. There is little left to mark their site.

TEMPLE OF MINDEN. Now called the Temple of Bellona; on a mound south of Victoria Gate (zone 461).

WINTER GARDEN. The central block of the newly constructed Temperate House (which, along with the two octagons, was completed in 1862) was originally known as the Winter Garden.

SOURCES OF PLANT RECORDS

1. Literature

Sources of all published records are abbreviated in the text as follows:

JB: the original checklist of the Flora Survey of 1873/4, page numbers only; *Journal of Botany* 13: 9–12, 42–49, 71–77 (1875); all other references to the Journal are accompanied by the volume number.

AS: the revised and updated checklist; *Bulletin of Miscellaneous Information, Kew, Additional Series* 5: 1–223 (1906).

S1–S36: Supplements to the original lists; editor/compiler and contents (P, vascular plants; B, bryophytes; C, other non-vascular cryptogams; F, fungi and lichens; A, animals) are noted below (Table 3) for interest:

Additional records have been sought and often located in the following publications.

Surrey, Kew and the London area:

BFS	J. A. Brewer, Flora of Surrey (1863)
SFS	C. E. Salmon, Flora of Surrey (1931)
LFS	J. E. Lousley, Flora of Surrey (1976)
FSSC	A. C. Lesley, Flora of Surrey Supplement and Checklist (1987)
FLA	R. M. Burton, Flora of the London Area (1983)
EAK	Ecological Appraisal Kew Lake and Pond; Tony Stones for Posford Duvivier (1999)
CHAM	Chamomile (*Chamaemelum nobile* L.) at the Royal Botanic Gardens, Kew; Hilary Wendt for English Nature and Plant Life (Jan. 2000).
DEA	D. E. Allen, The bramble florula of Queen's Cottage Grounds, Kew through a century and a quarter. The London Naturalist 85: 29–38 (2006).

Carpets of snowdrops in the Queen's Cottage Grounds

Table 3. The Supplements to Nicholson's checklists of the wild flora and fauna of Kew Gardens.

Suppl.	Reference	Editor	Contents
I	Bull. Misc. Inform., Kew 1906: 46–47 (1906)	Anon.	F
II	Bull. Misc. Inform., Kew 1907: 97–100 (1907)	Anon.	A
III	Bull. Misc. Inform., Kew 1907: 156–187 (1907)	Anon.	A
IV	Bull. Misc. Inform., Kew 1907: 238–244 (1907)	Anon.	F
V	Bull. Misc. Inform., Kew 1907: 282–283 (1907)	Anon.	A
VI	Bull. Misc. Inform., Kew 1907: 401–403 (1907)	Anon.	A
VII	Bull. Misc. Inform., Kew 1908: 120–127 (1908)	Anon.	P, B, A
VIII	Bull. Misc. Inform., Kew 1908: 272–283 (1908)	Anon.	A
IX	Bull. Misc. Inform., Kew 1909: 243–256 (1909)	Anon.	A
X	Bull. Misc. Inform., Kew 1909: 369–376 (1909)	Anon.	P, F
XI	Bull. Misc. Inform., Kew 1910: 79–84 (1910)	Anon.	A
XII	Bull. Misc. Inform., Kew 1911: 365–377 (1911)	Anon.	P, C, F, A
XIII	Bull. Misc. Inform., Kew 1912: 161–166 (1912)	Anon.	F
XIV	Bull. Misc. Inform., Kew 1913: 195–199 (1913)	Anon.	F
XV	Bull. Misc. Inform., Kew 1917: 73–76 (1917)	Anon.	A
XVI	Bull. Misc. Inform., Kew 1920: 212–217 (1920)	Anon.	P, B, F, A
XVII as 'XVI'	Bull. Misc. Inform., Kew 1922: 189–193 (1922)	Anon.	A
XVIII as 'XVII'	Bull. Misc. Inform., Kew 1936: 60–66 (1936)	Anon.	P, C, F, A
XIX as 'XVIII'	Bull. Misc. Inform., Kew 1938: 390–396 (1938)	Anon.	A
XX	Kew Bull. 3: 113–124 (1948)	H. K. Airy Shaw	F, A
XXI	Kew Bull. 4: 231–238 (1949)	H. K. Airy Shaw	P, F, A
XXII	Kew Bull. 7: 285–288 (1952)	H. K. Airy Shaw	P, F, A
XXIII	Kew Bull. 13: 295–301 (1958)	H. K. Airy Shaw	A
XXIV	Kew Bull. 15: 169–191 (1961)	H. K. Airy Shaw	P, C, F, A
XXV	Kew Bull. 16: 139–146 (1962)	H. K. Airy Shaw	A
XXVI	Kew Bull. 19: 391–397 (1965)	H. K. Airy Shaw	A
XXVII	Kew Bull. 20: 201–231 (1966)	H. K. Airy Shaw	F
XXVIII	Kew Bull. 21: 229–239 (1967)	H. K. Airy Shaw	A
XXIX	Kew Bull. 28: 387–409 (1973)	H. K. Airy Shaw	P, F, A
XXX	Kew Bull. 43: 437–451 (1988)	B. Verdcourt	P, A
XXXI	Kew Bull. 48: 169–184 (1993)	B. Verdcourt	P, F, A
XXXII	Kew Bull. 50: 645–657 (1995)	B. Verdcourt	P, A
XXXIII	Kew Bull. 55: 721–752 (2000)	B. Verdcourt	P, F, A
XXXIV	Kew Bull. 57: 1007–1022 (2002)	B. Verdcourt	A
XXXV	Kew Bull. 59: 639–649 (2004)	B. Verdcourt	P, A
XXXVI	Kew Bull. 64: 183–194 (2009)	B. Verdcourt	P, A

TCL	Transactions. City of London Entomological and Natural History Society (1880–1913) cont. as TLN
TLN	Transactions of the London Natural History Society (1914–1920) cont. as LN
LN/LN:S	The London Naturalist (1921–present); S = Supplement to the volume (with its own pagination)

United Kingdom in general:

BLRC	Botanical Locality Record Club Reports (1874–1887)
WBEC	Watson Botanical Exchange Club Reports (1884–1934)
TBEC	Thirsk Botanical Exchange Club Reports (1859–1866) cont. as LBEC
LBEC	London Botanical Exchange Club Reports (1867–1869) cont. as BEC
BEC	Botanical Exchange Club Reports (1870–1879) cont. as BECB
BECB	Botanical Exchange Club of the British Isles Reports (1879–1900) cont. as BECS
BECS	Botanical Exchange Club and Society of the British Isles Reports (1901–1913) cont. as BSEC
BSEC	Botanical Society and Exchange Club of the British Isles Reports (1914–1947) cont. as WAT
WAT	Watsonia (1949–present)
YB	Year Book of the Botanical Society of the British Isles (1949–1953) cont. as BSP
BSP	Botanical Society of the British Isles Proceedings (1954–1969) amalgamated with Watsonia Vol. 8
BSN	Botanical Society of the British Isles News (1972–present)

For further details relating to Botanical Exchange Club Reports see the Appendix.

2. Exsiccatae

Exsiccatae supporting, or supplementing, literature records are still being sought. At the time of writing only the Kew herbarium (K) has been systematically searched. By chance, several grass species collected by C. E. Hubbard, but not reported to Airy Shaw and therefore absent from the supplements, were discovered in the course of another project and this led to the search. It remains to be seen how many other species have been collected but not reported, and how many have found their way into other herbaria. Herbarium specimens that have been found are cited in chronological order under the same two headings as the observations.

Some of the bramble specimens are mounted on extra large sheets at Kew and are stored separately from those on standard sheets. Such specimens are designated (K*) in the list that follows.

The New Survey

The new survey began late in 1998 and is likely to continue indefinitely. The recording units employed are the numbered management zones of the Living Collections Department (now called HPE, Horticulture and Public Experience). There are 40 of these in North Arboretum, 44 in West Arboretum, 53 in South Arboretum and 16 in Herbaceous. There are a further twelve areas that have no official zone numbers, but some of these are not strictly within the Gardens. In all, 165 zones are being surveyed. Species lists for each zone are cumulative and no attempt is being made to make year by year comparisons. Species numbers per zone range from zero (Palm House Pond) to over 200 (Conservation Area with the lawns around the Herbarium close behind). Individual records are entered into a database and some 20,000 have already been processed. Changes in the landscape of the Gardens are constantly happening and in some cases species found for the first time in the survey have already been lost. The records, however, have not been removed from the database.

Some changes, like the demonstration wheatfield strips along the Broadwalk sown in late 1999 ready for the New Millenium, have been discontinued. A second wheatfield was created in the lawn to the south of the Lake at the end of Syon Vista in 2001. In subsequent years this area was sown with Opium Poppies, used for dumping sludge when the Lake was drained, and most recently put back to grass. In other areas mass planting of native bulbs (wild daffodil — although it transpires that these are in fact Tenby daffodil — wild fritillary etc.) is taking place, increasing the area of doubt about what to record and what to ignore. As stated before, Nicholson avoided most species that were only in cultivation but it is useful to know which native species that might appear to be wild in the Gardens were, in fact, introduced. Baker listed a number of species from 'Hort. Kew' and a number as 'Cult. Hort. Kew', but it is not clear whether the former were cultivated or not; some of them appear to be unlikely natives of the Gardens. It was too tempting to avoid including in the current Catalogue all British natives and archaeophytes that are or have been cultivated, in order to forewarn future generations of wild flora surveyors of the likely source of their records; such plants that have not yet been found wild are included in the text but on a blue background.

Layout of the Catalogue

The plant names used in this catalogue are those currently accepted by C. A. Stace, *New Flora of the British Isles, ed. 2* (1997), with some updates from Stace, *Field Flora of the British Isles* (1999) and in some cases the opinion of specialists at Kew and elsewhere. Names used in the original literature or herbarium sources, where different, are given as synonyms in italics beneath the accepted name. Occasionally the author of a combination used in the early literature could not be determined; the unknown author's name has therefore been substituted by 'auct. dub.' until it can be traced. Accepted English names, again, are mostly taken from Stace.

The list of taxa broadly follows the sequence used by Stace and set out in D. H. Kent, *List of Vascular Plants of the British Isles* (1992). Observations under each taxon are arranged chronologically. Square brackets [ ] in the citations have been used in several ways and indicate one of the following: my own comments; my updating of names from those in the published source; and text imported from elsewhere in the literature, particularly where there were joint observations. Context should make usage clear.

An asterisk (*) against the accepted name indicates taxa that are neither natives nor archaeophytes in the British flora. Entries on a blue background indicate that the species is or has been cultivated in the Gardens but has never been observed as a wild plant; this applies only to natives and archaeophytes of Great Britain. Such entries enclosed in square brackets **[ ]** indicate species thought to have been British natives at the time of their original cultivation but which are now regarded as aliens.

Where information is available details for each main entry are presented under the following headings:

First record. The year that the plant was first recorded growing wild in the Gardens. The date given in square brackets [ ] is that of the first known cultivation of the species at Kew.

Nicholson. George Nicholson's records from the 1873/4 survey and the 1906 revision transcribed in full from the original texts. Other records made by Nicholson are not included here.

References/Other references. All published records, other than Nicholson's surveys, in chronological order. Those from 'Kew Gardens' are placed before those from 'Kew Environs.' Individual records are separated by bullet points (•). Some early, but unpublished, records of my own can also be found here if they predate published records.

Exsicc. All known herbarium specimens of wild collected plants. As far as this draft of the catalogue is concerned these are almost exclusively from the Kew herbarium (K). Arrangement as for 'References' above.

Cult. Summary of the history of cultivation of the species in the Gardens. Historical literature records are shown as year of publication, observation by the author on the status of the taxon (if such is included), and an abbreviated reference as follows:

JH = John Hill, *Hortus Kewensis* (1768); this volume was reissued in 1769 with minor corrections and described as edition 2.
HK1 = W. Aiton, *Hortus Kewensis* (1789), 3 volumes.
HK2 = W. T. Aiton, *Hortus Kewensis*, ed. 2 (1810–1813), 5 volumes.
HKE = W. T. Aiton, *Epitome* (1814).

Herbarium records are shown as date of collection, locality in which grown (if known; otherwise entered as 'unspec[ified].') and herbarium code, e.g. (K). The labelling of cultivated material is uneven and often uninformative. Sometimes labels designed for cultivated specimens have been used for wild-collected plants — both from within and from outside the Gardens — which were subsequently filed in the herbarium under cultivated material. Records of taxa currently in cultivation are taken from Kew's Living Collections Database and are entered as 'In cult.' If there is no record in the database the entry will state 'Not currently in cult.' However, some plantings, especially those for naturalisation, do not bear accession numbers and do not appear in the database.

CURRENT STATUS. A brief observation on the presence and abundance of the taxon as noted in the new survey and some remarks on its probable history. These notes were correct at the time of writing, but details of individual taxa are likely to change on a daily basis as the new survey progresses. Many of these observations say 'No recent sightings' either because the taxon has not yet been refound or because it is genuinely extinct in the Gardens. The word 'extinct' is used only when there is absolute certainty that this is the case.

Note: species entries are numbered sequentially, but a few species were added after completion of the text. These are inserted in the appropriate places with a suffix added to the number of the preceding species.

ACKNOWLEDGEMENTS

My thanks are due to many colleagues at Kew for their support and assistance, not least for the many determinations undertaken on my behalf. Worthy of particular mention are Bernard Verdcourt, Peter Edwards, Sally Bidgood, Kaj Vollesen, Gemma Bramley, Nicholas Hind, Alan Paton and Mike Lock (all Herbarium); Brian Spooner (Mycology); John Lonsdale (HPE); Kate Pickard, Michele Losse, Kiri Ross-Jones and Mandy Ingram (Archives). Thanks are also due to Mark Pitman (formerly of LCD, North Arboretum) for his efforts to save *Salvia verbenaca* when its only remaining site, the grass around the old student allotments, was scheduled for restoration. Peter Crane (former Director of the Gardens), Nigel Taylor (Curator of Living Collections) and Simon Owens (former Keeper of the Herbarium) are acknowledged for their interest and support. My special thanks go to David Allen (Hampshire) whose detailed knowledge of Surrey brambles helped in the interpretation of early records from Kew Gardens, and for his kind offer to catalogue the existing brambles in the Gardens. Mention must also be made of invaluable help I received in 1990 and 1991 from the late Mike Mullin (formerly of the Natural History Museum, London); and of the interest in the wild plants and animals of Kew that I constantly shared with Rupert Hastings until his untimely death in 1993.

I must also thank those who have contributed to or taken an interest in the new survey. To those mentioned above, I should add: Barry Phillips (Wisley RHS Garden and former Surrey County Recorder) for help with surveying aquatics, willows and roses, Eric Clement (Hampshire) for help with aliens, and Sandra Bell (HPE) for organising the boat that made the aquatic survey possible; Sandra, again, for leading the latest effort to conserve *Salvia verbenaca*; Mark Bridger (formerly HPE), Simon Cole and Tony Hall (both HPE) and Colin Clubbe (Conservation) for their interest in conservation aspects arising from the survey; Joy Corbett (HPE) for plant sightings; and Angela Bond and Sharon MacDonald (Herbarium and ex-Herbarium respectively) for helping with the survey itself. Special mention must be made of Audrey Thorne (Spelthorne, Surrey) for painstakingly entering the survey and historical records into the wildlife database. I must also thank Dr Margaret Leigh (Ealing), Conservation Officer of the Royal Mid-Surrey Golf Club, for inviting me to survey the golf course and for giving me free access to the land; this enterprise has allowed me to extend the Kew Gardens survey into an area that retains much of its original grassland and was once an integral part of the Royal Estate.

Finally, I must thank Sandra Bell and Nigel Taylor for vaulable input into the Introduction; and acknowledge Kew's Publications team, Gina Fullerlove, John Harris, Lydia White, Ruth Linklater and Margaret Newman without whose considerable expertise this book would not have been possible.

The Catalogue

CLUBMOSSES

Lycopodiaceae

1. Huperzia selago (L.) Bernh. ex Schrank & Mart.
Lycopodium selago L.
fir clubmoss
Cult.: 1768 (JH: 432); 1789, Nat. of Britain (HK1, 3: 471); 1813, Nat. of Britain (HK2, 5: 495); 1814, Nat. of Britain (HKE: 324). Not currently in cult.

2. Lycopodiella inundata (L.) Holub
Lycopodium inundatum L.
marsh clubmoss
Cult.: 1768 (JH: 432); 1789, Nat. of Britain (HK1, 3: 471); 1813, Nat. of Britain (HK2, 5: 494); 1814, Nat. of Britain (HKE: 324). Not currently in cult.

3. Lycopodium clavatum L.
stag's-horn clubmoss
Cult.: 1768 (JH: 432); 1789, Nat. of Britain (HK1, 3: 471); 1813, Nat. of Britain (HK2, 5: 493); 1814, Nat. of Britain (HKE: 324); 1907, unspec. (K). Not currently in cult.

4. Lycopodium annotinum L.
interrupted clubmoss
Cult.: 1813, Nat. of Britain (HK2, 5: 494); 1814, Nat. of Britain (HKE: 324). Not currently in cult.

5. Diphasiastrum alpinum (L.) Holub
Lycopodium alpinum L.
alpine clubmoss
Cult.: 1768 (JH: 432); 1813, Nat. of Britain (HK2, 5: 493); 1814, Nat. of Britain (HKE: 324). Not currently in cult.

6. Diphasiastrum complanatum (L.) Holub
Issler's clubmoss
Cult.: 1930, unspec. (K). Not currently in cult.

Selaginellaceae

7. Selaginella selaginoides (L.) P.Beauv.
Lycopodium selaginoides L.
lesser clubmoss
Cult.: 1768 (JH: 432); 1813, Nat. of Britain (HK2, 5: 494); 1814, Nat. of Britain (HKE: 324). Not currently in cult.

8. *Selaginella kraussiana (Kunze) A.Braun
Krauss's clubmoss
First record: 1908 [1867].
References: **Kew Gardens.** 1908, *A.B.Jackson* (S7: 126): Occurs in a naturalised condition in the Rhododendron Dell and in the Rockery. • 1911, *Church* (S12: 375): Growing by clump of *Saxifraga caespitosa* over rocks in Rock garden.
Exsicc.: **Kew Gardens.** Naturalised in the Rhododendron Dell, on damp, mossy banks; 5 xii 1948; *P.G.Taylor* s.n. (K).
Cult.: 1867, unspec. (K); 1889, unspec. (K); 1961, House 26A (K). In cult. under glass.
Current status: Not seen for many years and probably extinct; there is no longer any suitable habitat for it either in the Rhododendron Dell or on the Rockery, both of which have been thoroughly refurbished since the last sightings.

Krauss's clubmoss (*Selaginella kraussiana*). Once a feature of damp places in the Rock Garden and Rhododendron Dell, but now extinct.

Isoetaceae

9. Isoetes lacustris L.
quillwort
Cult.: 1768 (JH: 431); 1789, Nat. of Wales & Scotland (HK1, 3: 470); 1813, Nat. of Britain (HK2, 5: 530); 1814, Nat. of Britain (HKE: 330). Not currently in cult.

HORSETAILS

Equisetaceae

10. Equisetum hyemale L.
rough horsetail
Cult.: 1768 (JH: 431); 1789, Nat. of Britain (HK1, 3: 455); 1813, Nat. of Britain (HK2, 5: 492); 1814, Nat. of Britain (HKE: 324). Not currently in cult.

11. Equisetum variegatum Schleich. ex F.Weber & D.Mohr
variegated horsetail
Cult.: 1813, Nat. of Scotland (HK2, 5: 492); 1814, Nat. of Scotland (HKE: 324). In cult. under glass.

12. Equisetum fluviatile L.
E. limosum var. *fluviatile* (L.) Asch.
E. limosum auct. non L.
water horsetail
First record: 1873/4 [1768].
Nicholson: **1873/4 survey** (JB: 77): **P.** Several large patches on the Richmond side of lake. • **1906 revision** (AS: 91): **A.** Lake.
Cult.: 1768 (JH: 431); 1789, Nat. of Britain (HK1, 3: 454, 455); 1813, Nat. of Britain (HK2, 5: 491, 492); 1814, Nat. of Britain (HKE: 324). Not currently in cult.
Current status: No recent sightings. Even before the Lake was emptied and refurbished this species had long since disappeared.

13. Equisetum arvense L.
field horsetail
First record: 1873/4 [1768].
Nicholson: **1873/4 survey** (JB: 77): **B.** Here and there in shrubberies near Old Lily House. **P.** Not frequent. • **1906 revision** (AS: 91): Here and there in shrubberies. Not frequent.
Cult.: 1768 (JH: 431); 1789, Nat. of Britain (HK1, 3: 454); 1813, Nat. of Britain (HK2, 5: 491); 1814, Nat. of Britain (HKE: 324). Not currently in cult.
Current status: Occasionally found but unexpectedly rare considering its reputation for persistence and its ability to spread far and wide in cultivated ground.

14. Equisetum sylvaticum L.
wood horsetail
Cult.: 1768 (JH: 431); 1789, Nat. of Britain (HK1, 3: 454); 1813, Nat. of Britain (HK2, 5: 492); 1814, Nat. of Britain (HKE: 324). Not currently in cult.

Field horsetail (*Equisetum arvense*). An aggressive and persistent weed surprisingly uncommon in the Gardens.

15. Equisetum palustre L.
marsh horsetail
Cult.: 1768 (JH: 431); 1789, Nat. of Britain (HK1, 3: 454); 1813, Nat. of Britain (HK2, 5: 492); 1814, Nat. of Britain (HKE: 324). Not currently in cult.

FERNS

Ophioglossaceae

16. Ophioglossum vulgatum L.
adder's-tongue
First record: 1873/4 [1768].
Nicholson: **1873/4 survey** (JB: 77): **P.** On a plot of ground a few yards square about 100 yards in a straight line from Railway Gate towards Brentford Ferry. • **1906 revision** (AS: 91): **A.** In open place in wood near Azalea garden.
Cult.: 1768 (JH: 430); 1789, Nat. of Britain (HK1, 3: 455); 1813, Nat. of Britain (HK2, 5: 495); 1814, Nat. of Britain (HKE: 324). In cult.
Current status: No recent sightings and extinct as a wild plant; it is still in cultivation, however, in the woodland garden. The 1873/4 locality is now lawn with planted trees and no longer a suitable habitat.

17. Botrychium lunaria (L.) Sw.
Osmunda lunaria L.
moonwort
Cult.: 1768 (JH: 430); 1789, Nat. of Britain (HK1, 3: 455); 1813, Nat. of Britain (HK2, 5: 496); 1814, Nat. of Britain (HKE: 325). Not currently in cult.

Osmundaceae

18. Osmunda regalis L.
royal fern
Cult.: 1768 (JH: 430); 1789, Nat. of Britain (HK1, 3: 456); 1813, Nat. of Britain (HK2, 5: 499); 1814, Nat. of Britain (HKE: 325); 1904, unspec. (K); 1915, unspec. (K); 1928, unspec. (K); 1936, unspec. (K); 1976, Filmy Fern House (Corridor) (K). In cult.

Adiantaceae

19. Cryptogramma crispa (L.) R.Br. ex Hook.
Osmunda crispa L.
Pteris crispa (L.) All.
parsley fern
Cult.: 1768 (JH: 430); 1789, Nat. of Britain (HK1, 3: 457); 1813, Nat. of Britain (HK2, 5: 520); 1814, Nat. of Britain (HKE: 328). Not currently in cult.

20. Adiantum capillus-veneris L.
maidenhair fern
Cult.: 1768 (JH: 429); 1789, Nat. of Britain (HK1, 3: 468); 1813, Nat. of Britain (HK2, 5: 526); 1814, Nat. of Britain (HKE: 329); 1869, unspec. (K); 1886, unspec. (K); 1887, unspec. (K); 1934, unspec. (K); 1935, unspec. (K); 1936, unspec. (K); 1953, Temperate House (K); 1956, unspec. (K); 1973, Lower Nursery (K). In cult. under glass.

Adder's-tongue (*Ophioglossum vulgatum*). Once known from two localities but now extinct. Cultivated plants can still be seen in the Woodland Garden.

Marsileaceae

21. Pilularia globulifera L.
pillwort
Cult.: 1768 (JH: 431); 1789, Nat. of Britain (HK1, 3: 470); 1813, Nat. of Britain (HK2, 5: 530); 1814, Nat. of Britain (HKE: 330); 1975, unspec. (K). Not currently in cult.

Hymenophyllaceae

22. Hymenophyllum tunbrigense (L.) Sm.
Trichomanes tunbrigense L.
Tunbridge filmy-fern
Cult.: 1768 (JH: 429); 1813, Nat. of Britain (HK2, 5: 530); 1814, Nat. of Britain (HKE: 329). Not currently in cult.

23. Trichomanes speciosum Willd.
T. brevisetum R.Br.
Killarney fern
CULT.: 1813, Nat. of Britain (HK2, 5: 529); 1814, Nat. of Britain (HKE: 329). Not currently in cult.

POLYPODIACEAE

24. Polypodium vulgare L.
polypody
FIRST RECORD: 2008 [1768].
CULT.: 1768 (JH: 428); 1789, Nat. of Britain (HK1, 3: 462); 1813, Nat. of Britain (HK2, 5: 504); 1814, Nat. of Britain (HKE: 326). In cult.
CURRENT STATUS: Rare; a recent colonist of the new river wall along Ferry Lane, on the side facing the river.

25. Polypodium cambricum L.
P. vulgare β *cambricum* (L.) Willd.
southern polypody
CULT.: 1768 (JH: 428); 1789, Nat. of Britain (HK1, 3: 463); 1813, Nat. of Britain (HK2, 5: 504); 1814, Nat. of Britain (HKE: 326). Not currently in cult.

DENNSTAEDTIACEAE

26. Pteridium aquilinum (L.) Kuhn
Pteris aquilina L.
bracken
FIRST RECORD: 1873/4 [1768].
NICHOLSON: **1873/4 survey** (JB: 77): **P.** A starved plant or two in wood near Winter Garden. **Q.** Several good-sized tracts. • **1906 revision** (AS: 91): **A.** Here and there in wood. **Q.** Common.
OTHER REFERENCES: **Kew Gardens.** 1999, *Stones* (EAK: 5): By Palm House Pond.
CULT.: 1768 (JH: 429); 1789, Nat. of Britain (HK1, 3: 458); 1813, Nat. of Britain (HK2, 5: 521); 1814, Nat. of Britain (HKE: 328); 1901, Tropical Fern House (K); 1902, in a greenhouse (K); 1903, in the garden (K); 1904, unspec. (K). Not currently in cult.
CURRENT STATUS: Occasional, mostly in the Conservation Area; it is not as common as might have been expected from such an aggressive weed.

THELYPTERIDACEAE

27. Thelypteris palustris Schott
Acrostichum thelypteris L.
Aspidium thelypteris (L.) Sw.
Polypodium thelypteris (L.) Weis
marsh fern
CULT.: 1768 (JH: 428); 1789, Nat. of Britain (HK1, 3: 465); 1813, Nat. of Britain (HK2, 5: 509); 1814, Nat. of Britain (HKE: 326); 1937, unspec. (K). In cult.

28. Phegopteris connectilis (Michx.) Watt
Polypodium phegopteris L.
beech fern
CULT.: 1768 (JH: 428); 1789, Nat. of Britain (HK1, 3: 464); 1813, Nat. of Britain (HK2, 5: 506); 1814, Nat. of Britain (HKE: 326). In cult.

29. Oreopteris limbosperma (Bellardi ex All.) Holub
Aspidium oreopteris Sw.
lemon-scented fern
CULT.: 1813, Nat. of Britain (HK2, 5: 509); 1814, Nat. of Britain (HKE: 326). Not currently in cult.

ASPLENIACEAE

30. Phyllitis scolopendrium (L.) Newman
Asplenium scolopendrium L.
Scolopendrium officinarum Sw.
hart's-tongue
FIRST RECORD: 1999 [1768].
REFERENCES: **Kew Gardens.** 1999, *Stones* (EAK: 5): By Palm House Pond.
CULT.: 1768 (JH: 429); 1789, Nat. of Britain (HK1, 3: 461); 1813, Nat. of Britain (HK2, 5: 518); 1814, Nat. of Britain (HKE: 328); 19th cent., unspec. (K). In cult.
CURRENT STATUS: Mostly in brickwork around the Palm House Pond and in damp corners of buildings. It has been around as a wild plant for a long time but not formally recorded until recently.

31. Asplenium adiantum-nigrum L.
black spleenwort
CULT.: 1768 (JH: 429); 1789, Nat. of Britain (HK1, 3: 462); 1813, Nat. of Britain (HK2, 5: 518); 1814, Nat. of Britain (HKE: 328). Not currently in cult.

32. Asplenium obovatum Viv.
Asplenium lanceolatum Huds.
lanceolate spleenwort
CULT.: 1813, Nat. of England (HK2, 5: 518); 1814, Nat. of England (HKE: 328). Not currently in cult.

33. Asplenium marinum L.
sea spleenwort
CULT.: 1768 (JH: 429); 1789, Nat. of Britain (HK1, 3: 462); 1813, Nat. of Britain (HK2, 5: 515); 1814, Nat. of Britain (HKE: 327). In cult. under glass.

Hart's-tongue fern (*Phyllitis scolopendrium*) on the wall of the Palm House Pond.

34. Asplenium trichomanes L.
maidenhair spleenwort
FIRST RECORD: 2001 [1768].
REFERENCES: **Kew Gardens.** 2001, *Cope* (S35: 649): On wall by Jodrell Laboratory (153).
CULT.: 1768 (JH: 429); 1789, Nat. of Britain (HK1, 3: 461); 1813, Nat. of Britain (HK2, 5: 516); 1814, Nat. of Britain (HKE: 327); 1935, unspec. (K). In cult.
CURRENT STATUS: On a shaded wall by the Jodrell Laboratory. It has a long history of cultivation and may have been growing wild for many years before finally being noticed. It was lost when the laboratory was extended.

35. Asplenium viride Huds.
green spleenwort
CULT.: 1789, Nat. of Britain (HK1, 3: 461); 1813, Nat. of Britain (HK2, 5: 516); 1814, Nat. of Britain (HKE: 327); 1939, unspec. (K). Not currently in cult.

36. Asplenium ruta-muraria L.
wall-rue
CULT.: 1768 (JH: 429); 1789, Nat. of Britain (HK1, 3: 462); 1813, Nat. of Britain (HK2, 5: 517); 1814, Nat. of Britain (HKE: 328). In cult.

37. Asplenium × alternifolium Wulfen
(septentrionale × trichomanes)
A. germanicum Weis
CULT.: 1813, Nat. of Scotland (HK2, 5: 516); 1814, Nat. of Scotland (HKE: 327). Not currently in cult.

38. Asplenium septentrionale (L.) Hoffm.
Acrostichum septentrionale L.
forked spleenwort
CULT.: 1768 (JH: 428); 1789, Nat. of Britain (HK1, 3: 457); 1813, Nat. of Britain (HK2, 5: 515); 1814, Nat. of Britain (HKE: 327). In cult. under glass.

[**39. Asplenium fontanum** (L.) Bernh.
Polypodium fontanum L.
Aspidium fontanum (L.) Sw.
CULT.: 1789, Nat. of Britain (HK1, 3: 463); 1813, Nat. of England (HK2, 5: 511); 1814, Nat. of England (HKE: 327)
Note. Listed in error as native.]

40. Ceterach officinarum Willd.
Asplenium ceterach L.
Grammitis ceterach (L.) Sw.
rustyback
CULT.: 1768 (JH: 429); 1789, Nat. of Britain (HK1, 3: 461); 1813, Nat. of Britain (HK2, 5: 503); 1814, Nat. of Britain (HKE: 326). Not currently in cult.

WOODSIACEAE

41. Athyrium filix-femina (L.) Roth
Polypodium filix-femina L.
Aspidium filix-femina (L.) Sw.
lady-fern
CULT.: 1768 (JH: 428); 1789, Nat. of Britain (HK1, 3: 465); 1813, Nat. of Britain (HK2, 5: 512); 1814, Nat. of Britain (HKE: 327); 1856, unspec. (K); 1866, unspec. (K); 1958, unspec. (K). In cult.

42. Athyrium distentifolium Tausch ex Opiz
Polypodium rhaeticum L.
alpine lady-fern
CULT.: 1768 (JH: 428); 1789, Nat. of England (HK1, 3: 465). In cult.
Note. Possibly misidentified and discarded from cultivation; listed as a native of England, but in fact a rare Scottish species.

43. Athyrium flexile (Newman) Druce
Newman's lady-fern
CULT.: In cult. under glass.

44. Gymnocarpium dryopteris (L.) Newman
Polypodium dryopteris L.
oak fern
CULT.: 1768 (JH: 428); 1789, Nat. of Britain (HK1, 3: 466); 1813, Nat. of Britain (HK2, 5: 506); 1814, Nat. of Britain (HKE: 326). In cult.

45. Gymnocarpium robertianum (Hoffm.) Newman
Polypodium calcareum Sm.
limestone fern
CULT.: 1813, Nat. of England (HK2, 5: 507); 1814, Nat. of England (HKE: 326). In cult. under glass.
Note. Known for many years on the London-bound platform of Kew Gardens Station, but recently removed during refurbishment.

46. Cystopteris fragilis (L.) Bernh.
Polypodium fragile L.
Aspidium fragile (L.) Sw.
Polypodium regium L.
Aspidium regium (L.) Sw.
A. dentatum Sw.
brittle bladder-fern
CULT.: 1768 (JH: 428); 1789, Nat. of Britain (HK1, 3: 466); 1813, Nat. of Britain (HK2, 5: 512 and: 513 as *regium*); Nat. of Scotland & Wales (ibid.: 511 as *dentatum*), 1814, Nat. of Britain (HKE: 327 and as *regium*), Nat. of Scotland (ibid.: 327 as *dentatum*); 1894, unspec. (K); 1936, unspec. (K). In cult. under glass.

47. Cystopteris dickieana R.Sim
Dickie's bladder-fern
CULT.: 1973, Alpine Dept. (K). In cult.

48. Woodsia ilvensis (L.) R.Br.
Acrostichum ilvense L.
Polypodium ilvense (L.) Vill.
oblong Woodsia
CULT.: 1768 (JH: 428); 1813, Alien (HK2, 5: 505); 1814, Alien (HKE: 326). Not currently in cult.
Note. Listed as an alien, but in fact known as a native since 1690.

49. Woodsia alpina (Bolton) Gray
Acrostichum ilvense auct. non L.
Polypodium hyperboreum Sw.
alpine Woodsia
CULT.: 1789, Nat. of Wales & Scotland (HK1, 3: 457); 1813, Nat. of Wales & Scotland (HK2, 5: 505); 1814, Nat. of Wales & Scotland (HKE: 326). Not currently in cult.

DRYOPTERIDACEAE

50. Polystichum setiferum (Forssk.) T.Moore ex Woyn.
soft shield-fern
CULT.: pre-1911, unspec. (K); 1936, unspec. (K). In cult.

51. Polystichum aculeatum (L.) Roth
Polypodium aculeatum L.
Aspidium aculeatum (L.) Sw.
A. lobatum Sw.
hard shield-fern
CULT.: 1768 (JH: 428); 1789, Nat. of Britain (HK1, 3: 465); 1813, Nat. of Britain (HK2, 5: 509), Nat. of England (ibid.: 510 as *lobatum*); 1814, Nat. of Britain (HKE: 326), Nat. of England (ibid.: 327 as *lobatum*). In cult.

52. Polystichum lonchitis (L.) Roth
Polypodium lonchitis L.
Aspidium lonchitis (L.) Sw.
holly-fern
CULT.: 1768 (JH: 428); 1789, Nat. of Britain (HK1, 3: 463); 1813, Nat. of Britain (HK2, 5: 507); 1814, Nat. of Britain (HKE: 326). In cult.

53. Dryopteris filix-mas (L.) Schott
Polypodium filix-mas L.
Aspidium filix-mas (L.) Sw.
Nephrodium filix-mas (L.) Rich.
male-fern
FIRST RECORD: 1873/4 [1768].
NICHOLSON: **1873/4 survey** (JB: 77): **P.** About a dozen large plants in wood midway between Engine House and Juniper Avenue. **Q.** Common. • **1906 revision**: Omitted.
CULT.: 1768 (JH: 428); 1789, Nat. of Britain (HK1, 3: 464); 1813, Nat. of Britain (HK2, 5: 510); 1814, Nat. of Britain (HKE: 326); pre-1820, unspec. (K); 1856, unspec. (K); 1937, unspec. (K); 1974, under oak in woodland garden, plot 166-02 (K). In cult.
CURRENT STATUS: Widely scattered in shady areas but not common. It may well have been planted or have naturalised from planted specimens in most of its stations although some populations are well away from known cultivated plants. It was certainly wild at the time of Nicholson's first survey but curiously omitted from the 1906 revision.

54. Dryopteris affinis subsp. **borreri** (Newman) Fraser-Jenk.
Nephrodium filix-mas var. *affine* Newman
scaly male-fern
FIRST RECORD: 1879.
NICHOLSON: **1906 revision** (AS: 91): **A.** Here and there in wood near pumping station. **Q.** Common
EXSICC.: **Kew Gardens.** [Unlocalised]; 1879; *Nicholson* (ABD, *fide* J.P.Pugh in WAT3: 59).
CURRENT STATUS: No recent sightings. It may still be in the Conservation Area but much of this has been inaccessible for many years and has therefore not been properly surveyed.

55. Dryopteris aemula (Aiton) Kuntze
hay-scented buckler-fern
CULT.: pre-1867, unspec. (K). In cult.

56. Dryopteris cristata (L.) A.Gray
Polypodium cristatum L.
Aspidium cristatum (L.) Sw.
crested buckler-fern
CULT.: 1768 (JH: 428); 1789, Nat. of Britain (HK1, 3: 464); 1813, Nat. of England (HK2, 5: 509); 1814, Nat. of England (HKE: 326). In cult. under glass.

57. Dryopteris carthusiana (Vill.) H.P.Fuchs
Aspidium spinulosum Sw.
Nephrodium spinulosum (D.F.Müll.) Strempel, nom. illegit.
narrow buckler-fern
FIRST RECORD: 1906 [1813].
NICHOLSON: **1906 revision** (AS: 91): Common in Queen's Cottage Grounds.
CULT.: 1813, Nat. of Britain (HK2, 5: 510); 1814, Nat. of Britain (HKE: 327); pre-1867, unspec. (K); 1936, unspec. (K). In cult.
CURRENT STATUS: No recent sightings. See observation under the preceding *D. affinis* subsp. *borreri*.

58. Dryopteris dilatata (Hoffm.) A.Gray
Aspidium dilatatum (Hoffm.) Sm.
Nephrodium dilatatum (Hoffm.) Fritsch
broad buckler-fern
FIRST RECORD: 1873/4 [1813].
NICHOLSON: **1873/4 survey** (JB: 77): **P.** A couple of plants with [*D. filix-mas*]. **Q.** Here and there. This and [*D. filix-mas* and *Pteridium aquilinum*] seem to be the only genuine native ferns of our Flora. Some two or three others grow about the Old Ruined Arch and Merlin's Cave, but as they occur nowhere else one may reasonably suppose them to have been planted at some time. • **1906 revision**: Omitted.
CULT.: 1813, Nat. of Britain (HK2, 5: 510); 1814, Nat. of Britain (HKE: 327); 20th cent., unspec. (K). In cult.
CURRENT STATUS: Still in the Conservation Area but has not been seen recently in any of the places where Nicholson thought it may have been planted.

59. Dryopteris expansa (C.Presl) Fraser-Jenk. & Jermy
D. assimilis S.Walker
northern buckler-fern
CULT.: 1946, unspec. (K). In cult. under glass.

BLECHNACEAE

60. Blechnum spicant (L.) Roth
Osmunda spicant L.
B. boreale Sw.
hard-fern
CULT.: 1768 (JH: 430); 1789, Nat. of Britain (HK1, 3: 456); 1813, Nat. of Britain (HK2, 5: 522); 1814, Nat. of Britain (HKE: 328). In cult.

AZOLLACEAE

61. *Azolla filiculoides Lam.
water fern
FIRST RECORD: 1999.
REFERENCES: **Kew Gardens.** 1999, *Cope* (S35: 649): In the Larch Pond in the Conservation Area (310) before it dried out.
CURRENT STATUS: In the dewpond in the Conservation Area before it dried out. The pond is to be reinstated at some time in the near future and the plant may reappear although it could prove to be a problem.

Water fern (*Azolla filiculoides*) an alien and often unwelcome species that grew in the pond in the Conservation Area but was apparently lost when the pond dried out.

CONIFERS

PINACEAE

62. Pinus sylvestris L.
Scots pine
CULT.: 1768 (JH: 450); 1789, Nat. of Scotland (HK1, 3: 366); 1813, Nat. of Scotland (HK2, 5: 314); 1814, Nat. of Scotland (HKE: 299). In cult.

CUPRESSACEAE

63. *Chamaecyparis lawsoniana (A.Murray) Parl.
Lawson's cypress
FIRST RECORD: 1958.
REFERENCES: **Kew Gardens.** 1958, *Milne-Redhead* (BSP4: 40; LN38: 19; S33: 743): Self-sown seedlings from a wall near Queen's Cottage.
CULT.: In cult.
CURRENT STATUS: No recent sightings; the plants were presumably removed to protect the wall from root damage.

64. Juniperus communis L.
juniper
a. subsp. **communis**
CULT.: 1768 (JH: 444); 1789, Nat. of Britain (HK1, 3: 414); 1813, Nat. of Britain (HK2, 5: 414); 1814, Nat. of Britain (HKE: 314). In cult.

b. subsp. **nana** (Hook.) Syme
J. communis var. *montana* Aiton
CULT.: 1789, Nat. of Britain (HK1, 3: 414); 1813, Nat. of Britain (HK2, 5: 415). In cult.

TAXACEAE

65. Taxus baccata L.
yew
FIRST RECORD: 1998 [1768].
REFERENCES: **Kew Gardens.** 1998, *Cope* (S33: 743): Self-sown seedlings in following:- North Arb.: (110, 122, 123), West Arb.: (214), Herbaceous: (158). • 1998, *Verdcourt* (S33: 743): self-sown seedlings on piles of stored rockery stone (by 116).
CULT.: 1768 (JH: 456); 1789, Nat. of Britain (HK1, 3: 415); 1813, Nat. of Britain (HK2, 5: 415); 1814, Nat. of Britain (HKE: 314). In cult.
CURRENT STATUS: Seedlings are widespread, usually near the parent tree, but are seldom allowed to persist. They were first noted in 1993 but not immediately reported. Doubtless seedlings have been around for as long as the species has been cultivated.

Yew (*Taxus baccata*). A cultivated tree that commonly seeds itself throughout the Gardens.

FLOWERING PLANTS

DICOTYLEDONS

Aristolochiaceae

[66. **Asarum europaeum** L.
asarabacca
Cult.: 1789, Nat. of England (HK1, 2: 124); 1811, Nat. of England (HK2, 3: 141); 1814, Nat. of England (HKE: 139).
Note. Listed in error as native.]

67. **Aristolochia clematitis** L.
birthwort
First record: 1877 [1768].
Exsicc.: **Kew Gardens.** Pleasure Grounds; 19 vii 1877; *G.Nicholson* 1130 (K).
Cult.: 1768 (JH: 328); 1789, Nat. of England (HK1, 3: 312); 1813, Nat. of England (HK2, 5: 228); 1814, Nat. of England (HKE: 286); 1933, unspec. (K); 1936, Herbaceous Ground (K). In cult.
Current status: This was not listed by Nicholson in his revised list so presumably it had already been lost by then. He seldom recorded cultivated plants, except when they appeared to have naturalised, but this is an unlikely wild species for Kew.

Nymphaeaceae

68. **Nymphaea alba** L.
Castalia speciosa Salisb.
white water-lily
First record: 1873/4 [1768].
Nicholson: **1873/4 survey** (JB: 10; SFS: 108): **Strip.** Several plants grow in moat near "Old Deer Park" just within the prescribed limits. All in lake have been planted. • **1906 revision** (AS: 74): In ha-ha bordering **Q**. Elsewhere planted.
Cult.: 1768 (JH: 229); 1789, Nat. of Britain (HK1, 2: 227); 1811, Nat. of Britain (HK2, 3: 292); 1814, Nat. of Britain (HKE: 165); 1878, unspec. (K); 1883, unspec. (K); 1887, Hardy Tank (K); 1944, Aquatic Garden (K). In cult.
Current status: There are no recent sightings of wild plants. Plants currently in cultivation in the Lake were sourced from the New Forest; it is in a pond on the golf course in the Old Deer Park and from its flower-colour it has naturalised from cultivation.

69. **Nuphar lutea** (L.) Sibth. & Sm.
Nymphaea lutea L.
yellow water-lily
First record: 1873/4 [1768].
Nicholson: **1873/4 survey** (JB: 10; SFS: 108): **Strip.** In company with [*Nymphaea alba*]. Lake specimens all planted. • **1906 revision** (AS: 74): In ha-ha bordering **Q**. Elsewhere planted.
Other references: **Kew Environs.** 1902, *Beeby* (SFS: 108): Thames between Kew and Richmond.
Exsicc.: **Kew Gardens.** Thames, opposite Sion House, Surrey; *W.T.Western* s.n. (K).
Cult.: 1768 (JH: 229); 1789, Nat. of Britain (HK1, 2: 227); 1811, Nat. of Britain (HK2, 3: 295); 1814, Nat. of Britain (HKE: 165); 1887, Hardy Tank (K); 1944, Aquatic Garden (K). In cult.
Current status: There are several patches in the Lake and the Palm House Pond, all planted long ago but now thoroughly naturalised. When the Lake was recently dredged and cleaned it returned spontaneously and has subsequently been augmented by plants sourced from the New Forest. There are a few other modern cultivated specimens, especially in the Banks Building Pond. It disappeared long ago from the river — probably as a result of strengthening of the banks — and from the Moat.

70. **Nuphar pumila** (Timm) DC.
least water-lily
Cult.: 1814, Nat. of Scotland (HKE: Add.). Not currently in cult.

Ceratophyllaceae

71. **Ceratophyllum demersum** L.
C. aquaticum var. *demersum* (L.) auct. dub.
rigid hornwort
First record: 1873/4 [1768].
Nicholson: **1873/4 survey** (JB: 73): Grows plentifully in the moat nearly the whole length of the Gardens. • **1906 revision** (AS: 87): **A.** Lake. **Strip.** Plentiful in the ha-ha, nearly its whole boundary length of gardens.
Exsicc.: **Kew Gardens.** Ha-ha, Kew; vi 1927; *Findlay* s.n. (K). **Kew Environs.** Thames, Kew; vii 1930; *Pearce* s.n. (K).
Cult.: 1768 (JH: 374); 1789, Nat. of Britain (HK1, 3: 351); 1813, Nat. of Britain (HK2, 5: 281); 1814, Nat. of Britain (HKE: 294). Not currently in cult.
Current status: In the Water Lily Pond, where it dominates. It was in the Lake until its recent refurbishment and appears to have found its own way back, probably with the help of waterfowl. It is no longer in the Moat.

72. **Ceratophyllum submersum** L.
soft hornwort
Cult.: 1789, Nat. of Britain (HK1, 3: 351); 1813, Nat. of Britain (HK2, 5: 282); 1814, Nat. of Britain (HKE: 294). Not currently in cult.

White water-lily (*Nymphaea alba*). Extinct as a wild plant but naturalised in a pond in the Old Deer Park.

RANUNCULACEAE

73. Caltha palustris L.

marsh-marigold

a. var. **palustris**

FIRST RECORD: 1873/4 [1768].

NICHOLSON: **1873/4 survey** (JB: 10): **Strip.** Along side of moat. Common. • **1906 revision** (AS: 74): Common along ha-ha, etc.

OTHER REFERENCES: **Kew Gardens.** 1890, *Bennett* (WBEC6: 2): Hort. Croydon, 1889. Root from Kew Gardens as the plant of Forster, but doubtful.

CULT.: 1768 (JH: 342); 1789, Nat. of Britain (HK1, 2: 273); 1811, Nat. of Britain (HK2, 3: 361); 1814, Nat. of Britain (HKE: 175); 1904, unspec. (K); 1925, unspec. (K); 1926, unspec. (K); 1944, Rock Garden (K). In cult.

CURRENT STATUS: By the Water Lily Pond, a station not listed in the LCD database for a known cultivated accession; despite having once been native in the immediate vicinity it is likely to have been planted by the pond.

b. var. **radicans** (T.F.Forst.) Hook.

Caltha radicans T.F.Forst.

CULT.: 1811, Nat. of Scotland (HK2, 3: 360); 1814, Nat. of Scotland (HKE: 175). Not currently in cult.

74. Trollius europaeus L.

globeflower

CULT.: 1768 (JH: 360); 1789, Nat. of Britain (HK1, 2: 271); 1811, Nat. of Britain (HK2, 3: 359); 1814, Nat. of Britain (HKE: 175); 1898, unspec. (K); 1902, unspec. (K); 1905, unspec. (K); 1930, unspec. (K); 1940, Herbarium Ground (K); 1961, Rock Garden (K); 1964, Herbaceous Dept. (K). In cult.

75. Helleborus foetidus L.

stinking hellebore

CULT.: 1768 (JH: 342); 1789, Nat. of England (HK1, 2: 272); 1811, Nat. of England (HK2, 3: 360); 1814, Nat. of England (HKE: 175); 1888, unspec. (K); 1924, unspec. (K); 1931, Herbaceous Beds (K); 1933, Herb. Ground (K); 1988, unspec. (K). In cult.

76. Helleborus viridis subsp. **occidentalis** (Reut.) Schiffn.

H. viridis s. lat.

green hellebore

CULT.: 1768 (JH: 342); 1789, Nat. of Britain (HK1, 2: 272); 1811, Nat. of Britain (HK2, 3: 360); 1814, Nat. of Britain (HKE: 175); 1888, unspec. (K); 1889, unspec. (K); 1896, unspec. (K); 1932, Herbaceous borders (K); 1935, unspec. (K); 1981, bed 156-06 (K); 1988, unspec. (K). In cult.

77. *Eranthis hyemalis (L.) Salisb.

Helleborus hyemalis L.

winter aconite

FIRST RECORD: 1873/4 [1768].

NICHOLSON: **1873/4 survey** (JB: 10; SFS: 104): **B.** Here and there in turf and beds skirting "Palace Grounds." • **1906 revision** (AS: 74): In turf in palace grounds, etc.

CULT.: 1768 (JH: 342); 1883, unspec. (K); 1924, unspec. (K); 1945, bed of lilacs near Main Gate (K). In cult.

CURRENT STATUS: A few recent sightings; the patch in a bed near Kew Palace has recently been grassed over but the plant can still be found in nearby flower-beds; there are also some near the Ice-house and several other spots in other parts of the Gardens. It was originally planted in most of its localities but in one or two places it seems to be spontaneous.

78. *Nigella damascena L.

love-in-a-mist

FIRST RECORD: 1983 [1768].

REFERENCES: **Kew Gardens.** 1983, *Hastings & Cope* (LN63: 143): Recently cleared plot near the herbarium building. • 1983/4, *Cope* (S31: 181): New student vegetable plots behind Hanover House.

CULT.: 1768 (JH: 343); 1789, Alien (HK1, 2: 248); 1811, Alien (HK2, 3: 326); 1814, Alien (HKE: 170); 1887, unspec. (K); 1918, unspec. (K); 1971, Jodrell (K). In cult.

CURRENT STATUS: Found outside the Jodrell Laboratory in 2006, on a heap of topsoil being stored in the Paddock in 2007, and in a flower-bed near the Temperate House in 2008, but otherwise is unrecorded since 1984.

79. Aconitum napellus L.

monk's-hood

CULT.: 1768 (JH: 346); 1789, Alien (HK1, 2: 245); 1811, Alien (HK2, 3: 323); 1814, Alien (HKE: 169); 1901, unspec. (K); 1903, unspec. (K); 1926, Herb. Exp. Ground (K); 1934, Arboretum (K); 1934, unspec. (K); 1936, Herbaceous Ground (K); 1998, bed 156-10 (K). In cult.

Note. Originally cultivated as an alien but found as a native in 1821.

80. *Consolida ajacis (L.) Schur

C. ambigua auct. non L.

Delphinium consolida L.

D. ajacis L.

larkspur

FIRST RECORD: 1877 [1768].

REFERENCES: **Kew Gardens.** 1983/4, *Cope* (S31: 181): New student vegetable plots behind Hanover House.

EXSICC.: **Kew Environs.** [Unlocalised]; viii 1877; *Baker* 43 (K). • Kew Green; 24 vi & 14 viii 1920; *Fraser* s.n. (K).

CULT.: 1768 (JH: 345); 1789, Nat. of England (HK1, 2: 242); 1811, Nat. of England (HK2, 3: 318); 1814, Nat. of England (HKE: 169); 1891, unspec. (K); 1998, bed 156-06 (K). Not currently in cult.

CURRENT STATUS: Very rare; it makes an occasional appearance in disturbed soil but never persists.

81. Actaea spicata L.

baneberry

CULT.: 1768 (JH: 172/10); 1789, Nat. of Britain (HK1, 2: 221); 1811, Nat. of Britain (HK2, 3: 286); 1814, Nat. of Britain (HKE: 164); 1980s, bed 165-06 (K). In cult.

82. Anemone nemorosa L.

wood anemone

FIRST RECORD: 2001 [1768].

REFERENCES: **Kew Gardens.** 2001, *Cope* (S35: 646): On the edge of Kew Green opposite Ferry Lane. It grows right on the edge of the Green and has probably escaped either from Kew Gardens or gardens nearby on Kew Green. It has subsequently been noted as naturalised in several parts of the Gardens. A pink-flowered variety occurs in lightly wooded grassland near the Temple of Bellona (461).

Wood anemone (*Anemone nemorosa*). A British native species that is not native in the Gardens; now widespread and looking very much at home, as here near the Bamboo Garden.

CULT.: 1768 (JH: 361); 1789, Nat. of Britain (HK1, 2: 256); 1811, Nat. of Britain (HK2, 3: 340); 1814, Nat. of Britain (HKE: 172); 1880, unspec. (K); 1894, unspec. (K); 1896, unspec. (K); 1915, unspec. (K); undated, rocks (K). In cult.

CURRENT STATUS: Probably originally cultivated, but now thoroughly naturalised and spreading. It is doubtful if Nicholson would have missed it as a wild plant and it is unlikely to have arrived on its own; as a result, it has never been formally recorded. A pink-flowered variety occurs in grass near the Temple of Bellona. A small population survived for a few years on the edge of Kew Green opposite Ferry Lane but has since disappeared.

[**83. Anemone apennina** L.

blue anemone

CULT.: 1789, Nat. of England (HK1, 2: 257); 1811, Nat. of England (HK2, 3: 340); 1814, Nat. of England (HKE: 172)

Note. Listed in error as native.]

84. *Anemone blanda Schott & Kotschy

Balkan anemone

FIRST RECORD: 2008 [1965].

CULT.: 1965, unspec. (K); 1970, Alpine House (K). In cult.

CURRENT STATUS: A small patch naturalised in the Conservation Area.

[85. **Anemone ranunculoides** L.
yellow anemone
CULT.: 1789, Nat. of England (HK1, 2: 257); 1811, Nat. of England (HK2, 3: 340); 1814, Nat. of England (HKE: 172).
Note. Listed in error as native.]

86. *Anemone coronaria L.
poppy anemone
FIRST RECORD: 2004 [1768].
CULT.: 1768 (JH: 361). In cult.
CURRENT STATUS: Naturalised in rough grass under trees near White Peaks and probably arrived as a contaminant of the supply of *Chionodoxa forbesii* that was used in a mass planting.

87. *Anemone trifolia subsp. **albida** (Mariz) Tutin
FIRST RECORD: 2002 [1811].
REFERENCES: **Kew Gardens.** 2002, *Cope* (S35: 646): Naturalised in rough grass under trees near Pagoda Vista (zone 354).
CULT.: 1811, Alien (HK2, 3: 340); 1814, Alien (HKE: 172). Not currently in cult.
CURRENT STATUS: Known only from the above locality where it forms a small but persistent patch.

88. Pulsatilla vulgaris Mill.
Anemone pulsatilla L.
pasqueflower
CULT.: 1768 (JH: 361); 1789, Nat. of England (HK1, 2: 255); 1811, Nat. of England (HK2, 3: 337); 1814, Nat. of England (HKE: 172); 1904, unspec. (K); 1911, unspec. (K); 1914, unspec. (K); 1950, Rock Garden (K); 1950, Chalk Garden (K); 1962, Herb. & Alpine Dept. (K); 1972, Herbaceous & Alpine Dept. (K). In cult.

89. Clematis vitalba L.
traveller's-joy
FIRST RECORD: 1998 [1768].
REFERENCES: **Kew Gardens.** 1998, *Cope* (S33: 743): North Arb.: behind Banks Building (zones 107 and 108); Lower Nursery, on the river side of the access road (zone 120); around Brentford Ferry car park; Herbaceous: by the wall along Kew Road (zone 158); Ferry Lane, opposite Herbarium.
CULT.: 1768 (JH: 340, 438); 1789, Nat. of England (HK1, 2: 260); 1811, Nat. of England (HK2, 3: 345); 1814, Nat. of England (HKE: 173); 1948, Arboretum (K). In cult.
CURRENT STATUS: All records are very recent (see above) although the species has been present for a long time. It is probably wild on the towpath outside the Gardens and has spread into various parts of the Gardens either from here or from the Conservation Area where it has been deliberately planted.

90. Ranunculus acris L.
meadow buttercup
FIRST RECORD: 1873/4 [1768].
NICHOLSON: **1873/4 survey** (JB: 10): Common. • **1906 revision** (AS: 74): Common.
EXSICC.: **Kew Gardens.** In field around the Herbarium; v 1928; *C.A.Smith* 6018 (K) • Herbarium field, in long grass in damp part of the field (near Buildings), a few very large plants with rather glaucous stems; 7 v 1945; *Ross-Craig & Sealy* 1050 (K, det. J.Rossiter 12 i 1954 as var. *steveni*).
CULT.: 1768 (JH: 187); 1789, Nat. of Britain (HK1, 2: 269); 1811, Nat. of Britain (HK2, 3: 357); 1814, Nat. of Britain (HKE: 175); 1895, unspec. (K); 1933, unspec. (K); 1936, unspec. (K); 1940, Herb Garden (K); 1951, Herbarium Ground (K). In cult.
CURRENT STATUS: Widespread but conspicuous only in the 'butterfly conservation' areas where the grass is infrequently mown and the plant allowed to reach maturity.

91. Ranunculus repens L.
creeping buttercup
FIRST RECORD: 1856 [1768].
NICHOLSON: **1873/4 survey** (JB: 10): Abundant in all the divisions. • **1906 revision** (AS: 74): Common.
OTHER REFERENCES: **Kew Gardens.** 1999, *Stones* (EAK: 2): By Lake.
EXSICC.: **Kew Gardens.** Hort. Kew.; 1856; *anon.* (K) • Thames-bank, Kew – Richmond, between Brentford Ferry and Isleworth Gate; on mud just below high-tide mark; 25 vi 1929; *Summerhayes & Turrill* s.n. (K) • Herbarium Exp. Ground; waste ground; etiolated from growing under inverted flower pot; 14 v 1930; *Summerhayes & Ballard* 465 (K). • Herbarium Exp. Ground; waste ground; normal plants; 14 v 1930; *Summerhayes & Ballard* 466 (K). **Kew Environs.** River bank near Kew Bridge; growing on old wall near top, at average high tide mark; 7 v 1928; *Summerhayes & Turrill* s.n. (K).
CULT.: 1768 (JH: 187); 1789, Nat. of Britain (HK1, 2: 269); 1811, Nat. of Britain (HK2, 3: 357); 1814, Nat. of Britain (HKE: 174); 1937, from a plant transplanted from Thames side near Kew Gardens in 1936 (K). In cult.
CURRENT STATUS: Ubiquitous in grassland.

92. Ranunculus bulbosus L.
bulbous buttercup
FIRST RECORD: 1873/4 [1768].
NICHOLSON: **1873/4 survey** (JB: 10): Equally common with [*R. acris* and *R. repens*]. • **1906 revision** (AS: 74): Common.
EXSICC.: **Kew Gardens.** River bank near Isleworth Gate; top zone on ± dry bank; 2 v 1928; *Turrill* s.n. (K). • [Unlocalised]; in short grassland, sandy soil; 10 v 1928; *Hubbard* s.n. (K). • In field round the Herbarium; v 1928; *C.A.Smith* 6016 (K). • Near the

Herbarium; in grasses; 16 vi 1928; *Hubbard* s.n. (K). • Near Tennis Courts by the old mulberry tree; 14 v 1931; *Turrill* s.n. (K). • Near *Juglans* tree, between Herbarium and Tennis Courts; in grassland; 19 v 1932; *Turrill* s.n. (K).
CULT.: 1768 (JH: 187); 1789, Nat. of Britain (HK1, 2: 268); 1811, Nat. of Britain (HK2, 3: 356); 1814, Nat. of Britain (HKE: 175); 1929, unspec. (K); 1933, unspec. (K); 1934, Herbarium Ground (K); 1936, Herbarium Ground (K); 1937, Herbarium Ground (K); 1937, unspec. (K); 1947, Herbarium Ground (K). Not currently in cult.
CURRENT STATUS: Widespread but rather uncommon; it is the least abundant of the three main buttercup species in the Gardens.

93. **Ranunculus sardous** Crantz

R. hirsutus Curtis
R. parvulus L.
hairy buttercup

CULT.: 1789, Nat. of England (HK1, 2: 268); 1811, Nat. of England (HK2, 3: 356, 357); 1814, Nat. of England (HKE: 174). Not currently in cult.

94. ***Ranunculus muricatus** L.

rough-fruited buttercup

FIRST RECORD: 1986 [1789].
REFERENCES: **Kew Gardens.** 1986, *Hastings* (LN65: 196; S33: 743): on disturbed ground.
CULT.: 1789, Alien (HK1, 2: 270); 1811, Alien (HK2, 3: 358); 1814, Alien (HKE: 174). In cult.
CURRENT STATUS: No recent sightings. A rare casual outside of Cornwall and only likely to return as an escape from cultivation.

95. **Ranunculus parviflorus** L.

small-flowered buttercup

CULT.: 1768 (JH: 187); 1789, Nat. of England (HK1, 2: 270); 1811, Nat. of England (HK2, 3: 358); 1814, Nat. of England (HKE: 174); 1924, unspec. (K); 1936, unspec. (K). Not currently in cult.

96. **Ranunculus arvensis** L.

corn buttercup

CULT.: 1768 (JH: 187); 1789, Nat. of Britain (HK1, 2: 269); 1811, Nat. of Britain (HK2, 3: 358); 1814, Nat. of Britain (HKE: 174); 1867, unspec. (K). Not currently in cult.
Note. Originally grown as a native, but now regarded as an archaeophyte.

97. **Ranunculus auricomus** L.

Goldilocks buttercup

FIRST RECORD: 1909 [1768].
REFERENCES: **Kew Gardens.** 1909, *Turrill* in *Rolfe & A.B.Jackson* (S10: 373): A large patch on the side of the path at the back of the Tropical Aroid House. **Kew Environs.** 1926, *Marquand* (BSEC8(4): 561): Meadow near the river, Kew; depauperate type.
EXSICC.: **Kew Gardens.** Behind Aroid House; v 1919; *Turrill* s.n. (K). • Under Walnut tree behind Aroid House; 17 vi 1919; *Turrill* s.n. (K). • Under Walnut tree behind Aroid House; 6 iv 1920; *Turrill* s.n. (K). • Under *Juglans* tree behind Aroid House No. 1; growing thickly in grass; 27 iv 1927; *Turrill* s.n. (K). • Back of Aroid House No. 1; in grass under *Juglans* tree; 1 vi 1927; *Turrill* s.n. (K). • Near herbarium; amongst grasses; 12 v 1929; *Hubbard* s.n. (K). • Near herbarium field; amongst grasses in patches; 12 v 1929; *Hubbard* s.n. (K). • Herbarium grounds; 22 v 1936; *Ross-Craig, Sealy & Burtt* 467 (K). • Near Herbarium Keeper's Garden; v 1947; *Hutchinson* s.n. (K). • Rear of Number One; growing under Celtis and Walnut tree, in shady position, soil sandy loam, *Luzula campestris* in same situation; 15 vi 1948; *O.J.Ward* 119 (K).
CULT.: 1768 (JH: 187); 1789, Nat. of Britain (HK1, 2: 266); 1811, Nat. of Britain (HK2, 3: 354); 1814, Nat. of Britain (HKE: 174); 1927, Herb. Expt. Ground (K); 1985, Alpine Dept. (K); 1986, Alpine Dept. (K). In cult.
CURRENT STATUS: Certainly extinct in the areas around the Aroid House (Nash Conservatory) and

Goldilocks buttercup (*Ranunculus auricomus*). Once known from near the Nash Conservatory but became extinct there in the 1980s; recently rediscoverd in another part of the Gardens.

the Herbarium since 1980 despite conservation efforts. One patch was recently discovered in long grass between the Water Lily House and the Cycad House.

98. Ranunculus sceleratus L.
celery-leaved buttercup
FIRST RECORD: 1873 [1768].
NICHOLSON: **1873/4 survey** (JB: 10): Two plants near end of lake. One near wall of moat at end of "Syon Vista," 1873. None seen in 1874. • **1906 revision** (AS: 74): **A.** By lake.
EXSICC.: **Kew Environs.** Rt. bank of Thames, near Kew Bridge, towards Richmond side; vii 1928; *C.A. Smith* 6081 (K).
CULT.: 1768 (JH: 187); 1789, Nat. of Britain (HK1, 2: 266); 1811, Nat. of Britain (HK2, 3: 354); 1814, Nat. of Britain (HKE: 174); 1933, Herbaceous Ground (K). Not currently in cult.
CURRENT STATUS: Still not at all common but can be found in the Gravel Pit in the Conservation Area and in the Moat adjacent to the Ancient Meadow.

99. Ranunculus lingua L.
greater spearwort
CULT.: 1768 (JH: 186); 1789, Nat. of Britain (HK1, 2: 265); 1811, Nat. of Britain (HK2, 3: 352); 1814, Nat. of Britain (HKE: 174). In cult.

100. Ranunculus flammula L.
lesser spearwort
CULT.: 1768 (JH: 186); 1789, Nat. of Britain (HK1, 2: 265); 1811, Nat. of Britain (HK2, 3: 351); 1814, Nat. of Britain (HKE: 174). Not currently in cult.

101. Ranunculus reptans L.
creeping spearwort
CULT.: 1811, Nat. of Britain (HK2, 3: 351); 1814, Nat. of Britain (HKE: 174). Not currently in cult.

102. Ranunculus ficaria L.
incl. *R. ficaria* subsp. *bulbilifer* Lambinon
lesser celandine
FIRST RECORD: 1873/4 [1768].
NICHOLSON: **1873/4 survey** (JB: 10): **B.** Plentiful under trees near "Grand Entrance" and "Cumberland Gate," also under the two large limes on which the mistletoe grows. • **1906 revision** (AS: 74): Plentiful in shady places.
EXSICC.: **Kew Gardens.** Under walnut-tree at back of Aroid House; 10 iii 1921; *Turrill* s.n. (K). • Herbarium grounds; 19 iv 1922, *anon.* (K). • Path between Herbarium & Tennis Courts; under Horse Chestnut tree with Pines on one side; 30 i 1928; *Summerhayes & Turrill* 3919 (K) & 3927 (K). By path between Herbarium & Tennis Courts; under Lime tree; 30 i 1928; *Turrill & Summerhayes* 3927 bis (K). • Near Office of Works; in grass; 9 ii 1928; *Summerhayes & Turrill* s.n. (K). • Near Office of

Lesser celandine (*Ranunculus ficaria*). An attractive spring flower in scattered localities but never common and often regarded as an unwelcome weed.

Works; under Horse Chestnut, with *Aegopodium*; 9 ii 1928; *Summerhayes & Turrill* 3926 (K). • Tow-path just outside Kew Gardens; in reed swamp, submerged at high-tide with *Caltha*, *Oenanthe crocata*, *Phalaris*; 19 iii 1928; *Summerhayes & Turrill* s.n. (K). • Office of Works; under lime among ivy; 2 iv 1928; *Summerhayes & Turrill* 3920 (K). • Near the Herbarium, under *Aesculus* tree; 7 iv 1928; *Summerhayes & Turrill* 3922 (K). • [Unlocalised]; on rubbish heap, growing under Sycamore among rather large leaved ivy; 2 iv 1928; *Summerhayes & Turrill* 3923 (K). • River-bank near Isleworth Gate; lowest zone, in water, tall plant in clump of *Oenanthe crocata* & *Phalaris*; 2 v 1928; *Turrill* s.n. (K). • River bank near Isleworth Gate; top zone, on ± dry bank; 2 v 1928; *Turrill* s.n. (K). • Just by Office of Works; under trees in grass, *Aegopodium* & ivy; 15 v 1928; *Summerhayes & Turrill* s.n. (K). • River-bank between Kew and Richmond near the Isleworth Gate; growing in *Phalaris-Oenanthe* marsh; 19 x 1928; *Summerhayes & Turrill* 3931 (K). • Near the Tennis Courts; under *Juglans* tree, amongst ivy; 22 x 1928; *Summerhayes & Turrill* 3930A (K). • [Unlocalised]; under Aesculus tree; 22 x 1928; *Summerhayes & Turrill* 3930B (K). • Near Herbarium; under lime tree; 22 x 1928; *Summerhayes & Turrill* 3930C (K). • Near the Office of Works; 22 x 1928; *Summerhayes & Turrill* 3930D (K). • Near Near Office of Works; in shade of trees with ivy; 10 v 1929; *Summerhayes & Turrill* 3925 (K). • Thames-bank, near Isleworth Gate; in *Phalaris* zone in front of *Petasites* zone; 25 vi 1929; *Summerhayes & Turrill* 3928 (K). • Tow Path by Kew Palace; 3 iv 1948; *O.J.Ward* 118 (K). **Kew Environs.** River-bank between Kew and Richmond; inter-tidal zone; 25 v 1928; *Summerhayes & Turrill* 3929 (K). • Kew – Richmond tow-path; found in hedgerows in shady position & in rather poor soil, associated with *Galium aparine* & *Urtica dioica*; 15 iv 1948; *Parker* s.n. (K).

CULT.: 1768 (JH: 187); 1789, Nat. of Britain (HK1, 2: 266); 1811, Nat. of Britain (HK2, 3: 353); 1814, Nat. of Britain (HKE: 174); 1905, unspec. (K); 1926, unspec. (K); 1929, grown in pot from natural seed collected near Office of Works, Kew, 1928, sown at once, germinated next spring (K); 1951, Herb. Expt. Ground (K); numerous cytology vouchers from the mid-1960s. In cult.

CURRENT STATUS: Scattered but never common; the subspecies of recent sightings have not yet been determined; those of Nicholson are unknown and much of the available herbarium material was collected either without the base or too early for any tubers to have formed. It is probable that both subsp. *ficaria* and subsp. *bulbilifer* occur in the Gardens.

[**103. Ranunculus hederaceus** L.

ivy-leaved crowfoot

CULT.: 1768 (JH: 187); 1789, Nat. of Britain (HK1, 2: 270); 1811, Nat. of Britain (HK2, 3: 358); 1814, Nat. of Britain (HKE: 174). Not currently in cult.]

104. Ranunculus aquatilis L.

R. heterophyllus Wigg.

common water-crowfoot

FIRST RECORD: 1931 [1768].

REFERENCES: **Kew Gardens.** 1931, *Fraser* (SFS: 91; S33: 743): In the ha-ha.

CULT.: 1768 (JH: 187); 1789, Nat. of Britain (HK1, 2: 270); 18111, Nat. of Britain (HK2, 3: 359); 1814, Nat. of Britain (HKE: 174). Not currently in cult.

CURRENT STATUS: No recent sightings. There is now no suitable habitat for it.

105. Ranunculus peltatus Schrank

pond water-crowfoot

FIRST RECORD: 1894.

EXSICC.: **Kew Environs.** Ditch by River Thames, Kew; 26 v 1894; *Hosking* s.n. (K).

CURRENT STATUS: No recent sightings. There is now no suitable habitat for it.

106. Ranunculus penicillatus subsp. **pseudofluitans** (Syme) S.Webster

R. pseudofluitans (Syme) Newbould ex Baker & Foggitt

R. pseudofluitans var. *penicillatus* (Dumort.) Hiern

stream water-crowfoot

FIRST RECORD: 1883.

REFERENCES: **Kew Environs.** 1883, *Crespigny* (BLRC2: 203; S33: 743): Ditch by the towing-path between Kew and Richmond. • 1931, *Wolley-Dod* (SFS: 93; S33: 743): Ditch between Kew Gardens and river.

CURRENT STATUS: No recent sightings. There is now no suitable habitat for it.

107. Ranunculus fluitans Lam.

R. peltatus var. *penicillatus* auct. *fide* JB: 77

river water-crowfoot

FIRST RECORD: 1867 [1789].

NICHOLSON: **1873/4 survey** (JB: 10): Moat and lake. Not common in latter place. • **1906 revision** (AS: 74): Lake and ha-ha.

EXSICC.: **Kew Gardens.** Ha-ha; 9 vii 1920; *Fraser* s.n. (K). **Kew Environs.** Kew; 1867, *Newbould* in *Herb. Watson* (K).

CULT.: 1789, Nat. of Britain (HK1, 2: 271); 1811, Nat. of Britain (HK2, 3: 359); 1814, Nat. of Britain (HKE: 174). Not currently in cult.

CURRENT STATUS: No recent sightings. The riverside habitat no longer exists and it has been lost from the Lake.

108. Ranunculus circinatus Sibth.

R. aquatilis var. b

fan-leaved water-crowfoot

FIRST RECORD: 1873/4 [1789].

NICHOLSON: **1873/4 survey** (JB: 77; SFS: 88): Moat by Kew Gardens. • **1906 revision** (AS: 74): Ha-ha.

OTHER REFERENCES: **Kew Gardens.** 1924, *Catchside* (WBEC 41: 285; LN30: S3): In the moat at Kew.

CULT.: 1789, Nat. of Britain (HK1, 2: 270); 1811, Nat. of Britain (HK2, 3: 359); 1814, Nat. of Britain (HKE: 174). Not currently in cult.

CURRENT STATUS: No recent sightings. There is now no suitable habitat for it.

[**109. Ranunculus gramineus** L.

CULT.: 1789, Alien (HK1, 2: 265); 1811, Nat. of Wales (HK2, 3: 352); 1814, Nat. of Wales (HKE: 174). Note. Listed in error as native.]

110. Adonis annua L.

A. autumnalis L.

pheasant's-eye

CULT.: 1789, Nat. of England (HK1, 2: 264); 1811, Nat. of Britain (HK2, 3: 350); 1814, Nat. of Britain (HKE: 173); 1911, unspec. (K); 1914, seedlings, unspec. (K); 1935, seedlings, unspec. (K). In cult. Note. Originally grown as a native, but now regarded as an archaeophyte.

111. Myosurus minimus L.
mousetail

FIRST RECORD: 2003 [1768].
REFERENCES: **Kew Gardens.** 2003, *Cope* (S35: 646): In bare ground along the Riverside Walk (213).
EXSICC.: **Kew Gardens.** • Riverside Walk; in recently cleared bare ground under trees by the boundary fence; 7 v 2003; *Cope* 704 (K).
CULT.: 1768 (JH: 184); 1789, Nat. of Britain (HK1, 1: 399); 1811, Nat. of Britain (HK2, 2: 199); 1814, Nat. of Britain (HKE: 85); 1932, Herbaceous Ground (K). In cult.
CURRENT STATUS: Extremely rare; a handful of plants appeared in the Riverside Walk in 2003 following clearance of rough vegetation. Whether or not these were the progeny of previously cultivated material has not been established. If they were, then they have either passed the last 70 years as seed (a long way from the likely source) or the plant has flowered regularly but has gone unnoticed. Neither possibility seems very likely but it is hard to believe that it could occur in the Gardens as a truly wild plant. By 2005 it had almost vanished under encroaching grass. In 2008, there was an unconfirmed report of it from the Rhododendron Dell during a weeding operation and the plant may have been lost before its identity could be confirmed. The habitat in the Gardens is not entirely typical and considerable management input will be required to maintain the population.

112. Aquilegia vulgaris L.
columbine

CULT.: 1768 (JH: 346); 1789, Nat. of Britain (HK1, 2: 247); 1811, Nat. of Britain (HK2, 3: 325); 1814, Nat. of Britain (HKE: 170); 1855, unspec. (K); 1856, unspec. (K); 1878, unspec. (K); 1882, unspec. (K). In cult.

113. Thalictrum flavum L.
common meadow-rue

CULT.: 1768 (JH: 340); 1789, Nat. of Britain (HK1, 2: 263); 1811, Nat. of Britain (HK2, 3: 349); 1814, Nat. of Britain (HKE: 173); 1846, unspec. (K); 1878, unspec. (K); 1882, unspec. (K); 1901, unspec. (K); 1916, unspec. (K); 1929, unspec. (K); 1988, Alpine & Herbaceous (K). In cult.

114. Thalictrum minus L.
T. majus Crantz
lesser meadow-rue

CULT.: 1768 (JH: 340); 1789, Nat. of Britain (HK1, 2: 262); 1811, Native of Britain (HK2, 3: 348), Nat. of England (ibid.: 347 as *majus*); 1814, Nat. of Britain (HKE: 173), Nat. of England (ibid.: as *majus*); 1879, unspec. (K); 1929, cult. in experimental plot (K); 1988, Herbaceous, bed 156-01 (K); 1988, Alpine & Herbaceous (K). In cult.

Mousetail (*Myosurus minimus*). A rare British native that unexpectedly appeared along the Riverside Walk in 2003. Its origin in the Gardens has never been resolved and careful management is required to maintain it.

115. Thalictrum alpinum L.
alpine meadow-rue

CULT.: 1768 (JH: 340); 1789, Nat. of Britain (HK1, 2: 261); 1811, Nat. of Britain (HK2, 3: 346); 1814, Nat. of Britain (HKE: 173). Not currently in cult.

BERBERIDACEAE

116. Berberis vulgaris L.
barberry

CULT.: 1768 (JH: 435); 1789, Nat. of Britain (HK1, 1: 479); 1811, Nat. of England (HK2, 2: 313); 1814, Nat. of England (HKE: 102); pre-1867, unspec. (K); 1880, Arboretum (K); 1881, unspec. (K); 1882, Arboretum (K); 1883, unspec. (K); 1888, Arboretum (K); 1891, unspec. (K); 1891, Berberis Dell (K); 1895, Arboretum (K); 1900, Arboretum (K); 1911, unspec. (K); 1916, Arboretum (K); 1921, unspec. (K); 1930, Arboretum (K); 1931, Arboretum (K); 1944, Berberis Dell (K); 1945, Berberis Dell (K); 1960, Arboretum (K); 1961, Arboretum (K); 1990, unspec. (K); 1996, bed 463-03 (K); 1997, bed 463-03 (K). In cult.

[**117. Epimedium alpinum** L.
barren-wort

CULT.: 1789, Alien (HK1, 1: 157); 1810, Nat. of England (HK2, 1: 260); 1814, Nat. of England (HKE: 35).
Note. Listed in error as native.]

PAPAVERACEAE

118. *Papaver pseudoorientale (Fedde) Medw.
oriental poppy
FIRST RECORD: 2001 [1984].
REFERENCES: **Kew Gardens.** 2001, *Cope* (S35: 646): Incorporated in the Robinsonian Meadow sown in front of the temporary Cycad House (224), and now naturalised.
CULT.: 1984, bed 156-13 (K); 1998, bed 156-11 (K); 1998, bed 156-13 (K). Not currently in cult.
CURRENT STATUS: Naturalised in the Robinsonian Meadow in front of 'Climbers & Creepers' and may persist for a short while.

119. Papaver somniferum L.
opium poppy
FIRST RECORD: 1983/4 [1768].
REFERENCES: **Kew Gardens.** 1983/4, *Cope* (S31: 181): New student vegetable plots behind Hanover House.
CULT.: 1768 (JH: 172/6); 1789, Nat. of England (HK1, 2: 224); 1811, Nat. of England (HK2, 3: 290); 1814, Nat. of England (HKE: 164); 1856, unspec. (K); 1876, unspec. (K); 1933, unspec. (K); 1936, Herbaceous Ground (K). In cult.
CURRENT STATUS: Occasional; it has largely disappeared from the Paddock where it was abundant in 1990. In 2004 a demonstration plot was sown at the end of Syon Vista where the Wheatfield had been in previous years. It is occasionally found elsewhere as a casual, mostly in its common pale lilac colour but sometimes in more exotic forms.

120. Papaver rhoeas L.
P. strigosum (Boenn.) Schur
common poppy
FIRST RECORD: 1873/4 [1768].
NICHOLSON: **1873/4 survey** (JB: 10): **B, P, Strip.** Here and there, not common. • **1906 revision** (AS: 74): Here and there, not common.
OTHER REFERENCES: **Kew Environs.** 1878, *G.Nicholson* (BEC18: 13): Kew.
EXSICC.: **Kew Environs.** [Unlocalised]; vii 1878; *G.Nicholson* 51b (K).
CULT.: 1768 (JH: 172/6); 1789, Nat. of Britain (HK1, 2: 224); 1811, Nat. of Britain (HK2, 3: 289); 1814, Nat. of Britain (HKE: 164); 1933, Arboretum (K); 1935, Herbarium Ground (K); 1936, Herbarium Ground (K); 1955, Herbarium Ground (K); 1964, Order Beds (K). In cult.
CURRENT STATUS: Still occurs in the Paddock and elsewhere; it was sown in the Wheatfields along the Broad Walk in 1999 and was abundant the following year. It was then sown into the new Wheatfield at the south end of the Lake in 2001. It is likely to come up almost anywhere in the Gardens from seed persisting in the soil.

121. Papaver dubium L.
long-headed poppy
FIRST RECORD: 1888 [1768].
REFERENCES: **Kew Gardens.** 1963, *J.L.Gilbert* (S29: 406): Near Herbarium cycle shed.
EXSICC.: **Kew Gardens.** Mound of earth from excavation for foundations of new wing; 14 x 1965; *Verdcourt* 4240 (K). **Kew Environs.** Thames Banks, Kew; 10 vi 1888; *Fraser* s.n. (K).
CULT.: 1768 (JH: 172/6); 1789, Nat. of Britain (HK1, 2: 224); 1811, Nat. of Britain (HK2, 3: 289); 1814, Nat. of Britain (HKE: 164); 1934, Herbarium Ground (K); 1973, S. Arboretum (K). In cult.
CURRENT STATUS: Still occurs occasionally but is not as abundant as it was in 1990 in what is left of the Paddock.

Common poppy (*Papaver rhoeas*). Widespread in flower-beds and often very abundant. It has been sown several times into 'wheatfield' displays and usually persists.

Long-headed poppy (*Papaver dubium*). Generally scarcer than common poppy but was abundant in the Paddock after its restoration in 1990; it has declined since.

122. Papaver hybridum L.
rough poppy
CULT.: 1768 (JH: 172/6); 1789, Nat. of England (HK1, 2: 223); 1811, Nat. of England (HK2, 3: 288); 1814, Nat. of England (HKE: 164); 1916, unspec. (K). Not currently in cult.
Note. Originally cultivated as a native, but now regarded as an archaeophyte.

123. Papaver argemone L.
prickly poppy
FIRST RECORD: 1966 [1768].
REFERENCES: **Kew Gardens.** 1990, *Cope* (S31: 182): Paddock, after restoration in 1990.
EXSICC.: **Kew Environs.** [Unlocalised]; 7 vii 1966; *Meikle* s.n. (K).
CULT.: 1768 (JH: 172/6); 1789, Nat. of Britain (HK1, 2: 223); 1811, Nat. of Britain (HK2, 3: 289); 1814, Nat. of Britain (HKE: 164); 1932, unspec. (K); 1967, Experimental Ground (K). Not currently in cult.
CURRENT STATUS: Not seen in the Paddock in recent years; it was first noted here in 1987 (unpubl.) and briefly reappeared after major disturbance. By far the rarest of the poppies in the Gardens.

124. Meconopsis cambrica (L.) Vig.
Papaver cambrica L.
Welsh poppy
CULT.: 1768 (JH: 172/6); 1789, Nat. of Wales (HK1, 2: 225); 1811, Nat. of England (HK2, 3: 290); 1814, Nat. of England (HKE: 165). Not currently in cult.

125. Glaucium flavum Crantz
Chelidonium glaucium L.
Glaucium luteum Crantz
yellow horned-poppy
FIRST RECORD: 2004 [1789].
CULT.: 1789, Nat. of Britain (HK1, 2: 223); 1811, Nat. of Britain (HK2, 3: 287); 1814, Nat. of Britain (HKE: 164); 1904, unspec. (K); 1910, unspec. (K); 1929, unspec. (K); 1932, Herbaceous Ground (K); 1975, bed 156-11 (K); 1998, bed 156-11 (K). In cult.
CURRENT STATUS: Grew from seed in topsoil being stored in the Paddock but did not persist.

[**126. Glaucium corniculatum** (L.) Rudolph
Chelidonium corniculatum L.
Glaucium phoenicium Crantz, nom. illegit.
red horned-poppy
CULT.: 1789, Nat. of England (HK1, 2: 223); 1811, Nat. of England (HK2, 3: 288); 1814, Nat. of England (HKE: 164).
Note. Listed in error as a native.]

[**127. Roemeria hybrida** (L.) DC.
Chelidonium hybridum L.
Glaucium violaceum Juss.
Violet Horned-poppy
CULT.: 1789, Nat. of Wales (HK1, 2: 223); 1811, Nat. of England (HK2, 3: 288); 1814, Nat. of England (HKE: 164).
Note. Listed in error as a native.]

128. Chelidonium majus L.
C. majus var. *laciniatum* (Mill.) Koch
greater celandine
FIRST RECORD: 1873/4 [1789].
NICHOLSON: **1873/4 survey** (JB: 10): **B** and **Pal.** Abundant in latter division. • **1906 revision** (AS: 74 and in SFS: 113): **B, P.** Abundant in latter division. Var. *laciniatum*: **B, P.** Here and there with the type.
EXSICC.: **Kew Gardens.** Near Office of Works; seedlings near mature plants, on bare ground beneath trees and shrub layer; 2 iv 1928; *Summerhayes & Turrill* s.n. (K). • Near Office of Works; in shade under trees & at base of wall; 6 v 1928; *Summerhayes & Turrill* s.n. (K). • Near Office of Works Dept.; against wall, in sandy soil, abundant; 16 vi 1928; *Hubbard* s.n. (K). • Near Office of Works Dept.; by edge of wall, common, 18 vi 1928;

Hubbard s.n. (K). • Grounds of Herbarium House; 11 vi 1958; *Dunk, Halliday & Uchlein* 146 (K). **Kew Environs.** Towpath between Kew & Richmond; in semi-shade under trees; 23 v 1944; *Souster* 68 (K).
CULT.: 1789, Nat. of Britain (HK1, 2: 222); 1811, Nat. of Britain (HK2, 3: 287); 1814, Nat. of Britain (HKE: 164); 1856, unspec. (K); 1897, unspec. (K); 1898, unspec. (K); 1904, unspec. (K); 1998, bed 156-14 (K). In cult.
CURRENT STATUS: Self-sown in several flower-beds and in neglected corners by buildings.

129. *Eschscholzia californica Cham.
Californian poppy
FIRST RECORD: 1932 [1883].
EXSICC.: **Kew Gardens.** On rubbish tip in lower nursery; 1 vii 1932; *Hubbard* s.n. (K).
CULT.: 1883, unspec. (K); 1894, unspec. (K); 1998, bed 156-11 (K). In cult.
CURRENT STATUS: Naturalised in a few scattered localities, especially on disturbed soil.

130. *Macleaya cordata (Willd.) R.Br.
five-seeded plume-poppy
FIRST RECORD: 2004.
CULT.: In cult.
CURRENT STATUS: Grew from seed in topsoil being stored in the Paddock, but did not persist.

[**131. Corydalis solida** (L.) Clairv.
Fumaria solida (L.) Mill.
F. bulbosa var. *solida* L.
bird-in-a-bush
CULT.: 1789, Alien (HK1, 3: 2); 1812, Nat. of England (HK2, 4: 240); 1814, Nat. of England (HKE: 219).
Note. Listed in error in later editions of *Hort. Kew.* as a native; correctly listed as a alien in the first edition.]

132. *Pseudofumaria lutea (L.) Borkh.
yellow corydalis
FIRST RECORD: 2000 [1789].
REFERENCES: **Kew Gardens.** 2000, *Cope* (S35: 646): Around the Marianne North Gallery. Subsequently near the Water Lily House (163) and the Administration Block.
CULT.: 1789, Alien (HK1, 3: 2); 1812, Nat. of England (HK2, 4: 240); 1814, Nat. of England (HKE: 219). Not currently in cult.
CURRENT STATUS: Naturalised in a few places, especially near buildings.

133. Ceratocapnos claviculata (L.) Lidén
Fumaria claviculata L.
climbing corydalis
CULT.: 1789, Nat. of Britain (HK1, 3: 3); 1812, Nat. of Britain (HK2, 4: 242); 1814, Nat. of Britain (HKE: 219). Not currently in cult.

134. Fumaria capreolata subsp. **babingtonii** (Pugsley) P.D.Sell
white ramping-fumitory
FIRST RECORD: 2002 [1789].
REFERENCES: **Kew Gardens.** 2002, *Cope* (S35: 646): Ancient Meadow (321).
CULT.: 1789, Nat. of Britain (HK1, 3: 3); 1812, Nat. of Britain (HK2, 4: 241); 1814, Nat. of Britain (HKE: 219); 1922, unspec. (K). Not currently in cult.
CURRENT STATUS: Extremely rare; so far there is just a single record from the Ancient Meadow near Isleworth Gate.

135. Fumaria muralis subsp. **boraei** (Jord.) Pugsley
common ramping-fumitory
FIRST RECORD: pre-1862 [1887].
REFERENCES: **Kew Gardens.** 1998, *Cope* (S33: 743): North Arb.: lawn behind Herbarium after disturbance during building works (113).
EXSICC.: **Kew Environs.** Kew [as *F. muralis*], *Borrer* in *Herb. Borrer* (K). • Kew [2 sheets as *F. glabella*], *Borrer* in *Herb. Borrer* (K). • Kew; vi 1870; *Baker* in *Herb. Watson* (K).
CULT.: 1887, unspec. (K). Not currently in cult.
CURRENT STATUS: Still to be found behind the Herbarium and appeared along the Broad Walk as a weed in the Wheatfields that were sown in 1999. Rather scattered elsewhere but still the commonest of the three fumitories. The specimens in the Borrer herbarium are undated, but are probably his own collections; they could have been collected as early as 1805 or as late as the early 1860s, but the bulk of his specimens seem to date from between 1829 and 1862 (the year he died).

136. Fumaria officinalis L.
common fumitory
FIRST RECORD: 1873/4 [1789].
NICHOLSON: **1873/4 survey** (JB: 10): **B, P, Q, Pal.** Common. • **1906 revision** (AS: 74): Common in all the divisions.
EXSICC.: **Kew Gardens.** East of bowling green; on recently turned-over soil; 28 v 1929; *Ballard* s.n. (K).
CULT.: 1789, Nat. of Britain (HK1, 3: 3); 1812, Nat. of Britain (HK2, 4: 241); 1814, Nat. of Britain (HKE: 219); 1936, Herbarium Ground (K). Not currently in cult.
CURRENT STATUS: Rare and widely scattered. The previous species is now much commoner.

137. Fumaria densiflora DC.
dense-flowered fumitory
CULT.: 1887, unspec. (K). Not currently in cult.

138. Fumaria parviflora Lam.
few-flowered fumitory
CULT.: 1812, Nat. of England (HK2, 4: 241); 1814, Nat. of England (HKE: 219). Not currently in cult.

PLATANACEAE

139. *Platanus × hispanica Mill. ex Münchh. (occidentalis × orientalis)
P. × acerifolia (Aiton) Willd.
P. × hybrida Brot.
London plane
FIRST RECORD: 1941 [1843–53].
REFERENCES: **Kew Gardens.** 1958, *Welch* (LFS: 238): Small plant nr. Brentford Gate. • 1990, *Cope* (S31: 182): Seedling in paddock, after restoration in 1990. **Kew Environs.** 1941, *Cooke* (LN34: S253): River wall between Kew and Richmond.
CULT.: 1843–1853, Arboretum (K); 1856, Palace Grounds (K); 1862, Herbarium Grounds (K); 1868, Hollow Walk, Pleasure Gnds (K); 1880, Arboretum (K); 1881, emormous tree overhanging Rhododendron Dell nr. Tulip Tree (K); 1881, small tree between *Pinus excelsa* & Atlas Cedar (K); 1881, in right side of road, Cumberland Gate (K); 1884, Arboretum (K); 1885, Arboretum (K); 1916, Rhod. Dell (K); 1917, Kew Green (K); 1919, Azalea Garden (K); 1919, Herbarium Ground (K); 1941, large tree near King Wm's Temple (K); 1944, Kew Green (K). In cult.
CURRENT STATUS: Has not persisted in the Paddock but there are other recent sightings of seedlings at the base of the wall of Wing A. No doubt these are derived from one or other of the two fine specimens on the edge of the Paddock near the boundary wall. Six or seven young trees were planted beyond the wall adjacent to Ferry Lane, and several other mature trees exist in various parts of the Gardens.

ULMACEAE

140. Ulmus glabra Huds.
U. campestris var. *glabra* (Huds.) auct. dub.
U. campestris var. *latifolia* auct. dub.
wych elm
CULT.: 1789, Nat. of Britain (HK1, 1: 319); 1814, Nat. of England (HKE: Add); 1858, centre tree in front of Herbarium (K); 1880, Arboretum (K); 1881, Arboretum (K); 1881, Palace Nursery (Ko; 1885, Arboretum (K); 1892, Arboretum (K); 1896, Arboretum (K); 1900, Arboretum (K); 1907, Arboretum (K); 1933, Elm Collection (K); 1936, Arboretum (K); 1937, Arboretum (K); 1938, unspec. (K); 1961, Arboretum (K); 1986, Arboretum (K). In cult.

141. Ulmus × elegantissima Horw. (glabra × plotii)
CULT.: 1900, Arboretum (K); 1958, unspec. (K); 1965, unspec. (K). Not currently in cult.

142. Ulmus × vegeta (Loudon) Ley (glabra × minor)
Huntingdon elm
CULT.: 1880, Arboretum (K); 1900, Arboretum (K); 1933, Arboretum (K); 1936, Arboretum (K); 1937, Arboretum (K); 1938, Arboretum (K); 1958, Arboretum (K). Not currently in cult.

143. Ulmus × hollandica Mill. (glabra × minor? × plotii)
U. campestris var. *fungosa* auct. dub.
U. suberosa Moench
Dutch elm
FIRST RECORD: 1910 [1789]
EXSICC.: **Kew Environs.** Towing path, Kew; 8 ix 1910; *Fraser* s.n. (K).
CULT.: 1789, Nat. of Britain (HK1, 1: 319); 1811, Nat. of England (HK2, 2: 107); 1814, Nat. of England (HKE: 72); 1880, Arboretum (K); 1881, Arboretum (K); 1892, Arboretum (K); 1900, Arboretum (K); 1910, Arboretum (K); 1911, Arboretum (K); 1929, Palace Lawn (K); 1932, on the lawn behind the Ice Well (K); 1933, Arboretum (K); 1936, Arboretum (K); 1937, Arboretum (K); 1938, Arboretum (K); 1949, Arboretum (K); 1955, Arboretum (K); 1958, Arboretum (K); 1960, unspec. (K). In cult.
CURRENT STATUS: Its status as a wild plant is not known, but there are still several plants in cultivation.

144. Ulmus procera Salisb.
U. campestris auct. angl.
English elm
FIRST RECORD: 1888 [1768].
EXSICC.: **Kew Environs.** Kew; iii 1888; *Gamble* 19839 (K).
CULT.: 1768 (JH: 458); 1789, Nat. of Britain (HK1, 1: 319); 1811, Nat. of England (HK2, 2: 107), Nat. of Britain (ibid.); 1814, Nat. of Britain (HKE: 72); 1880, Arboretum (K); 1881, unspec. (K); 1885, Arboretum (K); 1888, next Lodge in front of Herb. (K); 1910, unspec. (K); 1912, Arboretum (K); 1935, Arboretum (K); 1936, on riverside of Willow Colln. (K); 1936, near Kew Palace (K); 1936, Seven Sisters Lawn (K); 1936, Willow Colln., Queen's Cott. Gnd. (K); 1936, in nursery by Brentford Ferry Gate (K); 1936, by Grand Ave. (K); 1937, Willow Colln. (K); 1937, Seven Sisters Lawn (K); 1937, W. side of Ruined Arch (K); 1937, by Rhododendron Dell (K); 1937, W. of Cedar in Colln. (K); 1937, Riverside nursery by Brentford Gate (K); 1937, near *Quercus cerris* in Grand Ave. (K); 1938, by Brentford Gate (K); 1938, second of two trees S. of Victoria Gate (K); 1942, unspec. (K); 1945, by Grand Ave. (K);

1949, Arboretum (K); 1961, unspec. (K). Not currently in cult.
CURRENT STATUS: Numerous saplings occur in the Conservation Area originating from cultivated trees and first noticed only as recently as 1976.

145. Ulmus minor Mill.
a. subsp. **minor**
U. carpinifolia Gled.
small-leaved elm
FIRST RECORD: 1910 [1880].
EXSICC.: **Kew Environs.** Towing path, Kew; 8 ix 1910 and 16 iii 1911; *Fraser* s.n. (K).
CULT.: 1880, Arboretum (K); 1881, Arboretum (K); 1885, Arboretum (K); 1892, Arboretum (K); 1894, Arboretum (K); 1899, unspec. (K); 1900, Arboretum (K); 1901, Arboretum (K); 1911, unspec. (K); 1912, Arboretum (K); 1922, Palace Lawn (K); 1933, Arboretum (K); 1936, Arboretum (K); 1937, Arboretum (K); 1942, Arboretum (K); 1986, Arboretum (K). Not currently in cult.
CURRENT STATUS: Not known; there are no recent confirmed sightings.

b. subsp. **angustifolia** (Weston) Stace
U. campestris var. *stricta* auct. dub.
Cornish elm
CULT.: 1789, Nat. of Britain (HK1, 1: 319) 1936, Arboretum (K); 1937, Arboretum (K). In cult.

Hop (*Humulus lupulus*). Common on the riverbank and in the Old Deer Park, but rare in the Gardens.

c. subsp. **sarniensis** (C.K.Schneid.) Stace
Jersey elm
CULT.: 1880, Arboretum (K); 1881, Arboretum (K); 1892, Arboretum (K); 1894, Arboretum (Ko; 1900, Arboretum (K); 1936, Arboretum (K); 1937, Arboretum (K); 1938, Arboretum (K); 1945, Arboretum (K); 1949, Arboretum (K); 1958, unspec. (K). Not currently in cult.

146. Ulmus × viminalis Lodd. ex Loudon (minor × plotii)
CULT.: 1880, Arboretum (K); 1892, Arboretum (K); 1900, Arboretum (K); 1910, near Cart Road near Ash Collection (K); 1933, Arboretum (K); 1936, Arboretum (K); 1937, Arboretum (K); 1938, near *Quercus cerris* on Grand Walk (K); 1949, Arboretum (K); 1958, Arboretum (K). Not currently in cult.

147. Ulmus plotii Druce
Plot's elm
CULT.: 1899, unspec. (K); 1900, Arboretum (K); 1936, Arboretum (K); 1937, unspec. (K); 1969, Elm Collection, near Brentford Ferry Gate (K). In cult.

CANNABACEAE

148. *Cannabis sativa L.
hemp
FIRST RECORD: 2003 [1768].
CULT.: 1768 (JH: 420). Not currently in cult.
CURRENT STATUS: A couple of plants were found in a neglected corner of the Old Deer Park, possibly – but doubtfully – originating from birdseed. The site has since been cleared.

149. Humulus lupulus L.
hop
FIRST RECORD: 1884 [1768].
EXSICC.: **Kew Gardens.** Thames Bank, Kew – Richmond, near Isleworth Gate; top zone above *Petasites*; 16 viii 1928; *Summerhayes & Turrill* s.n. (K). • Arboretum; 13 ix 1934, *Dallimore* 14 (K). **Kew Environs.** Thames banks, Kew; 31 viii 1884; *Fraser* s.n. (K). • River-bank, Kew; viii 1929; *Pearce* s.n. (K).
CULT.: 1768 (JH: 421); 1789, Nat. of Britain (HK1, 3: 400); 1813, Nat. of Britain (HK2, 5: 386); 1814, Nat. of Britain (HKE: 309); 1934, Herbarium Ground (K); 1936, Herbaceous Ground (K); 1936, Herbarium Ground (K). In cult.
CURRENT STATUS: Still present on the riverbank but is very rare within the Gardens; on the golf course in the Old Deer Park there is quite a lot of it to be seen scrambling over trees and shrubs.

MORACEAE

150. *Ficus carica L.

fig

FIRST RECORD: 1943 [1768].
REFERENCES: **Kew Environs.** 1943, *Cooke* (LN34: S252; S33: 743): Thames side between Kew and Richmond.
CULT.: 1768 (JH: 441); 1940, Herb Garden Wall (K). In cult.
CURRENT STATUS: No recent sightings; it is probable that the tree was destroyed when the riverbank was strengthened in the late 1970s or early 1980s.

URTICACEAE

151. Urtica dioica L.
U. galeopsifolia Wierzb. ex Opiz.

common nettle

FIRST RECORD: 1873/4 [1789].
NICHOLSON: **1873/4 survey** (JB: 73): Common in back parts of shrubberies and in woods. • **1906 revision** (AS: 87): Common in the woods.
OTHER REFERENCES: **Kew Gardens.** 1999, *Stones* (EAK: 2 & 5): By Lake and Palm House Pond.
EXSICC.: **Kew Gardens.** Herbaceous ground; 15 vi 1937; *Howes* s.n. (K). • Arboretum; 12 ix 1938; *Dallimore* 108 (K). • Herbarium Ground; weed; 18 x 1938; *Blakelock* s.n. (K). • [Unlocalised]; 1 viii 1944; *Souster* 136 (K). **Kew Environs.** Kew Bridge; this huge nettle is something different from *U. dioica*; July [late 1800s]; *Mill* s.n. (K).
CULT.: 1789, Nat. of Britain (HK1, 3: 341); 1813, Nat. of Britain (HK2, 5: 263); 1814, Nat. of Britain (HKE: 291); 1922, unspec. (K). Not currently in cult.
CURRENT STATUS: Abundant throughout the Gardens. The non-stinging form, once separated as *U. galeopsifolia*, appeared in some quantity in the Paddock in 2007 and again in 2008 in topsoil being stored there.

152. Urtica urens L.

small nettle

FIRST RECORD: 1873/4 [1789].
NICHOLSON: **1873/4 survey** (JB: 73): Frequent wherever the soil gets turned. • **1906 revision** (AS: 87): Frequent wherever the soil gets turned.
EXSICC.: **Kew Gardens.** [Unlocalised]; on rubbish heap; viii 1931; *Hubbard* s.n. (K). • Mound of earth from excavation for foundations of new wing; 14 x 1965; *Verdcourt* 4239 (K). • Near tennis courts; rough ground; 18 vi 1969; *Ross-Craig* s.n. (K). • Temperate House; cultivated land; 19 x 1980; *C.M. Parker* 13 (K). • Order Beds; 15 ix 1981; *Borg* 11 (K). • Student plots; ix 1981; *Locke* 5 (K). **Kew Environs.** Thames banks, Kew; 9 ix 1894; *Fraser* s.n. (K). • Kew; 29 ix 1901; *Clarke* 498 (K). • Kew; 9 vi 1931; *Fraser* s.n. (K).

Fig (*Ficus carica*). Grew wild along the Thames between Kew and Richmond in the 1940s but was probably lost when the riverbank was strengthened in the late 1970s or early 1980s.

CULT.: 1789, Nat. of Britain (HK1, 3: 341); 1813, Nat. of Britain (HK2, 5: 263); 1814, Nat. of Britain (HKE: 291). Not currently in cult.
CURRENT STATUS: A common weed of flower-beds, allotments and disturbed ground throughout the Gardens.

153. *Urtica pilulifera L.
U. dodartii L.
U. pilulifera var. *dodartii* (L.) Asch.

Roman nettle

FIRST RECORD: 1861 [1768]
REFERENCES: **Kew Gardens.** 1990, *Cope* (S31: 182): Paddock, after restoration in 1990. **Kew Environs.** 1878, *G.Nicholson* (BEC18: 18; BSEC9: 692): The lower branches of several vigorous plants growing on rubbish-heaps at Kew [Bridge] had the deeply serrated leaves of typical *pilulifera*, whilst the upper part of the plant showed the entire or subentire leaves of var. *dodartii*.
EXSICC.: **Kew Environs.** Kew Bridge; vii 1861; *Churchill* s.n. (K). • Kew; waste ground; vii 1878; *G.Nicholson* 1151 (K).
CULT.: 1768 (JH: 378); 1789, Nat. of England (HK1,3: 340, as *U. dodartii*), Alien (HK1, 3: 340); 1813, Nat. of England (HK2, 5: 261, as *U. dodartii*), Alien (HK2, 5: 262); 1814, Nat. of England (HKE: 291, as *U. dodartii*), Alien (HKE: 291); 1894, unspec. (K); 1936, Herbaceous Ground (K). Not currently in cult.

Roman nettle (*Urtica pilulifera*). First seen in the Paddock after its restoration in 1990 but not seen again until 2007 when the Paddock had once again been disturbed.

CURRENT STATUS: This has only ever been known from the Paddock. It was recorded there in 1990 but did not appear again until 2007. The plant was identified as var. *dodartii* by J.M.Mullin but few authors bother to maintain the distinction.

154. Parietaria judaica L.
P. diffusa Mert. & W.D.J.Koch
P. officinalis auct. non L.
pellitory-of-the-wall
FIRST RECORD: 1852 [1789].
NICHOLSON: **1873/4 survey** (JB: 73): **B.** Walls about Herbarium. **P.** Old Ruined Arch. **Strip.** By side of towing-path and on wall of moat. • **1906 revision** (AS: 87): Old walls. **Strip.** Side of towing-path.
EXSICC.: **Kew Gardens.** Brick wall facing towpath; on ledge about 1 ft above ground; 14 ii 1928; *Summerhayes & Turrill* s.n. (K). • Towpath, Kew; growing on old brick wall at back of Kew Palace; 24 vii 1944; *Souster* 193 (K). **Kew Environs.** Wall at Kew; vi. 1852; *Syme* in *Herb. Watson* (K) • Wall at Kew; viii 1852; *Syme* in *Herb. Watson* (K). • Kew; shady ground; 1872; *Watson* in *Herb.* Watson (K). • Thames Banks, Kew; 15 ix 1883; *Fraser* s.n. (K). • Kew; 19 v 1935; *Wedgwood* s.n. (K).
CULT.: 1789, Nat. of Britain (HK1, 3: 429); 1813, Nat. of Britain (HK2, 5: 436); 1814, Nat. of Britain (HKE: 317); 1963, Herbarium Expt. Garden (K). Not currently in cult.
CURRENT STATUS: Widespread; common against walls and in paving and occasionally in dry ground under trees.

155. *Soleirolia soleirolii (Req.) Dandy
mind-your-own-business
FIRST RECORD: 1998.
REFERENCES: **Kew Gardens.** 1998, *Cope* (S33: 744): North Arb.: around the entrance to the Aroid House (117).
CULT.: In cult. under glass.
CURRENT STATUS: Uncommon; it regularly occurs outside the Nash Conservatory, in places on and around the Rockery and in cracks between paving blocks around Kew Palace.

JUGLANDACEAE

156. *Juglans regia L.
walnut
FIRST RECORD: 2003 [1768].
CULT.: 1768 (JH: 444); 1789, Alien (HK1, 3: 360); 1813, Alien (HK2, 5: 295); 1814, Alien (HKE: 296); 1881, Arboretum (K); 1883, unspec. (K); 1942, behind Aroid House (K). In cult.
CURRENT STATUS: Planted in various places in the Gardens and is now self-sowing along the towpath.

Mind-your-own-business (*Soleirolia soleirolii*). An inconspicuous plant of paving cracks and at the base of walls, and particularly common in the Rock Garden.

157. *Juglans nigra L.
black walnut
FIRST RECORD: 2003 [1768]
CULT.: 1768 (JH: 444); 1789, Alien (HK1, 3: 360); 1813, Alien (HK2, 5: 296); 1814, Alien (HKE: 296); 1862, unspec. (K); 1901, unspec. (K); 1930, Arboretum (K). In cult.
CURRENT STATUS: Planted in the Conservation Area and elsewhere in the Gardens and is now self-sowing along the towpath.

MYRICACEAE

158. Myrica gale L.
bog-myrtle
CULT.: 1768 (JH: 424, 448); 1789, Nat. of Britain (HK1, 3: 396); 1813, Nat. of Britain (HK2, 5: 379); 1814, Nat. of Britain (HKE: 308). Not currently in cult.

FAGACEAE

159. Fagus sylvatica L.
beech
FIRST RECORD: 1937 [1768].
REFERENCES: **Kew Gardens.** 1998, *Cope* (S33: 744): Self-sown seedlings in West Arb.: beech circle (242).
EXSICC.: **Kew Gardens.** Shrubbery, Herbarium; seedling; 24 v 1937; *Bullock* s.n. (K).
CULT.: 1768 (JH: 441); 1789, Nat. of Britain (HK1, 3: 362); 1813, Nat. of Britain (HK2, 5: 297); 1814, Nat. of Britain (HKE: 296); 1880, Arboretum (K); 1882, near Winter Garden (K); 1882, Holly Walk (K); 1883, Arboretum (K); 1884, Arboretum (K); 1885, Arboretum (K); 1887, unspec. (K); 1907, Arboretum (K); 1920, unspec. (K); 1922, unspec. (K); 1936, Arboretum (K); 1937, Arboretum (K); 1941, by Lion Gate (K); 1961, Arboretum (K). In cult.
CURRENT STATUS: Seedlings occur in the Beech Circle derived from trees that were originally planted. It was previously noted as spontaneous in 1980 (unpubl.) but doubtless seedlings have been around for as long as the species has been cultivated.

160. Castanea sativa Mill.
Fagus castanea L.
C. vesca Gaertn.
sweet chestnut
FIRST RECORD: 1999 [1768].
REFERENCES: **Kew Gardens.** 1999, *Cope* (S35: 648): Seedlings in wooded area between the Stable Yard and the Temperate House (354). Subsequently, seedlings have been found in numerous other places where the tree has been planted.

Beech (*Fagus sylvatica*). All mature trees have been planted but seedlings can be found in many parts of the Gardens.

CULT.: 1768 (JH: 441); 1789, Nat. of England (HK1, 3: 361); 1813, Nat. of England (HK2, 5: 298); 1814, Nat. of England (HKE: 296); 1880, Arboretum (K); 1881, on path from Pinetum by Lake (K); 1905, Arboretum (K). In cult.
CURRENT STATUS: Occasional seedlings can be found in the vicinity of cultivated trees but these are seldom allowed to grow to any size before being removed.

161. *Quercus ilex L.
Q. gramuntia L.
evergreen oak
FIRST RECORD: 1983/4 [1768].
REFERENCES: **Kew Gardens.** 1983/4, *Cope* (S31: 181): Seedling in new student vegetable plots behind Hanover House.
CULT.: 1768 (JH: 452); 1843–1853, unspec. (K); 1874, Pleasure Grounds (K); 1880, Arboretum (K); 1882, behind mess room (K); 1903, near Ruined Arch (K); 1905, Arboretum (K); 1907, Arboretum (K); 1910, unspec., tree blown down by wind 20 ii 1910 (K); 1913, Arboretum (K); 1921, Temperate House (K); 1927, Arboretum (K); 1935, Sion Vista (K); 1935, near Main Gate (K); 1978, unspec. (K). In cult.
CURRENT STATUS: Has not persisted behind Hanover House. There are no other recent sightings of seedlings despite how widely grown it is in the Gardens. There are mature trees on the edge of the golf course in the Old Deer Park close to the boundary of the Gardens.

162. Quercus petraea (Matt.) Liebl.
Q. pubescens Willd.
Q. robur var. *sessilis* Martyn
Q. robur auct. non L.
sessile oak

CULT.: 1789, Nat. of Britain (HK1, 3: 359); 1813, Nat. of Britain (HK2, 5: 294 as *robur*), Nat. of England (ibid.: as *pubescens*); 1814, Nat. of Britain (HKE: 296 as *robur*), Nat. of England (ibid.: as *pubescens*); 1880, Arboretum (K); 1884, Arboretum (K); 1888, Arboretum (K); 1911, Arboretum (K); 1978, Arboretum South, plot 216 (K). In cult.

163. Quercus robur L.
Q. pedunculata Ehrh. ex Hoffm.
Q. robur var. *pedunculata* (Ehrh. ex Hoffm.) Hook.f.
Q. escula L.
pedunculate oak

FIRST RECORD: 1990 [1768].
REFERENCES: **Kew Gardens.** 1990, *Cope* (S31: 182): Seedling in paddock, after restoration in 1990.
CULT.: 1768 (JH: 452); 1789, Nat. of Britain (HK1, 3: 359); 1813, Nat. of Britain (HK2, 5: 294); 1814, Nat. of Britain (HKE: 296); 1843-1853, Arboretum (K); 1874, Pleasure Grd. (K); 1880, Arboretum (K); 1881, unspec. (K); 1882, left side of walk from Railway Gate (K); 1885, Arboretum (K); 1887, Arboetum (K); 1905, Arboretum (K); 1910, Arboretum (K); 1919, unspec. (K); 1929, Arboretum (K); 1934, Arboretum (K); 1939, corner of ride, N.E. of Queen's Cottage (K); 1944, Arboretum (K); 1966, Arboretum (K); 1978, unspec. (K). In cult.
CURRENT STATUS: Seedlings occasionally appear in the vicinity of cultivated trees and have doubtless been doing so for the last two hundred years; they are seldom left to grow to any size before removal.

BETULACEAE

164. Betula pendula Roth
B. alba auct. non L.
B. alba var. *pendula* (Roth) W.T.Aiton
silver birch

FIRST RECORD: 1999 [1768].
REFERENCES: **Kew Gardens.** 1999, *Cope* (S35: 648): In lightly wooded area near Queen Charlotte's Cottage (325). Trees were largely ignored by Nicholson in his original survey of 1873–4 and most have subsequently gone unrecorded.
CULT.: 1768 (JH: 435); 1789, Nat. of Britain (HK1, 3: 336); 1813, Nat. of Britain (HK2, 5: 299); 1814, Nat. of Britain (HKE: 296); 1880, Arboretum (K); 1881, large tree with iron label nr. Tulip Tree (K); 1883, Arboretum (K); 1905, Arboretum (K); 1934, unspec. (K); 1949, Arboretum (K); 1962, Arboretum (K); 1963, Arboretum (K). In cult.

Silver birch (*Betula pendula*). Spontaneous in many parts of the Gardens, especially the Conservation Area, but probably from trees that were originally planted.

CURRENT STATUS: A significant component of the canopy in parts of the Conservation Area and elsewhere; surprisingly unrecorded as it freely seeds into its surroundings. Whether or not any of the trees are truly wild is a moot point.

165. Betula pubescens Ehrh.
downy birch

CULT.: 1880, Arboretum (K); 1881, Arboretum (K); 1904, Arboretum (K); 1913, Arboretum (K); 1935, Arboretum (K); 1962, Arboretum (K); 1963, Arboretum (K); 1969, unspec. (K). In cult.

166. Betula nana L.
dwarf birch

CULT.: 1768 (JH: 435); 1789, Nat. of Scotland (HK1, 3: 337); 1813, Nat. of Scotland (HK2, 5: 300); 1814, Nat. of Scotland (HKE: 296). In cult.

167. Alnus glutinosa (L.) Gaertn.
Betula alnus L.
B. alnus var. *glutinosa* L.
alder

FIRST RECORD: 1928 [1768].
REFERENCES: **Kew Gardens.** 1999, *Stones* (EAK: 2 & 5): Seedlings by Lake and Palm House Pond.
EXSICC.: **Kew Environs.** Kew, banks of Thames; 16 vi 1928; *Hubbard* s.n. (K). • Thames bank, Kew;

growing on top of wall; 9 viii 1928; *Summerhayes & Turrill* s.n. (K). • Thames bank, Kew – Richmond; at top of river wall, opposite Kew end of Syon House Grounds; 21 iii 1929; *Summerhayes & Turrill* s.n. (K). • Towpath, Kew; 26 iv 1944; *Souster* 42 (K). • Towpath, Kew; 7 v 1944; *Souster* 57 (K). • Towpath, Kew; 27 ii 1950; *Souster* 1067 (K).
CULT.: 1768 (JH: 435); 1789, Nat. of Britain (HK1, 3: 338); 1813, Nat. of Britain (HK2, 5: 258); 1814, Nat. of Britain (HKE: 291); 1843–1853, Arboretum (K); 1880, Arboretum (K); 1882, Arboretum (K); 1890, Arboretum (K); 1894, Arboretum (K); 1895, Lake (K); 1898, unspec. (K); 1900, Arboretum (K); 1904, Arboretum (K); 1920, Arboretum (K); 1931, Arboretum (K); 1938, Arboretum (K); 1962, Arboretum, at N.E. end of Lake (K); 1964, Arboretum, at N.E. end of Lake (K); 1965, Arboretum (K). In cult.
CURRENT STATUS: Still present on the towpath south of Brentford Gate; seedlings derived from cultivated trees occur in scattered localities in the Gardens but are seldom left for long before removal.

CORYLACEAE

168. Carpinus betulus L.
hornbeam
FIRST RECORD: 1931 [1768].
REFERENCES: **Kew Gardens.** 1983/4, *Cope* (S31: 181): Seedling in new student vegetable plots behind Hanover house.
EXSICC.: **Kew Environs.** Thames banks, Kew; 1 viii 1931; *Fraser* s.n. (K).
CULT.: 1768 (JH: 436); 1789, Nat. of Britain (HK1, 3: 362); 1813, Nat. of Britain (HK2, 5: 301); 1814, Nat. of Britain (HKE: 297); 1880, unspec. (K); 1881, side of Old Deer Park, Lion Walk (K); 1882, unspec. (K); 1883, unspec. (K); 1885, unspec. (K); 1938, Arboretum (K); 1965, Syon Vista (K); 1979, behind Fumigation Chamber, Herbarium Grounds (K). In cult.
CURRENT STATUS: Seedlings are frequently found behind Hanover House in the vicinity of the parent tree.

169. Corylus avellana L.
hazel
FIRST RECORD: 1920 [1768].
EXSICC.: **Kew Environs.** Kew Green; 29 v 1920; *Fraser* s.n. (K).
CULT.: 1768 (JH: 438); 1789, Nat. of Britain (HK1, 3: 363); 1813, Nat. of Britain (HK2, 5: 302); 1814, Nat. of Britain (HKE: 297); 1880, Arboretum (K); 1882, Arboretum (K); 1912, Arboretum (K); 1931, Arboretum (K); 1954, ornamental shrubbery opposite House No. 4 (K); 1978, plot 310 in S part nr Old Deer Park (K); 1978, plot 310, large group by the river bank (K); 1978, plot 310, large group in N part of area (K); 1978, plot 234 (K). In cult.
CURRENT STATUS: Seedlings in the Gardens derived from planted trees have been noted only since 2000, having previously been ignored. It is no longer found on Kew Green.

PHYTOLACCACEAE

170. *Phytolacca acinosa Roxb.
P. americana auct. non L.
P. esculenta auct. non Van Houtte
Indian pokeweed
FIRST RECORD: 1960 [1955].
REFERENCES: **Kew Gardens.** 1983/4, *Cope* (S31: 181): New student vegetable plots behind Hanover House. • **Kew Environs.** 1960, *Townsend* (LN40: 20): On the towpath at Kew. • 1987, *Leslie* (FSSC: 74): Plants from the well-known locality on Kew Green have been determined by EJC [Eric Clement] as *P. esculenta*.
EXSICC.: **Kew Environs.** Thames bank, near Kew; 1960; *Townsend* s.n. (K).
CULT.: 1955, unspec. (K). Not currently in cult.
CURRENT STATUS: Persistent by the Herbarium and in front of Hanover House; it is also in the Director's Garden and occasionally in one or two other places including the heap of stored topsoil in the Paddock. Plants in St Anne's churchyard, to which Leslie referred, have been noted since 1979.

Hazel (*Corylus avellana*). A cultivated tree that seeds easily into various parts of the Gardens, doubtless aided by birds and squirrels. [Photo: P.J.Cribb]

Indian pokeweed (*Phytolacca acinosa*). Well established around the Herbarium and often appears when soil is moved or heavily disturbed.

CHENOPODIACEAE

171. *Chenopodium ambrosioides L.

Mexican-tea

FIRST RECORD: 2002 [1768].

REFERENCES: **Kew Gardens.** 2002, *Cope* (S35: 648): Between the Rhododendron Dell and the Bamboo Garden (231). Probably from birdseed.

CULT.: 1768 (JH: 382); 1789, Alien (HK1, 1: 313); 1811, Alien (HK2, 2: 99); 1814, Alien (HKE: 71); 1894, unspec. (K); 1963, unspec. (K). In cult.

CURRENT STATUS: A weed in the Rhododendron Dell, doubtless originating from birdseed, and in disturbed ground in front of the Jodrell Laboratory, but not persistent in either locality.

172. Chenopodium bonus-henricus L.

good-King-Henry

CULT.: 1768 (JH: 382); 1789, Nat. of Britain (HK1, 1: 311); 1811, Nat. of Britain (HK2, 2: 98); 1814, Nat. of Britain (HKE: 71). In cult.

Note. Listed as native but now regarded as an archaeophyte.

173. Chenopodium glaucum L.

oak-leaved goosefoot

FIRST RECORD: 1918 [1768].

REFERENCES: **Kew Gardens.** 1918, *Worsdell fide Airy Shaw* (S29: 407 and see exsicc.): Kew Gardens. • 1929, *Summerhayes & Hubbard* fide *Airy Shaw* (S29: 407): Lower Nursery. • 1933, *Hubbard fide Airy Shaw* (S29: 407): Lower Nursery. • 1956, *Harrison fide Airy Shaw* (S29: 407): Melon Yard, in gravel and poor soil, and in pots. • 1963, *Airy Shaw* (S29: 407): Melon Yard, at the foot of the walls of the propagating houses. • 1960, *Welch* (LFS: 153): Weed in Kew Gardens.

EXSICC.: **Kew Gardens.** [Unlocalised]; 22 vii 1918; *Worsdell* s.n. (K). • Lower Nursery; on side of leaf heap; ix 1929; *Summerhayes & Hubbard* s.n. (K). • [Unlocalised]; on rubbish heap; ix 1929; *Hubbard* s.n. (K). • Lower Nursery; weed; ix 1933; *Hubbard* s.n. (K). • Edge of path outside cocoa quarantine house; in gravel and poor soil & in grape vine pots sunk in gravel bed; 31 v 1956; *Morrison* s.n. (K).

CULT.: 1768 (JH: 382); 1789, Nat. of England (HK1, 1: 313); 1811, Nat. of England (HK2, 2: 100); 1814, Nat. of England (HKE: 71). In cult.

CURRENT STATUS: No recent sightings.

174. Chenopodium rubrum L.

C. botryodes Sm.

red goosefoot

FIRST RECORD: 1917 [1768].

REFERENCES: **Kew Gardens.** 1963, *J.L.Gilbert* (S29: 407): Near Rock Garden.

EXSICC.: **Kew Gardens.** [Unlocalised]; waste place; 8 viii 1917; *Worsdell* s.n. (K). • [Unlocalised]; weed of cultivated land; 16 vii 1944; *Souster* 191 (K).

CULT.: 1768 (JH: 382); 1789, Nat. of Britain (HK1, 1: 312); 1811, Nat. of Britain (HK2, 2: 98); 1814, Nat. of Britain (HKE: 70), Nat. of England (ibid.: Add. as *botryodes*); 1902, unspec. (K); 1923, unspec. (K). Not currently in cult.

CURRENT STATUS: Rare; turns up occasionally on disturbed soil but has never persisted.

175. Chenopodium polyspermum L.

C. acutifolium Sm.

many-seeded goosefoot

FIRST RECORD: 1873/4 [1768].

NICHOLSON: **1873/4 survey** (JB: 72): **B.** A single plant in shrubbery near "Old Lily House." **P.** Two or three plants in the enclosure near filter-beds. • **1906 revision** (AS: 86): Not uncommon as a weed in cultivated ground.

EXSICC.: **Kew Gardens.** [Unlocalised]; waste place; 6 viii 1917; *Worsdell* s.n. (K). • [Unlocalised]; on waste ground; viii 1924; *Hubbard* s.n. (K). • Herb. Expt. Ground; weed; 17 ix 1928; *Summerhayes & Turrill* s.n. (K). • [Unlocalised]; on cultivated ground; ix 1928; *Hubbard* s.n. (K). • Near tennis courts; rubbish heap; 1 x 1934; *Bullock* s.n. (K). •

Allotments by Herbarium; weed; 10 ix 1942; *Melville* s.n. (K). **Kew Environs.** Towing path, Kew; 5 ix 1921; *Fraser* s.n. (K).
CULT.: 1768 (JH: 382); 1789, Nat. of England (HK1, 1: 314); 1811, Nat. of Britain (HK2, 2: 100); 1814, Nat. of Britain (HKE: 71, & Add. as *acutifolium*); 1922, unspec. (K); 1934, unspec. (K). Not currently in cult.
CURRENT STATUS: Widespread as a weed of disturbed ground and sometimes abundant.

176. Chenopodium vulvaria L.
stinking goosefoot
FIRST RECORD: 1779 [1768].
REFERENCES: **Kew Gardens.** 1985, *London Natural History Society* (LN65: 195; S33: 744): [Around what is now the Princess of Wales Conservatory], 1 plant.
EXSICC.: **Kew Environs.** Kew Bridge; 1779; *Goodenough* (K).
CULT.: 1768 (JH: 382); 1789, Nat. of Britain (HK1, 1: 314); 1811, Nat. of Britain (HK2, 2: 100); 1814, Nat. of Britain (HKE: 71); 1894, unspec. (K). In cult.
CURRENT STATUS: No recent sightings and almost certainly now extinct. This is the earliest recorded vascular plant from the Kew area.

Stinking goosefoot (*Chenopodium vulvaria*). Long extinct, but the first vascular plant to be recorded from the Kew area.

177. Chenopodium hybridum L.
maple-leaved goosefoot
CULT.: 1768 (JH: 382); 1789, Nat. of Britain (HK1, 1: 312); 1811, Nat. of Britain (HK2, 2: 99); 1814, Nat. of Britain (HKE: 70); 1898, unspec. (K). Not currently in cult.
Note. Listed as native but now regarded as an archaeophyte.

178. Chenopodium urbicum L.
upright goosefoot
FIRST RECORD: 1863 [1768].
EXSICC.: **Kew Gardens.** [Unlocalised]; waste places; viii 1917; *Worsdell* s.n. (K). **Kew Environs.** Kew Bridge; vii 1863; *Irvine* s.n. (K).
CULT.: 1768 (JH: 382); 1789, Nat. of Britain (HK1, 1: 311); 1811, Nat. of Britain (HK2, 2: 98); 1814, Nat. of Britain (HKE: 70). Not currently in cult.
CURRENT STATUS: No recent sightings.

179. Chenopodium murale L.
nettle-leaved goosefoot
FIRST RECORD: 1873/4 [1768].
NICHOLSON: **1873/4 survey** (JB: 72 and in SFS: 550): **P.** Here and there near "Engine House." • **1906 revision** (AS: 86): **A.** Here and there in dry places.
OTHER REFERENCES: **Kew Gardens.** 1969, *Brenan* (S29: 406; and see Exsicc.): In garden of Herbarium House, 55 Kew Green. Bird-seed alien.
EXSICC.: **Kew Gardens.** Garden of Herbarium House, 55 Kew Green; bird-seed alien, 18 viii 1969; *Brenan* s.n. (K).
CULT.: 1768 (JH: 382); 1789, Nat. of Britain (HK1, 1: 312); 1811, Nat. of Britain (HK2, 2: 98); 1814, Nat. of Britain (HKE: 70); 1896, unspec. (K); 1919, unspec. (K); 1980, unspec. (K). Not currently in cult.
CURRENT STATUS: No recent sightings.

180. Chenopodium ficifolium Sm.
C. serotinum auct. non L.
fig-leaved goosefoot
FIRST RECORD: 1949 [1768].
REFERENCES: **Kew Gardens.** 1958, *Brenan* and *Airy Shaw* (Independently) (S24: 188): On heap of earth by path behind Aroid House. **Kew Environs.** 1988, *Lock* (S32: 657): Kew Green, Dog Walk, cricket practice nets.
EXSICC.: **Kew Gardens.** Stable Yard, Arboretum; in long grass by maure heap; 2 vii 1949; *Souster* s.n. (K).
CULT.: 1768 (JH: 382); 1789, Nat. of Britain (HK1, 1: 312); 1811, Nat. of England (HK2, 2: 99); 1814, Nat. of England (HKE: 70). Not currently in cult.
CURRENT STATUS: Occasional in disturbed ground but not persistent.

181. *Chenopodium opulifolium Schrad. ex W.D.J.Koch & Ziz
grey goosefoot
FIRST RECORD: 2007.
CURRENT STATUS: Came up in some quantity on a heap of topsoil being stored in the Paddock and has persisted.

182. Chenopodium album L.
C. viride L.
fat-hen
FIRST RECORD: 1873/4 [1768].
NICHOLSON: **1873/4 survey** (JB: 72): Common everywhere. • **1906 revision** (AS: 86): Common everywhere.
EXSICC.: **Kew Gardens.** [Unlocalised]; 29 vii 1918; *Worsdell* s.n. (K). • Mound of earth from excavation for foundations of new wing; 14 x 1965; *Verdcourt* 4246 (K). • Syringa beds; viii 1980; *Thurman* 7 (K).
CULT.: 1768 (JH: 382); 1789, Nat. of Britain (HK1, 1: 312); 1811, Nat. of Britain (HK2, 2: 99); 1814, Nat. of Britain (HKE: 70); 1893, unspec. (K); 1894, unspec. (K); 1930, Herb. Expt. Ground (K). Not currently in cult.
CURRENT STATUS: A common weed of disturbed ground throughout the Gardens

[**183. Chenopodium suecicum** Murr.
Swedish goosefoot
CULT.: 1789, Nat. of Britain (HK1, 1: 312).
Note. Listed in error as native.**]**

184. *Chenopodium giganteum D.Don
tree spinach
FIRST RECORD: 1985.
REFERENCES: **Kew Gardens.** 1985, *London Natural History Society* (LN65: 195; S33: 744): [Around what is now the Princess of Wales Conservatory], many plants.
CURRENT STATUS: Came up in some quantity in 2006 on a heap of topsoil being stored behind the Herbarium; in 2007 it came up on a similar heap of soil, and in similar quantity, in the Paddock. It has persisted in the latter.

185. *Chenopodium foliosum (Moench) Asch.
Blitum virgatum L.
strawberry goosefoot
FIRST RECORD: 1881 [1768].
REFERENCES: **Kew Environs.** 1881, *Baker* (BEC1: 55; S33: 744): Surrey side of Thames, between Kew and Richmond.
CULT.: 1768 (JH: 369); 1914, unspec. (K); 1969, Temperate Dept. (K). Not currently in cult.
CURRENT STATUS: A single plant appeared in 2004 in topsoil stored in the Paddock. There has been no subsequent disturbance to the spot where the plant appeared and it has not yet returned.

186. *Atriplex hortensis L.
garden orache
FIRST RECORD: 2007 [1887].
CULT.: 1887, unspec. (K); 1933, unspec. (K). In cult.
CURRENT STATUS: Several plants appeared in 2007 on topsoil stored in the Paddock. Both var. *hortensis* and var. *rubra* (Crantz) Roth were present.

187. Atriplex prostrata Boucher ex DC.
A. deltoidea Bab.
A. hastata auct. non L.
A. smithii Syme
spear-leaved orache
FIRST RECORD: 1873/4 [1768].
NICHOLSON: **1873/4 survey** (JB: 72): **B** and **P.** Here and there on dug ground. **Pal.** Common on the rubbish thrown over the wall when moat was cleared out. • **1906 revision** (AS: 86): In company with the next named [also = *A. prostrata*] ... common on rubbish heaps and in bare places.
OTHER REFERENCES: **Kew Gardens.** 1909, *Turrill* in *Rolfe & A.B.Jackson* (S10: 373): Several plants in broken ground near Kew Palace.
EXSICC.: **Kew Environs.** Thames bank, Kew; viii 1888; *Gamble* 20219 (K). • Towpath, Kew; 24 vii 1944; *Souster* 194 (K).
CULT.: 1768 (JH: 385); 1789, Nat. of Britain (HK1, 3: 431). In cult.
CURRENT STATUS: Common as a weed of disturbed ground in the northern part of the Gardens.

Strawberry goosefoot (*Chenopodium foliosum*). Appeared in the Paddock in 2004, its first recorded sighting since 1881. It has not been seen again.

188. Atriplex littoralis L.
Atriplex marina L.
grass-leaved Orache
CULT.: 1768 (JH: 385); 1789, Nat. of Britain (HK1, 3: 431); 1813, Nat. of Britain (HK2, 5: 440); 1814, Nat. of Britain (HKE: 317); 1898, unspec. (K); 1922, unspec. (K). Not currently in cult.

189. Atriplex patula L.
A. erecta Huds.
A. angustifolia Sm.
common orache
FIRST RECORD: 1873/4 [1768].
NICHOLSON: **1873/4 survey** (JB: 72): **P.** In filter-bed enclosure, growing with [*Chenopodium murale*] and *C. polyspermum*. • **1906 revision** (AS: 86): Here and there in all the divisions.
EXSICC.: **Kew Environs.** Bank of the Thames at Kew; 16 viii 1888; *anon.* (?*Gamble*) s.n. (K).
CULT.: 1768 (JH: 385); 1789, Nat. of Britain (HK1, 3: 431); 1813, Nat. of Britain (HK2, 5: 439 and as *angustifolia*), Native of England (ibid.: 440 as *erecta*); 1814, Nat. of Britain (HKE: 317 and as *angustifolia*), Nat. of England (ibid.: as *erecta*); 1922, unspec. (K). In cult.
CURRENT STATUS: No recent sightings.

190. Atriplex laciniata L.
frosted orache
CULT.: 1768 (JH: 385); 1789, Nat. of Britain (HK1, 3: 431); 1813, Nat. of Britain (HK2, 5: 439); 1814, Nat. of Britain (HKE: 317); ?early 20th cent., unspec. (K). Not currently in cult.

191. Atriplex portulacoides L.
sea-purslane
CULT.: 1768 (JH: 385); 1789, Nat. of Britain (HK1, 3: 430); 1813, Nat. of Britain (HK2, 5: 437); 1814, Nat. of Britain (HKE: 317); 1906, Arboretum (K). In cult. under glass.

192. Atriplex pedunculata L.
pedunculate sea-purslane
CULT.: 1789, Nat. of England (HK1, 3: 431); 1813, Nat. of England (HK2, 5: 440); 1814, Nat. of England (HKE: 317). Not currently in cult.

193. Beta vulgaris L.
a. subsp. **vulgaris**
root beet
FIRST RECORD: 1933 [1768].
EXSICC.: **Kew Gardens.** Amongst grasses near Herbarium; 20 viii 1933; *Hubbard* s.n. (K).
CULT.: 1768 (JH: 383); 1933, Herb Garden (K); 1936, Herb Garden (K); 1937, Herbaceous Ground (K). In cult.
CURRENT STATUS: No recent sightings; likely to have occurred only as an escape from allotments.

b. subsp. **maritima** (L.) Arcang.
B. maritima L.
sea beet
CULT.: 1768 (JH: 383); 1789, Nat. of Britain (HK1, 1: 315); 1811, Nat. of Britain (HK2, 2: 102); 1814, Nat. of Britain (HKE: 71). In cult.

194. *Beta trigyna Waldst. & Kit.
caucasian beet
FIRST RECORD: 1919 [c. 2000].
EXSICC.: **Kew Gardens.** Field near Herbarium; 28 vii 1919; *Turrill* s.n. (K). • Herbarium meadow by river wall; 19 vi 1979; *Panter* s.n. (K).
CULT.: c. 2000, unspec. (K). In cult.
CURRENT STATUS: Very rare; two plants were noted by the boundary wall near Cumberland Gate but these are likely to have been lost when the student vegetable plots were moved in 2007. In 2008 it appeared in topsoil being stored in the Paddock.

195. Sarcocornia perennis (Mill.) A.J.Scott
Salicornia fruticosa (L.) L.
S. radicans Sm.
perennial glasswort
CULT.: 1810, Nat. of Britain (HK2, 1: 12); 1814, Nat. of Britain (HKE: 2), Nat. of England (ibid.: Add. as *radicans*). Not currently in cult.

196. Salicornia europaea L.
S. herbacea (L.) L.
common glasswort
CULT.: 1768 (JH: 366); 1789, Nat. of Britain (HK1, 1: 4); 1810, Nat. of Britain (HK2, 1: 12); 1814, Nat. of Britain (HKE: 2); 1926, unspec. (K). Not currently in cult.

197. Salicornia fragilis P.W.Ball & Tutin
S. procumbens Sm.
yellow glasswort
CULT.: 1814, Nat. of England (HKE: Add.). Not currently in cult.

198. Suaeda vera Forssk. ex J.F.Gmel.
Salsola fruticosa L.
shrubby sea-blite
CULT.: 1768 (JH: 383); 1789, Nat. of England (HK1, 1: 317); 1811, Nat. of England (HK2, 2: 105); 1814, Nat. of England (HKE: 71). In cult.

199. Suaeda maritima (L.) Dumort.
Chenopodium maritimum L.
annual sea-blite
CULT.: 1768 (JH: 382); 1789, Nat. of Britain (HK1, 1: 314); 1811, Nat. of Britain (HK2, 2: 101); 1814, Nat. of Britain (HKE: 71). Not currently in cult.

200. Salsola kali L.
prickly saltwort
CULT.: 1768 (JH: 383); 1789, Nat. of Britain (HK1, 1: 316); 1811, Nat. of Britain (HK2, 2: 102); 1814, Nat. of Britain (HKE: 71). Not currently in cult.

AMARANTHACEAE

201. *Amaranthus retroflexus L.
common amaranth
FIRST RECORD: 1950 [1768].
REFERENCES: **Kew Gardens.** 1983, *Hastings* (LN63: 143): Many disturbed places in Kew Gardens. • 1983/4, *Cope* (S31: 181): New student vegetable plots behind Hanover House. • 1985, *London Natural History Society* (LN65: 195): Around what is now the Princess of Wales Conservatory.
EXSICC.: **Kew Gardens.** [Unlocalised]; cultivated land; 29 viii 1950; *Souster* 1191 (K).
CULT.: 1768 (JH: 384); 1789, Alien (HK1, 3: 349); 1813, Alien (HK2, 5: 276); 1814, Alien (HKE: 293); 1953, Herb. Expt. Ground (K). In cult.
CURRENT STATUS: Common around the student allotments behind Hanover House before the land was redeveloped, and frequently found in other highly disturbed areas.

Common amaranth (*Amaranthus retroflexus*). A common plant in heavily disturbed soil.

202. *Amaranthus hybridus L.
green amaranth
FIRST RECORD: 1985 [1768].
REFERENCES: **Kew Gardens.** 1985, *London Natural History Society* (LN65: 195; S33: 744): [Around what is now the Princess of Wales Conservatory].
CULT.: 1768 (JH: 384); 1789, Alien (HK1, 3: 349); 1813, Alien (HK2, 5: 275); 1814, Alien (HKE: 293); 1876, unspec. (K); 1893, unspec. (K); 1969, unspec. (K). Not currently in cult.
CURRENT STATUS: The purple-leaved form recently appeared in one of the kidney-shaped beds along the Broad Walk; otherwise very occasional in disturbed ground and doubtless a relict of former cultivation.

203. *Amaranthus hypochondriacus L.
prince's-feather
FIRST RECORD: 2007 [1768].
CULT.: 1768 (JH: 384); 1789, Alien (HK1, 3: 350); 1813, alien (HK2, 5: 277); 1814, alien (HKE: 293); 1893, unspec. (K). In cult. under glass.
CURRENT STATUS: Several plants grew from seed in topsoil stored in the Paddock.

203.1. *Amaranthus deflexus L.
perennial pigweed
FIRST RECORD: 2008 [1813].
CULT.: 1813, alien (HK2, 5: 274); 1814, Alien (HKE: 293); 1976, under glass (K). In cult.
CURRENT STATUS: Established in a small flower-bed adjacent to White Peaks.

204. *Amaranthus blitum L.
A. lividus L.
Guernsey pigweed
FIRST RECORD: 1985 [1789].
REFERENCES: **Kew Gardens.** 1985, *Burton* (FSSC: 11; LN65: 195; S33: 744): [Around what is now the Princess of Wales Conservatory].
CULT.: 1789, Nat. of England (HK1, 3: 348); 1813, Nat. of England (HK2, 5: 274); 1814, Nat. of England (HKE: 293). Not currently in cult.
CURRENT STATUS: An occasional weed in the Order Beds and presumably a relict of former cultivation.

205. *Amaranthus graecizans L.
short-tepalled pigweed
FIRST RECORD: 1998 [1789].
REFERENCES: **Kew Gardens.** 1998, *Cope* (S33: 744): North Arb.: lawn behind Herbarium after disturbance during building works (113).
CULT.: 1789, Alien (HK1, 3: 346); 1813, Alien (HK2, 5: 271); 1814, Alien (HKE: 273); 1979, Alpine (K). Not currently in cult.

Spring-beauty (*Claytonia perfoliata*). An occasional weed of shady places that was once much more common and described, as recently as 1983, as 'uncontrollable in Kew Gardens'.

CURRENT STATUS: Persistent in the lawn at the back of the Herbarium, at least until it was restored in 1999; it has not been seen anywhere since. It may return at the end of the present building works when the lawn will be given a chance to recover from serious damage.

PORTULACACEAE

206. *Portulaca oleracea* L.
purslane
FIRST RECORD: 1998 [1768].
REFERENCES: **Kew Gardens.** 1998, *Cope* (S33: 744): North Arb.: Paddock (104) and lawn behind Herbarium (113), ground disturbed during building works.
CULT.: 1768 (JH: 197). In cult. under glass.
CURRENT STATUS: A regular occurrence in the Herbarium lawn in the 1990s but now declining and perhaps extinct in most of its sites.

207. *Claytonia perfoliata* Donn ex Willd.
Montia perfoliata (Donn ex Willd.) Howell
spring-beauty
FIRST RECORD: 1873/4 [1811].
NICHOLSON: **1873/4 survey** (JB: 12): **Pal.** Common on the gravel walks behind House No. 1. • **1906 revision** (AS: 77): Not uncommon. A troublesome weed.
OTHER REFERENCES: **Kew Gardens.** 1936, *Cooke* (LN30: S40): Weed in flower beds. • 1950, *Welch* (LN30: S40): Still abundant. • 1983, *Burton* (FLA: 38): Uncontrollable at Kew Gardens.
EXSICC.: **Kew Gardens.** [Unlocalised]; iv 1897; *Collett* s.n. (K). • [Unlocalised]; beneath beech tree; 15 v 1932; *Hubbard* s.n. (K). • [Unlocalised]; summer annual, a common weed on cultivated land; 24 vii 1944; *Souster* 123 (K). • Near Kew Palace; weed in flower-bed; 15 iv 1994; *N.P. Taylor* 1699 (K). **Kew Environs.** Kew; vi 1929; *Pearce* s.n. (K).
CULT.: 1811, Alien (HK2, 2: 54); 1814, Alien (HKE: 64); 1951, Arboretum (K). Not currently in cult.
CURRENT STATUS: Widespread, but seldom common and can no longer be considered a serious weed.

208. *Montia fontana* L. subsp. ***chondrosperma*** (Fenzl) Walters
M. lamprosperma Cham.
blinks
FIRST RECORD: 1873/4 [1789].
NICHOLSON: **1873/4 survey** (JB: 12 and in SFS: 194): **P.** In the turf round Winter Garden; in beds at end of "Syon Vista." • **1906 revision** (AS: 77): **A:** In the turf round temperate house; in turf and in beds at end of "Syon Vista."

Blinks (*Montia fontana*). A tiny spring-ephemeral known only from Kew Green, opposite the Herbarium, and a patch of *Festuca filiformis* grassland north of the Rhododendron Dell. Both sites are vulnerable.

OTHER REFERENCES: **Kew Gardens.** 1991, *Hastings* (S31: 182): Kew Green (R.B.G. side), abundant. • 1996, *London Natural History Society* (LN76: 196): Kew Green.
EXSICC.: **Kew Gardens.** W. of Temperate House; minute annual growing in lawn, dry situation; 25 iv 1944; *Souster* 38 (K).
CULT.: 1789, Nat. of Britain (HK1, 1: 123); 1810, Nat. of Britain (HK2, 1: 183); 1814, Nat. of Britain (HKE: 26). Not currently in cult.
CURRENT STATUS: A large population on Kew Green (RBG side) and some north of the Rhododendron Dell are all that are currently known. Both sites are vulnerable due to soil enrichment by grass clippings. The subspecies of early cultivated plants are not known.

209. *Calandrinia ciliata (Ruíz & Pav.) DC.
red-maids
FIRST RECORD: 2008.
CURRENT STATUS: Several plants came up in bare ground from which a small tree had been removed.

CARYOPHYLLACEAE

210. Arenaria serpyllifolia L.
thyme-leaved sandwort
a. subsp. **serpyllifolia**
FIRST RECORD: 1853 [1768].
NICHOLSON: **1873/4 survey** (JB: 12): Everywhere. • **1906 revision** (AS: 76): Everywhere.
EXSICC.: **Kew Gardens.** Botanic Garden Wall; vii 1853; *Brocas* 178 (K).
CULT.: 1768 (JH: 203); 1789, Nat. of Britain (HK1, 2: 101); 1811, Nat. of Britain (HK2, 3: 99); 1814, Nat. of Britain (HKE: 133); 1949, Herbarium Ground (K). Not currently in cult.
CURRENT STATUS: Declining and now very rare; found only near the river at the end of Syon Vista during the current survey.

b. subsp. **leptoclados** (Rchb.) Nyman
A. serpyllifolia var. *leptoclados* Rchb.
FIRST RECORD: 1873/4 [1936].
NICHOLSON: **1873/4 survey** (JB: 12): **Pal.** Kitchen garden ground. • **1906 revision** (AS: 76): Here and there.
CULT.: 1936, Herbarium Ground (K); 1949, Herbarium Ground (K). Not currently in cult.
CURRENT STATUS: No recent sightings.

211. Arenaria norvegica Gunnerus
Arctic sandwort
CULT.: 1922, unspec. (K). Not currently in cult.

212. *Arenaria balearica L.
mossy sandwort
FIRST RECORD: Probably between 1779 and 1805.
EXSICC.: **Kew Environs.** Kew; May [probably between 1779 and 1805]; *Goodenough* s.n. (K).
CULT.: In cult.
CURRENT STATUS: No recent sightings.

213. Moehringia trinervia (L.) Clairv.
Arenaria trinervia L.
three-nerved sandwort
FIRST RECORD: 1873/4 [1789].
NICHOLSON: **1873/4 survey** (JB: 12): **P.** Moist shady places near Richmond Wall, also in all the woods. • **1906 revision** (AS: 76): In all the woods.
CULT.: 1789, Nat. of Britain (HK1, 2: 101); 1811, Nat. of Britain (HK2, 3: 98); 1814, Nat. of Britain (HKE: 133). Not currently in cult.
CURRENT STATUS: This plant occurs along the Riverside Walk and was recently found by the Lake; it can still be found beside the wall along Kew Road but otherwise is rather rare and missing from likely places in the woodland areas.

214. Honckenya peploides (L.) Ehrh.
Arenaria peploides L.
sea sandwort
CULT.: 1768 (JH: 203); 1789, Nat. of Britain (HK1, 2: 100); 1811, Nat. of Britain (HK2, 3: 98); 1814, Nat. of Britain (HKE: 133). In cult.

215. Minuartia verna (L.) Hiern
Arenaria verna L.
A. laricifolia With.
spring sandwort
CULT.: 1789, Nat. of England (HK1, 2: 102); Nat. of Britain (ibid.: 102 as *laricifolia*), 1811, Nat. of Britain (HK2, 3: 100); 1814, Nat. of Britain (HKE: 133). Not currently in cult.

216. Minuartia hybrida subsp. **tenuifolia** (L.) Kerguélen
Arenaria tenuifolia L.
fine-leaved sandwort
FIRST RECORD: 1906 [1768].
NICHOLSON: **1906 revision** (AS: 76): **A.** On ground at top of wall of ha-ha facing river.
OTHER REFERENCES: **Kew Gardens.** 1931, *Wright* (SFS: 186): Thames side, on wall of Kew Gardens.
CULT.: 1768 (JH: 203); 1789, Nat. of England (HK1, 2: 102); 1811, Nat. of England (HK2, 3: 100); 1814, Nat. of England (HKE: 133). Not currently in cult.
CURRENT STATUS: No recent sightings; it has gone from the riverside wall top.

217. Minurtia sedoides (L.) Hiern
Cherleria sedoides L.
cyphel
CULT.: 1789, Nat. of Scotland (HK1, 2: 103); 1811, Nat. of Scotland (HK2, 3: 101); 1814, Nat. of Scotland (HKE: 133). Not currently in cult.

218. Stellaria nemorum L.
wood stitchwort
CULT.: 1789, Nat. of Britain (HK1, 2: 99); 1811, Nat. of Britain (HK2, 3: 96); 1814, Nat. of Britain (HKE: 133). Not currently in cult.

219. Stellaria media (L.) Vill.
Alsine media L.
common chickweed
FIRST RECORD: 1873/4 [1768].
NICHOLSON: **1873/4 survey** (JB: 12): Everywhere in beds etc. Also in turf whenever it becomes rather bare. • **1906 revision** (AS: 76): Everywhere in beds etc. Also in turf whenever it becomes rather bare.
EXSICC.: **Kew Gardens.** [Unlocalised]; weed on cultivated ground; 25 v 1929; *Hubbard* s.n. (K). • [Unlocalised]; weed on cultivated ground; 10 v 1933; *Hubbard* s.n. (K). • [Unlocalised]; weed of cultivated land; 6 v 1944; *Souster* 55 (K).
CULT.: 1768 (JH: 199); 1789, Nat. of Britain (HK1, 1: 378); 1811, Nat. of Britain (HK2, 2: 175); 1814, Nat. of Britain (HKE: 82). Not currently in cult.
CURRENT STATUS: Almost ubiquitous in the Gardens.

220. Stellaria pallida (Dumort.) Crép.
lesser chickweed
FIRST RECORD: 1982.
REFERENCES: **Kew Gardens.** 1982, *Hastings* (S31: 182): Kew Green (R.B.G. side). **Kew Environs.** pre-2000, *Bartlett* (LN76: 196): Richmond Golf Course.
CURRENT STATUS: Rare. It may still be on the Green where it was first found but is seriously threatened by mulching of the bare soil at the base of trees that it favours; it has not been seen for several years.

221. Stellaria holostea L.
greater stitchwort
CULT.: 1789, Nat. of Britain (HK1, 2: 100); 1811, Nat. of Britain (HK2, 3: 96); 1814, Nat. of Britain (HKE: 133). In cult.

222. Stellaria palustris Retz.
S. glauca With.
marsh stitchwort
CULT.: 1811, Nat. of Britain (HK2, 3: 97); 1814, Nat. of Britain (HKE: 133). Not currently in cult.

Lesser chickweed (*Stellaria pallida*). First recorded on Kew Green in 1982 but has not been seen since 2005; the habitat was lost to mulching of the soil around the tree under which it grew.

223. Stellaria graminea L.
S. scapigera Willd.
lesser stitchwort
FIRST RECORD: 1873/4 [1789].
NICHOLSON: **1873/4 survey** (JB: 12): **B.** Frequent in turf on most of the lawns. **P.** Plentiful. • **1906 revision:** Omitted.
EXSICC.: **Kew Gardens.** [Unlocalised]; growing in rough grass; 3 vi 1944; *Souster* 79 (K).
CULT.: 1789, Nat. of Britain (HK1, 2: 100); 1811, Nat. of Britain (HK2, 3: 97), Nat. of Scotland (ibid.: as *scapigera*); 1814, Nat. of Britain (HKE: 133), Nat. of Scotland (ibid.: as *scapigera*). Not currently in cult.
CURRENT STATUS: Widespread, and particularly abundant in the Conservation Area and Ancient Meadow. It can be found in almost any area of long grass except in the northern part of the Gardens. Strangely overlooked in Nicholson's revised checklist.

224. Stellaria uliginosa Murray
bog stitchwort
FIRST RECORD: 1852 [1811].
EXSICC.: **Kew Environs.** Meadow at the foot of Kew Bridge; v 1852; *Stevens* s.n. (K).
CULT.: 1811, Nat. of Britain (HK2, 3: 97); 1814, Nat. of Britain (HKE: 133). Not currently in cult.
CURRENT STATUS: No recent sightings. It is very unlikely that there is now anywhere for it grow.

[225. Holosteum umbellatum L.
jagged chickweed
CULT.: 1789, Nat. of England (HK1, 1: 123); 1810, Nat. of England (HK2, 1: 183); 1814, Nat. of Britain (HKE: 26).
Note. Listed in error as native.**]**

226. Cerastium cerastoides (L.) Britton
Stellaria cerastoides L.
starwort mouse-ear
CULT.: 1811. Nat. of Scotland (HK2, 3: 97); 1814, Nat. of Scotland (HKE: 133). Not currently in cult.

227. Cerastium arvense L.
field mouse-ear
FIRST RECORD: 1873/4 [1768].
NICHOLSON: **1873/4 survey** (JB: 11 and in SFS: 179): **P.** Open dry places. Makes beautiful masses on the top of wall facing river. **Strip.** Frequent in towing-path towards Brentford Ferry. • **1906 revision** (AS: 76): **A.** Open dry places. In quantity on top of wall of ha-ha facing river.
CULT.: 1768 (JH: 203); 1789, Nat. of Britain (HK1, 2: 119); 1811, Nat. of Britain (HK2, 3: 135); 1814, Nat. of Britain (HKE: 138). In cult.
CURRENT STATUS: No recent sightings.

228. Cerastium alpinum L.
alpine mouse-ear
CULT.: 1768 (JH: 203); 1811, Nat. of Britain (HK2, 3: 136); 1814, Nat. of Britain (HKE: 138). Not currently in cult.

229. Cerastium arcticum Lange
C. latifolium auct. non L.
Arctic mouse-ear
CULT.: 1789, Nat. of Britain (HK1, 2: 120); 1811, Nat. of Britain (HK2, 3: 137); 1814, Nat. of Britain (HKE: 138). Not currently in cult.

230. Cerastium fontanum Baumg.
C. triviale Link
C. vulgatum L. 1762, non L. 1755
C. fontanum subsp. *holosteoides* (Fries) Salmon, Ommering & de Voogd.
common mouse-ear
FIRST RECORD: 1873/4 [1768].
NICHOLSON: **1873/4 survey** (JB:11): Everywhere. Common in beds and shrubberies, also in turf. • **1906 revision** (AS: 76): Common in beds and shrubberies, also in turf.
EXSICC.: **Kew Gardens.** [Unlocalised]; among short grasses, on sandy soil; 10 v 1928, *Hubbard* s.n. (K). • Herbarium field; amongst grasses; v 1933; *Hubbard* s.n. (K). • [Unlocalised]; a rather infrequent weed of cultivated land & sometimes growing in grass; 5 v 1944; *Souster* 53 (K). **Kew Environs.** Bank of Thames near Kew Gardens; growing so that it is submerged by every tide; 5 v 1928; *Summerhayes* 332 (K). • Kew, Thames bank; on river-wall, from crevices, near top; 13 vii 1928; *Turrill* s.n. (K).
CULT.: 1768 (JH: 203); 1789, Nat. of Britain (HK1, 2: 119); 1811, Nat. of Britain (HK2, 3: 135); 1814, Nat. of Britain (HKE: 138); 1938, Herb. Ground (K). Not currently in cult.
CURRENT STATUS: Widespread and very common. The two specimens cited above from 'Kew Environs' are filed in the Herbarium under subsp. *holosteoides*.

231. Cerastium glomeratum Thuill.
C. viscosum auct. non L.
sticky mouse-ear
FIRST RECORD: 1873/4 [1768].
NICHOLSON: **1873/4 survey** (JB: 11): Abundant and very typical on dug ground among the oak and other collections at end of "Syon Vista." • **1906 revision** (AS: 76): Abundant in dry places in all the divisions.
EXSICC.: **Kew Gardens.** Field near the Herbarium; 6 v 1924; *Turrill* s.n. (K).
CULT.: 1768 (JH: 203); 1789, Nat. of Britain (HK1, 2: 119); 1811, Nat. of Britain (HK2, 3: 135); 1814, Nat. of Britain (HKE: 138); 1937, Herbarium Ground (K); 1950, Herb. Exp. Grd. (K). Not currently in cult.
CURRENT STATUS: Widespread and fairly common, often in the same places as the preceding.

232. Cerastium diffusum Curtis
C. tetrandrum Curtis
dwarf mouse-ear
CULT.: 1811, Nat. of Scotland (HK2, 3: 135); 1814, Nat. of Scotland (HKE: 138); 1950, Herb. Exp. Grd. (K). Not currently in cult.

233. Cerastium semidecandrum L.
little mouse-ear
FIRST RECORD: 1987 [1768].
REFERENCES: **Kew Gardens.** 1987, *Hastings* (S31: 182): Kew Green (R.B.G. side).
CULT.: 1768 (JH: 203); 1789, Nat. of Britain (HK1, 2: 119); 1811, Nat. of Britain (HK2, 3: 135); 1814, Nat. of Britain (HKE: 138). Not currently in cult.
CURRENT STATUS: Wall-top along Ferry Lane near Hanover House but seriously threatened by current building works. It has not been seen on Kew Green for several years.

234. Myosoton aquaticum (L.) Moench
Cerastium aquaticum L.
Stellaria aquatica (L.) Scop.
water chickweed
FIRST RECORD: 1873/4 [1768].
NICHOLSON: **1873/4 survey** (JB: 12): **Strip.** A couple of plants about 100 yards on the Brentford side of Isleworth Gate. • **1906 revision** (AS: 76): **Strip:** Near Isleworth Gate.
OTHER REFERENCES: **Kew Environs.** [Unspecified dates], *anon.* (LN30: S37): Many scattered records by the Thames from Chertsey to Kew.
EXSICC.: **Kew Environs.** Kew; viii 1888; *Gamble* 20217 (K).
CULT.: 1768 (JH: 203); 1789, Nat. of Britain (HK1, 2: 120); 1811, Nat. of Britain (HK2, 3: 137); 1814, Nat. of Britain (HKE: 138). Not currently in cult.
CURRENT STATUS: No recent sightings; it was last seen on the towpath in 1975.

235. Moenchia erecta (L.) P.Gaertn., B.Mey. & Scherb.
upright chickweed
CULT.: 1789, Nat. of Britain (HK1, 1: 173); 1810, Nat. of Britain (HK2, 1: 282); 1814, Nat. of Britain (HKE: 39). Not currently in cult.

[236. Bufonia tenuifolia L.
CULT.: 1810, Nat. of England (HK2, 1: 275); 1814, Nat. of England (HKE: 37).
Note. Listed in error as native.**]**

237. Sagina nodosa (L.) Fenzl
Spergula nodosa L.
knotted pearlwort
FIRST RECORD: 1883 [1768].
NICHOLSON: **1906 revision**: Not recorded.
EXSICC.: **Kew Gardens.** Bank of Thames, opposite Isleworth; vii 1883; *Fraser* s.n. (K). • Thames bank near Isleworth Gate, on river wall about 3 stones from top, above normal high tides; 9 vii 1929; *Summerhayes* s.n. (K). **Kew Environs.** On bank of Thames between Kew & Richmond, in wall of river; viii 1929; *Hubbard* s.n. (K).
CULT.: 1768 (JH: 201); 1789, Nat. of Britain (HK1, 2: 121); 1811, Nat. of Britain (HK2, 3: 138); 1814, Nat. of Britain (HKE: 138). Not currently in cult.
CURRENT STATUS: No recent sightings.

238. Sagina subulata (Sw.) C.Presl
Spergula subulata Sw.
heath Pearlwort
CULT.: 1811, Nat. of Britain (HK2, 3: 138); 1814, Nat. of Britain (HKE: 138). Not currently in cult.

239. Sagina saginoides (L.) H. Karst,
Spergula saginoides L.
alpine pearlwort
CULT.: 1789, Nat. of England (HK1, 2: 121); 1811, Nat. of Scotland (HK2, 3: 138); 1814, Nat. of Scotland (HKE: 138). Not currently in cult.

240. Sagina procumbens L.
procumbent pearlwort
FIRST RECORD: 1873/4 [1768].
NICHOLSON: **1873/4 survey** (JB: 12): **B, P** and **Pal.** Not confined to shade. Often met with in the most open places such as Pagoda Avenue, the dry slopes about Palm House, etc. • **1906 revision** (AS: 76): Everywhere. Not unfrequent in dry places in turf.
EXSICC.: **Kew Environs.** Kew; 5 v 1894; *Hosking* s.n. (K). • Kew Road, Richmond; 16 vi 1923; *Fraser* s.n. (K). • Kew, River bank; on old wall of concrete with *Cerastium vulgatum*, *Nasturtium*, *Marchantia* etc.; 26 vi 1928; *Summerhayes & Turrill* s.n. (K).
CULT.: 1768 (JH: 172/8); 1789, Nat. of Britain (HK1, 1: 172); 1810, Nat. of Britain (HK2, 1: 281); 1814, Nat. of Britain (HKE: 39). Not currently in cult.
CURRENT STATUS: Widespread in the Gardens and rather commoner than the next species.

241. Sagina apetala Ard.
S. ciliata Fr.
S. erecta Lam. [as *S. recta*]
annual pearlwort
FIRST RECORD: 1873/4 [1768].
NICHOLSON: **1873/4 survey** (JB: 12): '*S. apetala*': **B** and **P:** A common weed. '*S. ciliata*': **B.** Walks near Palm House, growing with [*S. apetala*]. Plentiful at foot of wall on Kew Road from the "Cumberland Gate" to "Melon Yard Gate." • **1906 revision** (AS: 76): '*S. apetala*': Everywhere. '*S. ciliata*': Here and there as a weed on walks.
EXSICC.: **Kew Environs.** Kew Road, Richmond; 24 vi 1899; *Fraser* s.n. (K).
CULT.: 1768 (JH: 172/8); 1789, Nat. of England (HK1, 1: 172); 1810, Nat. of Britain (HK2, 1: 281); 1814, Nat. of Britain (HKE: 39). Not currently in cult.
CURRENT STATUS: Widespread in the Gardens but not common.

242. Sagina maritima Don
sea pearlwort
CULT.: 1814, Nat. of Ireland (HKE: Add.). Not currently in cult.
Note. Listed as a native of Ireland, but in fact found all around the coasts of Great Britain.

243. Scleranthus perennis L.
perennial knawell
CULT.: 1768 (JH: 372); 1789, Nat. of England (HK1, 2: 84); 1811, Nat. of England (HK2, 3: 73); 1814, Nat. of England (HKE: 130); 1922, unspec. (K); 1950, Director's Garden (K). Not currently in cult.

244. Scleranthus annuus L.
annual knawell
FIRST RECORD: 19th cent. [1768].
EXSICC.: **Kew Gardens.** [Unlocalised]; 19th cent. (K).
CULT.: 1768 (JH: 372); 1789, Nat. of Britain (HK1, 2: 83); 1811, Nat. of Britain (HK2, 3: 73); 1814, Nat. of Britain (HKE: 130). In cult.
CURRENT STATUS: No recent sightings.

245. Corrigiola litoralis L.
strapwort
CULT.: 1789, Nat. of England (HK1, 1: 377); 1811, Nat. of England (HK2, 2: 173); 1814, Nat. of England (HKE: 82); 1882, unspec. (K); 1922, unspec. (K); 1932, unspec. (K); 1938, Herbarium Ground (K). Not currently in cult.

246. *Corrigiola telephiifolia Pourr.
FIRST RECORD: 1947.
REFERENCES: **Kew Gardens.** 1947, *Ross-Craig* (S21: 236, and see Exsicc.): Two or three plants in turf where old beech tree had recently been removed at E. end of birch collection, flowering 17 x 1947.
EXSICC.: **Kew Gardens.** Near E. end of birch collection; turf where old beech had been felled; 17 x 1947; *Ross-Craig* s.n. (K).
CULT.: 1932, unspec. (K). Not currently in cult.
CURRENT STATUS: No recent sightings.

247. Herniaria glabra L.
smooth rupturewort
CULT.: 1768 (JH: 372); 1789, Nat. of England (HK1, 1: 310); 1811, Nat. of England (HK2, 2: 97); 1814, Nat. of England (HKE: 70). Not currently in cult.

[**248. Herniaria hirsuta** L.
hairy rupturewort
CULT.: 1789, Nat. of England (HK1, 1: 311); 1811, Nat. of England (HK2, 2: 97); 1814, Nat. of England (HKE: 70).
Note. Listed in error as native.]

249. Illecebrum verticillatum L.
coral-necklace
CULT.: 1768 (JH: 382); 1789, Nat. of England (HK1, 1: 290); 1811, Nat. of England (HK2, 2: 60); 1814, Nat. of England (HKE: 65); 1922, unspec. (K); 1924, unspec. (K). Not currently in cult.

250. Polycarpon tetraphyllum (L.) L.
four-leaved allseed
CULT.: 1789, Nat. of England (HK1, 1: 123); 1810, Nat. of England (HK2, 1: 184); 1814, Nat. of England (HKE: 26). Not currently in cult.

251. Spergula arvensis L.
Spergula sativa Boenn, nom. illegit.
corn spurrey
FIRST RECORD: 1873/4 [1768].
NICHOLSON: **1873/4 survey** (JB: 12): **B.** Casual in flower-beds. **P.** Common in beds and bare turf about lake. • **1906 revision** (AS: 77): Two forms of this species occur, *S. arvensis* and *S. sativa*. The latter is the more abundant.
EXSICC.: **Kew Gardens.** Collected in cultivated bed near Lily Pond; uncommon weed in these gardens; 29 viii 1950; *Souster* 1190 (K).
CULT.: 1768 (JH: 201); 1789, Nat. of Britain (HK1, 2: 120); 1811, Nat. of Britain (HK2, 3: 137); 1814, Nat. of Britain (HKE: 138); Undated, Herbarium Ground (K). Not currently in cult.
CURRENT STATUS: No recent sightings.

[**252. Spergula morisonii** Boreau
S. pentandra auct. non L.
pearlwort spurrey
CULT.: 1789, Nat. of England (HK1, 2: 121); 1811, Nat. of England (HK2, 3: 138); 1814, Nat. of England (HKE: 138).
Note. Listed in error as native.]

253. Spergularia rupicola Lebel ex Le Jol.
rock sea-spurrey
CULT.: 1950, Herb. Exp. Grd. (K); 1955, Herb. Expt. Grnd. (K). In cult.

254. Spergularia media (L.) C.Presl
Arenaria media L.
greater sea-spurrey
CULT.: 1789, Nat. of England (HK1, 2: 102); 1950, Herbarium Experimental Ground (K). Not currently in cult.

255. Spergularia marina (L.) Griseb.
Arenaria marina L.
A. rubra var. *marina* L.
lesser sea-spurrey
CULT.: 1789, Nat. of Britain (HK1, 2: 101); 1811, Nat. of Britain (HK2, 3: 99); 1814, Nat. of Britain (HKE: 133); 1922, unspec. (K); 1923, unspec. (K). Not currently in cult.

256. Spergularia rubra (L.) J. & C.Presl
Arenaria rubra L.
A. rubra var. *campestris* L.
A. fastigiata Sm.
sand spurrey
First record: 1873/4 [1768].
Nicholson: **1873/4 survey** (JB: 12): Everywhere. On most of the walks in the Pleasure Grounds, where it is common among turf in the dryer places. • **1906 revision** (AS: 77): Everywhere. Walks and amongst turf in the dryer places.
Exsicc.: **Kew Environs.** Kew Green; vii 1883; *Fraser* s.n. (K). • Kew Green; in sandy soil, in bare places in turf, fairly common; 17 vi 1928; *Hubbard* s.n. (K).
Cult.: 1768 (JH: 203); 1789, Nat. of Britain (HK1, 2: 101); 1811, Nat. of Britain (HK2, 3: 99), Nat. of Scotland (ibid.: 101 as *fastigiata*); 1814, Nat. of Britain (HKE: 133), Nat. of Scotland (ibid.: as *fastigiata*); 1950, Kew Green (K). [Very unlikely to have been cultivated in this location since it was already a common wild plant here]. Not currently in cult.
Current status: Common on Kew Green (non-RBG side); it was once common in the gravel footpath across the Green, but was destroyed when the path was relaid in 2000 and later sealed; it is common in the short turf nearby; it also occurs in cracks in hardstanding around the Banks Building although this site is currently being refurbished.

257. Lychnis flos-cuculi L.
ragged-robin
First record: 1873/4 [1768].
Nicholson: **1873/4 survey** (JB: 11): **B.** A few plants on the mound of the temple of Aeolus. **Pal.** Three or four in the corner close to "Princess's Gate." **Strip.** Two or three opposite "Brentford Docks." • **1906 revision** (AS: 76): **B.** Near Temple of Aeolus. **P.** Near palace. **Strip.** By side of ha-ha.
Cult.: 1768 (JH: 192); 1789, Nat. of Britain (HK1, 2: 117); 1811, Nat. of Britain (HK2, 3: 133); 1814, Nat. of Britain (HKE: 137); 1934, unspec. (K). In cult.
Current status: No recent sightings.

258. Lychnis viscaria L.
sticky catchfly
Cult.: 1768 (JH: 193); 1789, Nat. of Britain (HK1, 2: 117); 1811, Nat. of Britain (HK2, 3: 133); 1814, Nat. of Britain (HKE: 137); 1897, unspec. (K); 1931, unspec. (K); 1970, unspec. (K); 1972, unspec. (K). In cult.

259. Lychnis alpina L.
alpine catchfly
Cult.: 1789, Alien (HK1, 2: 117); 1811, Nat. of Scotland (HK2, 3: 134); 1814, Nat. of Scotland (HKE: 137). Not currently in cult.
Note. Originally listed as an alien but found as a native in 1811.

Corncockle (*Agrostemma githago*). Known as a wild plant in 1906 but was lost until it was resown into 'wheatfield' displays in 1999 and 2001. It occasionally reappears as relicts from these sowings.

260. Agrostemma githago L.
Lychnis githago (L.) Scop.
corncockle
First record: 1873/4 [1768].
Nicholson: **1873/4 survey** (JB: 11): Came up about lake in newly-sown grass, 1873. • **1906 revision** (AS: 76): Here and there as weed in plantations and flower-beds.
Cult.: 1768 (JH: 194); 1789, Nat. of Britain (HK1, 2: 115); 1811, Nat. of Britain (HK2, 3: 131); 1814, Nat. of Britain (HKE: 137); 1950, Order Beds (K). In cult.
Current status: Extinct as a wild plant; it was sown in 1999 in the Wheatfields along the Broad Walk and was abundant in 2000. A few plants came up subsequently in grass but were mown off before setting seed. More were sown into the Wheatfield at the south end of the Lake in 2001. It still occasionally occurs from seed dispersed from the original sowings but is not viable in the soil for very long and is likely to disappear in due course.

261. Silene nutans L.
Nottingham catchfly
CULT.: 1768 (JH: 195); 1789, Nat. of England (HK1, 2: 94); 1811, Nat. of Britain (HK2, 3: 89); 1814, Nat. of Britain (HKE: 132); 1918, unspec. (K); 1967, unspec. (K); 1969, unspec. (K); 1970, unspec. (K); 1972, unspec. (K); 1973, unspec. (K); 1974, bed 156-23 (K). Not currently in cult.

262. Silene otites (L.) Wibel
Cucubalus otites L.
Spanish catchfly
CULT.: 1768 (JH: 195); 1789, Nat. of England (HK1, 2: 93); 1811, Nat. of England (HK2, 3: 86); 1814, Nat. of England (HKE: 132); 1897, unspec. (K); 1969, unspec. (K); 1972, unspec. (K). Not currently in cult.

263. Silene vulgaris (Moench) Garcke
S. inflata Sm.
S. cucubalus Wibel
S. cucubalus var. *puberula* Hook.f.
Cucubalus behen L.
bladder campion
FIRST RECORD: 1873/4 [1768].
NICHOLSON: **1873/4 survey** (JB: 11): **B.** A casual flower-bed weed. **P.** Frequent about lake and elsewhere. A single plant of the var. *puberula* grows in wood in front of "Engine House." • **1906 revision** (AS: 76 and in SFS: 171): A casual weed. Frequent about lake and elsewhere. Var. *puberula*: **A.** In wood near pumping station.
EXSICC.: **Kew Gardens.** Herbarium field; amongst grass near to the river wall; 30 vi 1927, *Turrill* s.n. (K).
CULT.: 1768 (JH: 195); 1789, Nat. of Britian (HK1, 2: 92); 1811, Nat. of Britain (HK2, 3: 83); 1814, Nat. of Britain (HKE: 132); 1918, unspec. (K); 1969, unspec. (K); 1970, Jodrell Glass (K); 1971, Jodrell Glass (K); 1972, unspec. (K).1973, unspec. (K). In cult.
CURRENT STATUS: No recent sightings.

264. Silene uniflora Roth
Silene maritima With.
sea campion
CULT.: 1811, Nat. of Britain (HK2, 3: 88); 1814, Nat. of Britain (HKE: 132); 1971, Jodrell Glass (K). In cult.

265. Silene acaulis (L.) Jacq.
moss campion
CULT.: 1789, Nat. of Britain (HK1, 2: 99); 1811, Nat. of Britain (HK2, 3: 96); 1814, Nat. of Britain (HKE: 132); 1909, unspec. (K). In cult. under glass.

266. *Silene armeria L.
sweet-William catchfly
FIRST RECORD: 1926 [1768].
EXSICC.: **Kew Gardens.** Weed in *Aster* beds; viii 1926, *Hubbard* s.n. (K).
CULT.: 1768 (JH: 195); 1789, Nat. of England (HK1, 2: 98); 1811, Nat. of England (HK2, 3: 94); 1814, Nat. of England (HKE: 132); 1968, Decorative Dept. (K); 1970, Glasshouse (K). Not currently in cult.
CURRENT STATUS: No recent sightings.

267. Silene noctiflora L.
night-flowering catchfly
FIRST RECORD: 1990 [1789].
REFERENCES: **Kew Gardens.** 1990, *Cope* (S31: 182): Paddock, after restoration in 1990.
CULT.: 1789, Nat. of England (HK1, 2: 96); 1811, Nat. of England (HK2, 3: 92); 1814, Nat. of England (HKE: 131); 1916, unspec. (K); 1970, unspec. (K); 1971, Jodrell Glass (K). Not currently in cult.
CURRENT STATUS: No recent sightings; it has not reappeared in the Paddock since 1990.

268. Silene latifolia subsp. **alba** (Miller) Greuter & Burdet
Lychnis vespertina Sibth.
white campion
FIRST RECORD: 1873/4 [1933].
NICHOLSON: **1873/4 survey** (JB: 11): **P.** Many plants in young plantations between Pagoda Avenue and Richmond Road. • **1906 revision** (AS: 76): Here and there in shrubberies and plantations.
CULT.: 1933, unspec. (K); 1934, unspec. (K); 1969, Greenhouse (K); 1970, Jodrell Glass (K); 1971, Jodrell Glass (K). In cult. under glass.
CURRENT STATUS: A rare weed of flower-beds and shrubberies. It has, despite its rarity, managed to hybridise with *S. dioica* (*S.* × *hampeana* q.v.).

269. Silene dioica (L.) Clairv.
Lychnis diurna Sibth.
L. dioica L.
red campion
FIRST RECORD: 1873/4 [1768].
NICHOLSON: **1873/4 survey** (JB: 11): **P** and **Q.** Not uncommon in the woods. • **1906 revision** (AS: 76): **Q.** Common in the woods.
EXSICC.: **Kew Gardens.** Queen's Cottage; [c. 1914]; *Flippance* s.n. (K). • Queen's Cottage Grounds; damp glade in woodland; 6 vi 1950; *Ross-Craig & Sealy* 1639 (K).
CULT.: 1768 (JH: 193); 1789, Nat. of Britain (HK1, 2: 118); 1811, Nat. of Britain (HK2, 3: 134); 1814, Nat. of Britain (HKE: 137); 1898, unspec. (K); 1969, unspec. (K); 1972, unspec. (K). In cult.
CURRENT STATUS: In most of the wooded areas and occasionally in long grass throughout the Gardens.

270. Silene × hampeana Meusel & K.Werner (dioica × latifolia)
FIRST RECORD: 1998.
REFERENCES: **Kew Environs.** 1998, *Cope* (S33: 744): Ferry Lane, roadside verge opposite Herbarium.
CURRENT STATUS: A single sighting along Ferry Lane and a more recent one (2004) behind the Herbarium. Most of the pale pink-flowered campions in the Gardens are not this hybrid.

271. *Silene coeli-rosea (L.) Godr.
rose-of-heaven
FIRST RECORD: 1990 [1789].
REFERENCES: **Kew Gardens.** 1990, *Cope* (S31: 182): Paddock, after restoration in 1990.
CULT.: 1789, Alien (HK1, 2: 116); 1811, Alien (HK2, 3: 132); 1814, Alien (HKE: 137). In cult.
CURRENT STATUS: No recent sightings; it has not reappeared in the Paddock since 1990.

272. Silene gallica L.
S. quinquevulnera L.
S. gallica var. *quinquevulnera* (L.) Mert. & Koch
S. anglica L.
S. lusitanica L.
small-flowered catchfly
FIRST RECORD: 1880 [1768].
NICHOLSON: **1906 revision**: Not recorded.
OTHER REFERENCES: **Kew Gardens.** 1880, *Baker* (BECB1(2): 40): Thames bank opposite Sion House between Kew and Richmond [both varieties]. • 1990, *Cope* (S31: 182): Paddock, after restoration in 1990 [both varieties].
CULT.: 1768 (JH: 195); 1789, Nat. of England (HK1, 2: 93 as *anglica*, 94 as *quinquevulnera*); 1811, Nat. of Britain (HK2, 3: 87 as *anglica*), Nat. of England (ibid.: as *quinquevulnera*); 1814, Nat. of Britain (HKE: 131 as *anglica*), Nat. of England (ibid.: as *quinquevulnera*); 1919, unspec. (K); 1938, Herbarium Ground (K); 1952, Herbarium Ground (K). In cult.
CURRENT STATUS: No recent sightings. Plants found in the Paddock in 1990 were likely to have been descendants of specimens cultivated in the Herbarium Experimental Ground.

273. Silene conica L.
sand catchfly
CULT.: 1789, Alien (HK1, 2: 96); 1811, Nat. of England (HK2, 3: 91); 1814, Nat. of England (HKE: 131); 1950, Order Beds (K); 1970, Jodrell Glass (K); 1971, Jodrell Glass (K). In cult. under glass.

[274. Silene conoidea L.
CULT.: 1789, Nat. of England (HK1, 2: 95); 1811, Alien (HK2, 3: 90); 1814, Alien (HKE: 131).
Note. The 1789 record is almost certainly the result of confusion between this species and *S. conica*.]

275. *Silene bellidifolia Juss. ex Jacq.
S. vespertina Retz.
FIRST RECORD: 1880 [1970].
NICHOLSON: **1906 revision**: Not recorded.
OTHER REFERENCES: **Kew Gardens.** 1880, *Baker* (BECB1(2): 40; S33: 744): Thames bank opposite Sion House, midway between Kew and Richmond.
CULT.: 1970, unspec. (K); 1971, Jodrell Glass (K). Not currently in cult.
CURRENT STATUS: No recent sightings.

276. *Silene colorata Poir.
FIRST RECORD: 1990 [1970].
REFERENCES: **Kew Gardens.** 1990, *Cope* (S31: 182): Paddock, after restoration in 1990.
CULT.: 1970, Jodrell Glass (K). Not currently in cult.
CURRENT STATUS: No recent sightings; it has not reappeared in the Paddock since 1990.

277. *Cucubalus baccifer L.
berry catchfly
FIRST RECORD: 19th Cent. [1768]
NICHOLSON: **1873/4 survey**: Not recorded. • **1906 revision**: Not recorded.
EXSICC.: **Kew Gardens.** 'Hort. Bot. Kew.' [label is 19th Cent.; no other information]; *anon.* (K).
CULT.: 1768 (JH: 195); 1789, Nat. of England (HK1, 2: 91); 1811, Nat. of England (HK2, 3: 83); 1814, Nat. of England (HKE: 131); 1934, unspec. (K). Not currently in cult.
CURRENT STATUS: At the end of the wall between the Rockery and the Order Beds, and in the Director's garden. It is vulnerable in the area of the Rockery because of enthusiastic weeding and had not been seen for a couple of years before making a return.

278. Saponaria officinalis L.
soapwort
FIRST RECORD: 1999 [1768].
REFERENCES: **Kew Gardens.** 1999, *Cope* (S35: 646): An extensive patch in long grass on the edge of the Conservation Area (310).
CULT.: 1768 (JH: 194); 1789, Nat. of England (HK1, 2: 86); 1811, Nat. of England (HK2, 3: 76); 1814, Nat. of England (HKE: 130); 1950, Herbarium Field (K). In cult.
CURRENT STATUS: There are one or two patches naturalised on the edge of the Conservation Area but these were almost certainly originally planted.

278.1. *Saponaria cerastoides Fisch. ex C.A.Mey.
FIRST RECORD: 2008 [1836].
CULT.: 1836, unspec. (K); 1916, unspec. (K); 1930, unspec. (K); 1973, Herb. & Alpine Dept. (K). Not currently in cult.
CURRENT STATUS: Two or three plants were found on a small heap of compost in the Lower Nursery.

279. Petrorhagia nanteuilii (Burnat) P.W.Ball & Heywood
childing pink
FIRST RECORD: 1990 [1930].
REFERENCES: **Kew Gardens.** 1990, *Cope* (S31: 182): Paddock, after restoration in 1990.
CULT.: 1930, Herb. Expt. Ground (K). Not currently in cult.
CURRENT STATUS: No recent sightings; it has not reappeared in the Paddock since 1990. See note under the following species. The 1930 record above has been confirmed in the herbarium.

280. Petrorhagia prolifera (L.) P.W.Ball & Heywood
Dianthus prolifer L.
proliferous pink
CULT.: 1768 (JH: 205); 1789, Nat. of England (HK1, 2: 89); 1811, Nat. of England (HK2, 3: 79); 1814, Nat. of England (HKE: 130). In cult.
Note. This and the previous species are difficult to distinguish and it is likely that the records all apply to the same taxon, although it is uncertain which this should be. Aiton used the vernacular name 'Proliferous Pink' for the species in cultivation and it has been known as a British plant since 1650.

281. Dianthus gratianopolitanus Vill.
D. glaucus Huds.
D. caesius Sm.
cheddar pink
CULT.: 1768 (JH: 205); 1789, Nat. of Britain (HK1, 2: 89); 1811, Nat. of Britain (HK2, 3: 82); 1814, Nat. of Britain (HKE: 131). In cult.

[282. Dianthus caryophyllus L.
clove pink
CULT.: 1789, Nat. of England (HK1, 2: 89); 1811, Nat. of England (HK2, 3: 79); 1814, Nat. of England (HKE: 130).
Note. Listed in error as native.]

283. Dianthus deltoides L.
maiden pink
FIRST RECORD: 1906 [1768].
NICHOLSON: **1906 revision** (AS: 76 and in SFS: 169): **A.** In turf of Pagoda Vista, but does not flower on account of its being constantly mown down. In Larch collection does flower. Slopes of temperate house.
OTHER REFERENCES: **Kew Gardens.** 1983, *Burton* (FLA: 32): [There is no longer] suitable territory for it ... at Kew Gardens.
CULT.: 1768 (JH: 205); 1789, Nat. of Britain (HK1, 2: 89); 1811, Nat. of Britain (HK2, 3: 80); 1814, Nat. of Britain (HKE: 131); 1921, unspec. (K); 1932, unspec. (K); 1939, Herbarium Ground (K); 1951, Herbarium Ground (K); 1952, Herbarium Ground (K); 1953, Herbarium Ground (K); 1954, Herbarium Ground (K); 1955, Herbarium Ground (K); 1958, Herbarium Ground (K); 1959, Herbarium Ground (K). In cult.
CURRENT STATUS: Long extinct as a wild plant in the Gardens. Because of close mowing of the turf it was never a prominent plant and seldom managed to flower; as a result the seed bank in the soil has probably been exhausted.

284. Dianthus armeria L.
Deptford pink
CULT.: 1768 (JH: 205); 1789, Nat. of England (HK1, 2: 88); 1811, Nat. of England (HK2, 3: 79); 1814, Nat. of England (HKE: 130); 1950, Herbarium Ground (K); 1952, Herb. Exp. Gdn (K). In cult.

[285. Dianthus arenarius L.
stone pink
CULT.: 1789, Nat. of Britain (HK1, 2: 91).
Note. Listed in error as native.]

Maiden pink (*Dianthus deltoides*). At one time this was a component of turf along Pagoda Vista and near the Temperate House, but it has been extinct for many years. [Photo: B.R.Tebbs]

Polygonaceae

286. Persicaria bistorta (L.) Samp.
Polygonum bistorta L.
bistort
First record: 2003 [1768].
Cult.: 1768 (JH: 330); 1789, Nat. of Britain (HK1, 2: 30); 1811, Nat. of Britain (HK2, 2: 417); 1814, Nat. of Britain (HKE: 117); 1933, Arboretum (K). In cult.
Current status: By the Lake in small quantity and probably originally planted.

287. Persicaria vivipara (L.) Ronse Decr.
Polygonum viviparum L.
alpine bistort
Cult.: 1768 (JH: 330); 1789, Nat. of Britain (HK1, 2: 30); 1811, Nat. of Britain (HK2, 2: 417); 1814, Nat. of Britain (HKE: 117). In cult.

288. Persicaria amphibia (L.) Gray
Polygonum amphibium L.
amphibious bistort
First record: 1873/4 [1768].
Nicholson: **1873/4 survey** (JB: 73): **Strip.** Several plants in moat. • **1906 revision** (AS: 86): **Strip.** In ha-ha.
Other references: **Kew Gardens.** 1999, *Stones* (EAK: 2): By Lake.
Cult.: 1768 (JH: 330); 1789, Nat. of Britain (HK1, 2: 31); 1811, Nat. of Britain (HK2, 2: 417); 1814, Nat. of Britain (HKE: 117); 1940, Aquatic Garden (K). In cult.
Current status: Occasional around the Lake and along the towpath, the latter only as the terrestrial form. There are no early records from within the Gardens suggesting that it has been planted by the Lake. It is no longer in the Moat.

289. Persicaria maculosa Gray
Polygonum persicaria L.
redshank
First record: 1870 [1768].
Nicholson: **1873/4 survey** (JB: 72): **B, P** and **Q.** Common. • **1906 revision** (AS: 86): Common everywhere.
Other references: **Kew Gardens.** 1999, *Stones* (EAK: 2): By Lake. **Kew Environs.** 1932, *Lousley* (LN34: S239): Between Kew and Richmond.
Exsicc.: **Kew Gardens.** [Unlocalised]; on rubbish heap; viii 1931; *Hubbard* s.n. (K). • Allotments; vii 1942; *Hutchinson* s.n. (K). • Allotments by the Herbarium; weed; 10 ix 1942; *Melville* s.n. (K). **Kew Environs.** Kew; vii 1870; *Baker* in *Herb. Watson* (K). • Kew; 4 viii 1871; *Hooker* s.n. (K). • Tow Path, Richmond to Kew; soil sandy gravel; 1 viii 1948; *O.J.Ward* 107 (K). • Thames Embankment, Kew; 23 viii 1950; *Naylor* 427 (K).
Cult.: 1768 (JH: 330); 1789, Nat. of Britain (HK1, 2: 31); 1811, Nat. of Britain (HK2, 2: 418); 1814, Nat. of Britain (HKE: 117). Not currently in cult.
Current status: A common weed of disturbed ground throughout the Gardens.

290. Persicaria lapathifolia (L.) Gray
Polygonum lapathifolium L.
P. maculatum (Gray) Bab.
P. nodosum Pers.
P. nodosum var. *erectum* Rouy
pale persicaria
First record: 1873/4 [1789].
Nicholson: **1873/4 survey** (JB: 73 and in SFS: 562): **B** and **P.** Fairly common. • **1906 revision** (AS: 86): Fairly common in all the divisions.
Other references: **Kew Environs.** 1931–32, *Lousley* (LN34: S240): [var. *erectum*] Between Kew and Richmond.
Exsicc.: **Kew Gardens.** River-bank Kew – Richmond, Richmond side of Isleworth Gate; on thrown-up mud; 16 viii 1928; *Summerhayes & Turrill* 3332 (K), 3333 (K). **Kew Environs.** Kew; viii 1878; *Baker* 1121 (K). • Kew; 25 vii 1902; *Worsdell* s.n. (K).
Cult.: 1789, Nat. of England (HK1, 2: 30); 1811, Nat. of England (HK2, 2: 417); 1814, Nat. of England (HKE: 117); 1897, unspec. (K). Not currently in cult.
Current status: There has been only a single recent occurrence on heavily disturbed ground at the edge of the Stable Yard, so its decline since Nicholson's time has been quite marked.

291. Persicaria hydropiper (L.) Spach
Polygonum hydropiper L.
water-pepper
First record: 1873/4 [1768].
Nicholson: **1873/4 survey** (JB: 72): **P.** Here and there at edge of lake. **Strip.** Very abundant by side of river. • **1906 revision** (AS: 86): **A.** Here and there along edge of lake. **Strip.** Very abundant by ha-ha and river.
Exsicc.: **Kew Gardens.** River-bank Kew – Richmond, Richmond side of Isleworth Gate; on thrown-up mud; 16 viii 1928; *Summerhayes & Turrill* 3269 (K). **Kew environs.** Thames bank, Kew; 1883; *Fraser* s.n. (K).
Cult.: 1768 (JH: 330); 1789, Nat. of Britain (HK1, 2: 31); 1811, Nat. of Britain (HK2, 2: 418); 1814, Nat. of Britain (HKE: 117). Not currently in cult.
Current status: Recently found as a weed in the Azalea beds along the Broad Walk, but otherwise it has not been seen for many years.

292. Persicaria mitis (Schrank) Opiz ex Assenov
Polygonum laxiflorum Weihe
tasteless water-pepper
FIRST RECORD: 1885.
REFERENCES: **Kew Environs.** 1934, *Robbins* (LN14: 75 &: S84; LN34: S240; S33: 744): Thames bank between Kew and Richmond, plentifully.
EXSICC.: **Kew Environs.** Thames banks, Kew; 1885; *Nicholson* s.n. (K). • By the Thames below Richmond; 7 viii 1927; *Britton* 3229 (K).
CURRENT STATUS: Still occurs on the river bank adjacent to Ferry Lane but no longer in any quantity.

293. Persicaria minor (Huds.) Opiz
Polygonum minus Huds.
small water-pepper
FIRST RECORD: 1873/4 [1789].
NICHOLSON: **1873/4 survey** (JB: 72 and in SFS: 560): **P.** Very common round edge of lake wherever the Junci, etc., leave it room. • **1906 revision** (AS: 86): **A.** Common about lake, etc.
OTHER REFERENCES: **Kew Gardens.** 1907, *A.B.Jackson* (LN34: S240): By the lake at Kew.
CULT.: 1789, Nat. of England (HK1, 2: 31); 1811, Nat. of England (HK2, 2: 418); 1814, Nat. of England (HKE: 117). Not currently in cult.
CURRENT STATUS: Recently found as a weed in the Azalea beds along the Broad Walk. It has not been seen near the Lake for many years.

294. Koenigia islandica L.
Iceland-purslane
CULT.: 1789, Alien (HK1, 1: 123); 1810, Alien (HK2, 1: 183); 1814, Alien (HKE: 26); 1926, unspec. (K). Not currently in cult.
Note. Originally grown as an alien and not discovered as a native of Britain until 1934.

295. *Fagopyrum esculentum Moench
Polygonum fagopyrum L.
buckwheat
FIRST RECORD: 2003 [1768].
REFERENCES: **Kew Gardens.** 2003, *Cope* (S35: 648): Component of the Wheatfield sown into the south end of Syon Vista (261).
CULT.: 1768 (JH: 330); 1789, Nat. of England (HK1, 2: 33); 1811, Nat. of England (HK2, 2: 421); 1814, Nat. of England (HKE: 117); 1894, unspec. (K). Not currently in cult.
CURRENT STATUS: Sown into the Wheatfield at the south end of the Lake in 2001. It formed a population to one side and until the turf replacing this patch is fully established it may reappear from time to time.

296. Polygonum maritimum L.
sea knotgrass
CULT.: 1768 (JH: 330); 1789, Nat. of England (HK1, 2: 32). Not currently in cult.

297. Polygonum arenastrum Boreau
equal-leaved knotgrass
FIRST RECORD: 1998.
REFERENCES: **Kew Gardens.** 1998, *Cope* (S33: 744): North Arb.: lawn behind Herbarium after disturbance during building works.
CURRENT STATUS: Increased rapidly in 2003 especially in the new lawn in front of the Orangery, probably as a contaminant of the seed, but has subsequently declined. There are one or two other scattered records of similar origin.

298. Polygonum aviculare L.
knotgrass
FIRST RECORD: 1873/4 [1768].
NICHOLSON: **1873/4 survey** (JB: 72): Common at path-edges in bare places. • **1906 revision** (AS: 86): Common in bare places everywhere.
EXSICC.: **Kew Gardens.** [Unlocalised]; annual weed of waste land; 1 viii 1944; *Souster* 134 (K).
CULT.: 1768 (JH: 330); 1789, Nat. of Britain (HK1, 2: 32); 1811, Nat. of Britain (HK2, 2: 419); 1814, Nat. of Britain (HKE: 117); 1899, unspec. (K). Not currently in cult.
CURRENT STATUS: A common weed of disturbed ground throughout the Gardens.

299. *Polygonum cognatum Meisn.
P. alpestre C.A.Mey.
Indian knotgrass
FIRST RECORD: 1872.
REFERENCES: **Kew Environs.** 1872, *Naylor* (JB10: 338): Has been noticed abundantly by Mr Naylor for the last two years in the meadow near Kew Bridge, so often referred to as a locality for introduced species. • 1872, *Naylor* fide *Nicholson* (BEC18: 18; S33: 744): On the Surrey side of the Thames, near Kew Bridge, where it was first gathered by Mr Naylor in 1872. • 1887, *F.H.Ward* (WBEC4: 12; S33: 744): Naturalised on Kew Green. • 1917, *A.B. Jackson* (LN34: S241; S33: 744): Westerly Wier, Kew Green. The plant persisted here from before 1872 until it was destroyed by the construction of the tennis courts in 1923. • 1954, *Bangerter & Welch* (LN34: 62; S33: 744): It seems to have disappeared from our area, where it was known near Kew Green for fifty years up to 1923.
EXSICC.: **Kew Environs.** Kew Green, in quantity on a piece of ground here which at high tide is covered by the Thames; 1872; *Naylor* in *Herb. Watson* (K). • Many roots of it in the summer of 1872, on a plot of ground, on which was thrown [sic], close by the

bridge at Kew; *Watson* in *Herb. Watson* (K). • Westerley Ware, Kew; vii 1877; *Nicholson* s.n. (K). • Westerley Ware, Kew; 1 vii 1886; *Fraser* s.n. (K). • Established on Westerley Ware, 1872 et ante, still maintining its ground; 23 viii 1888; *Fraser* s.n. (K). •Westerley Weir near Kew Green; 17 viii 1918; *Fraser* s.n. (K). • Westerley Ware, Kew; the plant grew here before 1872 but was destroyed by the making of tennis courts in 1923, see *J. Bot.* 1892: 338; 6 vi 1922; *Fraser* s.n. (K).
CURRENT STATUS: No recent sightings. The tennis courts on what was Westerley Ware are probably those on the far side of Kew Bridge and therefore out of our area. It has been a long time since Kew Green was flooded at high tide.

300. *Fallopia japonica (Houtt.) Ronse Decr.
Japanese knotweed
FIRST RECORD: 1998 [1881].
REFERENCES: **Kew Gardens.** 1998, *Cope* (S33: 744): North Arb.: edge of Paddock near boundary wall (104); behind Banks Building (108); West Arb.: beech circle (242).
CULT.: 1881, unspec. (K); 1883, unspec. (K). In cult.
CURRENT STATUS: Rare but persistent; it frequently pushes its way through asphalt at the edge of the Herbarium car park but never achieves much height.

301. *Fallopia baldschuanica (Regel) Holub
Russian-vine
FIRST RECORD: 2001 [1895].
REFERENCES: **Kew Gardens.** 2001, *Cope* (S35: 648): Scrambling over trees and shrubs near the Oxenhouse Gate (327).
CULT.: 1895, unspec. (K); 1896, unspec. (K); 1904, unspec. (K); 1905, Arboretum (K); 1919, unspec. (K). Not currently in cult.
CURRENT STATUS: Scrambling over trees near the Oxenhouse Gate, doubtless as a relict of former cultivation.

302. Fallopia convolvulus (L.) Á.Löve
Polygonum convolvulus L.
black-bindweed
FIRST RECORD: 1873/4 [1768].
NICHOLSON: **1873/4 survey** (JB: 72): **P.** A plant or two in shrubbery at end of "Syon Vista." Common among newly-planted shrubs bordering Old Deer Park. • **1906 revision** (AS: 86): Here and there in all the divisions.
OTHER REFERENCES: **Kew Gardens.** 1931, *Fraser* (SFS: 557): Kew Gardens.
EXSICC.: **Kew Gardens.** Near the Herbarium; weed; 31 vii 1928; *Summerhayes & Turrill* 3335 (K). • [Unlocalised]; Kew Gardens [see references above]; 10 vi 1931; *Fraser* s.n. (K). • By Herbarium; 10 ix 1942; *Melville* s.n. (K). **Kew Environs.** Kew; viii 1878; *G. Nicholson* s.n. (K). • Kew; 20 ix 1886; *Fraser* s.n. (K).
CULT.: 1768 (JH: 330); 1789, Nat. of Britain (HK1, 2: 33); 1811, Nat. of Britain (HK2, 2: 421); 1814, Nat. of Britain (HKE: 117); 1881, unspec. (K); 1908, unspec. (K). Not currently in cult.
CURRENT STATUS: An occasional weed in disturbed ground and especially abundant in the last two or three years in topsoil stored in the Paddock.

303. Rumex acetosella L.
sheep's sorrel
FIRST RECORD: 1873/4 [1768].
NICHOLSON: **1873/4 survey** (JB: 72): Very much more plentiful than [*Rumex acetosa*]. • **1906 revision** (AS: 87): Very common everywhere.
CULT.: 1768 (JH: 162); 1789, Nat. of Britain (HK1, 1: 487); 1811, Nat. of Britain (HK2, 2: 323); 1814, Nat. of Britain (HKE: 104). Not currently in cult.
CURRENT STATUS: Widespread in short grass throughout the Gardens and often very abundant, particularly where the ground is more acid as on Kew Green.

Black-bindweed (*Fallopia convolvulus*). Not common but always seems to appear when there is major disturbance of soil.

Water dock (*Rumex hydrolapathum*). A conspciuous feature of the Lake margin, where it is possibly native, and in the Conservation Area where it probably isn't.

304. Rumex acetosa L.
common sorrel
FIRST RECORD: 1873/4 [1768].
NICHOLSON: **1873/4 survey** (JB: 72): Fairly common in each division. • **1906 revision** (AS: 87): Fairly common in each division.
CULT.: 1768 (JH: 162); 1789, Nat. of Britain (HK1, 1: 487); 1811, Nat. of Britain (HK2, 2: 323); 1814, Nat. of Britain (HKE: 104). Not currently in cult.
CURRENT STATUS: Common in the longer grass throughout the Gardens.

305. Rumex pseudoalpinus Höfft.
monk's-rhubarb
Cult.: In cult. (1969) possibly under the old name *R. alpinus* auct.

306. Rumex hydrolapathum Huds.
R. aquaticus auct. non L.
water dock
FIRST RECORD: 1873/4 [1789].
NICHOLSON: **1873/4 survey** (JB: 72 and in SFS: 570): **P.** Here and there round lake. **Strip.** Not uncommon. • **1906 revision** (AS: 87): **A.** Here and there round lake. **Strip.** Not uncommon.
OTHER REFERENCES: **Kew Gardens.** 1999, *Stones* (EAK: 2): By Lake.
CULT.: 1789, Nat. of England (HK1, 1: 483); 1811, Nat. of Britain (HK2, 2: 320); 1814, Nat. of Britain (HKE: 104). In cult.
CURRENT STATUS: Rare; it can be found by the Lake and in the Conservation Area although it has certainly been introduced in the latter

307. *Rumex cristatus DC.
Greek dock
FIRST RECORD: 1918.
REFERENCES: **Kew Gardens.** 1998, *Cope* (S33: 744): North Arb.: Paddock (104). **Kew Environs.** 1938, *Lousley* (LN34: S245; S33: 744): Thames bank, Kew.
EXSICC.: **Kew Gardens.** Queen's Cottage Grounds; 29 vii 1918; *Worsdell* s.n. (K). **Kew Environs.** River embankment, Kew Bridge; 18 vii 1932; [*illegible initials*] s.n. (K).
CURRENT STATUS: Recently appeared in the Paddock but has otherwise been absent or overlooked for many years. It has not been refound on the riverbank or in the Conservation Area.

308. *Rumex patientia L.
patience dock
FIRST RECORD: 1944 [1768].
REFERENCES: **Kew Gardens.** 1990, *Hastings* (S31: 183): Several at Beech Clump. **Kew Environs.** 1944, *Lousley* (LN24: 7): At frequent intervals up the river to above Kew Bridge.
CULT.: 1768 (JH: 162); 1789, Alien (HK1, 1: 482); 1811, Alien (HK2, 2: 318); 1814, Alien (HKE: 103). Not currently in cult.
CURRENT STATUS: No recent sightings.

309. Rumex crispus L.
curled dock
a. subsp. **crispus**
FIRST RECORD: 1873/4 [1768].
NICHOLSON: **1873/4 survey** (JB: 72): **B.** A few plants at edge of pond. **P.** Borders of shrubberies near lake. • **1906 revision** (AS: 87): Borders of shrubberies, etc.
CULT.: 1768 (JH: 162); 1789, Nat. of Britain (HK1, 1: 482); 1811, Nat. of Britain (HK2, 2: 318); 1814, Nat. of Britain (HKE: 103). Not currently in cult.
CURRENT STATUS: Fairly common in shady places and in long grass.

b. subsp. **littoreus** (J.Hardy) Akeroyd
R. crispus var. *triangulatus* Syme
FIRST RECORD: 1877.
EXSICC.: **Kew Environs.** Kew; waste ground; vii 1877; *Nicholson* s.n. (K).
CURRENT STATUS: No recent sightings.

c. subsp. **uliginosus** (Le Gall) Akeroyd
R. crispus var. *uliginosus* Le Gall
R. elongatus Guss.
FIRST RECORD: 1920.
REFERENCES: **Kew Environs.** 1923, *Pugsley* (WBEC40: 259): Thames bank at Kew, Surrey, 7 vii, 1923. These specimens are from the tidal mudbanks of the Thames near Kew, where the plant occurs with [*Glyceria maxima*], [*Schoenoplectus* × *carinatus*], and other estuarine species. It is noteworthy that it seems confined to the mud, and does not grow on the waste ground above the river-wall, where other Docks abound. It was first recorded as an introduced British plant by Trimen in "Journ. Bot." XI, p.237 (1873), from specimens collected lower down the river on the Surrey Bank between Hammersmith and Putney Bridges. The plant has thus maintained itself by the Thames for at least fifty years, and apparently extends as far above London as the tide. • 1933, *Lousley* (WBEC50: 232; LN34: S242): Estuarine mud ... above Kew Bridge. This plant was first recorded from the Thames banks by Trimen in *Journ. Bot.* 1873, p.327, and had been noted from various stations between Kew and Putney at intervals during the last sixty years. This year I found it in three different localities between Kew and Mortlake, and judging from my previous experience in searching for the plant the present summer must have been extraordinarily favourable to it. It is noteworthy that, whereas *R. elongatus* is entirely restricted to the estuarine mud, hybrids of it with other Docks appear on the towing path above. • 1944, *Lousley* (LN24: 6): I gathered it at Kew Bridge and Mortlake in 1933, but observed it at the former station for several years after this.
EXSICC.: **Kew Environs.** By the Thames, Kew; 1 vii 1920; *Britton* 2203 (K). • Estuarine mud above Kew Bridge; 1 vii 1933; *Lousley* s.n. (K).
CURRENT STATUS: No recent sightings. Very little now grows on the mud, and the hybrid *Schoenoplectus* mentioned above has been extinct for a long time.

310. Rumex × **pratensis** Mert. & W.D.J.Koch (crispus × obtusifolius)
R. acutus auct. non L.
CULT.: 1789, Nat. of Britain (HK1, 1: 484); 1811, Nat. of Britain (HK2, 2: 320); 1814, Nat. of Britain (HKE: 104). Not currently in cult.

311. Rumex conglomeratus Murray
R. paludosus With.
clustered dock
FIRST RECORD: 1873/4 [1789].
NICHOLSON: **1873/4 survey** (JB: 72): **P** and **Strip.** Very common about lake and near river. **B.** Here and there by side of pond. • **1906 revision** (AS: 87): **A.** Very common about lake and near river.
OTHER REFERENCES: **Kew Gardens.** 1999, *Stones* (EAK: 2 & 5): By Lake and Palm House Pond.
EXSICC.: **Kew Gardens.** Thames-bank, Kew – Richmond; in *Petasites* zone at outer (inland) edge, near Isleworth Gate; 25 vi 1929; *Summerhayes & Turrill* 3058 (K). **Kew Environs.** Thames bank, Kew; top of bank, not normally flooded; 13 viii 1928; *Summerhayes & Turrill* 3059 (K). • Thames banks, Kew; 1 viii 1931; *Fraser* s.n. (K).
CULT.: 1789, Nat. of Britain (HK1, 1: 482). Not currently in cult.
CURRENT STATUS: Known from several spots around the Lake and elsewhere.

312. Rumex × **knafii** Čelak. (conglomeratus × maritimus)
FIRST RECORD: 1928.
EXSICC.: **Kew Environs.** Thames bank, Kew; lowest zone on mud, completely submerged at high tides; 13 vii 1928; *Turrill* s.n. (K).
CURRENT STATUS: No recent sightings. The zoning of vegetation on the tidal mud has long since been lost to strengthening of the banks and most of the vegetation itself has gone.

313. Rumex sanguineus L.
wood dock
FIRST RECORD: 1880 [1768].
NICHOLSON: **1906 revision** (AS: 87): Here and there in dry places.
OTHER REFERENCES: **Kew Gardens.** 1915, *A.B.Jackson* (BSEC4: 369): Seedlings, ex hort. Kew.
EXSICC.: **Kew Gardens.** [Unlocalised]; grass under trees; 26 viii 1918; *Worsdell* s.n. (K). • Thames-bank; at outer (inland) edge of *Petasites* zone, near Isleworth Gate; 25 vi 1929; *Summerhayes & Turrill* 3067 (K). **Kew Environs.** Thames side, Kew; vi 1880; *Baker* 1099b (K). • Side of tow path – between Kew & Richmond bridges; vii 1928; *C.A. Smith* 6083 (K).
CULT.: 1768 (JH: 162); 1789, Nat. of England (HK1, 1: 482); 1811, Nat. of England (HK2, 2: 318); 1814, Nat. of England (HKE: 103). In cult.
CURRENT STATUS: Common throughout the Gardens in shady places.

314. Rumex pulcher L.
R. pulcher subsp. *woodsii* (De Not.) Arcang.
R. pulcher subsp. *divaricatus* (L.) Murb.
fiddle dock
FIRST RECORD: 1867 [1768].
NICHOLSON: **1873/4 survey** (JB: 72): **B.** A plant or two near wall close to Grand Entrance. • **1906 revision** (AS: 87 and in SFS: 567): **P.** Here and there in dry places.
OTHER REFERENCES: **Kew Gardens.** 1972, *Airy Shaw* (S29: 406): One plant by new fumigation chamber E of Wing C. • 1986, *Hastings* (LN65: 196): Kew Gardens. • 1995, *Cope* (S32: 657): Kew Green, near Administration Block etc. **Kew Environs.** Pre-1908, rep. 1938, *Nicholson* in Herb. LD (BSEC12: 138; S33: 744): Kew. • 1908, *A.B.Jackson* (LN34: S244): Kew Green. • 1945, *Lousley* (LN24:5; S33: 744): Kew, doubtless on waste ground. • 1980–81, *Hastings* (LN61: 102): Kew Green.
EXSICC.: **Kew Gardens.** Kew Green, Bird Cage Walk, south end; amongst grass; 7 viii 1946; *Turrill* s.n. (K). **Kew Environs.** Kew; ii 1867; *Hemsley* s.n. (K). • Kew Green; i 1923; *Turrill* s.n. (K).
CULT.: 1768 (JH: 162); 1789, Nat. of Britain (HK1, 1: 484); 1811, Nat. of Britain (HK2, 2: 321); 1814, Nat. of Britain (HKE: 104). Not currently in cult.
CURRENT STATUS: Barely survives on Kew Green on either side of Birdcage Walk. Mowing of the grass, on both sides of Birdcage Walk, has inhibited flowering and subsequent replenishment of the seed bank. Three individuals were noted in 2008.

315. Rumex obtusifolius L.
broad-leaved dock

a. var. **obtusifolius**
R. friesii Gren. & Godr.
R. obtusifolius subsp *agrestis* (Wallr.) Rech.
FIRST RECORD: 1873/4 [1768].
NICHOLSON: **1873/4 survey** (JB: 72): **B.** Here and there in shrubberies. **P.** Common on border of wood nearly the whole length of Syon Vista. • **1906 revision** (AS: 86): Here and there in shrubberies and along border of wood whole length of Syon Vista.
OTHER REFERENCES: **Kew Environs.** 1875, *G.Nicholson* (BEC16: 24): Banks of Thames, Kew. Of one of Mr Nicholson's specimens Dr Boswell says: "I should call this *R. friesii* collected before the fruit petals were matured." There are other specimens from Mr Nicholson, however, which seem certainly *R. sylvestris*. – T.R.A.Briggs.
EXSICC.: **Kew Gardens.** Queen's Cottage Grounds; 29 vii 1918; *Worsdell* s.n. (K). • Thames-bank, Kew – Richmond; in *Petasites* zone near Isleworth Gate, at outer edge of zone; 25 vi 1929; *Summerhayes & Turrill* 3066 (K). **Kew Environs.** Thames bank, Kew; 13 vii 1928; *Turrill* s.n. (K). • Thames bank, Kew; near the top of the bank, flooded at high tide; 13 viii 1928; *Summerhayes & Turrill* 3063 (K). • Thames bank, Kew; top of bank, not normally flooded; 13 viii 1928; *Summerhayes & Turrill* 3064 (K). • River embankment, Kew Bridge; 19. vii. 1932; [*illegible initials*] s.n. (K).
CULT.: 1768 (JH: 162); 1789, Nat. of Britain (HK1, 1: 484); 1811, Nat. of Britain (HK2, 2: 320); 1814, Nat. of Britain (HKE: 104). Not currently in cult.
CURRENT STATUS: Abundant throughout the Gardens in long grass and shady places.

b. var. **microcarpus** Dierb.
R. obtusifolius var. *sylvestris* (Wallr.) Koch
R. obtusifolius subsp. *transiens* (Simonk.) Rech.f.
R. sylvestris Wallr.
FIRST RECORD: 1873/4.
NICHOLSON: **1873/4 survey** (JB: 72): On the **Strip** this variety is much more common than [var. *obtusifolius*], and nearly as common as *R. conglomeratus*. I have never been able to find a single plant away from the river. • **1906 revision**: Not recorded.
OTHER REFERENCES: **Kew Environs.** 1875, *G.Nicholson* (BEC16: 24): Banks of Thames, Kew. Of one of Mr Nicholson's specimens Dr Boswell says: "I should call this *R. friesii* collected before the fruit petals were matured." There are other specimens from Mr Nicholson, however, which seem certainly *R. sylvestris*. – T.R.A.Briggs. • 1877, *Baker* (BEC18: 9): Kew, Surrey. Racemes much more slender than in type, very few flowers fertilised, enlarged petals subentire or faintly toothed. Several tufts intermixed with ordinary form. "This, with the few fertilised flowers, has much the look of a hybrid dock." – T.R.A.Briggs. • 1907, *Linton* (JB45: 296; BECS2: 590): [I] think that the *R. sylvestris* from Kew riverside, though I have only a poor specimen, is [*R. conglomeratus* × *obtusifolius*]. • 1945, *Lousley* (LN24: 5): On both sides of the Thames up to just above Kew Bridge.
EXSICC.: **Kew Environs.** Thames bank near Kew Bridge; 30 vii 1940; *Brenan* 6445 (K).
CURRENT STATUS: Known only from the towpath on either side of Brentford Gate; the records may refer to var. *transiens* but no attempt has been made to distinguish between these critical taxa.

316. Rumex × dufftii Hausskn. (obtusifolius × sanguineus)
FIRST RECORD: 1983.
REFERENCES: **Kew Environs.** 1983, *Hastings* (LN63: 143; S33: 744): Thames path outside Kew Gardens.
CURRENT STATUS: No recent sightings.

317. Rumex palustris Sm.
marsh dock
FIRST RECORD: 1951 [1811].
REFERENCES: **Kew Environs.** 1951, *Boniface* (LN34: S244; S33: 744): River-wall between Kew and Richmond, six plants.

CULT.: 1811, Nat. of England (HK2, 2: 320); 1814, Nat. of England (HKE: 104). Not currently in cult.
CURRENT STATUS: No recent sightings. The river wall has been repaired and renovated several times since the 1950s.

318. Rumex maritimus L.
golden dock
FIRST RECORD: 2008 [1768].
CULT.: 1768 (JH: 162); 1789, Nat. of Britain (HK1, 1: 484); 1811, Nat. of Britain (HK2, 2: 319); 1814, Nat. of Britain (HKE: 104). Not currently in cult.
CURRENT STATUS: Recently found, for the first time, on the golf course in the Old Deer Park.

319. Oxyria digyna (L.) Hill
Rumex digynus L.
mountain sorrel
CULT.: 1768 (JH: 158, 162); 1789, Nat. of Britain (HK1, 1: 486); 1811, Nat. of Britain (HK2, 2: 322); 1814, Nat. of Britain (HKE: 104); 1841, unspec. (K); 1926, unspec. (K); 1931, unspec. (K); 1971, Alpine & Herbaceous Dept. (K). Not currently in cult.

PLUMBAGINACEAE

320. Limonium vulgare Mill.
Statice limonium L.
common sea-lavender
CULT.: 1768 (JH: 183); 1789, Nat. of England (HK1, 1: 383); 1811, Nat. of England (HK2, 2: 180); 1814, Nat. of England (HKE: 83). In cult.

321. Limonium bellidifolium (Gouan) Dumort.
Statice reticulata auct. angl, non L.
matted sea-lavender
CULT.: 1789, Nat. of England (HK1, 1: 384); 1811, Nat. of England (HK2, 2: 182); 1814, Nat. of England (HKE: 83). Not currently in cult.

322. Limonium binervosum (G.E.Sm.) C.E.Salmon
rock sea-lavender
CULT.: 1934, Herbaceous Ground (K). Not currently in cult.

323. Armeria maritima Willd.
Statice armeria L.
thrift
CULT.: 1768 (JH: 77); 1789, Nat. of Britain (HK1, 1: 383); 1811, Nat. of Britain (HK2, 2: 179); 1814, Nat. of Britain (HKE: 83). In cult.

PAEONIACEAE

[324. Paeonia mascula (L.) Mill.
P. corallina Retz.
peony
CULT.: 1811, Nat. of England (HK2, 3: 315); 1814, Nat. of England (HKE: 168).
Note. Originally grown as a native, but has been known as an introduction since 1803.**]**

ELATINACEAE

325. Elatine hydropiper L.
eight-stamened waterwort
CULT.: 1768 (JH: 172/9); 1811, Nat. of England (HK2, 2: 245); 1814, Nat. of England (HKE: 118). Not currently in cult.

CLUSIACEAE

326. Hypericum androsaemum L.
tutsan
FIRST RECORD: 1999 [1769].
REFERENCES: **Kew Gardens.** 1999, *Cope* (S35: 646): Conservation Area (310). Subsequently in various other places in the Gardens; doubtless a relic of former cultivation and probably spread by birds.
CULT.: 1769 (JH: 205*); 1789, Nat. of Britain (HK1, 3: 103); 1812, Nat. of Britain (HK2, 4: 422); 1814, Nat. of Britain (HKE: 242); 1882, Arboretum (K); 1890, Rockery (K); 1894, Arboretum (K); 1932, Arboretum (K); 1934, Herb Garden (K); 1951, Order Beds (K); 1959, Arboretum Nursery (K); 1961, Order Beds (K); 1983, unspec. (K). In cult.
CURRENT STATUS: Thoroughly naturalised from cultivated plants in several places and may be increasing.

327. Hypericum perforatum L.
H. × *desetangsii* Lamotte (maculatum × perforatum)*
H. dubium × *perforatum*
perforate St John's-wort
FIRST RECORD: 1873/4 [1769].
NICHOLSON: **1873/4 survey** (JB: 12 and in SFS: 197): **P.** Common on wall facing river. **Pal.** Frequent except on kitchen garden ground. • **1906 revision** (AS: 77): **A.** Common on wall facing river. **P.** Frequent in turf.
OTHER REFERENCES: **Kew Gardens.** 1941, *Sandwith* (BSEC13: 53; S31: 182): [refering to *H.* × *desetangsii*] In at least two spots in shade in the wild part of Kew Gardens, apparently a survival of the native flora, 1941 and subsequent (only *H. perforatum* and *H. humifusum* are recorded in *The Wild Fauna amd Flora* of the Gardens, published in 1906, and its numerous supplements).

Perforate St John's-wort (*Hypericum perforatum*). Widespread, but uncommon, in rough grassland.

CULT.: 1769 (JH: 205*); 1789, Nat. of Britain (HK1, 3: 106); 1812, Nat. of Britain (HK2, 4: 426); 1814, Nat. of Britain (HKE: 242); 1900, unspec. (K); 1916, unspec. (K*); 1931, Herb. Exp. Ground (K); 1959, Arboretum Nursery (K); 1961, Order Beds (K); 1961, Order Beds (K*); 1980, unspec. (K). In cult.
CURRENT STATUS: Fairly widespread; a number of specimens recorded as this species seem to be *H.* × *desetangsii*, but the two sometimes grow in proximity and they are known to introgress. In practice they are very hard to distinguish and are treated together in this Catalogue. The second parent is absent and has never been recorded in the wild although it has been cultivated on and off since 1812.

328. Hypericum maculatum Crantz
H. dubium Leers
imperforate St John's-wort

a. subsp. **maculatum**
CULT.: 1812, Nat. of Britain (HK2, 4: 426); 1814, Nat. of Britain (HKE: 242); 1895, unspec. (K); 1982, unspec. (K); 1984, unspec. (K). In cult.

b. subsp. **obtusiusculum** (Tourlet) Hayek
CULT.: 1894, unspec. (K); 1922, unspec. (K); 1927, unspec. (K); 1951, unspec. (K); 1959, Arboretum Nursery (K). Not currently in cult.

329. Hypericum tetrapterum Fr.
H. quadrangulum L.
square-stalked St John's-wort
CULT.: 1769 (JH: 205*); 1789, Nat. of Britain (HK1, 3: 106); 1812, Nat. of Britain (HK2, 4: 426); 1814, Nat. of Britain (HKE: 242); 1922, unspec. (K). In cult.

330. Hypericum humifusum L.
trailing St John's-wort
FIRST RECORD: 1873/4 [1769].
NICHOLSON: **1873/4 survey** (JB: 12 and in SFS: 199): **B.** In dry places on the lawns this beautiful little plant often occurs. **P.** Here and there about lake and Winter Garden. • **1906 revision** (AS: 77): Not uncommon in turf in dry places.
EXSICC.: **Kew Gardens.** Growing occasionally as weed in nursery & apparently introduced with loam; 12 vii 1944; *Souster* 111 (K).
CULT.: 1769 (JH: 205*); 1789, Nat. of Britain (HK1, 3: 106); 1812, Nat. of Britain (HK2, 4: 427); 1814, Nat. of Britain (HKE: 242); 1951, unspec. (K); 1959, Arboretum Nursery (K). Not currently in cult.
CURRENT STATUS: No recent sightings and probably now extinct.

331. Hypericum pulchrum L.
slender St John's-wort
CULT.: 1769 (JH: 206*); 1789, Nat. of Britain (HK1, 3: 108); 1812, Nat. of Britain (HK2, 4: 429); 1814, Nat. of Britain (HKE: 242); 1922, unspec. (K); 1951, unspec. (K); Arboretum Nursery (K). In cult.

332. Hypericum hirsutum L.
hairy St John's-wort
CULT.: 1769 (JH: 205*); 1789, Nat. of Britain (HK1, 3: 107); 1812, Nat. of Britain (HK2, 4: 428); 1814, Nat. of Britain (HKE: 242); 1961, Order Beds (K). Not currently in cult.

333. Hypericum montanum L.
pale St John's-wort
CULT.: 1769 (JH: 205*); 1789, Nat. of Britain (HK1, 3: 107); 1812, Nat. of Britain (HK2, 4: 428); 1814, Nat. of Britain (HKE: 242); 1894, unspec. (K); 1949, Herbarium Ground (K). Not currently in cult.

334. Hypericum elodes L.
marsh St John's-wort
CULT.: 1769 (JH: 206*); 1789, Nat. of Britain (HK1, 3: 108); 1812, Nat. of Britain (HK2, 4: 428); 1814, Nat. of Britain (HKE: 242); 1934, unspec. (K); 1951, unspec. (K). Not currently in cult.

335. *Hypericum annulatum Moris
FIRST RECORD: 2004 [1961].
CULT.: 1961, Order Beds (K); 1961, Rock Garden (K). Not currently in cult.
CURRENT STATUS: Grew from seed in topsoil stored in the Paddock and behind the Princess of Wales Conservatory with several plants in both locations. The area behind the Conservatory has now been renovated and the plant has not reappeared in the Paddock.

[336. **Hypericum barbatum** Jacq.
CULT.: 1812, Nat. of Scotland (HK2, 4: 427); 1814, Nat. of Scotland (HKE: 242).
Note. Listed in error as native.]

TILIACEAE

337. **Tilia platyphyllos** Scop.
T. europaea var. *corallina* (Sm.) auct. dub.
large-leaved lime
CULT.: 1789, Nat. of Britain (HK1, 2: 229); 1811, Nat. of Britain (HK2, 3: 299); 1881, Arboretum, Old Collection (K); 1881, near Railway Gate (K); 1882, unspec. (K); 1884, Arboretum, Old Collection (K); 1889, unspec. (K). In cult.
Note: It was presumably planted on the towpath near Brentford Gate.

338. **Tilia cordata** Mill.
T. parvifolia Ehrh. ex Hoffm.
T. europaea var. *parvifolia* (Ehrh. ex Hoffm.) auct. dub.
small-leaved lime
CULT.: 1789, Nat. of Britain (HK1, 2: 229); 1811, Nat. of England (HK2, 3: 299); 1814, Nat. of England (HKE: 166). In cult.

339. **Tilia × vulgaris** Hayne (cordata × platyphyllos)
T. × europaea auct. non L.
lime
FIRST RECORD: 1928 [1768].
EXSICC.: **Kew Gardens.** 1st lime tree after the planes from the Herbarium end of Bird-Cage walk; 6 vii 1928; *Turrill* s.n. (K). • Nr. Herbarium; ?cult.; 1929; *Sealy* s.n. (K). • Kew Green; with galls of *Cercidonyia tilicola*; 7 vii 1941, *Melville* s.n. (K). **Kew Environs.** Tow-path, Richmond; specimen probably planted; 8 vi 1938; ?*Perkins* s.n. (K).
CULT.: 1768 (JH: 457); 1789, Nat. of Britain (HK1, 2: 229); 1811, Nat. of Britain (HK2, 3: 299); 1814, Nat. of Britain (HKE: 166); 1880, Arboretum (K); 1881, Arboretum, Old Collection (K); 1881, Palm House end of Lake (K); 1884, near Cumberland Gate (K); 1884, Arboretum (K); 1901, near Palace Gates (K); 1905, Arboretum (K); 1919, unspec. (K); 1928, Arboretum (K); 1929, unspec. (K); 1935, Arboretum (K); 1943, Arboretum (K); 1963, Arboretum (K); 1964, by Flagstaff (K); 1980, plot 162 (K); 1980, plot 451 (K); 1980, plot 462 (K); 1980, plot 463 (K); 1980, plot 482 (K). In cult.
CURRENT STATUS: Still on Kew Green by Birdcage Walk where it was planted by George IV when part of the Green was enclosed. It is probable that all trees in the vicinity of the Gardens have been planted and there is, as yet, no sign of self-seeding.

MALVACEAE

340. **Malva moschata** L.
musk-mallow
FIRST RECORD: 1873/4 [1768].
NICHOLSON: **1873/4 survey** (JB: 12): **Pal.** Here and there on the sloping bank facing Botanic Garden and Pleasure Grounds. **P.** About lake; also on the wood side of "Syon Vista." • **1906 revision** (AS: 77): **P, A.** About lake; also on the wood side of "Syon Vista."
CULT.: 1768 (JH: 212); 1789, Nat. of Britain (HK1, 2: 450); 1812, Nat. of Britain (HK2, 4: 217); 1814, Nat. of Britain (HKE: 215); 1881, unspec. (K); 1887, unspec. (K). In cult.
CURRENT STATUS: Rare; it regularly occurs near the Brick Pit in the Conservation Area and may be spreading.

[341. **Malva alcea** L.
greater musk-mallow
CULT.: 1789, Nat. of England (HK1, 2: 450); 1812, Alien (HK2, 4: 217); 1814, Alien (HKE: 215).
Note. Originally listed in error as native but this corrected in later editions.]

Musk mallow (*Malva moschata*). Never common and less so now than formerly, but regularly appears in the Conservation Area near the Brick Pit.

342. Malva sylvestris L.
M. mauritiana L.
common mallow
FIRST RECORD: 1873/4 [1768].
NICHOLSON: **1873/4 survey** (JB: 12): Plentiful. • **1906 revision** (AS: 77): Abundant.
OTHER REFERENCES: **Kew Environs.** 1950, B. Welch (LN31: S44): Towpath between Richmond and Kew [white-flowered plants].
EXSICC.: **Kew Environs.** Kew; form with small white flowers; vi 1877; Baker s.n. (K). • Kew; [seedling only]; 9 vi 1931; *Fraser* s.n. (K).
CULT.: 1768 (JH: 212); 1789, Nat. of Britain (HK1, 2: 449); 1812, Nat. of Britain (HK2, 4: 216); 1814, Nat. of Britain (HKE: 215); 1887, unspec. (K); 1930, unspec. (K); 1932, unspec. (K); 1935, Herbarium Ground (K); 1936, unspec. (K); 1938, Herbarium Ground (K). In cult.
CURRENT STATUS: Widespread but seldom common; three colour-forms (pink, pale blue and white) regularly came up behind the Herbarium until the lawn was resown in 1999 after building works. All three varieties were in cultivation in the 1930s in the Herbarium Experimental Ground and the plants now growing are presumably their descendants. More recent building works for the new Herbarium extension have severely damaged the site.

343. *Malva nicaeensis All.
French mallow
FIRST RECORD: 1877 [1930].
EXSICC.: **Kew Environs.** Kew; viii 1877; *Baker* s.n. (K).
CULT.: 1930, unspec. (K). Not currently in cult.
CURRENT STATUS: No recent sightings.

344. *Malva parviflora L.
least mallow
FIRST RECORD: 1872 [1768].
EXSICC.: **Kew Gardens.** Weed in path of Herbarium Ground; 22 vii 1953; *Blakelock* s.n. (K). **Kew Environs.** Waste ground by Kew Bridge; ix 1872; *Watson* in *Herb. Watson* (K).
CULT.: 1768 (JH: 212); 1915, unspec. (K); 1952, Herbarium Ground (K); 1953, Herbarium Ground (K). Not currently in cult.
CURRENT STATUS: No recent sightings.

345. *Malva pusilla Sm.
M. rotundifolia L.
M. rotundifolia var. *pusilla* auct. non (Sm.) Aiton
small mallow
FIRST RECORD: 1873/4 [1768].
NICHOLSON: **1873/4 survey** (JB: 12 and in SFS: 205): **B.** A few plants on most of the lawns. **P.** Now and then near lake. • **1906 revision** (AS: 77): **B:** Here and there on most of the lawns. **A:** Now and then near lake.
OTHER REFERENCES: **Kew Gardens.** 1990, *Cope* (S31: 182): Paddock, after restoration in 1990. **Kew Environs.** 1902, *Beeby* (SFS: 205): Thames bank between Kew and Richmond.
CULT.: 1768 (JH: 212); 1789, Nat. of Britain (HK1, 2: 449); 1812, Nat. of Britain (HK2, 4: 216); 1814, Nat. of Britain (HKE: 215). Not currently in cult.
CURRENT STATUS: No recent sightings; it has not reappeared in the Paddock since 1990.

346. Malva neglecta Wallr.
M. rotundifolia auct. non L.
dwarf mallow
FIRST RECORD: 1953 [1934].
REFERENCES: **Kew Gardens.** 1953, *Meikle* (S32: 656): Weed in former Herbarium Experimental Ground.
CULT.: 1934, unspec. (K). Not currently in cult.
CURRENT STATUS: Until the latest phase of building works this species was a regular component of the grass behind the Herbarium but the population may have been destroyed by disturbance; it has more recently been found on slopes around the Secluded Garden and in grass near Kew Palace.

347. *Malva verticillata L.
M. crispa L.
Chinese mallow
FIRST RECORD: 1943 [1768].
REFERENCES: **Kew Gardens.** 1943, *Welch* (S32: 656; LN31: S44): Kew Gardens, among potatoes. **Kew Environs.** 1943, *Welch* (LN31: S44): Waste ground near Kew Bridge.
CULT.: 1768 (JH: 212); 1789, Alien (HK1, 2: 499); 1812, Alien (HK2, 4: 217); 1814, Alien (HKE: 215); 1836, unspec. (K); 1969, unspec. (K). Not currently in cult.
CURRENT STATUS: No recent sightings. It seems only to have occurred as a weed of the allotments created on lawns in the Gardens during the Second World War.

348. Lavatera arborea L.
tree-mallow
CULT.: 1768 (JH: 211); 1789, Nat. of Britain (HK1, 2: 450); 1812, Nat. of Britain (HK2, 4: 219); 1814, Nat. of Britain (HKE: 216); 1898, unspec. (K); 1930, Herb. Expt. Ground (K). Not currently in cult.

349. Althaea officinalis L.
marsh mallow
CULT.: 1768 (JH: 213); 1789, Nat. of Britain (HK1, 2: 445); 1812, Nat. of Britain (HK2, 4: 207); 1814, Nat. of Britain (HKE: 215); 1880, unspec. (K); 1923, Cambridge Cottage Grounds (K); 1929, unspec. (K); 1932, unspec. (K); 1951, Order Beds (K). In cult.

350. *Urocarpidium peruvianum (L.) Krapov.
FIRST RECORD: 1983 [1934].
REFERENCES: **Kew Gardens.** 1983, *Cope & Hastings* (FSSC: 112; LN63: 143): Recently cleared plot near the herbarium buiding. • 1983/4, *Cope* (S31: 181): New student vegetable plots behind Hanover House.
CULT.: 1934, unspec. (K); 1970, Order Beds (K). Not currently in cult.
CURRENT STATUS: No recent sightings; it has not reappeared behind Hanover House since 1983.

DROSERACEAE

351. Drosera rotundifolia L.
round-leaved sundew
CULT.: 1768 (JH: 191); 1789, Nat. of Britain (HK1, 1: 389); 1811, Nat. of Britain (HK2, 2: 189); 1814, Nat. of Britain (HKE: 84); ?1963, unspec. (K). In cult. under glass.

352. Drosera anglica Huds.
D. longifolia L.
great sundew
CULT.: 1768 (JH: 191); 1811, Nat. of England (HK2, 2: 189); 1814, Nat. of England (HKE: 84). In cult. under glass.

353. Drosera intermedia Hayne
oblong-leaved sundew
CULT.: 1789, Nat. of Britain (HK1, 1: 389); 1811, Nat. of Britain (HK2, 2: 189); 1814, Nat. of Britain (HKE: 84). In cult. under glass.

CISTACEAE

354. Tuberaria guttata (L.) Fourr.
Cistus guttatus L.
spotted rock-rose
CULT.: 1768 (JH: 204); 1789, Nat. of England (HK1, 2: 237); 1811, Nat. of England (HK2, 3: 309); 1814, Nat. of England (HKE: 167); 1920, unspec. (K). Not currently in cult.

355. Helianthemum nummularium (L.) Mill.
Cistus helianthemum L.
C. serpyllifolius L.
C. surrejanus L.
C. tomentosus Sm.
common rock-rose
CULT.: 1768 (JH: 204); 1789, Nat. of Britain (HK1, 2: 239 as *helianthemum*), Nat. of England (ibid.: 238 as *surrejanus*); 1811, Nat. of Britain (HK2, 3: 312 as *helianthemum*), Nat. of England (ibid.: 310 as *surrejanus*); 1814, Nat. of Britain (HKE: 167 as *helianthemum*), Nat. of England (ibid.: as *surrejanus*); Nat. of Scotland (ibid.: Add. as *tomentosus*). In cult.

356. Helianthemum apenninum (L.) Mill.
Cistus appeninus L.
C. polifolius L.
C. salicifolius Huds.
C. ledifolius Georgi
white rock-rose
CULT.: 1768 (JH: 204); 1789, Nat. of England (HK1, 2: 238, 239), Alien (ibid.: 237, 239); 1811, Nat. of England (HK2, 3: 310, 313), Alien (ibid.); 1814, Nat. of England (HKE: 167), Alien (ibid.); 1921, Arboretum (K); 1971, loc. unspec. (K). In cult.

357. Helianthemum oelandicum subsp. **incanum** (Willk.) G.López
Cistus anglicus L.
C. canus L.
C. marifolius L.
hoary rock-rose
CULT.: 1768 (JH: 204); 1789, Nat. of England (HK1, 2: 237 as *anglicus*), Alien (ibid.: as *marifolius*); 1811, Nat. of England (HK2, 3: 309); 1814, Alien (HKE: 167); 1969, Rock Garden (K). Not currently in cult.

VIOLACEAE

358. Viola odorata L.
sweet violet
FIRST RECORD: 1873/4 [1768].
NICHOLSON: **1873/4 survey** (JB: 11): **Pal.** In turf among young trees. • **1906 revision** (AS: 76): **P.** Plentiful under trees near Brentford Ferry, also between palace and herbarium.
CULT.: 1768 (JH: 315); 1789, Nat. of Britain (HK1, 3: 289); 1811, Nat. of Britain (HK2, 2: 45); 1814, Nat. of Britain (HKE: 63). In cult.
CURRENT STATUS: Recently noticed in the meadow by the Main Gate, where it is steadily increasing, and this seems to be its only extant locality.

359. Viola hirta L.
hairy violet
CULT.: 1768 (JH: 315); 1789, Nat. of England (HK1, 3: 289); 1811, Nat. of England (HK2, 2: 45); 1814, Nat. of England (HKE: 62); 1909, unspec. (K). Not currently in cult.

Common dog-violet (*Viola riviniana*). Rarely occurs along rides in the Conservation Area.

360. Viola riviniana Rchb.
common dog-violet
First record: 2001.
References: **Kew Gardens.** 2001, *Cope* (S35: 646): Conservation Area (310).
Cult.: In cult.
Current status: Along rides in the Conservation Area.

361. Viola reichenbachiana Jord. ex Boreau
early dog-violet
Cult.: In cult.

362. Viola canina L.
heath dog-violet
a. subsp. **canina**
First record: 1873/4 [1768].
Nicholson: **1873/4 survey** (JB: 11): **P.** Here and there in turf and beds near Pagoda, also on slopes near lake. This presents all the characters of the typical *V. canina*, at least so do all my dried specimens. • **1906 revision** (AS: 76): **A.** Here and there in turf near pagoda, also on slopes near lake.
Cult.: 1768 (JH: 315); 1789, Nat. of Britain (HK1, 3: 289); 1811, Nat. of Britain (HK2, 2: 46); 1814, Nat. of Britain (HKE: 63); 1976, bed 144-01 (K). Not currently in cult.
Current status: No recent sightings.

b. subsp. **montana** (L.) Hartm.
V. montana L.
Cult.: 1768 (JH: 315). Not currently in cult.

363. Viola lactea Sm.
pale dog-violet
Cult.: 1811, Nat. of England (HK2, 2: 46); 1814, Nat. of England (HKE: 63); 1921, unspec. (K). Not currently in cult.
Note. Not discovered as a wild plant in Britain until 1798.

364. Viola palustris L.
marsh violet
Cult.: 1768 (JH: 315); 1789, Nat. of Britain (HK1, 3: 289); 1811, Nat. of Britain (HK2, 2: 45); 1814, Nat. of Britain (HKE: 62). Not currently in cult.

365. Viola lutea Huds.
V. grandiflora auct. non L.
mountain pansy
Cult.: 1789, Nat. of Britain (HK1, 3: 291); 1811, Nat. of Britain (HK2, 2: 48); 1814, Nat. of Britain (HKE: 63); 1896, unspec. (K); 1949, Herbarium Experimental Ground (K); 1977, 'location 7' (K). Not currently in cult.

366. Viola tricolor L.
wild pansy
First record: 1873/4 [1768].
Nicholson: **1873/4 survey** (JB: 11): **B.** A flower-bed weed. **P.** Now and then near Winter Garden. • **1906 revision** (AS: 76): **B:** A flower-bed weed.
Cult.: 1768 (JH: 315); 1789, Nat. of Britain (HK1, 3: 291); 1811, Nat. of Britain (HK2, 2: 48); 1814, Nat. of Britain (HKE: 62); 1861, unspec. (K); 1948, Experimental Ground (K); 1985, Herbaceous (K). In cult.
Current status: An occasional weed of flower-beds.

367. Viola arvensis Murray
V. tricolor var. *arvensis* (Murray) DC.
field pansy
First record: 1873/4 [1948].
Nicholson: **1873/4 survey** (JB: 11): Occurs here and there with [*V. tricolor*]. • **1906 revision** (AS: 76): **B.** This occurs here and there with [*V. tricolor*].
Cult.: 1948, unspec. (K); 1949, Experimental Ground (K); 1950, Experimental Ground (K). In cult.
Current status: An increasingly common weed that may be have recently been boosted by further introduction as a contaminant of grass seed.

368. Viola × contempta Jord. (*arvensis* × *tricolor*)
CULT.: 1948, unspec. (K); 1950, Herbarium Experimental Ground (K); 1954, unspec. (K). Not currently in cult.

369. Viola kitaibeliana Schult.
dwarf pansy
CULT.: 1951, Herbarium Ground (K). Not currently in cult.

TAMARICACEAE

[370. Tamarix gallica L.
tamarisk
CULT.: 1789, Alien (HK1, 1: 375); 1811, Nat. of England (HK2, 2: 172); 1814, Nat. of England (HKE: 81).
Note. Listed in error as native.]

FRANKENIACEAE

371. Frankenia laevis L.
sea-heath
CULT.: 1768 (JH: 192); 1789, Nat. of England (HK1, 1: 480); 1811, Nat. of England (HK2, 2: 315); 1814, Nat. of England (HKE: 103). In cult.

[372. Frankenia pulverulenta L.
CULT.: 1789, Nat. of England (HK1, 1: 480); 1811, Nat. of England (HK2, 2: 316); 1814, Nat. of England (HKE: 103).
Note. Listed in error as native.]

CUCURBITACEAE

373. Bryonia dioica Jacq.
B. alba L.
white bryony
FIRST RECORD: 1873/4 [1768].
NICHOLSON: **1873/4 survey** (JB: 44): **Q.** A plant or two near the "Cottage." • **1906 revision** (AS: 80): A few plants near the Queen's Cottage.
OTHER REFERENCES: **Kew Gardens.** 1971, *Airy Shaw* (S29: 407): Two or three plants in shrubberies and copses near tennis courts by Kew Palace, May 1962, and in hedge (since destroyed) separating Herbarium grounds from garden of Hanover House, Aug. 1971. Only recorded from near Queen's Cottage in 1906.
EXSICC.: **Kew Gardens.** Queen's Cottage; [c. 1914]; *Flippance* s.n. (K). • Near Tennis Courts; over bushes; vi 1932; *Turrill* s.n. (K). • Holly Hedge near Tennis Courts; 9 vi 1939; *Turrill* s.n. (K).
CULT.: 1768 (JH: 144); 1789, Nat. of Britain (HK1, 3: 384); 1813, Nat. of Britain (HK2, 5: 348); 1814, Nat. of Britain (HKE: 303); 1933, Herb Garden (K); 1936, Herbaceous Ground (K); 1939, Herbarium Ground (K); 1943, by old yard (K); 1946, Herb. Ground (K); 1948, Herbarium Ground (K); 1968, unspec. (K); 1971, plot 156.28 (K). In cult.
CURRENT STATUS: Widespread in shrubberies.

LOASACEAE

374. *Blumenbachia hieronymi Urban
FIRST RECORD: 1985.
REFERENCES: **Kew Gardens.** 1985, *London Natural History Society* (LN65: 195; S33: 745): [Around what is now the Princess of Wales Conservatory]. • 1986, *Grenfell* (BSN42: 18; S33: 745): On dumped soil in Kew Gardens reported in 1985 firstly by Dr S.O'Donnell and later by J.B.Latham, C.G.Hanson and others: in view of the locality, however, its status must, at best, be considered doubtful.
CURRENT STATUS: No recent sightings.

White bryony (*Bryonia dioica*). Britain's only native member of the cucumber family occurs in many parts of the Gardens, usually scrambling over shrubs.

SALICACEAE

[375. Populus alba L.
white poplar
CULT.: 1768 (JH: 450); 1789, Nat. of England (HK1, 3: 405); 1813, Nat. of Britain (HK2, 5: 395); 1814, Nat. of Britain (HKE: 310).
Note. Listed in error as native.**]**

[376. Populus × canescens (Aiton) Sm. (alba × tremula)
grey poplar
P. alba var. canescens Aiton
CULT.: 1789, Nat. of England (HK1, 3: 405); 1813, Nat. of England (HK2, 5: 395); 1814, Nat. of England (HKE: 310).
Note. Listed in error as native.**]**

377. Populus tremula L.
aspen
CULT.: 1768 (JH: 450); 1789, Nat. of Britain (HK1, 3: 405); 183, Nat. of Britain (HK2, 5: 395); 1814, Nat. of Britain (HKE: 310); 1843–1853, Arboretum (K); 1880, Arboretum (K); 1881, West of Carpinus Avenue (K); 1892, Arboretum (K); 1894, Arboretum (K); 1896, unspec. (K); 1915, Arboretum (K); 1918, Arboretum (K); 1931, Arboretum (K). In cult.

378. Populus nigra subsp. **betulifolia** (Pursh) Dippel
P. nigra auct. non L.
black-poplar
FIRST RECORD: 1928 [1768]
EXSICC.: **Kew Environs.** Thames bank, Kew; top of bank, not normally flooded; 13 viii 1928; *Summerhayes & Turrill* 3062 (K). • Thames banks, Kew; 29 viii 1930; *Fraser* s.n. (K). • Thames banks, Kew; 14 ix 1930; *Fraser* s.n. (K). • Thames banks, Kew; 15 iv 1931; *Fraser* s.n. (K).
CULT.: 1768 (JH: 450); 1789, Nat. of Britain (HK1, 3: 405); 1813, Nat. of Britain (HK2, 5: 396); 1814, Nat. of Britain (HKE: 311); 1843-1853, Arboretum (K); 1881, Arboretum (K); 1884, Arboretum (K); 1907, unspec. (K); 1914, Arboretum (K); 1930, Arboretum (K); 1931, Arboretum (K); 1935, Arboretum (K); 1951, Arboretum (K); 1961, Arboretum (K); 1964, Arboretum (K); 1965, Arboretum (K). In cult.
CURRENT STATUS: Known from mature trees that were possibly planted long ago on the towpath south of Brentford Gate. Likely also to have planted on the golf course in the Old Deer Park. A species of local conservation interest.

379. Salix pentandra L.
S. hermaphroditica L.
bay willow
CULT.: 1768 (JH: 454); 1789, Nat. of Britain (HK1, 3: 389), Nat. of England (ibid.: as *hermaphroditica*); 1813, Nat. of Britain (HK2, 5: 353); 1814, Nat. of Britain (HKE: 304); 1843–1853, Arboretum (K); 1880, Arboretum (K); 1927, unspec. (K); 1928, unspec. (K); 1936, Arboretum (K); 1937, Arboretum (K); 1939, unspec. (K). In cult.

380. Salix fragilis L.
a. var. **fragilis**
S. decipiens Hoffm.
crack-willow
FIRST RECORD: 1906 [1768].
NICHOLSON: **1906 revision** (AS: 87): **Strip.** Not uncommon by towing-path and river.
EXSICC.: **Kew Environs.** By the Thames between Kew & Richmond; 17 v 1910; *Fraser* s.n. (K). • Thames-bank near Kew; 1912, *Elwes & Henry* s.n. (K). • Riverside, Kew; v 1917; *Worsdell* s.n. (K). • Bank of R. Thames, Kew; 28 v 1933; *Hubbard* s.n. (K). • Side of R. Thames, Kew; 10 v 1948 (catkins), 31 x 1948 (leaves); *Meikle* 1506 (K). • Side of Thames near Kew Palace; 10 v 1948 (catkins), 31 x 1948 (leaves); *Meikle* 1511 (K).
CULT.: 1768 (JH: 454); 1789, Nat. of Britain (HK1, 3: 390); 1813, Nat. of Britain (HK2, 5: 355), Nat. of England (ibid.: as *decipiens*); 1814, Nat. of Britain (HKE: 305), Nat. of England (ibid.: as *decipiens*); 1843–1853, Arboretum (K); 1880, Arboretum (K); 1931, Arboretum (K); 1936, Arboretum (K); 1945, S. side of Lake (K). Not currently in cult.
CURRENT STATUS: Still present on the towpath.

b. var. **russelliana** (Sm.) W.D.J.Koch
Bedford willow
FIRST RECORD: 2003 [1813].
CULT.: 1813, Nat. of England (HK2, 5: 353); 1814, Nat. of England (HKE: 304). In cult.
CURRENT STATUS: Planted in the Conservation Area but now seems to be colonising the towpath.

381. Salix × meyeriana Rostk. ex Willd. (fragilis × pentandra)
Shiny-leaved Willow
CULT.: 1880, Arboretum (K); 1936, Arboretum (K). In cult.

382. Salix alba L.
a. var. **alba**
white willow
FIRST RECORD: 2003 [1789].
CULT.: 1789, Nat. of Britain (HK1, 3: 393); 1813, Nat. of Britain (HK2, 5: 365); 1814, Nat. of Britain (HKE: 306); 1880 & 1881, by Lake (K); 1895, unspec. (K); 1896, unspec. (K); 1936, N.E. side of Lake (K). In cult.
CURRENT STATUS: Planted in the Conservation Area and now colonising the towpath.

b. var. **caerulea** (Sm.) Dumort.
S. caerulea Sm.
cricket-bat willow
FIRST RECORD: 1949 [1813].
EXSICC.: **Kew Environs.** Side of Thames near Kew; 1949; *Meikle* s.n. (K).
CULT.: 1813, Nat. of England (HK2, 5: 365); 1814, Nat. of England (HKE: 306); 1843–53, Arboretum (K); 1880, Arboretum (K); 1906, unspec. (K); 1931, Seven Sisters Lawn (K). In cult.
CURRENT STATUS: Planted in the Conservation Area but to date has not been refound by the river.

c. var. **vitellina** (L.) Stokes
S. vitellina L.
FIRST RECORD: 1912 [1768]
EXSICC.: **Kew Environs.** Tow path above Kew; 21 vii 1912; *Britton* 770 (K). • Towing path, Kew; 11 v 1927 (leaves), 10 viii 1927; *Fraser* 653 (K).
CULT.: 1768 (JH: 454); 1789, Nat. of England (HK1, 3: 389); 1813, Nat. of England (HK2, 5: 355); 1814, Nat. of England (HKE: 305); 1880, Arboretum (K); 1936, Arboretum (K). In cult.
CURRENT STATUS: No recent sightings.

383. *Salix alba L. × **triandra** L.
FIRST RECORD: 1926 [1931].
REFERENCES: **Kew Environs.** 1926, *Fraser* (LN34: S257): Towpath, Kew.
CULT.: 1931, Arboretum (K). Not currently in cult.
CURRENT STATUS: If the tree on the towpath was correctly identified it must have escaped from the gardens since the hybrid is not otherwise known to occur in Britain.

384. *Salix × **sepulcralis** Simonk. (alba × babylonica)
weeping willow
FIRST RECORD: 1926 [1880].
REFERENCES: **Kew Environs.** 1930, *Fraser & Lousley* (LN34: S257; S33: 745): Thames bank near Kew Gardens.
EXSICC.: **Kew Environs.** By the ha-ha, Kew; 4 iv 1926 (catkins), 16 vii 1926 & 8 x 1926 (leaves); *Fraser* 602 (K). • Thames banks, Kew; 19 x 1932 (leaves), 18 iv 1933 (catkins); *Fraser* 787 (K).
CULT.: 1880, Arboretum (K); 1881, Arboretum (K); 1906, banks of Lake (K); 1907, Arboretum (K); 1936, N. side of Lake (K); 1936, South side of Lake (K); N.E. end of Lake (K). In cult.
CURRENT STATUS: No recent sightings; last noted on the towpath in 1981.

385. Salix × **rubens** Schrank (alba × fragilis)
hybrid crack-willow
CULT.: 1880, Arboretum (K); 1892, west end of Lake (K); 1897 & 1898, unspec. (K); 1936, Arboretum (K); 1961, unspec. (K). In cult.
Note: Present in the Consevration Area where it has been planted. It is represented by nothovar. *basfordiana* Scaling ex Salter, one of whose parents is *S. alba* var. *vitellina* (L.) Stokes which has not recently been recorded in the Gardens.

386. Salix × **ehrhartiana** Sm. (alba × pentandra)
Ehrhart's willow
CULT.: 1929, unspec. (K); 1936, unspec. (K). Not currently in cult.

387. Salix triandra L.
S. amygdalina L.
S. undulata Ehrh.
almond willow
FIRST RECORD: 1872 [1789].
NICHOLSON: **1906 revision** (AS: 87): **Strip.** Common by towing-path and river ... the prevailing willow of the Thames banks within our limits.
OTHER REFERENCES: **Kew Gardens.** 1880, *Baker* (BECB1: 35, and see Exsicc.): Hort. Kew. **Kew Environs.** 1934, *Fraser* (LN14: S90): Towpath, Kew.
EXSICC.: **Kew Gardens.** [Unlocalised]; viii 1880; *Baker* 1134 (K). **Kew Environs.** Thames side between Kew and Richmond; ix 1872; *Baker* in *Herb. Watson* (K). • Thames banks, Kew; 22 iv 1905 (catkins), 12 viii 1905 (leaves); *Fraser* 226 (K).
CULT.: 1789, Nat. of Britain (HK1, 3: 389); 1813, Nat. of Britain (HK2, 5: 352, 353); 1814, Nat. of Britain (HKE: 304); 1867, unspec. (*Herb. Watson* K); 1880, Arboretum (K); 1884, Arboretum (K); 1895, Lower Nursery (K); 1896, unspec. (K); 1936, Arboretum (K). Not currently in cult.
CURRENT STATUS: No recent sightings.

388. Salix × **mollissima** Hoffm. ex Elwert (triandra × viminalis)
S. × *hippophaifolia* Thuill.
S. × *lanceolata* Sm.
sharp-stipuled willow
FIRST RECORD: 1926 [1813].
REFERENCES: **Kew Environs.** 1930, *Fraser & Lousley* (LN34: S257; S33: 745): Above Kew Bridge.
EXSICC.: **Kew Environs.** Thames banks, Kew; 9 vii 1926 (leaves), 8 iv 1927 (catkins); *Fraser* 582 (K). • Thames bank near Kew Bridge; 16 viii 1930; *Lousley* s.n. (K). • Side of Thames near Kew; 20 iv 1949 (catkins), 10 viii 1949 (leaves); *Meikle* 1582 (K).
CULT.: 1813, Nat. of England (HK2, 5: 352, 365); 1814, Nat. of England (HKE: 304, 306); 1843-1853, Arboretum (K); 1867, unspec. (K); 1880, Arboretum (K); 1898, unspec. (K); 1899, unspec. (K); 1924, unspec. (K); 1927, unspec. (K); 1936, SW end of Lake (K). In cult.
CURRENT STATUS: No recent sightings; nothovars. *hippophaifolia* and *undulata* have both been recorded.

389. Salix purpurea L.
S. helix L.
S. lambertiana Sm.
purple willow
FIRST RECORD: 1918 [1768].
REFERENCES: **Kew Environs.** 1918, *Bishop* (LN34: S257; S33: 745): By river Thames between Kew and Richmond. • 1934, *anon.* (LN14: S90; S33: 745): Thames bank between Kew and Richmond.
CULT.: 1768 (JH: 454, 455); 1789, Nat. of Britain (HK1, 3: 390); 1813, Nat. of England (HK2, 5: 356 and as *lambertiana*), Nat. of Britain (ibid.: as *helix*); 1814, Nat. of England (HKE: 305 and as *lambertiana*), Nat. of Britain (ibid.: as *helix*); 1843-53, Arboretum (K); 1867, unspec. (*Herb. Watson* K); 1880, Arboretum (K); 1884, Arboretum (K); 1895, unspec. (K); 1897, unspec. (K); 1899. unspec. (K); 1926, unspec. (K); 1929, unspec. (K); 1931, Arboretum (K); 1933, unspec. (K); 1936, Arboretum (K). Not currently in cult.
CURRENT STATUS: No recent sightings. It has not been possible to separate records of subsp. *lambertiana* from those of subsp. *purpurea*.

390. Salix × doniana G.Anderson ex Sm. (purpurea × repens)
CULT.: 1843–53, Arboretum (K); 1880, Arboretum (K); 1936, Arboretum (K). In cult.

391. Salix × rubra Huds. (purpurea × viminalis)
green-leaved willow
FIRST RECORD: 1914 [1813].
REFERENCES: **Kew Environs.** 1931/2, *Fraser* (BSEC10: 450; LN34: S257; S33: 745): Thames bank, Kew.
EXSICC.: **Kew Gardens.** Isleworth Ferry Gate, Kew; 1 ix 1926 (leaves), 25 iii 1927 (catkins); *Fraser* 632 (K). **Kew Environs.** Towing path, eastern boundary of Kew; 26 iii 1914 (catkins), 4 viii 1914 (leaves); *Fraser* 409 (K). • Thames banks, Kew; 22 iii 1924 (catkins), 19 vii 1924 (leaves); *Fraser* 476 (K). • Thames banks, Kew; 27 ii 1926 (catkins), 10 vii 1926 (leaves); *Fraser* 587 (K). • Thames banks, Kew; 28 ii 1926 (catkins), 9 vii 1926 (leaves); *Fraser* 588 (K). • Thameside near Kew Gardens; 18 iv 1930; *Fraser* s.n. (K). • Thames banks, Kew; 30 iii 1931 (catkins), 25 ix 1932 (leaves); *Fraser* 753 (K).
CULT.: 1813, Nat. of England (HK2, 5: 357); 1814, Nat. of England (HKE: 305); 1880, Arboretum (K); 1884, Arboretum (K); 1898, unspec. (K); 1905, Arboretum (K); 1929, unspec. (K); 1930, unspec. (K); 1931, Arboretum (K); 1932, unspec. (K); 1936, Arboretum (K). In cult.
CURRENT STATUS: Still to be found in the Conservation Area, where it has been planted, but is no longer on the towpath.

392. Salix × forbyana Sm. (purpurea × viminalis × cinerea)
fine osier
CULT.: 1813, Nat. of England (HK2, 5: 357); 1814, Nat. of England (HKE: 305). Not currently in cult.

393. *Salix daphnoides Vill.
European violet-willow
FIRST RECORD: 1910 [1843–53].
EXSICC.: **Kew Gardens.** River-side, Kew; iv 1910; *Turrill* s.n. (K). • By the ha-ha, Kew; 27 ii 1926 (catkins), 7 viii 1926 (leaves); *Fraser* 585 (K). • Isleworth Ferry Gate; 18 iii 1927 (catkins), 8 vii 1927 (leaves); *Fraser* 585 (K).
CULT.: 1843–53, Arboretum (K); 1880, Arboretum (K); 1884, Arboretum (K); 1931, Arboretum (K); 1936, Arboretum (K); 1956, Herbarium Ground (K). Not currently in cult.
CURRENT STATUS: No recent sightings.

394. *Salix acutifolia Willd.
Siberian violet-willow
FIRST RECORD: 1923 [1867].
EXSICC.: **Kew Gardens.** By the ha-ha, Kew; 12 ix 1923 (leaves), 11 iii 1924 (catkins); *Fraser* 467 (K).
CULT.: 1867, unspec. (*Herb. Watson* K); 1927, unspec. (K); 1928, unspec. (K); 1929, unspec. (K); 1931, Arboretum (K); 1935, Arboretum (K); 1936, Arboretum (K); 1936, Nursery (K). In cult.
CURRENT STATUS: No recent sightings.

395. Salix viminalis L.
osier
FIRST RECORD: 1917 [1768].
REFERENCES: **Kew Environs.** 1930, *Fraser & Lousley* (LN34: S258; S33: 745): Thames bank near Kew.
EXSICC.: **Kew Environs.** Riverside, Kew; v 1917; *Worsdell* s.n. (K). • Thames bank, Kew; 14 iii 1945; *Melville* s.n. (K).
CULT.: 1768 (JH: 455); 1789, Nat. of Britain (HK1, 3: 393); 1813, Nat. of Britain (HK2, 5: 364); 1814, Nat. of Britain (HKE: 306); 1843–53, Arboretum (K); 1880, Arboretum (K); 1895, Lower Nursery (K); 1897, unspec. (K); 1931, Arboretum (K); 1936, Arboretum (K); 1937, Lake side (K). In cult.
CURRENT STATUS: Planted in the Conservation Area and last seen on the towpath in 1982. It can still be found, however, on the golf course in the Old Deer Park.

[**396. Salix × calodendron** Wimm. (caprea × cinerea × viminalis)
S. acuminata Sm.
Holme willow
CULT.: 1813, Nat. of Britain (HK2, 5: 364); 1814, Nat. of Britain (HKE: 306).
Note. Listed in error as native.]

397. Salix caprea L.
S. aquatica Lam.
goat willow
FIRST RECORD: 1899 [1768].
EXSICC.: **Kew Environs.** Towing path, Old Deer Park, Richmond; 29 iii 1899 (catkins), 29 v 1899, 5 vii 1900, 26 vii 1900 (leaves); *Fraser* 28 (K).
CULT.: 1768 (JH: 455); 1789, Nat. of Britain (HK1, 3: 393); 1813, Nat. of Britain (HK2, 5: 364 and: 363 as *aquatica*); 1814, Nat. of Britain (HKE: 306, and: 305 as *aquatica*); 1843–53, Arboretum (K); 1880, Arboretum (K); 1894, Arboretum (K); 1897, unspec. (K); 1931, Arboretum (K); 1936, unspec. (K). In cult.
CURRENT STATUS: Plants, possibly wild, have been found in the Conservation Area, but it has not been seen on the towpath since 1976.

398. Salix × reichardtii A.Kern. (caprea × cinerea)
FIRST RECORD: 2003.
CULT.: 1843–53, Arboretum (K). Not currently in cult.
CURRENT STATUS: Plants can still be found in the Conservation Area and are possibly self-sown from trees once cultivated.

399. Salix × latifolia J.Forbes (caprea × myrsinifolia)
CULT.: 1843–53, Arboretum (K); 1884, Arboretum (K). Not currently in cult.

400. Salix × sericans Tausch ex A.Kern. (caprea × viminalis)
S. laurina auct. non Sm.
broad-leaved osier
FIRST RECORD: 1976 [1813].
REFERENCES: **Kew Gardens.** 1976, *Cope* (unpubl.): towpath.
CULT.: 1813. Nat. of England (HK2, 5: 354); 1814, Nat. of England (HKE: 304); 1843–53, Arboretum (K); 1880, Arboretum (K); 1884, Arboretum (K); 1893, Lake (K); 1897, unspec. (K); 1898, unspec. (K); 1899, unspec. (K); 1905, Arboretum (K); 1931, Arboretum (K); 1932, unspec. (K); 1936, Arboretum (K); 1937, Arboretum (K). In cult.
CURRENT STATUS: No recent sightings; it is likely that the plants were removed when the riverbank was strengthened in the late 1970s.

401. Salix cinerea L.
grey willow
a. subsp. **cinerea**
FIRST RECORD: 2003 [1813].
EXSICC.: **Kew Environs.** Riverside, Kew; v 1917; *Worsdell* s.n. (K).
CULT.: 1813, Nat. of Britain (HK2, 5: 360); 1814, Nat. of Britain (HKE: 305); 1843–53, Arboretum (K); 1880, Arboretum (K); 1897, unspec. (K); 1899, unspec. (K); 1931, Arboretum (K); 1937, Arboretum (K). Not currently in cult.
CURRENT STATUS: There are plants in the Conservation Area, but the subspecies does not appear in the current Living Collections Database; it is therefore possible that the plants found are spontaneous. It has not been refound on the riverside.

b. subsp. **oleifolia** (Sm.) Macreight
S. oleifolia Sm.
FIRST RECORD: 2003 [1813].
CULT.: 1813, Nat. of Britain (HK2, 5: 363); 1814, Nat. of Britain (HKE: 305); 1980, Rock Garden (K). In cult.
CURRENT STATUS: In the Conservation Area where it has almost certainly been planted, and on the towpath where it may be native.

402. Salix cinerea L. × **myrsinifolia** Salisb.
CULT.: 1843–53, Arboretum (K); 1880, Arboretum (K); 1897, unspec. (K); 1899, unspec. (K); 1907, Arboretum (K); 1909, unspec. (K); 1927, unspec. (K); 1929, unspec. (K); 1930, unspec. (K); 1931, unspec. (K); 1933, unspec. (K); 1936, Arboretum (K). Not currently in cult.

403. Salix × laurina Sm. (cinerea × phylicifolia)
CULT.: 1843–53, Arboretum (K); 1880, Arboretum (K); 1881, unspec. (K); 1894, Arboretum (K); 1897, unspec. (K); 1907, Queens Cott. Grounds (K); 1923, Arboretum (K); 1931, Arboretum (K); 1936, Arboretum (K); 1978, Arboretum South (K); 1980, by the Lake (K). Not currently in cult.

404. Salix × pontederiana Willd. (cinerea × purpurea)
CULT.: 1843–53, Arboretum (K); 1880, Arboretum (K); 1881, unspec. (K); 1884, Arboretum (K); 1898, unspec. (K); 1901, Arboretum (K); 1931, Arboretum (K). Not currently in cult.

405. Salix × smithiana Willd. (cinerea × viminalis)
silky-leaved osier
CULT.: 1843–53, Arboretum (K); 1880, Arboretum (K); 1881, Arboretum (K); 1884, Arboretum (K); 1899, unspec. (K); 1931, Arboretum (K). Not currently in cult.
Note: This has been planted on the golf course in the Old Deer Park, but is no longer grown in the Gardens.

406. Salix aurita L.
eared willow
CULT.: 1768 (JH: 455); 1789, Nat. of Britain (HK1, 3: 391); 1813, Nat. of Britain (HK2, 5: 363); 1814, Nat. of Britain (HKE: 305). In cult.

407. Salix × stipularis Sm. (aurita × caprea × viminalis)
CULT.: 1813, Nat. of England (HK2, 5: 365); 1814, Nat. of England (HKE: 306); 1933, unspec. (K). Not currently in cult.

408. Salix × multinervis Döll. (aurita × cinerea)
CULT.: 1843–53, Arboretum (K); 1880, Arboretum (K); 1881, Arboretum (K); 1936, Arboretum (K); 1940, Nursery (K); 1978, Arboretum South (K). In cult.

409. Salix × grahamii Borrer ex Baker (aurita × herbacea × repens)
CULT.: 1936, Nursery (K). Not currently in cult.

410. Salix × coriacea J.Forbes (aurita × myrsinifolia)
CULT.: 1843–53, Arboretum (K); 1880, Arboretum (K); 1881, Arboretum (K). Not currently in cult.

411. Salix × ludificans F.B.White (aurita × phylicifolia)
CULT.: 1881, Arboretum (K); 1936, Arboretum (K). Not currently in cult.

412. Salix × sesquitertia F.B.White (aurita × phylicifolia × purpurea)
CULT.: 1899, unspec. (K). Not currently in cult.

413. Salix × ambigua Ehrh. (aurita × repens)
CULT.: 1843–53, Arboretum (K); 1898 & 1901, unspec. (K); 1899 & 1900, unspec. (K); 1951, Herbarium Ground (K). Not currently in cult.

414. Salix × fruticosa Döll (aurita × viminalis)
CULT.: 1899, unspec. (K). Not currently in cult.

415. Salix myrsinifolia Salisb.
S. andersoniana Sm.
S. cotinifolia Sm.
S. forsteriana Sm.
S. hirta Sm.
S. nigricans Sm.
S. rupestris Donn ex Sm.
dark-leaved willow
FIRST RECORD: Pre-1948 [1813].
REFERENCES: **Kew Environs.** 1976, *Lousley* (LFS: 245; S33: 745): Known for many years on the Thames bank outside Kew Gardens, but destroyed when the embankment was rebuilt about 1948. Garden escape.
CULT.: 1813, Nat. of Britain (HK2, 5: 363 as *cotinifolia*), Nat. of England (ibid.: 354 as *nigricans* and: 362 as *hirta*), Nat. of Scotland (ibid.: 358 as *rupestris* and: 359 as *andersoniana* and *forsteriana*); 1814, Nat. of Britain (HKE: 306 as *cotinifolia*), Nat. of England (ibid.: 304 as *nigricans* and: 305 as *hirta*), Nat. of Scotland (ibid.: 305 as *andersoniana*, *forsteriana* and *rupestris*); 1843–53, Arboretum (K); 1867, unspec. (*Herb. Watson* K); 1870, unspec. (*Herb. Watson* K); 1880, Arboretum (K); 1881, Arboretum (K); 1897, unspec. (K); 1898, unspec. (K); 1908, unspec. (K); 1909, unspec. (K); 1932, unspec. (K); 1933, unspec. (K); 1936, Arboretum (K); 1937, Arboretum (K); 1980, by the Lake (K). In cult.
CURRENT STATUS: No recent sightings. Lousley's record implies that it is now extinct.

416. Salix × tetrapla Walker (myrsinifolia × phylicifolia)
S. andersoniana Sm. × *phylicifolia* L.
S. nigricans Sm. × *phylicifolia* L.
S. tenuifolia Sm.
FIRST RECORD: 1925 [1813].
REFERENCES: **Kew Gardens.** 1925, *Catchside* (WBEC42: 351; S33: 745): Bank of ha-ha. **Kew Environs.** 1925, *Lousley* (LN34: S259; S33: 745): Banks of Thames between Kew and Richmond. Introduced. • 1976, *Lousley* (LFS: 245): Known for many years on the Thames bank outside Kew Gardens, but destroyed when the embankment was rebuilt about 1948. Garden escape.
CULT.: 1813, Nat. of Britain (HK2, 5: 354); 1814, Nat. of Britain (HKE: 304). Not currently in cult.
CURRENT STATUS: No recent sightings. Lousley's record implies that it is now extinct.

417. Salix phylicifolia L.
S. croweana Sm.
S. dicksoniana Sm.
tea-leaved willow
CULT.: 1813, Nat. of Scotland (HK2, 5: 354 and: 362 as *dicksoniana*), Nat. of England (ibid.: 357 as *croweana*); 1814, Nat. of Scotland (HKE: 304 and: 305 as *dicksoniana*), Nat. of England (ibid.: 305 as *crowenana*); 1843–53, Arboretum (K); 1880, Arboretum (K); 1884, Arboretum (K); 1897, unspec. (K); 1905, unspec. (K); 1936, Nursery (K); 1940, Upper Nursery (K). Not currently in cult.

418. Salix phylicifolia L. × **repens** L.
CULT.: 1905, Arboretum (K); 1936, Arboretum (K). In cult.

419. Salix repens L.
creeping wllow
a. var. **repens**
S. adscendens Sm.
S. fusca L.
S. parvifolia Sm.
S. prostrata Sm.
S. rosmarinifolia auct. non L.

CULT.: 1768 (JH: 455); 1789, Nat. of Britain (HK1, 3: 392), Nat. of England (ibid.: as *fusca* and *rosmarinifolia*); 1813, Nat. of Britain (HK2, 5: 361 and as *fusca*,: 362 as *prostrata* and *rosmarinifolia*), Nat. of England (ibid.: 361 as *adscendens* and *parvifolia*); 1814, Nat. of Britain (HKE: 305 and as *fusca*, *prostrata* and *rosmarinifolia*), Nat. of England (ibid.: as *adscendens* and *parvifolia*); 1843–53, Arboretum (K); 1880, Arboretum (K); 1896, unspec. (K); 1897, unspec. (K); 1907, Arboretum (K); 1936, Arboretum (K). In cult.

b. var. **argentea** (Sm.) Wimm. & Grab.
S. argentea Sm.
CULT.: 1813, Nat. of Britain (HK2, 5: 361); 1814, Nat. of Britain (HKE: 305). In cult.

420. Salix × **friesiana** Andersson (repens × viminalis)
CULT.: 1843–53, Arboretum (K); 1978, Arboretum South (K). Not currently in cult.

421. Salix lapponum L.
S. arenaria L.
S. glauca auct. non L.
S. stuartiana Sm.
downy willow
CULT.: 1768 (JH: 455); 1789, Nat. of Scotland (HK1, 3: 392), Nat. of Britain (ibid.: as *arenaria*); 1813, Nat. of Scotland (HK2, 5: 360); 1814, Nat. of Scotland (HKE: 305 and as *glauca*,: Add. as *stuartiana*); 1900, unspec. (K); 1920, Rock Garden (K). Not currently in cult.

422. Salix × **boydii** E.F.Linton (lapponum × reticulata)
CULT.: 1927, unspec. (K). Not currently in cult.

423. Salix lanata L.
S. sphacelata Sommerf.
woolly willow
CULT.: 1789, Nat. of Scotland (HK1, 3: 391); 1813, Nat. of Scotland (HK2, 5: 363); 1814, Nat. of Scotland (HKE: 306); 1880, Arboretum (K); 1920, Rock Garden (K); 1925, unspec. (K). Not currently in cult.

424. Salix arbuscula L.
S. carinata Sm.
S. prunifolia Sm.
S. vacciniifolia Sm.
S. venulosa Sm.
mountain willow
CULT.: 1813, Nat. of Scotland (HK2, 5: 357, 358); 1814, Nat. of Scotland (HKE: 305); 1880, Arboretum (K); 1899, unspec. (K). Not currently in cult.

425. Salix × **pseudospuria** Rouy (arbuscula × lapponum)
CULT.: 1907, Arboretum (K). Not currently in cult.

426. Salix myrsinites L.
whortle-leaved willow
CULT.: 1789, Nat. of Scotland (HK1, 3: 390); 1813, Nat. of Scotland (HK2, 5: 358); 1814, Nat. of Scotland (HKE: 305). In cult.

427. Salix herbacea L.
dwarf willow
CULT.: 1768 (JH: 455); 1789, Nat. of Britain (HK1, 3: 390); 1813, Nat. of Britain (HK2, 5: 359); 1814, Nat. of Britain (HKE: 305); 1940, Upper Nursery (K). In cult.

428. Salix × **sadleri** Syme (herbacea × lanata)
CULT.: 1936, by King William's Temple (K). Not currently in cult.

429. Salix × **sobrina** F.B.White (herbacea × lapponum)
CULT.: 1899, unspec. (K); 1932, Arboretum (K). Not currently in cult.

430. Salix × **cernua** E.F.Linton (herbacea × repens)
CULT.: 1965, Exp. Ground (K). Not currently in cult.

431. Salix reticulata L.
net-leaved willow
CULT.: 1768 (JH: 455); 1789, Nat. of Britain (HK1, 3: 391); 1813, Nat. of Britain (HK2, 5: 359); 1814, Nat. of Britain (HKE: 305). Not currently in cult.

[432. Salix hastata L.
S. malifolia Sm.
CULT.: 1813, Nat. of England (HK2, 5: 357); 1814, Nat. of England (HKE: 305).
Note. Listed in error as native.]

[433. Salix petiolaris Sm.
slender willow
Cult.: 1813, Nat. of Britain (HK2, 5: 355); 1814, Nat. of Britain (HKE: 305).
Note. Listed in error as native.]

Brassicaceae

434. *Sisymbrium strictissimum L.

perennial rocket

First record: 1873 [1768].
Nicholson: **1873/4 survey:** Not recorded. • **1906 revision:** Not recorded.
Other references: **Kew Gardens.** 1987, *Leslie* (FSSC: 102; S33: 745): Although first recorded from the churchyard on Kew Green, in 1940, there are specimens in CGE from 'near Kew Herbarium' in 1873. **Kew Environs.** 1940, *D.H.Kent* (LN30: S19; S33: 745): Kew Churchyard. • 1982, *Clement* (BSN30: 10; S33: 745): Kew Green, 1873. And it is still (1981) abundant there, along the base of the N. wall of St. Ann's Churchyard, spreading by seed and rhizomes.
Exsicc.: **Kew Gardens.** Near Kew Herbarium; 1873; *leg. dub.* (CGE, n.v.). **Kew Environs.** Kew, St. Anne's Churchyard; common outside church wall; 11 vi 1979; *Verdcourt* 5325 (K).
Cult.: 1768 (JH: 172/22); 1789, Alien (HK1, 2: 392); 1812, Alien (HK2, 4: 114); 1814, Alien (HKE: 202). Not currently in cult.
Current status: Still in the churchyard. The whole of the churchyard was cleared and refurbished recently and the population of this species was greatly reduced as a result, but it is not under immediate threat.

435. *Sisymbrium irio L.

London-rocket

First record: Between 1881 and 1924 [1768].
Nicholson: **1906 revision:** Not recorded.
Other references: **Kew Gardens.** [Between 1881 and 1924], *Britten* (SFS: 135; S33: 745): Weed in Kew Gardens.
Exsicc.: **Kew Gardens.** [Unlocalised]; on rubbish heap; vii 1928; *Hubbard* s.n. (K).
Cult.: 1768 (JH: 172/22); 1789, Nat. of England (HK1, 2: 392); 1812, Nat. of England (HK2, 4: 114); 1814, Nat. of England (HKE: 202). Not currently in cult.
Current status: No recent sightings.

436. *Sisymbrium volgense M.Bieb. ex E.Fourn.

Russian mustard

First record: 1934.
References: **Kew Gardens.** 1934, *A.K.Jackson, Sandwith & Sprague* (WAT12: 311; S33: 745, and see Exsicc.).
Exsicc.: **Kew Gardens.** Herbarium field; weed; 23 vii 1934; *Sprague* s.n. (K). • Herbarium field nr. new wing; on recently disturbed ground; 30 vii 1934; *A.K.Jackson & Sandwith* s.n. (K).
Current status: No recent sightings.

Flixweed (*Descurainea sophia*). First noted in 2003 and always small and inconspicuous. Some very large specimens can be found in the Old Deer Park.

437. *Sisymbrium orientale L.

eastern rocket

First record: 1990.
References: **Kew Gardens.** 1990, *Cope* (S31: 182): Paddock, after restoration in 1990.
Current status: It has not reappeared in the Paddock but has recently been found outside the Jodrell Laboratory.

438. Sisymbrium officinale (L.) Scop.

Erysimum officinale L.

hedge mustard

First record: 1873/4 [1768].
Nicholson: **1873/4 survey** (JB: 10): Easily found in all the divisions. • **1906 revision** (AS: 75): Common in all the divisions.
Exsicc.: **Kew Environs.** Kew Road, near Lion Gate; grassy bank; 18 ix 1930; *Turrill* s.n. (K). • Kew; waste ground near R. Thames; 28 v 1933; *Hubbard* s.n. (K). • Richmond, opposite Old Deer Park, Kew Road; grass bank; 19 vi 1935; *Turrill* s.n. (K). • Richmond, opposite Old Deer Park, Kew Road; ditch bank; 15 viii 1935; *Turrill* s.n. (K).

CULT.: 1768 (JH: 172/18); 1789, Nat. of Britain (HK1, 2: 393); 1812, Nat. of Britain (HK2, 4: 111); 1814, Nat. of Britain (HKE: 202). Not currently in cult.
CURRENT STATUS: Widespread and common throughout the Gardens

439. Sisymbrium austriacum Jacq.
FIRST RECORD: Pre-1875.
EXSICC.: **Kew Environs.** Kew Bridge; [pre-1875]; *Irvine* s.n. (K).
CURRENT STATUS: No recent sightings.

440. Descurainea sophia (L.) Prantl
Sisymbrium sophia L.
flixweed
FIRST RECORD: 2003 [1768]
REFERENCES: **Kew Gardens.** 2003, *Cope* (S35: 646): Several plants growing up through mulch around the base of the Stone Pine (131).
CULT.: 1768 (JH: 172/22); 1789, Nat. of Britain (HK1, 2: 391); 1812, Nat. of Britain (HK2, 4: 114); 1814, Nat. of Britain (HKE: 202). Not currently in cult.
CURRENT STATUS: Several plants were found around the base of the Stone Pine in 2003, and in an adjacent bed in 2004, but they were weeded out before they flowered. It is not uncommon in front of the Aroid House and in 2004 it was in some quantity. The latter are subject to mowing and may not persist. Several very large plants were noted on the golf course in the Old Deer Park.

441. Alliaria petiolata (M.Bieb.) Cavara & Grande
Sisymbrium alliaria (L.) Scopoli
Erysimum alliaria L.
garlic mustard
FIRST RECORD: 1873/4 [1768].
NICHOLSON: **1873/4 survey** (JB: 10): **B.** Behind "Rockwork." **Pal** and **P.** Here and there. • **1906 revision** (AS:75): **A.** Ice house mound. **P, Q.** Here and there.
CULT.: 1768 (JH: 172/18); 1789, Nat. of Britain (HK1, 2: 393); 1812, Nat. of Britain (HK2, 4: 117); 1814, Nat. of Britain (HKE: 203). In cult. (2006). Not currently in cult.
CURRENT STATUS: Widespread but not common.

442. Arabidopsis thaliana (L.) Heynh.
Arabis thaliana L.
thale cress
FIRST RECORD: 1873/4 [1768].
NICHOLSON: **1873/4 survey** (JB: 10): **P.** Common about the Winter Garden and elsewhere. • **1906 revision** (AS: 75): Common as weed in shrubberies, etc.
EXSICC.: **Kew Gardens.** On heap of ashes near Herb. Expt. Ground, 12 v 1927, *Hubbard* s.n. (K). • Common weed in Arboretum nursery, 12 iv 1933, *Hubbard* s.n. (K). • Herbarium field, weed on 'fallow' part of cultivated ground, 31 iii 1945, *Ross-Craig & Sealy* 1042 (K). • Herbarium field, on 'fallow' part of cultivated ground, 25 iv 1945, *Ross-Craig & Sealy* 1049 (K).
CULT.: 1768 (JH: 172/20); 1789, Nat. of Britain (HK1, 2: 399); 1812, Nat. of Britain (HK2, 4: 106); 1814, Nat. of Britain (HKE: 201). Not currently in cult.
CURRENT STATUS: Widespread and often very abundant as a weed of flower-beds.

443. Isatis tinctoria L.
woad
CULT.: 1768 (JH: 172/17); 1789, Nat. of England (HK1, 2: 407); 1812, Nat. of England (HK2, 4: 78); 1814, Nat. of England (HKE: 197); 1933, Herb Garden (K); 1936, Herbaceous Ground (K); 1970, Herbaceous Dept. (K). In cult.
Note. Originally grown as a native but now regarded as an archaeophyte.

Garlic Mustard (*Alliaria petiolata*). Occasional, but important as the food-plant of the Orange-tip butterfly.

444. Erysimum cheiranthoides L.
Cheiranthus erysimoides Huds.
treacle-mustard

FIRST RECORD: 1873/4 [1789].
NICHOLSON: **1873/4 survey** (JB: 10): A common weed in beds and edges of shrubberies. Frequent on towing-path. • **1906 revision** (AS: 75): A common weed in beds and edges of shrubberies. **Strip.** Common on towing path.
CULT.: 1789, Nat. of Britain (HK1, 2: 394), Nat. of England (ibid.: as *erysimoides*); 1812, Nat. of Britain (HK2, 4: 115); 1814, Nat. of Britain (HKE: 202); 1969, Herbaceous Ground (K). Not currently in cult.
CURRENT STATUS: No recent sightings.

445. Erysimum cheiri (L). Crantz
Cheiranthus cheiri L.
C. fruticulosa L.
wallflower

CULT.: 1768 (JH: 172/19); 1789, Nat. of Britain (HK1, 2: 395); 1812, Alien (HK2, 4: 118); 1814, Alien (HKE: 203), Nat. of Britain (ibid.: as *fruticulosus*); 1906, unspec. (K); 1936, Keeper's Garden, abnormal form (K). In cult.

[446. Hesperis matronalis L.
H. inodora auct. non L.
dame's-violet

CULT.: 1789, Alien (HK1, 2: 398), Nat. of England (ibid.: 3: 496 as *inodora*); 1812, Nat. of England (HK2 4: 122); 1814, Nat. of England (HKE: 204).
Note. Listed in error as native.**]**

447. *Malcolmia maritima (L.) W.T.Aiton
Cheiranthus maritimus L.
Wilckia maritima (L.) Halácsy
Virginia stock

FIRST RECORD: 1877 [1768].
NICHOLSON: **1906 revision:** Not recorded.
OTHER REFERENCES: **Kew Gardens.** 1880, *Baker* (BECB1(2): 40; S33: 745): Thames bank opposite Sion House, midway between Kew and Richmond. **Kew Environs.** 1877, *Druce* (BSEC9(5): 683; S33: 745): Richmond and Kew.
CULT.: 1768 (JH: 172/19); 1789, Alien (HK1, 2: 395); 1812, Alien (HK2, 4: 121); 1814, Alien (HKE: 203). Not currently in cult.
CURRENT STATUS: No recent sightings.

448. *Malcolmia africana (L.) R.Br.
Hesperis africana L.
African stock

FIRST RECORD: 1932 [1768].
REFERENCES: **Kew Gardens.** 1932, *Hubbard* (BSN60: 43; S33: 745; and see Exsicc.): Weed in newly sown grass [near Herbarium].
EXSICC.: **Kew Gardens.** [Unspec.]; weed in recently sown grass; 28 vi 1932; Hubbard s.n. (K). • Weed in recently sown grass near Herbarium; 29 vi 1932; *Hubbard* s.n. (K).
CULT.: 1768 (JH: 172/19); 1812, Alien (HK2, 4: 121); 1814, Alien (HKE: 203). Not currently in cult.
CURRENT STATUS: No recent sightings.

[449. Matthiola incana (L.) W.T.Aiton
Cheiranthus incanus L.
hoary stock

CULT.: 1789, Alien (HK1, 2: 395); 1812, Nat. of England (HK2, 4: 119); 1814, Nat. of England (HKE: 203).
Note. Listed in error as native.**]**

450. Matthiola sinuata (L.) W.T.Aiton
Cheiranthus sinuatus L.
sea stock

CULT.: 1789, Nat. of England (HK1, 2: 397); 1812, Nat. of England (HK2, 4: 120); 1814, Nat. of England (HKE: 203). Not currently in cult.

451. Barbarea vulgaris W.T.Aiton
Erysimum barbarea L.
winter-cress

FIRST RECORD: 1873/4 [1768].
NICHOLSON: **1873/4 survey** (JB: 10): **P.** Vicinity of lake. **Strip.** Common. • **1906 revision** (AS: 75): **A.** About lake. **Strip.** Common.
EXSICC.: **Kew Gardens.** Bank of River Thames between Isleworth Gate & Isleworth Ferry; 23 vii 1948; *Ross-Craig & Sealy* 1572 (K). **Kew Environs.** Kew; grass bank; 26 v 1897; *Worsdell* s.n. (K).
CULT.: 1768 (JH: 172/18); 1789, Nat. of Britain (HK1, 2: 393); 1812, Nat. of Britain (HK2, 4: 109); 1814, Nat. of Britain (HKE: 201); 1934, unspec. (K). In cult.
CURRENT STATUS: Very rare and sporadic.

452. *Barbarea stricta Andrz.
small-flowered winter-cress

FIRST RECORD: 1873/4.
NICHOLSON: **1873/4 survey** (JB: 10; SFS: 123): **P.** Some half-dozen plants near lake. **Strip.** Occurs now and then with [*B. vulgaris*]. • **1906 revision** (AS: 75): **A.** Near lake. **Strip.** Not uncommon.
OTHER REFERENCES: **Kew Environs.** 1878, *anon.* (BEC18: 13): Very plentiful this year along the Surrey side of the Thames between Richmond and Kew. • 1915, *C.S.Nicholson* (LN30: S14): Bank of River Thames near Kew Gardens. • 1925, *Lousley* (LN30: S14): Towing path between Kew and Richmond. • 1929, *anon.* (LN8: S11): By R. Thames, near Kew Gardens. • 1977, *Pankhurst* and 1981, *Hastings* (FSSC: 17; LN61: 103): Reported again, from Kew towpath.

EXSICC.: **Kew Environs.** Kew; grass bank; 26 v 1897; *Worsdell* s.n. (K). • Surrey side of Thames above Kew; vi 1898; *Baker* 107 (K). • By the Thames between Kew & Richmond; mixed with *B. vulgaris*; 6 vi 1907; *A.B.Jackson* s.n. (K). • Thames Banks, Kew; 18 vi 1909; *Fraser* s.n. (K). • Thames Banks, Kew; 12 vii 1918; *Fraser* s.n. (K). • Between Kew and Richmond; on river bank; 20 v 1948; *Taylor* s.n. (K).
CURRENT STATUS: By ponds in the Old Deer Park, on the edge of the golf course.

453. *Barbarea intermedia Boreau
medium-flowered winter-cress
FIRST RECORD: 2001.
REFERENCES: **Kew Gardens.** 2001, *Cope* (S35: 646): Herbarium grounds near the allotments.
CURRENT STATUS: Rare and unpredictable.

[454. Barbarea verna (Mill.) Asch.
B. praecox (Sm.) R.Br.
American winter-cress
CULT.: 1812, Nat. of England (HK2, 4: 109); 1814, Nat. of England (HKE: 201).
Note. Listed in error as native.**]**

455. Rorippa nasturtium-aquaticum (L.) Hayek
Sisymbrium nasturtium-aquaticum L.
S. nasturtium Thunb.
Nasturtium officinale W.T.Aiton
water-cress
FIRST RECORD: 1873/4 [1768].
NICHOLSON: **1873/4 survey** (JB: 10): **Strip.** A few plants near moat. • **1906 revision** (AS: 74): **Strip.** Here and there along ha-ha.
EXSICC.: **Kew Gardens.** Towpath between Kew & Richmond; in muddy riverside opposite Isleworth; 2 viii 1948; *Souster* 924 (K). **Kew Environs.** Thames bank, Kew; from crevices of new stone wall, half-way down; 2 viii 1928; *Turrill* s.n. (K).
CULT.: 1768 (JH: 172/22); 1789, Nat. of Britain (HK1, 2: 388); 1812, Nat. of Britain (HK2, 4: 110); 1814, Nat. of Britain (HKE: 202). Not currently in cult.
CURRENT STATUS: Still to be found in the ditch alongside the towpath.

456. Rorippa palustris (L.) Besser
Nasturtium palustre (L.) DC.
N. terrestre With.
Sisymbrium terrestre (With.) R.Br.
marsh yellow-cress
FIRST RECORD: 1873/4 [1789].
NICHOLSON: **1873/4 survey** (JB: 11; SFS: 120): **P.** About lake. No so common as [*R. sylvestris*]. • **1906 revision** (AS: 74): **A.** About lake. Not so common as [*R. sylvestris*].
EXSICC.: **Kew Gardens.** Herbarium Exp. Ground; waste ground; normal form; 14 v 1930; *Summerhayes & Ballard* 467 (K). • Herbarium Exp. Ground; waste ground; etiolated from growing under inverted flower-pot; 14 v 1930; *Summerhayes & Ballard* 468 (K). • Kew, towpath by River Thames at intervals from Brentford to Isleworth Gates; river bank, growing among stone blocks which are used to face the bank, sometimes in fair quantity & growing with R. amphibia which is common; 11 vi 1947; *Ross-Craig & Sealy* 1382 (K). **Kew Environs.** Thames banks, Kew; 31 viii 1884; *Fraser* s.n. (K). • River bank, Kew; vii 1888; *Gamble* 20214 (K). • Kew; 9 vii 1893; *Clarke* 47558A (K). • Kew; on banks of Thames; 16 vi 1928; *Hubbard* s.n. (K). • Kew, Thames bank; growing on concrete wall in cracks, submerged at high tides; 13 vii 1928; *Turrill* s.n. (K). • Rt. bank of Thames – between Richmond & Kew; vii 1928; *C.A.Smith* 6095 (K). • Thames bank, Kew; from crevices of new stone wall, half-way down; 2 viii 1928; *Turrill* s.n. (K). • Towpath between Kew & Richmond; growing between stones in sea wall, subject to immersion in salt water at times; 18 vii 1948; *Souster* 883 (K).
CULT.: 1789, Nat. of England (HK1, 2: 389); 1812, Nat. of Britain (HK2, 4: 110); 1814, Nat. of Britain (HKE: 202). Not currently in cult.
CURRENT STATUS: Reappeared in 2004, after an absence of 20 years, in topsoil stored in the Paddock.

457. Rorippa sylvestris (L.) Besser
Nasturtium sylvestre (L.) R.Br.
Sisymbrium sylvestre L.
creeping yellow-cress
FIRST RECORD: 1873/4 [1768].
NICHOLSON: **1873/4 survey** (JB: 11; SFS: 120): **P.** Plentiful near edge of lake. **Strip.** Here and there. • **1906 revision** (AS: 74): **A.** Abundant near edge of lake. **Strip.** Here and there.
EXSICC.: **Kew Gardens.** Near the Herbarium; weed; 31 vii 1928; *Summerhayes & Turrill* 2946 (K). • Herbarium Experiment Ground; in waste ground; 30 viii 1928; *Summerhayes* 403 (K). • [Unlocalised]; troublesome perennial weed; 7 vi 1949; *Souster* 991 (K). **Kew Environs.** Thames banks, Kew; v 1899; *Fraser* s.n. (K). • Kew river bank; ix 1918; *Gamble* 30655 (K).
CULT.: 1768 (JH: 172/22); 1789, Nat. of Britain (HK1, 2: 389); 1812, Nat. of Britain (HK2, 4: 110); 1814, Nat. of Britain (HKE: 202); 1971, unspec. (K). Not currently in cult.
CURRENT STATUS: Rare; known only from a few localities in the northern end of the Gardens and one from the Lake.

458. Rorippa amphibia (L.) Besser
Nasturtium amphibium (L.) R.Br.
Sisymbrium amphibium L.
great yellow-cress
FIRST RECORD: 1863 [1768].
NICHOLSON: **1873/4 survey** (JB: 11; SFS: 121): **P.** Here and there round lake. **Strip.** Abundant. • **1906 revision** (AS: 75): **A.** Here and there about lake. **Strip.**
OTHER REFERENCES: **Kew Gardens.** 1863, *J.D.Salmon* (BFS: 22): Ditches by Kew Botanic Gardens. • 1999, *Stones* (EAK: 2): By Lake.
EXSICC.: **Kew Environs.** Thames banks, near Kew; 15 ix 1883, *Fraser* s.n. (K). • Thames banks, Kew; 27 viii 1909; *Fraser* s.n. (K). • Kew, River bank near Kew Bridge; growing on old river-wall, below average high-water mark; 7 v 1928; *Summerhayes & Turrill* s.n. (K). • Kew Gardens, on bank of R. Thames; v 1933; *Hubbard* s.n. (K). • Kew, Thames Bank; in mud; 6 vi 1939; *Hubbard* 9892 (K).
CULT.: 1768 (JH: 172/22); 1789, Nat. of Britain (HK1, 2: 389); 1812, Nat. of Britain (HK2, 4: 110); 1814, Nat. of Britain (HKE: 202). Not currently in cult.
CURRENT STATUS: Still in the ditch alongside the towpath but is no longer to be found by the Lake.

Coralroot (*Cardamine bulbifera*). A scarce British native cultivated in the Woodland Garden; it has subsequently naturalised and seems to be spreading.

459. Rorippa × erythrocaulis Borbás (amphibia × palustris)
Thames yellow-cress
FIRST RECORD: 1909.
REFERENCES: **Kew Gardens.** 1998, *Cope* (S33: 745): West Arb.: Riverside Walk (214). **Kew Environs.** 1916, *Britton* (BSEC4(6): 555; S33: 745, and see Exsicc.): By R. Thames, above Kew. • 1978, *Wild Flower Society* (FLA: 23; S33: 745): Recently seen by Kew Gardens.
EXSICC.: **Kew Environs.** By the Thames between Kew and Richmond; 13 vii 1909; *Britton* s.n. (K). • Thames banks above Kew Bridge; 30 vii 1909; *Fraser* s.n. (K). • By the Thames above Kew; 22 vii 1915; *Britton* s.n. (K). • Stone retaining wall of R. Thames near the Herbarium; with putative parents; 19 ix 1941; *Sandwith & Hoyle* 6667 (K).
CURRENT STATUS: This rather distinctive hybrid still occurs near the Brentford Gate along the Riverside Walk and was recently found near the Herbarium. Whether the Kew specimens are tetraploid (and fertile) or triploid (and sterile) is uncertain, but the original determination was made from apparently sterile plants. In the absence of recent records of *R. palustris* the hybrid may eventually disappear despite being perennial.

460. Armoracia rusticana P.Gaertn., B.Mey. & Scherb.
Cochlearia armoracia L.
A. lapathifolia Gilib.
horse-radish
FIRST RECORD: 1928 [1768].
EXSICC.: **Kew Gardens.** Outside Herbarium near Experimental Ground; 1 iii 1948; *Ross-Craig* s.n. (K). **Kew Environs.** Thames bank, Kew; top zone on mud with *Oenanthe* & *Phalaris*; 13 viii 1928; *Summerhayes & Turrill* s.n. (K).
CULT.: 1768 (JH: 172/14); 1789, Nat. of England (HK1, 2: 378); 1812, Nat. of England (HK2, 4: 90); 1814, Nat. of England (HKE: 199); 1926, unspec. (K); 1933, unspec. (K); 1936, Herb Gdn. (K); 1936, Herbaceous Ground (K). In cult.
CURRENT STATUS: Still surviving on the towpath though no longer in zoned vegetation. Recently noted on the golf course in the Old Deer Park.

461. Cardamine bulbifera (L.) Crantz
Dentaria bulbifera L.
coralroot
FIRST RECORD: 2002 [1768].
REFERENCES: **Kew Gardens.** 2002, *Cope* (S35: 646): Woodland Garden (166). A number of plants are scattered throughout the Woodland Garden (also 154, 159). Doubtless it was originally cultivated here but is now thoroughly naturalised.
CULT.: 1768 (JH: 172/17); 1789, Nat. of England (HK1, 2: 386); 1812, Nat. of England (HK2, 4: 101); 1814, Nat. of England (HKE: 201); 1905, unspec. (K). In cult.

Cuckooflower (*Cardamine pratensis*). A plant of damp grassland that has recently increased in the Gardens.

CURRENT STATUS: Woodland Garden in several spots where it has naturalised from plants that were originally cultivated.

462. Cardamine amara L.

large bitter-cress

FIRST RECORD: 1873/4 [1768].

NICHOLSON: **1873/4 survey** (JB: 10; SFS: 127): **Strip.** Plentiful by moat on both sides of Isleworth Gate. • **1906 revision** (AS: 75): By ha-ha. Abundant.

OTHER REFERENCES: **Kew Gardens.** • 1946, *Welch* (LN30: S15): Banks of Kew Gardens Moat near Isleworth Gate. Probably lost through disturbance, 1949. • 1976, *Lousley* (LFS: 126): By the moat of Kew Gardens until an exceptionally high tide destroyed it in March 1949. • 1981, *Hastings* (LN61: 103): Near the south end of Kew Gardens where it was evidently not eliminated by a flood tide in 1949 as stated by Lousley (1976: 126). **Kew Environs.** 1929, *anon.* (LN8: S11): Near Kew.

CULT.: 1768 (JH: 172/22); 1789, Nat. of Britain (HK1, 2: 388); 1812, Nat. of Britain (HK2, 4: 103); 1814, Nat. of Britain (HKE: 201). Not currently in cult.

CURRENT STATUS: No recent sightings.

463. Cardamine pratensis L.

C. pratensis var. *dentata* Rchb.

cuckooflower

FIRST RECORD: 1873/4 [1768].

NICHOLSON: **1873/4 survey** (JB: 10): Common in all the divisions. Less frequent in Botanic Garden proper than elsewhere. • **1906 revision** (AS: 75): Common in all the divisions. Var. *dentata*: **P.** Abundant in low ground between palace and herbarium.

OTHER REFERENCES: **Kew Gardens.** 1999, *Stones* (EAK: 2): By Lake.

EXSICC.: **Kew Gardens.** Near the Palace; damp places in long grass; 13 iv 1945; *Ross-Craig & Sealy* 1043 (K). • Side of ha-ha; growing in sandy soil with *Ranunculus ficaria*; 24 iii 1948; *Senogles* 11 (K). **Kew Environs.** Towpath between Kew & Richmond; favouring moist positions; 26 iv 1944; *Souster* 43 (K). • Tow-path, Kew – Richmond; found growing in a damp hedgerow on good loam, in association with *Urtica dioica* & *Ranunculus ficaria*; 5 iv 1948; *Parker* s.n. (K).

CULT.: 1768 (JH: 172/22); 1789, Nat. of Britain (HK1, 2: 388); 1812, Nat. of Britain (HK2, 4: 103); 1814, Nat. of Britain (HKE: 201); 1929, Herb. Expt. Ground (K). In cult.

CURRENT STATUS: Widespread and was increasing in the Herbarium lawn until the current building works began; this location was for a time buried under a mound of topsoil

[464. **Cardamine parviflora** L.
CULT.: 1789, Nat. of England (HK1, 2: 388).
Note. Listed in error as native.]

465. Cardamine impatiens L.

narrow-leaved bitter-cress

FIRST RECORD: 1873/4 [1768].

NICHOLSON: **1873/4 survey** (JB: 10; SFS: 130): **P.** An apetalous form of this plant grows very abundantly under the trees close by the wall from the "Unicorn Gate" to opposite the "Douglas Spar." Also about "Merlin's Cave." In the "Flora of Surrey," the neighbourhood of Godalming is given as the only known locality in the county. • **1906 revision** (AS: 75): An apetalous form grows abundantly under trees by wall skirting Kew Road, opposite flagstaff.

EXSICC.: **Kew Gardens.** Ex horto; raised from Kew seed [from] C.B.G.; 22 vi 1902; *Lowne* s.n. (K).

CULT.: 1768 (JH: 172/22); 1789, Nat. of Britain (HK1, 2: 387); 1812, Nat. of Britain (HK2, 4: 104); 1814, Nat. of Britain (HKE: 200). Not currently in cult.

CURRENT STATUS: Only in and around the Brick Pit in the Conservation Area where several fine plants are regularly seen.

466. Cardamine flexuosa With.
C. sylvatica Link
wavy bitter-cress
FIRST RECORD: 1873/4.
NICHOLSON: **1873/4 survey** (JB: 10; SFS: 129): **B** and **P.** Most frequent near lake. • **1906 revision** (AS: 75): **A.** Most frequent near lake.
EXSICC.: **Kew Gardens.** [Unlocalised]; cult. ground; 10 v 1928; *Hubbard* s.n. (K). • Thames bank, Kew – Richmond, Richmond side of Isleworth Gate; middle zone; 16 viii 1928; *Summerhayes & Turrill* s.n. (K). **Kew Environs.** Thames bank, Kew; from mud near top of bank to halfway down on steep slope; 2 viii 1928; *Turrill* s.n. (K). • Towpath between Kew & Richmond; favouring moist positions; 26 iv 1944; *Souster* 44 (K).
CURRENT STATUS: A widespread weed that can be abundant where it grows.

467. Cardamine hirsuta L.
C. bellidifolia auct. non L.
hairy bitter-cress
FIRST RECORD: 1873/4 [1768].
NICHOLSON: **1873/4 survey** (JB: 10; SFS: 128): **P** and **Pal.** Moist shady places. • **1906 revision** (AS: 75): Moist shady places in all the divisions.
EXSICC.: **Kew Gardens.** [Unlocalised]; common annual weed of cultivated land; 24 iv 1944; *Souster* 33 (K).
CULT.: 1768 (JH: 172/21, 22); 1789, Nat. of Britain (HK1, 2: 388); 1812, Nat. of Britain (HK2, 4: 104); 1814, Nat. of Britain (HKE: 200), Nat. of Scotland (ibid.: Add. as *bellidifolia*); 1970, Temperate Dept. (K). Not currently in cult.
CURRENT STATUS: An occasional weed that can be abundant where it grows.

468. *Cardamine corymbosa Hook.f.
New Zealand bitter-cress
FIRST RECORD: 1998.
REFERENCES: **Kew Gardens.** 1998, *Lock* (S33: 745): North Arb.: small circular bed by main entrance to the Banks Building (110).
CURRENT STATUS: Persistent where it was first noted and thus far resisting all attempts at eradication. Recently fleetingly seen by the Herbarium.

469. Arabis petraea (L.) Lam.
Cardamine petraea L.
A. hispida L.
northern rock-cress
CULT.: 1768 (JH: 172/21); 1812, Nat. of Britain (HK2, 4: 106); 1814, Nat. of Britain (HKE: 201). Not currently in cult.

470. Arabis glabra (L.) Bernh.
Turritis glabra L.
A. perfoliata Lam.
tower mustard
FIRST RECORD: 1805 [1768].
NICHOLSON: **1873/4 survey**: Not recorded. • **1906 revision**: Not recorded
OTHER REFERENCES: **Kew Gardens.** 1805, *Borrer* (BFS: 20; SFS: 125; S33: 745): Kew Garden wall.
EXSICC.: **Kew Environs.** Outside of Kew Garden wall; 1852; *Syme* in *Herb. Watson* (K).
CULT.: 1768 (JH: 172/20); 1789, Nat. of England (HK1, 2: 400); 1812, Nat. of England (HK2, 4: 109); 1814, Nat. of England (HKE: 201). In cult. under glass.
CURRENT STATUS: Probably extinct but it is not clear on which wall it might have been growing. Some of the boundary wall of the estate is original, but

New Zealand bitter-cress (*Cardamine corymbosa*). A recently and inadvertently imported species that is threatening to become a serious problem in Botanic Gardens. Thus far under control at Kew.

some, especially along Ferry Lane, has been rebuilt. The plant may even have been growing on the internal boundary wall of the botanic garden, but this had mostly been demolished by the time of Syme's record. A piece of original wall still stands between the Rock Garden and the Order Beds.

[471. **Arabis turrita** L.
tower cress
CULT.: 1789, Nat. of England (HK1, 2: 400); 1812, Nat. of England (HK2, 4: 107); 1814, Nat. of England (HKE: 201).
Note. Listed in error as native.]

472. **Arabis alpina** L.
alpine rock-cress
CULT.: 1768 (JH: 172/20); 1789, Alien (HK1, 2: 399); 1812, Alien (HK2, 4: 105); 1814, Alien (HKE: 201); 1855, unspec. (K); 1856, unspec. (K); 1946, Herbaceous Dept., Order Beds (K). Not currently in cult.
Note. Not discovered in Britain until 1887; all cultivated material before this time would therefore have been of European origin.

473. **Arabis hirsuta** (L.) Scop.
Turritis hirsuta L.
A. ciliata R.Br.
hairy rock-cress
CULT.: 1768 (JH: 172/20); 1789, Nat. of Britain (HK1, 2: 401); 1812, Nat. of Britain (HK2, 4: 107), Nat. of Ireland (ibid.: as *ciliata*); 1814, Nat. of Britain (HKE: 201), Nat. of Ireland (ibid.: as *ciliata*). Not currently in cult.

474. **Arabis scabra** All.
A. hispida auct. non L.
A. stricta Huds., nom. illegit.
Bristol rock-cress
CULT.: 1789, Nat. of England (HK1, 2: 400); Nat. of England (HK2, 4: 106); 1814, Nat. of England (HKE: 201). In cult.

475. ***Lunaria annua** L.
honesty
FIRST RECORD: 1998 [1768].
REFERENCES: **Kew Gardens.** 1998, *Cope* (S33: 745): North Arb.: wild area by Main Gate (132); West Arb.: Riverside Walk (213, 216).
CULT.: 1768 (JH: 172/11); 1789, Alien (HK1, 2: 385); 1812, Alien (HK2, 4: 98); 1814, Alien (HKE: 200). In cult.
CURRENT STATUS: Occasionally self-sown in flower-beds or persisting in neglected corners. It is quite widespread but never common.

476. ***Alyssum alyssoides** (L.) L.
A. calycinum L.
small Alison
FIRST RECORD: Pre-1875 [1768].
EXSICC.: **Kew Environs.** Kew Bridge, very near to a farm; July (pre 1875); *Mill* (K).
CULT.: 1768 (JH: 172/14); 1966, unspec. (K). In cult.
CURRENT STATUS: No recent sightings.

477. ***Alyssum simplex** Rudolfi
A. minus Rothm.
FIRST RECORD: 1972.
REFERENCES: **Kew Gardens.** 1972, *Airy Shaw* (S29: 406): A single plant near new fumigation chamber, E. of Wing C of Herbarium.
CURRENT STATUS: No recent sightings.

478. ***Berteroa incana** (L.) DC.
Alyssum incanum L.
hoary Alison
FIRST RECORD: 1878 [1768].
NICHOLSON: **1906 revision:** Not recorded.
EXSICC.: **Kew Gardens.** On walks of Queens Palace grounds; 13 vi 1878; ?*Hooker* s.n. (K). • Amongst grasses near herbarium; 8 viii 1933; *Hubbard* s.n. (K).
CULT.: 1768 (JH: 172/14). In cult.
CURRENT STATUS: Very rare; a single plant was seen outside the Jodrell Laboratory in 2004.

479. ***Lobularia maritima** (L.) Desv.
Clypeola maritima L.
Alyssum maritimum (L.) Lam.
A. halimifolium L.
sweet Alison
FIRST RECORD: 1932 [1768].
EXSICC.: **Kew Gardens.** [Unlocalised], weed on rubbish heap; 1 vii 1932; *Hubbard* s.n. (K).
CULT.: 1768 (JH: 172/14); 1812, Nat. of England (HK2, 4: 95); 1814, Nat. of England (HKE: 199); 1944, Herbaceous Dept. ("Order Beds") (K). Not currently in cult.
CURRENT STATUS: No recent sightings.

480. **Draba aizoides** L.
yellow whitlowgrass
CULT.: 1789, Alien (HK1, 2: 372); 1812, Nat. of Wales (HK2, 4: 92); 1814, Nat. of Wales (HKE: 199); 1894, unspec. (K); 1987 unspec. (K). In cult.
Note. The species was grown at Kew from European material before it was known as a wild plant in Great Britain (1804).

481. **Draba norvegica** Gunnerus
D. rupestris W.T. Aiton
rock whitlowgrass
CULT.: 1812, Nat. of Scotland (HK2, 4: 91); 1814, Nat. of Scotland (HKE: 199). Not currently in cult.

482. Draba incana L.
hoary whitlowgrass
CULT.: 1768 (JH: 172/11); 1789, Nat. of Britain (HK1, 2: 372); 1812, Nat. of Britain (HK2, 4: 91); 1814, Nat. of Britain (HKE: 199); 1894, unspec. (K); 1922, unspec. (K); 1974, bed 154-07 (K). In cult.

483. Draba muralis L.
wall whitlowgrass
CULT.: 1768 (JH: 172/11); 1789, Nat. of England (HK1, 2: 372); 1812, Nat. of England (HK2, 4: 92); 1814, Nat. of England (HKE: 199). Not currently in cult.

484. Erophila verna (L.) DC.
Draba verna L.
E. vulgaris DC.
common whitlowgrass
FIRST RECORD: 1871 [1768].
NICHOLSON: **1873/4 survey** (JB: 11): **B** and **P.** Extremely common on walks, in flower-beds, etc. • **1906 revision** (AS: 75): Extremely common on walks, in flower-beds, etc.
EXSICC.: **Kew Gardens.** Syon Vista; on newly turned up ground; 3 v 1871; *Britten* s.n. (K). • Arboretum Nursery; 1922 & 1925; *Hubbard* s.n. (K). • Arboretum Nursery; abundant; 8 iv 1925; *Hubbard* s.n. (K). • Arboretum Nursery; 8 i 1930; *anon.* s.n. (K). • Arboretum Nursery; common weed; 12 iv 1933; *Hubbard* s.n. (K). • Arboretum Nursery; weed; 17 iii 1945; *Souster* 195 (K).
CULT.: 1768 (JH: 172/11); 1789, Nat. of Britain (HK1, 2: 372); 1812, Nat. of Britain (HK2, 4: 91); 1814, Nat. of Britain (HKE: 199). Not currently in cult.
CURRENT STATUS: Rare; between Kew Palace and the Herbarium, especially between cobbles and paving stones. It is surprisingly uncommon since it is abundant in suitable places outside the Gardens.

485. Cochlearia anglica L.
English scurveygrass
CULT.: 1768 (JH: 172/13); 1789, Nat. of Britain (HK1, 2: 378); 1812, Nat. of Britain (HK2, 4: 89); 1814, Nat. of Britain (HKE: 199). Not currently in cult.

486. Cochlearia officinalis L.
a. subsp. **officinalis**
common scurvygrass
CULT.: 1768 (JH: 172/13); 1789, Nat. of Britain (HK1, 2: 377); 1812, Nat. of Britain (HK2, 4: 89); 1814, Nat. of Britain (HKE: 199); 1939, Herb Garden (K); 1949, Herbarium Ground (K). Not currently in cult.

b. subsp. **scotica** (Druce) P.S.Wyse Jacks.
C. groenlandica auct. non L.
CULT.: 1768 (JH: 172/13); 1789, Nat. of Britain (HK1 2: 378). Not currently in cult.

487. Cochlearia danica L.
Danish scurvygrass
CULT.: 1768 (JH: 172/13); 1789, Nat. of Britain (HK1, 2: 378); 1812, Nat. of Britain (HK2, 4: 90); 1814, Nat. of Britain (HKE: 199); 1948, Herbarium Ground (K); 1949, Herbarium Ground (K). Not currently in cult.

488. *Camelina sativa (L.) Crantz
Myagrum sativum L.
gold-of-pleasure
FIRST RECORD: 1873/4 [1768].
NICHOLSON: **1873/4 survey** (JB: 11 and in SFS: 138): **P.** Several plants came up near lake in newly-sown grass, 1873. Has not been seen since. • **1906 revision**, G. Nicholson (AS: 75): **A.** Here and there in cultivated ground.
CULT.: 1768 (JH: 172/10); 1789, Nat. of Britain (HK1, 2: 369); 1812, Nat. of Britain (HK2, 4: 93); 1814, Nat. of Britain (HKE: 199). Not currently in cult.
CURRENT STATUS: No recent sightings; it has not reappeared in the Paddock since 1990.

489. *Neslia paniculata (L.) Desv.
Myagrum paniculatum L.
ball mustard
FIRST RECORD: 1880 [1768].
NICHOLSON: **1906 revision:** Not recorded.
EXSICC.: **Kew Gardens.** Surrey side of Thames opposite Sion House; vii 1880; *Baker* s.n. (K).
CULT.: 1768 (JH: 172/10). Not currently in cult.
CURRENT STATUS: No recent sightings.

490. Capsella bursa-pastoris (L.) Medik.
Thlaspi bursa-pastoris L.
Bursa patagonica Druce
shepherd's-purse
FIRST RECORD: 1873/4 [1768].
NICHOLSON: **1873/4 survey** (JB: 11): Everywhere. • **1906 revision** (AS: 75): Everywhere.
OTHER REFERENCES: **Kew Environs.** 1923, *Todd* (BSEC7(1): 169): Kew.
EXSICC.: **Kew Gardens.** Near the Herbarium; iii 1922; *Turrill* s.n. (K). • [Unlocalised]; on cultivated sandy soil; 16 vi 1928; *Hubbard* s.n. (K). • Herb. Expt. Ground; weed; 10 vi 1929; *Turrill* s.n. (K). • Thames-bank, near Isleworth Gate, Kew – Richmond; in *Petasites* zone under *Petasites* leaves; 25 vi 1929; *Summerhayes & Turrill* s.n. (K). • Near the Tennis Courts; waste ground; vi 1932; *Turrill* s.n. (K). • [Unlocalised]; common weed in cultivated ground; 1 x 1942; *Melville* s.n. (K). • [Unlocalised]; common

annual weed of cultivated ground; 14 iv 1944; *Souster* 16 (K). • Herbarium Expt. Gd.; 31 iii 1945; *Ross-Craig & Sealy* 1041 (K). • By roadside in front of Herbarium House, 55 The Green; weed; 3 vi 1948; *Turrill* s.n. (K). • Margin of Kew Green, by Bird Cage Walk; 3 iv 1948; *Turrill* s.n. (K). **Kew Environs.** Thames bank, Kew; on bank of mud near the top; 2 viii 1928; *Turrill* s.n. (K).
CULT.: 1768 (JH: 172/13); 1789, Nat. of Britain (HK1, 2: 377); 1812, Nat. of Britain (HK2, 4: 81); 1814, Nat. of Britain (HKE: 197); 1887, unspec. (K); 1933, unspec. (K); 1934, Arboretum (K). Not currently in cult.
CURRENT STATUS: Widespread and often abundant in flower-beds and disturbed soil.

491. Hornungia petraea (L.) Rchb.
Lepidium petraeum L.
Hutchinsia petraea (L.) R.Br.
Hutchinsia
CULT.: 1768 (JH: 172/12); 1789, Nat. of Britain (HK1, 2: 373); 1812, Nat. of England (HK2, 4: 82); 1814, Nat. of England (HKE: 198). Not currently in cult.

492. Teesdalia nudicaulis (L.) R.Br.
Iberis nudicaulis L.
shepherd's cress
FIRST RECORD: 1906 [1768].
NICHOLSON: **1906 revision** (AS: 75 and in SFS: 150): On and near wall of ha-ha facing river.
CULT.: 1768 (JH: 172/14); 1789, Nat. of Britain (HK1, 2: 380); 1812, Nat. of Britain (HK2, 4: 83); 1814, Nat. of Britain (HKE: 198); 1922, unspec. (K). Not currently in cult.
CURRENT STATUS: No recent sightings.

493. Thlaspi arvense L.
Field penny-cress
FIRST RECORD: 1873/4 [1768].
NICHOLSON: **1873/4 survey** (JB: 11; SFS: 148): **P.** Some 8 or 10 plants near Winter Garden, 3 or 4 near lake. • **1906 revision** (AS: 75): **A.** Here and there in cultivated ground.
EXSICC.: **Kew Gardens.** Lower Nursery; on rubbish heap; 14 vi 1928; *Hubbard* s.n. (K). • Near the Herbarium; 20ft; vi 1928; *C.A. Smith* 6073 (K). • Herb. Expt. Ground; weed; 10 vi 1929; *Turrill* s.n. (K). • [Unlocalised]; amongst recently sown grass; 22 vi 1932; *Hubbard* s.n. (K).
CULT.: 1768 (JH: 172/13); 1789, Nat. of Britain (HK1, 2: 375); 1812, Nat. of Britain (HK2, 4: 80); 1814, Nat. of Britain (HKE: 197). In cult.
CURRENT STATUS: Rare; only a few recent sightings. At one time it was abundant on slopes behind Kew Palace, but is now very sporadic.

494. Thlaspi perfoliatum L.
perfoliate penny-cress
CULT.: 1768 (JH: 172/13); 1789, Alien (HK1, 2: 376); 1812, Nat. of England (HK2, 4: 81); 1814, Nat. of England (HKE: 197). Not currently in cult.

495. Thlaspi caerulescens J.&C.Presl
T. alpestre L. non Jacq.
alpine penny-cress
CULT.: 1789, Nat. of England (HK1, 2: 377); 1812, Nat. of England (HK2, 4: 81); 1814, Nat. of England (HKE: 197). Not currently in cult.

[496. Thlaspi montanum L.
CULT.: 1789, Nat. of England (HK1, 2: 376); 1812, Alien (HK2, 4: 82); 1814, Alien (HKE: 197). Note. Originally grown as a native, but this corrected in later editions.]

497. Iberis amara L.
wild candytuft
CULT.: 1768 (JH: 172/14); 1789, Nat. of England (HK1, 2: 380); 1812, Nat. of England (HK2, 4: 84); 1814, Nat. of England (HKE: 198); 1887, unspec. (K). In cult.

498. *Lepidium sativum L.
garden cress
FIRST RECORD: 1925 [1768].
REFERENCES: **Kew Gardens.** 1990, *Cope* (S31: 182): Paddock, after restoration in 1990.
EXSICC.: **Kew Gardens.** Aster beds; vii 1925; *anon.* s.n. (K).
CULT.: 1768 (JH: 172/12); 1789, Alien (HK1, 2: 373); 1812, Alien (HK2, 4: 89); 1814, [Cult.] (HKE: 198); 1897, unspec. (K). Not currently in cult.
CURRENT STATUS: A rare weed; seen only once since 1990.

499. Lepidium campestre (L.) W.T.Aiton
Thlaspi campestre L.
field pepperwort
CULT.: 1768 (JH: 172/13); 1789, Nat. of Britain (HK1, 2: 376); 1812, Nat. of Britain (HK2, 4: 88); Nat. of Britain (HKE: 198). Not currently in cult. Note. Originally grown as a native but now regarded as an archaeophyte.

500. Lepidium heterophyllum Benth.
L. smithii Hook.
Thlaspi hirtum Lam.
Smith's pepperwort
FIRST RECORD: 1847 [1768].
NICHOLSON: **1873/4 survey** (JB: 11; SFS: 147): **P.** Frequent all about lake, also in turf and on waste ground near the Winter Garden. • **1906 revision** (AS: 75): **A.** Frequent all about lake, also in turf, near winter garden.

OTHER REFERENCES: **Kew Gardens.** 1956, *Welch* (LFS: 121; LN36: 12; LN36: S338): Persists on slopes around the Temperate House. Known since 1945. • 1983, *Burton* (FLA: 17): Only the [record] for Kew Gardens represents a colony known over a long period.
EXSICC.: **Kew Gardens.** Round the Temperate House; in turf of grassy slopes; 23 v 1921; *Turrill* s.n. (K). • Round Temperate House; on grassy slopes, short turf on sandy slopes with *Rumex acetosella* etc.; 27 v 1946; *Ross-Craig & Sealy* 1077 (K). **Kew Environs.** Near Kew; vi 1847; *Stevens* s.n. (K).
CULT.: 1768 (JH: 172/13); 1789, Nat. of Wales (HK1, 2: 376); 1814, Nat. of Britain (HKE: Add.). Not currently in cult.
CURRENT STATUS: No recent sightings; certainly extinct around the Temperate House.

501. *Lepidium virginicum L.

least pepperwort

FIRST RECORD: 1927 [1768].
EXSICC.: **Kew Gardens.** On rubbish heap in lower nursery; 31 v 1927; *Hubbard* s.n. (K). • [Unlocalised]; amongst recently sown grass; 22 vi 1932; *Hubbard* s.n. (K). **Kew Environs.** Kew; vii 1929; *Irvine* s.n. (K).
CULT.: 1768 (JH: 172/12); 1789, Alien (HK1, 2: 375); 1812, Alien (HK2, 4: 89); 1814, Alien (HKE: 198); 1920, unspec. (K). Not currently in cult.
CURRENT STATUS: No recent sightings.

502. Lepidium ruderale L.

narrow-leaved pepperwort

FIRST RECORD: 1873/4 [1768].
NICHOLSON: **1873/4 survey** (JB: 11; SFS: 146): **B.** A single plant near Museum No. 1 (A plentiful crop of this appeared in 1873 on Kew Green, by sides of road near Church; also by sides of road leading from Kew to Richmond. This year [1874] only a few specimens have been seen together). • **1906 revision** (AS: 75). **A.** Here and there on dry ground, by walks, etc.
EXSICC.: **Kew Gardens.** [Unlocalised]; amongst recently sown grass, 1 vii 1932, *Hubbard* s.n. (K).
CULT.: 1768 (JH: 172/12); 1789, Nat. of England (HK1, 2: 375); 1812, Nat. of Britain (HK2, 4: 87); 1814, Nat. of Britain (HKE: 198). Not currently in cult.
CURRENT STATUS: No recent sightings.

503. Lepidium latifolium L.

dittander

CULT.: 1768 (JH: 172/12); 1789, Nat. of Britain (HK1, 2: 374); 1812, Nat. of Britain (HK2, 4: 85); 1814, Nat. of Britain (HKE: 198). Not currently in cult.

504. *Lepidium graminifolium L.

L. iberis L.

tall pepperwort

FIRST RECORD: 1872 [1768].
REFERENCES: **Kew Environs.** 1872, *Watson* (BEC15: 10; BSEC9(5): 684; S33: 745): Many plants of it on waste ground by Kew Bridge, Surrey, 1872, and garden, 1874. Root brought from waste ground near Kew Bridge.
CULT.: 1768 (JH: 172/12); 1812, Alien (HK2, 4: 86); 1814, Alien (HKE: 198). In cult.
CURRENT STATUS: No recent sightings.

505. *Lepidium draba L.

Cardaria draba (L.) Desv.

hoary cress

FIRST RECORD: 1933 [1812].
REFERENCES: **Kew Gardens.** 1990, *Cope* (S31: 182): Paddock, after restoration in 1990.
EXSICC.: **Kew Gardens.** Just outside the Herbarium S. end of Wing A; 31 v 1945; *?Ross-Craig* s.n. (K). **Kew Environs.** Near R. Thames; waste ground; 28 v 1933; *Hubbard* s.n. (K).
CULT.: 1812, Alien (HK2, 4: 86); 1814, Alien (HKE: 198); 1940, Herbaceous Ground (K). In cult.
CURRENT STATUS: Increasingly common around the Herbarium and in the Paddock but is rarely seen elsewhere.

506. Coronopus squamatus (Forssk.) Asch.

C. procumbens Gilib.
C. ruellii All.
Cochlearia coronopus L.

swine-cress

FIRST RECORD: 1931 [1768].
EXSICC.: **Kew Environs.** Kew; 10 vi 1931; *Fraser* s.n. (K).
CULT.: 1768 (JH: 172/13); 1789, Nat. of Britain (HK1, 2: 378); 1812, Nat. of Britain (HK2, 4: 76); 1814, Nat. of Britain (HKE: 197). In cult.
CURRENT STATUS: No recent sightings. The single herbarium record comprises only seedlings whose identity is not absolutely certain.

507. *Coronopus didymus (L.) Sm.

Lepidium didymum L.
Senebiera didyma (L.) Pers.

lesser swine-cress

FIRST RECORD: 1849 [1789].
Nicholson: **1873/4 survey** (JB: 11): Everywhere. A most troublesome weed. I have most carefully looked for [*C. squamatus*], but have never been able to find it within our present limits. • **1906 revision** (AS:75): Everywhere. A troublesome weed.
EXSICC.: **Kew Gardens.** On cult. sandy soil; 14 vi 1928; *Hubbard* s.n. (K). • Round the Herbarium; waste places; vi 1928; *C.A.Smith* 6060 (K). • Herbarium Experl. Ground; on loose soil cultivated previous year; 2 vii 1928; *Summerhayes & Hubbard*

s.n. (K). • Herbarium Exper. Ground; 4 vii 1928; *Summerhayes* s.n. (K). • On rubbish heap; 1 vii 1932; *Hubbard* s.n. (K) • Behind tennis courts; waste ground; 22 vi 1936; *Sealy & Burtt* 506 (K). • [Unlocalised]; frequent annual weed; 24 vii 1944; *Souster* 125 (K). • Mounds of earth from excavation for foundations of new wing; 14 x 1965; *Verdcourt* 4243 (K). **Kew Environs.** Kew Churchyard; vii 1849; *Taylor* 57 (K). • Kew; 20 viii 1925; *Wilson* s.n. (K). • Thames embankment, Kew; *Lewin* s.n. (K).
CULT.: 1789, Nat. of England (HK1, 2: 374); 1812, Nat. of England (HK2, 4: 76); 1814, Nat. of England (HKE: 197). Not currently in cult.
CURRENT STATUS: Almost ubiquitous in the Gardens; abundant in both grassland and flowerbeds.

508. Subularia aquatica L.
awlwort
CULT.: 1768 (JH: 172/12); 1789, Nat. of Wales & Scotland (HK1, 2: 371); 1812, Nat. of Britain (HK2, 4: 91); 1814, Nat. of Britain (HKE: 199). Not currently in cult.

509. *Conringia orientalis (L.) Dumort.
Brassica orientalis L.
Erysimum orientale (L.) Crantz
hare's-ear mustard
FIRST RECORD: 1963 [1768].
REFERENCES: **Kew Gardens.** 1963, *Meikle* (S29: 406): One plant behind Herbarium staff room.
CULT.: 1768 (JH: 172/20); 1789, Nat. of England (HK1, 2: 401); 1812, Nat. of England (HK2, 4: 117); 1814, Nat. of England (HKE: 202). Not currently in cult.
CURRENT STATUS: No recent sightings; the building referred to has long since been demolished and the land redeveloped.

510. Diplotaxis tenuifolia (L.) DC.
Sinapis tenuifolia (L.) R.Br.
Brassica muralis auct. non (L.) Huds.
perennial wall-rocket
FIRST RECORD: 1951 [1789].
EXSICC.: **Kew Gardens.** Edge of Kew Green near Main Gate to Gardens; 1 x 1951; *Turrill & Meikle* s.n. (K).
CULT.: 1789, Nat. of England (HK1, 2: 402); 1812, Nat. of England (HK2, 4: 128); 1814, Nat. of England (HKE: 205); 1923, unspec. (K). In cult.
CURRENT STATUS: A rare weed with just a few widely scattered records. It was, however, abundant around the staff allotments behind Hanover House in 2005, but this site has now been cleared for building works.

511. *Diplotaxis muralis (L.) DC.
Sisymbrium murale L.
Sinapis muralis (L.) R.Br.
annual wall-rocket
FIRST RECORD: 1872 [1768].
NICHOLSON: **1873/4 survey** (JB: 10 and in SFS: 142): **Strip.** On, and by the sides of, the towing-path. • **1906 revision** (AS: 75): **Strip.** On, and by the side of, the towing-path.
EXSICC.: **Kew Gardens.** On rubbish-tip, 6 vi 1939, *Hubbard* 9893 (K). • Mound of earth from excavation for foundations of new wing; 14 x 1965; *Verdcourt* 4242 (K). **Kew Environs.** Waste ground by Kew Bridge; ix 1872, *Watson* in *Herb. Watson* (K). • Thames Bank, Kew; viii 1888; *Gamble* 20221 (K).
CULT.: 1768 (JH: 172/22); 1812, Nat. of England (HK2, 4: 128); 1814, Nat. of England (HKE: 205). Not currently in cult.
CURRENT STATUS: A rare weed in the Gardens.

512. Brassica oleracea L.
cabbage
CULT.: 1768 (JH: 172/21); 1789, Nat. of England (HK1, 2: 402); 1812, Nat. of England (HK2, 4: 123); 1814, Nat. of England (HKE: 204); 1937, Herbaceous Ground (K); 1942, unspec. (K). In cult.

[513. Brassica napus subsp. **oleifera** (DC.) Metzg.
B. napus L. s. lat.
oil-seed rape
CULT.: 1789, Nat. of Britain (HK1, 2: 401); 1812, Nat. of Britain (HK2, 4: 123); 1814, Nat. of Britain (HKE: 204). Not currently in cult.
Note. Both Kew specimens cited were cultivated as 'Indian Rape' under the name *B. napus* var. *dichotoma*, but the exact interpretation of this name has not been verified. The plant was originally thought to be native but has always been introduced as a crop plant.**]**

514. Brassica rapa L.
a. subsp. **campestris** (L.) A.R.Clapham
B. rapa subsp. *sylvestris* (Lam.) Janch.
B. campestris L.
bargeman's cabbage
FIRST RECORD: 1852 [1789].
NICHOLSON: **1873/4 survey** (JB: 10): **Q** and **Strip.** Many plants in both localities. • **1906 revision** (AS: 75): **Q, Strip.** Many plants in both divisions.
OTHER REFERENCES: **Kew Environs.** 1932, *Lousley* (LN30: S21): Thames side between Kew and Richmond.
EXSICC.: **Kew Gardens.** Bank of River Thames near Isleworth Gate; 24 v 1947; *Burtt* 1372 (K). **Kew Environs.** Thames side a little above Kew; v-vii

1852; *Syme* in *Herb. Watson* (K). • Tow-path, Kew – Richmond; growing on gravelly bank in association with *Sisymbrium officinale*; 13 vi 1948; *Parker* s.n. (K). • Ha-ha, Old Deer Park, Richmond (seedlings); 19 iv 1933; *Fraser* s.n. (K). • Ha-ha, Old Deer Park, Richmond (seedlings); 6 viii 1934; *Fraser* s.n. (K).
CULT.: 1789, Nat. of England (HK1, 2: 401); 1812, Nat. of England (HK2, 4: 125); 1814, Nat. of England (HKE: 204); 1920, unspec. (K). Not currently in cult.
CURRENT STATUS: Very rare; just a couple of recent sightings. Possibly derived from birdseed.

[b. subsp. **rapa**
turnip
CULT.: 1789, Nat. of England (HK1, 2: 401); 1812, Nat. of England (HK2, 4: 123); 1814, Nat. of England (HKE: 204).
Note. Listed in error as native, but only ever an escape from cultivation.**]**

515. *Brassica juncea (L.) Czernj.
Chinese mustard
FIRST RECORD: 1980 [1917].
REFERENCES: **Kew Environs.** 1980, *Hastings* (FSSC: 18; S33: 745): Towpath by Kew Gardens.
EXSICC.: Cult. only.
CULT.: 1917, unspec. (K); 1929, unspec. (K); 1931, unspec. (K); 1934, unspec. (K); 1936, unspec. (K); 1969, Herbaceous Ground (K). In cult.
CURRENT STATUS: Rare; just one or two recent sightings within the Gardens.

516. Brassica nigra (L.) W.D.J.Koch
Sinapis nigra L.
black mustard
FIRST RECORD: 2001 [1768].
REFERENCES: **Kew Gardens.** 2001, *Cope* (S35: 646): Near the temporary Cycad House (223).
EXSICC.: Cult. only.
CULT.: 1768 (JH: 172/23); 1789, Nat. of Britain (HK1, 2: 403); 1812, Nat. of Britain (HK2, 4: 127); 1814, Nat. of Britain (HKE: 205); 1921, unspec. (K); 1936, Herb. Gdn. (K). Not currently in cult.
CURRENT STATUS: A very rare weed.

517. Sinapis arvensis L.
charlock
FIRST RECORD: 1983/4 [1768].
REFERENCES: **Kew Gardens.** 1983/4, *Cope* (S31: 181): New student vegetable plots behind Hanover House.
CULT.: 1768 (JH: 172/23); 1789, Nat. of Britain (HK1, 2: 403); 1812, Nat. of Britain (HK2, 4: 125); 1814, Nat. of Britain (HKE: 205). Not currently in cult.
CURRENT STATUS: Rare; it has not been seen recently behind Hanover House but it has appeared sporadically in one or two other places.

518. Sinapis alba L.
white mustard
FIRST RECORD: 1990 [1768].
REFERENCES: **Kew Gardens.** 1990, *Cope* (S31: 182): Paddock, after restoration in 1990.
CULT.: 1768 (JH: 172/23); 1789, Nat. of Britain (HK1, 2: 403); 1812, Nat. of Britain (HK2, 4: 126); 1814, Nat. of Britain (HKE: 205); 1934, unspec. (K). In cult.
CURRENT STATUS: Briefly grew as a weed in the Wheatfield sown in 2001 at the south end of the Lake and occasionally turns up elsewhere. It was first noted on the towpath in 1975 but has not been seen there since.

519. *Erucastrum gallicum (Willd.) O.Schulz
hairy rocket
FIRST RECORD: 1985.
REFERENCES: **Kew Gardens.** 1985, *London Natural History Society* (LN65: 196; S33: 745): [Around what is now the Princess of Wales Conservatory].
CURRENT STATUS: No recent sightings.

520. Coincya monensis (L.) Greuter & Burdet
Sisymbrium monense L.
Brassica monensis (L.) Huds.
Isle of Man cabbage
CULT.: 1768 (JH: 172/22); 1789, Nat. of Britain (HK1, 2: 390); 1812, Nat. of Britain (HK2, 4: 124); 1814, Nat. of Britain (HKE: 205); 1921, unspec. (K). Not currently in cult.

521. *Hirschfeldia incana (L.) Lagr.-Fossat
H. incana var. *hirta* (Bab.) O.E.Schulz
hoary mustard
FIRST RECORD: 1969.
REFERENCES: **Kew Gardens.** 1969, *Brenan* (S29: 406 and see Exsicc.).
EXSICC.: **Kew Gardens.** Garden of Herbarium House, 55 Kew Green; bird-seed alien; 18 viii 1969; *Brenan* s.n. (K).
CURRENT STATUS: Abundant in the Wheatfield at the south end of the Lake in 2001; scattered elsewhere but seems to be increasing.

522. *Carrichtera annua (L.) DC.
Vella annua L.
cress rocket
FIRST RECORD: 1965 [1768].
REFERENCES: **Kew Gardens.** 1965, *Airy Shaw* (S29: 406): One plant at foot of 'Marquand memorial window,' N. end of Wing C of Herbarium.
EXSICC.: **Kew Gardens.** Herbarium Ground; 15 xi 1965; *anon.* (K).
CULT.: 1768 (JH: 172/11); 1789, Nat. of England (HK1, 2: 370); 1812, Nat. of England (HK2, 4: 79); 1814, Nat. of England (HKE: 197). Not currently in cult.
CURRENT STATUS: No recent sightings.

523. Cakile maritima Scop.
Bunias cakile L.
sea rocket
CULT.: 1789, Nat. of Britain (HK1, 2: 406); 1812, Nat. of Britain (HK2, 4: 71); 1814, Nat. of Britain (HKE: 196). In cult. (2006). Not currently in cult.

524. *Rapistrum rugosum (L.) J.P.Bergeret
bastard cabbage
FIRST RECORD: Pre-1875.
EXSICC.: **Kew Environs.** Kew Bridge; July [pre-1875]; *Irvine* s.n. (K).
CURRENT STATUS: No recent sightings.

525. Crambe maritima L.
sea-kale
CULT.: 1768 (JH: 172/16); 1789, Nat. of Britain (HK1, 2: 407); 1812, Nat. of Britain (HK2, 4: 72); 1814, Nat. of Britain (HKE: 196); 1936, Herbaceous Ground (K); 1937, Herbaceous Ground (K); 1947, Order Beds (K); 1998, bed 156-19 (K). In cult.

526. Raphanus raphanistrum L.
a. subsp. **raphanistrum**
wild radish
FIRST RECORD: 1990 [1768].
REFERENCES: **Kew Gardens.** 1990, *Cope* (S31: 182): Paddock, after restoration in 1990.
CULT.: 1768 (JH: 172/21); 1789, Nat. of Britain (HK1, 2: 405); 1812, Nat. of Britain (HK2, 4: 129); 1814, Nat. of Britain (HKE: 205); 1938, unspec. (K). Not currently in cult.
CURRENT STATUS: Rare; it has not been seen in the Paddock since 1990 but turns up sporadically in other parts of the Gardens and on the golf course in the Old Deer Park.

b. subsp. **maritimus** (Sm.) Thell.
sea radish
CULT.: 1812, Nat. of Britain (HK2, 4: 129); 1814, Nat. of Britain (HKE: 205). Not currently in cult.

527. *Raphanus sativus L.
garden radish
FIRST RECORD: 2001 [1789].
REFERENCES: **Kew Gardens.** 2001, *Cope* (S35: 646): A single plant by the Lake (265), probably from birdseed.
CULT.: 1789, Alien (HK1, 2: 405); 1812, Alien (HK2, 4: 129); 1814, Alien (HKE: 205). In cult.
CURRENT STATUS: Casual; possibly a relict of former cultivation but its remote location near the Lake suggests that it is more likely to have come from birdseed.

RESEDACEAE

528. Reseda luteola L.
weld
FIRST RECORD: 1873/4 [1768].
NICHOLSON: **1873/4 survey** (JB: 11 and in SFS: 153): **P.** Plentiful on all the ground bordering lake. • **1906 revision** (AS: 75): **A.** Formerly plentiful on all the ground bordering lake.
CULT.: 1768 (JH: 221); 1789, Nat. of Britain (HK1, 2: 131); 1811, Nat. of Britain (HK2, 3: 153); 1814, Nat. of Britain (HKE: 141). Not currently in cult.
CURRENT STATUS: In a rough area near 'Climbers & Creepers', in the Paddock and on the golf course in the Old Deer Park close to the boundary of the Gardens.

529. Reseda lutea L.
wild mignonette
FIRST RECORD: 1873/4 [1768].
NICHOLSON: **1873/4 survey** (JB: 11): **B.** A single plant near clump of trees between "Old Lily House" and fence of Pleasure Garden, 1873. • **1906 revision** (AS: 75 and in SFS: 152): **A.** Here and there in waste places.
CULT.: 1768 (JH: 221); 1789, Nat. of Britain (HK1, 2: 132); 1811, Nat. of Britain (HK2, 3: 154); 1814, Nat. of Britain (HKE: 141); 1922, unspec. (K); 1933, Herb. Expt. Ground (K). In cult.
CURRENT STATUS: Only known as a weed in the Stable Yard and on the golf course in the Old Deer Park.

Wild mignonette (*Reseda lutea*). No longer to be found in the Gardens, but still occurs in the Old Deer Park.

Empetraceae

530. Empetrum nigrum L.
crowberry
CULT.: 1768 (JH: 419, 440); 1789, Nat. of Britain (HK1, 3: 394); 1813, Nat. of Britain (HK2, 5: 366); 1814, Nat. of Britain (HKE: 306); 1881, Arboretum (K); 1890, Arboretum (K); 1905, Canal Beds (K). Not currently in cult.

Heather (*Calluna vulgaris*). Once fairly common near the Queen's Cottage but has been lost for many years. It still occurs in the Old Deer Park.

Ericaceae

531. Loiseleuria procumbens (L.) Desv.
Azalea procumbens L.
trailing azalea
CULT.: 1789, Nat. of Scotland (HK1, 1: 204); 1810, Nat. of Britain (HK2, 1: 320); 1814, Nat. of Britain (HKE: 45). In cult. under glass.

532. Phyllodoce caerulea (L.) Bab.
Menziesia caerulea (L.) Sw.
blue heath
CULT.: 1814, Nat. of Scotland (HKE: Add.); 1906, Arboretum (K). In cult. under glass.
Note. Not discovered in Britain until 1812.

533. Andromeda polifolia L.
bog-rosemary
CULT.: 1768 (JH: 145); 1789, Nat. of Britain (HK1, 2: 68); 1811, Nat. of Britain (HK2, 3: 53); 1814, Nat. of Britain (HKE: 126); 1843–1853, Arboretum (K); 1880, Arboretum (K); 1881, Arboretum (K); 1905, Arboretum (K); 1976, unspec. (K). In cult.

534. Arctostaphylos uva-ursi (L.) Spreng.
Arbutus uva-ursi L.
bearberry
CULT.: 1768 (JH: 145, 434); 1789, Nat. of Britain (HK1, 2: 72); 1811, Nat. of Britain (HK2, 3: 57); 1814, Nat. of Britain (HKE: 127); 1880, Arboretum (K); 1884, Arboretum (K); 1913, unspec. (K). Not currently in cult.

535. Arctostaphylos alpinus (L.) Spreng.
Arbutus alpinus L.
Arctic bearberry
CULT.: 1768 (JH: 434); 1789, Nat. of Britain (HK1, 2: 72); 1811, Nat. of Scotland (HK2, 3: 56); 1814, Nat. of Scotland (HKE: 127). Not currently in cult.

536. Calluna vulgaris (L.) Hull
Erica vulgaris L.
heather
FIRST RECORD: 1873/4 [1768].
NICHOLSON: **1873/4 survey** (JB: 48): **P.** Several tufts in the turf by side of Cedar Avenue. Another more than a yard long near Larch collection. Here and there in wood. **Q.** Not uncommon in the open turf near "Cottage." • **1906 revision** (AS: 84): **A.** Here and there in open places. **Q.** Common in the open turf near Queen's Cottage.
CULT.: 1768 (JH: 441); 1789, Nat. of Britain (HK1, 2: 14); 1811, Nat. of Britain (HK2, 2: 387); 1814, Nat. of Britain (HKE: 111); 1880, Arboretum (K); 1897, unspec. (K); 1935, unspec. (K). In cult.
CURRENT STATUS: No longer found in the Gardens but is still in the Old Deer Park on the golf course.

537. Erica ciliaris L.
Dorset heath
CULT.: 1768 (JH: 441); 1789, Alien (HK1, 2: 21); 1811, Alien (HK2, 2: 394); 1814, Alien (HKE: 111); 1880, Arboretum (K); 1889, unspec. (K); 1899, unspec. (K). In cult.
Note. Originally listed as an alien; not discovered as a native until 1829.

538. Erica tetralix L.
cross-leaved heath
CULT.: 1768 (JH: 441); 1789, Nat. of Britain (HK1, 2: 18); 1811, Nat. of Britain (HK2, 2: 393); 1814, Nat. of Britain (HKE: 111); 1880, Arboretum (K); 1905, Arboretum (K). In cult.

539. Erica cinerea L.
bell heather
CULT.: 1768 (JH: 441); 1789, Nat. of Britain (HK1, 2: 19); 1811, Nat. of Britain (HK2, 2: 392); 1814, Nat. of Britain (HKE: 111); 1880, Arboretum (K); 1905, Arboretum (K). In cult.

540. Erica vagans L.
Cornish heath
CULT.: 1811, Nat. of Cornwall (HK2, 2: 367); 1814, Nat. of Cornwall (HKE: 111); 1884, Arboretum (K). In cult.

541. Vaccinium oxycoccus L.
cranberry
CULT.: 1768 (JH: 457); 1789, Nat. of Britain (HK1, 2: 13); 1811, Nat. of Britain (HK2, 2: 359); 1814, Nat. of Britain (HKE: 110); 1936, Arboretum (K). Not currently in cult.

542. Vaccinium vitis-idaea L.
cowberry
CULT.: 1768 (JH: 457); 1789, Nat. of Britain (HK1, 2: 13); 1811, Nat. of Britain (HK2, 2: 359); 1814, Nat. of Britain (HKE: 110); 1880, Arboretum (K); 1906, Arboretum (K); 1989, Arboretum (K). In cult.

543. Vaccinium uliginosum L.
bog bilberry
CULT.: 1789, Nat. of Britain (HK1, 2: 10); 1811, Nat. of Britain (HK2, 2: 356); 1814, Nat. of Britain (HKE: 110); 1880, unspec. (K); 1895, Arboretum (K). Not currently in cult.

544. Vaccinium myrtillus L.
bilberry
CULT.: 1768 (JH: 457); 1789, Nat. of Britain (HK1, 2: 10); 1811, Nat. of Britain (HK2, 2: 355); 1814, Nat. of Britain (HKE: 110); 1905, Arboretum (K). Not currently in cult.

PYROLACEAE

545. Pyrola minor L.
P. rosea Sm.
common wintergreen
CULT.: 1768 (JH: 189); 1789, Nat. of Britain (HK1, 2: 74); 1811, Nat. of Britain (HK2, 3: 58); 1814, Nat. of Britain (HKE: 127 as *minor*), Nat. of England (ibid.: Add. as *rosea*); 1925, unspec. (K). In cult. under glass.

546. Pyrola media Sw.
intermediate wintergreen
CULT.: 1814, Nat. of England (HKE: Add.). Not currently in cult.
Note. Discovered in Britain in 1807.

547. Pyrola rotundifolia L.
round-leaved wintergreen
CULT.: 1768 (JH: 189); 1789, Nat. of Britain (HK1, 2: 73); 1811, Nat. of Britain (HK2, 3: 58); 1814, Nat. of Britain (HKE: 127); 1977, plot 166-03 (K). Not currently in cult.

548. Orthilia secunda (L.) House
Pyrola secunda L.
serrated wintergreen
CULT.: 1768 (JH: 189); 1789, Nat. of Britain (HK1, 2: 74); 1811, Nat. of Britain (HK2, 3: 58); 1814, Nat. of Britain (HKE: 127). Not currently in cult.

549. Monesis uniflora (L.) A.Gray
Pyrola uniflora L.
one-flowered wintergreen
CULT.: 1789, Alien (HK1, 2: 74); 1811, Nat. of Scotland (HK2, 3: 59); 1814, Nat. of Scotland (HKE: 127). Not currently in cult.
Note. Discovered in Britain in 1793.

MONOTROPACEAE

550. Monotropa hypopitys L.
yellow bird's-nest
CULT.: 1768 (JH: 228, 318). Not currently in cult.

DIAPENSIACEAE

551. Diapensia lapponica L.
diapensia
CULT.: 1810, Alien (HK2, 1: 303); 1814, Alien (HKE: 42). In cult. under glass.
Note. Not discovered in Britain until 1951.

Primulaceae

552. Primula vulgaris Huds.
P. acaulis (L.) Hill
primrose
FIRST RECORD: 1873/4 [1768].
NICHOLSON: **1873/4 survey** (JB: 71): **Pal.** A few plants in turf midway between Herbarium and Palace. • **1906 revision** (AS: 84): A few plants in turf in herbarium grounds.
CULT.: 1768 (JH: 135); 1789, Nat. of Britain (HK1, 1: 192); 1810, Nat. of Britain (HK2, 1: 307); 1814, Nat. of Britain (HKE: 42); 1859, unspec. (K); 1895, unspec. (K); 1896, unspec. (K); 1909, unspec. (K); 1916, unspec. (K); 1925, unspec. (K); 1927, unspec. (K); 1971, Alpine & Herb. Dept. (K); 1976, unspec. (K). In cult.
CURRENT STATUS: Extinct in the localities cited above but not uncommon in the Conservation Area where it was probably introduced. In the *New Atlas* it is said to be alien in TQ17, but this may be questionable; Nicholson did not suggest that it is anything but native in the Palace Grounds.

553. Primula elatior (L.) Hill
oxslip
CULT.: 1789, Nat. of Britain (HK1, 1: 193); 1810, Nat. of Britain (HK2, 1: 307); 1814, Nat. of Britain (HKE: 42); 1887, unspec. (K); 1895, unspec. (K); 1957, Herbarium Ground (K). In cult.
Note. Still cultivated in the Rock Garden.

Primrose (*Primula vulgaris*). Extinct where it used to be native in the Herbarium grounds, but extensive patches can be found naturalised in the Conservation Area.

554. Primula veris L.
cowslip
P. officinalis (L.) Hill
FIRST RECORD: 1873/4 [1768].
NICHOLSON: **1873/4 survey** (JB: 71): In company with [*Primula vulgaris* — in turf midway between Herbarium and Palace], but rather more common. • **1906 revision** (AS: 84): In company with [*Primula vulgaris* — in turf in herbarium grounds], but more common.
CULT.: 1768 (JH: 135); 1789, Nat. of Britain (HK1, 1: 193); 1810, Nat. of Britain (HK2, 1: 307); 1814, Nat. of Britain (HKE: 42); 1920, unspec. (K); 1927, Herb. Expt. Ground (K); 1929, unspec. (K); 1930, unspec. (K); 1931, Herb. Expt. Ground (K); 1932, Herb. Expt. Ground (K); 1934, unspec. (K); 1948, Herbarium Ground (K). In cult.
CURRENT STATUS: For comments see *Primula vulgaris* above.

555. Primula farinosa L.
bird's-eye primrose
CULT.: 1768 (JH: 135); 1789, Nat. of Britain (HK1, 1: 193); 1810, Nat. of Britain (HK2, 1: 308); 1814, Nat. of Britain (HKE: 42); 1895, unspec. (K); 1924, unspec. (K); 1930s, unspec. (K). In cult. under glass.

556. Primula scotica Hook.
Scottish primrose
Cult.: In cult. under glass.

557. Hottonia palustris L.
water-violet
CULT.: 1768 (JH: 132); 1789, Nat. of England (HK1, 1: 197); 1810, Nat. of England (HK2, 1: 313); 1814, Nat. of Britain (HKE: 43). Not currently in cult.

558. *Cyclamen hederifolium Aiton
C. europaeum auct. non L.
sowbread
FIRST RECORD: 1982 [1768].
REFERENCES: **Kew Gardens.** 1982, *Cope* (unpubl.): a patch photographed along Pagoda Vista.
CULT.: 1768 (JH: 138); 1789, Alien (HK1, 1: 196); 1810, Nat. of England (HK2, 1: 311); 1814, Nat. of England (HKE: 43); 1978, unspec. (K). In cult.
CURRENT STATUS: Naturalised under trees along Pagoda Vista and in grass between Oxenhouse Gate and the Conservation Area.

559. *Cyclamen coum Mill.
eastern sowbread
FIRST RECORD: 2004 [1927].
CULT.: 1927, unspec. (K); 1961, Herbaceous Dept. (K). In cult.
CURRENT STATUS: Naturalised on a grassy knoll beside the Broad Walk away from known areas of cultivation.

560. Lysimachia nemorum L.
yellow pimpernel
CULT.: 1768 (JH: 129); 1789, Nat. of Britain (HK1, 1: 200); 1810, Nat. of Britain (HK2, 1: 316); 1814, Nat. of Britain (HKE: 44); 1991, Tem-TH21 (K). In cult. (2006). Not currently in cult.

561. Lysimachia nummularia L.
creeping-Jenny
FIRST RECORD: 1873/4 [1768].
NICHOLSON: **1873/4 survey** (JB: 72): **Pal.** Frequent in the short grass under wall between Palace and Herbarium. **Strip.** Several good patches in company with *Scutellaria galericulata*. • **1906 revision** (AS: 84 and in SFS: 446): Frequent in damp soils in herbarium and palace grounds. **Strip.** Near Isleworth Gate.
CULT.: 1768 (JH: 129); 1789, Nat. of Britain (HK1, 1: 200); 1810, Nat. of Britain (HK2, 1: 316); 1814, Nat. of Britain (HKE: 44). In cult.
CURRENT STATUS: Rare; mostly in the southern half of the Gardens.

562. Lysimachia vulgaris L.
yellow loosestrife
FIRST RECORD: 1873/4 [1768].
NICHOLSON: **1873/4 survey** (JB: 72): {This has been planted wherever it occurs within our limits. It is, however, very common in an "ait" in the "Old Deer Park," not far from here.}. • **1906 revision** (AS: 84): This has been planted wherever it occurs within our limits.
OTHER REFERENCES: **Kew Gardens.** 1999, *Stones* (EAK: 2): By Lake. **Kew environs.** 1918, *Bishop* (LN33: S184): Bank of Thames between Kew and Richmond. • 1934, *anon.* (LN13: S68): Between Kew and Richmond. • 1943, *Cooke* (LN33: S184): Bank of Thames between Kew and Richmond. • 1950, *D.H.Kent* (LN33: S184): Bank of Thames between Kew and Richmond.
CULT.: 1768 (JH: 129); 1789, Nat. of Britain (HK1, 1: 198); 1810, Nat. of Britain (HK2, 1: 314); 1814, Nat. of Britain (HKE: 44); 1883, unspec. (K); 1909, unspec. (K); 1934, Arboretum (K); 1936, Herb Gdn. (K); 1993, Melon Yard (K). In cult.
CURRENT STATUS: Rare; there have recently been a single sighting in the Conservation Area and one on the Lake margin. It was originally planted within the Gardens but may well be native on the riverbank.

563. Lysimachia thyrsiflora L.
tufted loosestrife
CULT.: 1768 (JH: 129); 1789, Nat. of England (HK1, 1: 199); 1810, Nat. of England (HK2, 1: 314); 1814, Nat. of Britain (HKE: 44); 1907, unspec. (K). Not currently in cult.

564. Trientalis europaea L.
chickweed-wintergreen
CULT.: 1768 (JH: 153); 1789, Nat. of Britain (HK1, 1: 493); 1811, Nat. of Britain (HK2, 2: 333); 1814, Nat. of Britain (HKE: 106); 1922, unspec. (K). Not currently in cult.

565. Anagallis tenella (L.) L.
Lysimachia tenella L.
bog pimpernel
CULT.: 1768 (JH: 129); 1789, Nat. of Britain (HK1, 1: 201); 1810, Nat. of Britain (HK2, 1: 317); 1814, Nat. of Britain (HKE: 44); 1923, unspec. (K). Not currently in cult.

566. Anagallis arvensis L.
a. subsp. **arvensis**
A. latifolia L.
scarlet pimpernel
FIRST RECORD: 1873/4 [1768].
NICHOLSON: **1873/4 survey** (JB: 72): **B.** On soil heaps and in flower-beds. **P.** Common about lake. • **1906 revision** (AS: 84): **B.** On soil heaps and in flower-beds. **A.** Common about lake.
EXSICC.: **Kew Gardens.** Hort. Kew (Wild Fl.), Arboretum; 17 vii 1909; *Turrill* s.n. (K). • Herb. Expt. Ground; as a weed; rather large blue flowers; 18 ix 1924; *Turrill* s.n. (K). • L. Nursery; on rubbish heap; 20 vi 1928; *Hubbard* s.n. (K). • [Unlocalised]; 30 v 1931, *Fraser* s.n. (K). • Arboretum; 26 vi 1933; *Dallimore* s.n. (K). • Herb. Gdn.; 1 vii 1936; *anon.* s.n. (K). • [Unlocalised]; summer annual weed of cultivated land; corollas brick-red; 29 vi 1950; *Souster* 1129 (K).
CULT.: 1768 (JH: 135); 1789, Nat. of Britain (HK1, 1: 201); 1810, Nat. of Britain (HK2, 1: 316); 1814, Nat. of Britain (HKE: 44); 1942, unspec. (K). In cult.
CURRENT STATUS: Uncommon as a flower-bed weed. Three colour varieties (red, blue and mauve) regularly occurred in the lawn behind the Herbarium until its restoration in 1999, after which it was not seen again. The blue form is still occasionally seen elsewhere, but it is not to be confused with subsp. *foemina* and records need to be checked.

b. subsp. **foemina** (Mill.) Schinz & Thell.
blue pimpernel
CULT.: 1814, Nat. of England (HKE: Add.); 1959, Herbarium Ground (K). Not currently in cult.
NOTE: Where this plant was cultivated in 1959 is also the locality of the blue variety of subsp. *arvensis*. The identity of both needs to be checked.

567. Anagallis minima (L.) E.H.L.Krause
Centunculus minimus L.
chaffweed
CULT.: 1769 (JH: 121) 1789, Nat. of Britain (HK1, 1: 155); 1810, Nat. of Britain (HK2, 1: 257); 1814, Nat. of Britain (HKE: 35). Not currently in cult.

568. Glaux maritima L.
sea-milkwort
CULT.: 1768 (JH: 329); 1789, Nat. of Britain (HK1, 1: 291); 1811, Nat. of Britain (HK2, 2: 62); 1814, Nat. of Britain (HKE: 65). Not currently in cult.

569. Samolus valerandi L.
brookweed
CULT.: 1768 (JH: 129); 1789, Nat. of Britain (HK1, 1: 227); 1810, Nat. of Britain (HK2, 1: 365); 1814, Nat. of Britain (HKE: 50). Not currently in cult.

GROSSULARIACEAE

570. Ribes rubrum L.
R. sylvestre (Lam.) Mert. & W.D.J.Koch
red currant
FIRST RECORD: 1946 [1768].
REFERENCES: **Kew Environs.** 1946, *Welch* (LN32: S117; S33: 745): Riverside, Richmond to Kew.
CULT.: 1768 (JH: 453); 1789, Nat. of Britain (HK1, 1: 279); 1811, Nat. of Britain (HK2, 2: 40); 1814, Nat. of Britain (HKE: 61); 1843-1853, Arboretum (K); 1891, unspec. (K); 1977, unspec. (K); 1992, unspec. (K); 1989, plot 338.02 (K). In cult.
CURRENT STATUS: No recent sightings.

571. Ribes spicatum E.Robson
R. petraeum auct. non Wulf.
downy currant
CULT.: 1811, Nat. of England (HK2, 2: 40, 41); 1814, Nat. of England (HKE: 62); 1993, unspec. (K). In cult.

572. *Ribes nigrum L.
black currant
FIRST RECORD: 1943 [1768].
REFERENCES: **Kew Environs.** 1943, *Welch* (LN32: S117): Riverside, Kew.
CULT.: 1768 (JH: 453); 1789, Nat. of Britain (HK1, 1: 279); 1811, Nat. of Britain (HK2, 2: 41); 1814, Nat. of Britain (HKE: 62); 1843–1853, Arboretum (K); 1880, Arboretum (K); 1881, Arboretum (K); 1887, unspec. (K); 1892, Arboretum (K); 1895, Arboretum (K); 1905, Nursery (K); 1913, Herbarium Ground (K); 1927, unspec. (K); 1982, unspec. (K); 1992, unspec. (K). In cult.
CURRENT STATUS: No recent sightings.

573. Ribes alpinum L.
mountain currant
CULT.: 1768 (JH: 453); 1789, Nat. of Britain (HK1, 1: 279); 1811, Nat. of Britain (HK2, 2: 41); 1814, Nat. of Britain (HKE: 62); 1843–1853, Arboretum (K); 1882, Arboretum (K); 1884, Arboretum (K); 1891, unspec. (K); 1894, Arboretum (K); 1898, unspec. (K); 1905, Arboretum (K); 1949, Arboretum (K); 1965, Arboretum (K). In cult.

574. *Ribes uva-crispa L.
R. grossularia L.
gooseberry
FIRST RECORD: 2001 [1768].
REFERENCES: **Kew Gardens.** 2001, *Cope* (S35: 647): Along the river wall near the Banks Building (108). Presumably bird-sown from nearby allotments.
CULT.: 1768 (JH: 453); 1789, Alien (HK1, 1: 280); 1811, Nat. of England (HK2, 2: 42); 1814, Nat. of England (HKE: 62); 1880, Arboretum (K); 1892, unspec. (K); 1894, Arboretum (K); 1913, Arboretum (K); 1965, unspec. (K); 1977, unspec. (K). Not currently in cult.
CURRENT STATUS: Probably bird-sown, against the river wall near the Banks Building Pond.

Gooseberry (*Ribes uva-crispa*). Once found against the boundary wall near the Banks Building; probably bird-sown.

CRASSULACEAE

575. Crassula tillaea Lest.-Garl.
Tillaea muscosa L.
mossy stonecrop

CULT.: 1789, Nat. of England (HK1, 1: 173); 1810, Nat. of England (HK2, 1: 282); 1814, Nat. of England (HKE: 39). In cult. under glass.

576. Umbilicus rupestris (Salis.) Dandy
Cotyledon lutea Huds.
C. umbilicus L.
navelwort

CULT.: 1768 (JH: 152); 1789, Nat. of Britain (HK1, 2: 107 as *umbilicus*), Nat. of England (ibid.: as *lutea*); 1811, Nat. of Britain (HK2, 3: 110 as *umbilicus*), Nat. of England (ibid.: as *lutea*); 1814, Nat. of Britain (HKE: 134 as *umbilicus*), Nat. of England (ibid.: as *lutea*); 1978, Herbarium Ground (K). In cult. under glass.

[**577. Sempervivum tectorum** L.
house-leek

CULT.: 1789, Nat. of Britain (HK1, 2: 148); 1811, Nat. of Britain (HK2, 3: 172); 1814, Nat. of Britain (HKE: 143).
Note. Listed in error as native.]

578. Sedum rosea (L.) Scop.
Rhodiola rosea L.
roseroot

CULT.: 1768 (JH: 419); 1789, Nat. of Britain (HK1, 3: 407); 1813, Nat. of Britain (HK2, 5: 397); 1814, Nat. of Britain (HKE: 311). In cult.

579. Sedum telephium L.
orpine

FIRST RECORD: 2001 [1768].
EXSICC.: **Kew Environs.** St. Anne's Church; wall with *Rubus* etc.; 14 xi 2001, *Verdcourt* 5528 (K).
CULT.: 1768 (JH: 209); 1789, Nat. of Britain (HK1, 2: 108); 1811, Nat. of Britain (HK2, 3: 111); 1814, Nat. of Britain (HKE: 135); 1955, Experimental Ground (K); 1977, unspec. (K); 1980, unspec. (K). In cult.
CURRENT STATUS: Persistent from an original planting in the churchyard but nowhere truly wild.

[**580. Sedum rupestre** L.
S. glaucum Waldst. & Kit.
S. reflexum L.
reflexed stonecrop

CULT.: 1789, Nat. of Britain (HK1, 2: 110 as *reflexum*), Nat. of England (ibid.: as *rupestre*); 1811, Nat. of Britain (HK2, 3: 113 as *reflexum*), Nat. of England (ibid.: 114 as *rupestre*); 1814, Nat. of Britain (HKE: 135 as *reflexum*), Nat. of England (ibid.: as *reflexum* and: Add. as *glaucum*).
Note. Listed in error as native.]

Biting stonecrop (*Sedum acre*). Once common on the boundary wall and in nearby turf but now confined to the riverbank near Kew Bridge.

581. Sedum forsterianum Sm.
rock stonecrop

CULT.: 1811, Nat. of Wales (HK2, 3: 114); 1814, Nat. of Wales (HKE: 135). Not currently in cult.

582. Sedum acre L.
biting stonecrop

FIRST RECORD: 1873/4 [1768].
NICHOLSON: **1873/4 survey** (JB: 44): **P.** On the wall and in the turf near it, the whole length of the river boundary. **Strip.** Here and there in the turf by towing-path. • **1906 revision** (AS: 79): On the wall and in the turf near it, the whole length of the river boundary. **Strip.** Here and there along towing-path.
CULT.: 1768 (JH: 210); 1789, Nat. of Britain (HK1, 2: 111); 1811, Nat. of Britain (HK2, 3: 115); 1814, Nat. of Britain (HKE: 135); 1933, Herbarium Ground (K); 1955, herb. ground (K); 1979, unspec. (K). In cult.
CURRENT STATUS: Still on the river embankment near Kew Bridge. It is reported in the *Atlas* to have been introduced in TQ17 although it is native in Surrey.

[**583. Sedum sexangulare** L.
tasteless stonecrop

CULT.: 1789, Nat. of England (HK1, 2: 111); 1811, Nat. of England (HK2, 3: 115); 1814, Nat. of England (HKE: 135).
Note. Listed in error as native.]

584. Sedum album L.
white stonecrop
First record: 1999 [1768].
References: **Kew Environs.** 1999, *Cope* (S35: 647): Bank of the Thames near Kew Bridge.
Cult.: 1768 (JH: 210); 1789, Nat. of England (HK1, 2: 111); 1811, Nat. of England (HK2, 3: 114); 1814, Nat. of England (HKE: 135); 1936, Herbarium Ground (K); 1977, Herbarium Ground (K). In cult. under glass.
Current status: In quantity on the riverbank towards Kew Bridge.

585. Sedum anglicum Huds.
English stonecrop
Cult.: 1789, Nat. of England (HK1, 2: 111); 1811, Nat. of Britain (HK2, 3: 115); 1814, Nat. of Britain (HKE: 135); 1932, Herbarium Ground (K). Not currently in cult.

[586. Sedum dasyphyllum L.
thick-leaved stonecrop
Cult.: 1789, Nat. of England (HK1, 2: 110); 1811, Nat. of England (HK2, 3: 113); 1814, Nat. of England (HKE: 135).
Note. Listed in error as native.]

587. Sedum villosum L.
hairy stonecrop
Cult.: 1768 (JH: 210); c. 1779–1805, unspec. (K); 1789, Nat. of Britain (HK1, 2: 112); 1811, Nat. of Britain (HK2, 3: 115); 1814, Nat. of Britain (HKE: 135). Not currently in cult.
Note. The earliest record is from the Goodenough herbarium and is undoubtedly from a cultivated plant.

Saxifragaceae

588. Saxifraga hirculus L.
marsh saxifrage
Cult.: 1789, Nat. of England (HK1, 2: 81); 1811, Nat. of England (HK2, 3: 69); 1814, Nat. of England (HKE: 129). In cult.

589. Saxifraga sibthorpii Boiss.
First record: 2008 [1969].
Cult.: In cult.
Current status: Found as a weed in the Rhododendron Dell. Very similar to the more frequently recorded *S. cymbalaria* (celandine saxifrage) in this country, but has reflexed sepals.

590. Saxifraga nivalis L.
alpine saxifrage
Cult.: 1768 (JH: 189); 1789, Nat. of Britain (HK1, 2: 79); 1811, Nat. of Britain (HK2, 3: 67); 1814, Nat. of Britain (HKE: 129). Not currently in cult.

591. Saxifraga stellaris L.
starry saxifrage
Cult.: 1768 (JH: 189); 1789, Nat. of Britain (HK1, 2: 78); 1811, Nat. of Britain (HK2, 3: 66); 1814, Nat. of Britain (HKE: 129). In cult.

[592. Saxifraga umbrosa L.
Pyrenean saxifrage
Cult.: 1789, Nat. of Britain (HK1, 2: 79); 1811, Nat. of Britain (HK2, 3: 67); 1814, Nat. of Britain (HKE: 129).
Note. Listed in error as native.]

593. Saxifraga oppositifolia L.
purple saxifrage
Cult.: 1768 (JH: 189); 1789, Nat. of Britain (HK1, 2: 80); 1811, Nat. of Britain (HK2, 3: 68); 1814, Nat. of Britain (HKE: 129). In cult. under glass.

594. Saxifraga aizoides L.
yellow saxifrage
Cult.: 1768 (JH: 189); 1789, Nat. of England (HK1, 2: 81); 1811, Nat. of Britain (HK2, 3: 69); 1814, Nat. of Britain (HKE: 129). Not currently in cult.

Meadow saxifrage (*Saxifraga granulata*). At one time abundant on the river side of the Gardens but now only in rare isolated patches; best seen in the meadow near the Main Gate.

595. Saxifraga rivularis L.
highland saxifrage
CULT.: 1811, Nat. of Scotland (HK2, 3: 70); 1814, Nat. of Scotland (HKE: 129). Not currently in cult.
Note. Discovered in Britain in 1792.

596. Saxifraga cernua L.
drooping saxifrage
CULT.: 1811, Nat. of Scotland (HK2, 3: 69); 1814, Nat. of Scotland (HKE: 129). Not currently in cult.
Note. Discovered in Britain in 1794.

597. Saxifraga granulata L.
meadow saxifrage
FIRST RECORD: 1851 [1768].
NICHOLSON: **1873/4 survey** (JB: 44): So common as to give, when in flower, quite a colour to the whole river length of the gardens. **Pal.** Plentiful (A few plants of the double form grow near the Palace. – Mr Lynch). • **1906 revision** (AS: 79): So common as to give, when in flower, quite a colour to the whole river length of the area.
OTHER REFERENCES: **Kew Gardens.** 1976, *Lousley* (LFS: 205): Still persists in the turf of Kew Gardens. • 1983, *Burton* (FLA: 77): Preserved ... by deliberate conservation measures as in Kew Gardens.
EXSICC.: **Kew Gardens.** Hort. Kew; 1856; *anon.* s.n. (K). • Grass just north of Main (Kew Green) Gate; 13 vi 1928; *Turrill* s.n. (K). • Near tennis courts; in grasses; v 1933; *Hubbard* s.n. (K). **Kew Environs.** Pastures about Kew; v 1851; *Prior* s.n. (K).
CULT.: 1768 (JH: 190); 1789, Nat. of Britain (HK1, 2: 81); 1811, Nat. of Britain (HK2, 3: 69); 1814, Nat. of Britain (HKE: 129); 1978, Alpine Nursery (K). In cult.
CURRENT STATUS: Still reasonably abundant in the meadow near the Main Gate but has declined over the rest of the Gardens and is probably extinct in most of its old localities. There were some magnificent specimens behind the cricket pavillion on Kew Green, but a path was mown through them and they have not thus far recovered.

598. Saxifraga hypnoides L.
S. elongata Panz. ex Ser.
S. hirta Donn
S. moschata D.Don
S. palmata Sm
S. platypetala Sm.
mossy saxifrage
CULT.: 1768 (JH: 190); 1789, Nat. of Britain (HK1, 2: 83); 1811, Nat. of Britain (HK2, 3: 72), Nat. of England (ibid.: 71 as *moschata*), Nat. of Wales (ibid.: 72 as *palmata*); 1814, Nat. of Britain (HKE: 129), Nat. of England (ibid.: 129 as *moschata* and: Add. as *platypetala*), Nat. of Wales (ibid.: 129 as *palmata*), Nat. of Scotland (ibid.: Add. as *elongata* & *hirta*); 1911, unspec. (K). In cult. under glass.

Rue-leaved saxifrage (*Saxifraga tridactylites*). Once known from old walls, gravel paths and flower-beds, but last recorded over a century ago.

599. Saxifraga rosacea Moench
Irish saxifrage

a. subsp. **rosacea**
CULT.: 1976, Alpine Dept. (K). In cult.

b. subsp. **hartii** (D.A.Webb) D.A.Webb
Cult.: In cult. under glass.

600. Saxifraga cespitosa L.
tufted saxifrage
CULT.: 1789, Nat. of England (HK1, 2: 82); 1811, Nat. of Wales (HK2, 3: 71); 1814, Nat. of Wales (HKE: 129). Not currently in cult.

601. Saxifraga tridactylites L.
S. granulata auct. (see SFS: 310)
rue-leaved saxifrage
FIRST RECORD: 1873/4 [1768].
NICHOLSON: **1873/4 survey** (JB: 44): **Pal.** A few plants on wall near Herbarium. **B.** As a weed in the gravel and beds near Museum No. 2. • **1906 revision** (AS: 79 and in SFS: 310; the record here is given under *S. granulata* but clearly refers to *S. tridactylites*): On old wall near herbarium. **A.** On wall of ha-ha.
CULT.: 1768 (JH: 190); 1789, Nat. of Britain (HK1, 2: 82); 1811, Nat. of Britain (HK2, 3: 71); 1814, Nat. of Britain (HKE: 128). Not currently in cult.
CURRENT STATUS: No recent sightings; the old wall near the Herbarium (running parallel to the river) has been demolished and rebuilt. The plant has not been found on the old wall still standing along Ferry Lane, nor on any other in the Gardens.

[602. **Saxifraga trifurcata** Schrader
S. pedatifida auct. non
CULT.: 1811, Nat. of Scotland (HK2, 3: 70); 1814, Nat. of Scotland (HKE: 129).
Note. Listed in error as native.]

603. ***Tellima grandiflora** (Pursh) Lindl.
fringe-cups
FIRST RECORD: 1977 [1926].
REFERENCES: **Kew Gardens.** 1977, *Cope* (unpubl.). • 2001, *Cope* (S35: 647): Naturalised in the Conservation Area (310). Subsequently in several other places near the Lake (252, 255) and just outside the Conservation Area (323).
CULT.: 1926, Rock Garden (K). In cult.
CURRENT STATUS: Established as an escape in several spots in the Conservation Area, on the edge of the Rhododendron Dell and near the Stable Yard.

604. **Chrysosplenium oppositifolium** L.
opposite-leaved golden-saxifrage
CULT.: 1768 (JH: 370); 1789, Nat. of Britain (HK1, 2: 77); 1811, Nat. of Britain (HK2, 3: 64); 1814, Nat. of Britain (HKE: 128). Not currently in cult.

605. **Chrysosplenium alternifolium** L.
alternate-leaved golden-saxifrage
CULT.: 1768 (JH: 370); 1789, Nat. of Britain (HK1, 2: 77); 1811, Nat. of Britain (HK2, 3: 64); 1814, Nat. of Britain (HKE: 128). Not currently in cult.

606. **Parnassia palustris** L.
grass-of-Parnassus
CULT.: 1768 (JH: 190); 1789, Nat. of Britain (HK1, 1: 381); 1811, Nat. of Britain (HK2, 2: 177); 1814, Nat. of Britain (HKE: 82); 1977, unspec. (K). Not currently in cult.

ROSACEAE

607. ***Sorbaria tomentosa** (Lindl.) Rehder
Himalayan sorbaria
FIRST RECORD: 2002 [1898].
REFERENCES: **Kew Environs.** 2002, *Hounsome* (LN82: 256): On a wall by the river at Kew.
CULT.: 1885, Arboretum (K); 1898, unspec. (K); 1899, unspec. (K); 1905, bed near Economic House (K). In cult.
CURRENT STATUS: No recent sightings.

[608. **Spiraea salicifolia** L.
Bridewort
CULT.: 1789, Alien (HK1, 2: 197); 1811, Nat. of Britain (HK2, 3: 254); 1814, Nat. of Britain (HKE: 158).
Note. Listed in error as native.]

609. **Filipendula vulgaris** Moench
Spiraea filipendula L.
dropwort
CULT.: 1768 (JH: 214); 1789, Nat. of Britain (HK1, 2: 199); 1811, Nat. of Britain (HK2, 3: 256); 1814, Nat. of Britain (HKE: 157); 1933, Herb. Ground (K). In cult.

610. **Filipendula ulmaria** (L.) Maxim.
Spiraea ulmaria L.
meadowsweet
FIRST RECORD: 1873/4 [1768].
NICHOLSON: **1873/4 survey** (JB: 43): Common by side of moat. These are the only wild ones. Those about lake and pond have all been planted. • **1906 revision** (AS: 78): Common by ha-ha; elsewhere planted.
EXSICC.: **Kew Gardens.** By Lily Lake; 22 vi 1929, *Lousley* s.n. (K). **Kew Environs.** Thames bank, Kew; growing on mud of high bank, upper zone; 7 viii 1928; *Summerhayes & Turrill* s.n. (K). • Tow path between Richmond and Kew Bridges; vii 1928; *C.A.Smith* 6102 (K).
CULT.: 1768 (JH: 214); 1789, Nat. of Britain (HK1, 2: 199); 1811, Nat. of Britain (HK2, 3: 257); 1814, Nat. of Britain (HKE: 157); 1936, Herb Gdn. (K). In cult.
CURRENT STATUS: Occurs only by the Lake and in the Conservation Area where Nicholson reports that it is not truly wild. It has not been seen recently along the towpath adjacent to the Gardens.

611. **Rubus chamaemorus** L.
cloudberry
CULT.: 1768 (JH: 216); 1789, Nat. of Britain (HK1, 2: 211); 1811, Nat. of Britain (HK2, 3: 270); 1814, Nat. of Britain (HKE: 161). Not currently in cult.

612. **Rubus saxatilis** L.
stone bramble
CULT.: 1768 (JH: 216); 1789, Nat. of Britain (HK1, 2: 210); 1811, Nat. of Britain (HK2, 3: 270); 1814, Nat. of Britain (HKE: 161). In cult.

613. **Rubus arcticus** L.
Arctic bramble
CULT.: 1768 (JH: 216); 1789, Alien (HK1, 2: 210); 1811, Nat. of Scotland (HK2, 3: 270); 1814, Nat. of Scotland (HKE: 161). Not currently in cult.
Note. Believed extinct in Britain since 1841.

614. **Rubus idaeus** L.
raspberry
FIRST RECORD: 1866 [1768].
NICHOLSON: **1873/4 survey**: Not recorded. • **1906 revision** (AS: 78): A few clumps in Queen's Cottage grounds.

Meadowsweet (*Filipendula ulmaria*). Once common along the Moat but now confined to the Lake and the Conservation Area where it has been planted.

OTHER REFERENCES: **Kew Gardens.** 1909, *Rolfe & A.B. Jackson* (S10: 370): Nicholson records a few clumps in Queens' Cottage grounds, but these now seem to have disappeared: at all events we could not find any. • 1911, *Rolfe* (S12: 375): In an enumeration of the Rubi growing indigenously at Kew, it was remarked [S10: 370] that a few clumps of *Rubus idaeus*, recorded growing in the Queen's Cottage grounds by the late Mr G.Nicholson, "now seem to have disappeared; at all events we could not find any." A large clump of it has since been pointed out to me by Mr C.P.Raffill."
EXSICC.: **Kew Gardens.** [Unlocalised]; viii 1866; *Baker* s.n. (K*).
CULT.: 1768 (JH: 216, 454); 1789, Nat. of Britain (HK1, 2: 209); 1811, Nat. of Britain (HK2, 3: 267); 1814, Nat. of Britain (HKE: 161). Not currently in cult.
CURRENT STATUS: Rare; one clump was found in the southern part of the Gardens and this is all that is currently known. It is a considerable distance from Nicholson's original locality, now lost, but may have been bird-sown from cultivated specimens close by near the southern end of the Temperate House.

615. *Rubus loganobaccus L.H.Bailey
loganberry
FIRST RECORD: 1963.
EXSICC.: **Kew Environs.** Kew Bridge; Thames bank; 1 viii 1963; *Neumann* R55 (K).
CURRENT STATUS: No recent sightings.

616 – 661. Rubus fruticosus L. agg. (species marked thus: ¶)
brambles
INDET RECORDS: 1999, *Stones* (EAK: 5): By Palm House Pond.
INDET EXSICC.: **Kew Gardens.** Queen's Gardens Grove; 27 vii 1908; *E.G.Gilbert* 481 (K). • Grown in Kew Gardens as *R. imbricatus*; 1910; *E.G.Gilbert* s.n. (K). • Queen's Cottage Grounds; 30 vii 1963; *Neumann* R41, R47 (all K). **Kew Environs.** Ferry Lane, Old Deer Park; 31 vii 1963; *Neumann* R 56, R57, R58, R59, R60 (all K). • Kew; Thames bank; 8 vii 1963; *Neumann* R4 (K).
CULT.: 1768 (JH: 216, 454); 1789, Nat. of Britain (HK1, 2: 210); 1811, Nat. of Britain (HK2, 3: 269); 1814, Nat. of Britain (HKE: 161); 1931, Keepers' Garden (K). Not currently in cult.
CURRENT STATUS: Brambles are found throughout the Gardens, but to date few of them have been named to microspecies. What follows has been pieced together from herbarium specimens and literature reports with some field observations. See DEA for further details.

616. Rubus nessensis W.Hall
R. suberectus G.Anderson ex Sm.
FIRST RECORD: Pre-1916 [1814].
EXSICC.: **Kew Environs.** Kew; pre-1916; *E.G.Gilbert* s.n. (K).
CULT.: 1814, Nat. of Britain (HKE: Add.); 1954, Herb. Expt. Grnd. (K). Not currently in cult.
CURRENT STATUS: No recent sightings.

617. *Rubus laciniatus Willd.
cut-leaved bramble
FIRST RECORD: 1998 [1932].
REFERENCES: **Kew Gardens.** 1998, *Cope* (S33: 745): West Arb.: Beech circle (242).
CULT.: 1932, unspec. (K). Not currently cult.
CURRENT STATUS: There are several recent records; the plant is of ornamental value, although viciously prickly, and would be a relict of former cultivation wherever it grows. It is abundant on the banks of the ditch between the Gardens and the Brentford Gate car park.

618. Rubus lentiginosus Lees
FIRST RECORD: 1909.
EXSICC.: **Kew Gardens.** Grown in Kew Gardens; 21 vii 1909; *E.G.Gilbert* s.n. (K).
CURRENT STATUS: Although the label of the specimen says 'grown in Kew Gardens' it is very unlikely that this plant was cultivated. No recent sightings.

619. Rubus viridescens (Rogers) T.A.W.Davis
R. thyrsoideus var. *viridescens* Rogers
FIRST RECORD: 1869.
NICHOLSON: **1873/4 survey**: Not recorded. • **1906 revision**: Not recorded.
EXSICC.: **Kew Gardens.** [Unlocalised]; ix 1869; *Baker* s.n. (K* sub *thyrsoideus* var. *viridescens*). • Grown in Kew Gardens; 30 viii 1909; *E.G.Gilbert* s.n. (K).
CURRENT STATUS: No recent sightings. Included here is a specimen labelled '*R. thyrsoideus*', the identity of which has not been established.

620. Rubus cardiophyllus Lefèvre & P.J.Müll.
R. cordifolius auct. non Bloxam
R. rhamnifolius auct. non Weihe & Nees
FIRST RECORD: 1866.
NICHOLSON: **1873/4 survey**: Not recorded. • **1906 revision**: Not recorded.
EXSICC.: **Kew Gardens.** [Unlocalised]; viii 1866; *Baker* s.n. (K* sub *cordifolius*). • [Unlocalised]; viii. 1866; *Baker* s.n. (K* sub *rhamnifolius*) • [unlocalised]; viii 1878; *Nicholson* s.n. (K). • Queen's Cottage Grounds; 30 vii 1963; *Neumann* R35 (K).
CURRENT STATUS: No recent sightings.

621. Rubus dumnoniensis Bab.
FIRST RECORD: 1906.
NICHOLSON: **1906 revision** (AS: 78; SFS: 266): **Q.** Here and there.
CURRENT STATUS: More likely to have been *R. cardiophyllus* Lef. & P.J.Mueller (DEA: 33) although one specimen has been renamed (DEA: 33) as *R. armipotens.* No recent sightings.

622. Rubus milfordensis Edees
?*R. villicaulis* auct. non Köhler ex Weihe & Nees
FIRST RECORD: 1910.
EXSICC.: **Kew Gardens.** [Unlocalised]; 13 vii 1910; *E.G.Gilbert* s.n. (K).
CURRENT STATUS: Uncertain; the record is based on a specimen at K labelled '*R. villicaulis*' but this determination has not been confirmed.

623. Rubus polyanthemus Lindeb.
R. pulcherrimus Neuman
FIRST RECORD: 1906.
NICHOLSON: **1906 revision** (AS: 78; SFS: 266): **Q.** Abundant.
OTHER REFERENCES: **Kew Gardens.** 1909, *Rolfe & A.B. Jackson* (S10: 370 and see Exsicc.): Abundant in Queen's Cottage grounds. Probably the commonest of the Kew brambles, and one of the handsomest when in flower. Aptly named.
EXSICC.: **Kew Gardens.** Queen's Cottage Grounds; abundant; 4 viii 1908; *A.B.Jackson & Rolfe* s.n. (K*). • [Unlocalised]; 12 vii 1911; *E.G.Gilbert* s.n. (K). • Queen's Cottage Grounds; 30 vii 1963, *Neumann* R40 (K). • Queen's Cottage Grounds; 30 vii 1963, *Neumann* R48 (K).
CURRENT STATUS: Still present in the conservation Area.

624. Rubus polyanthemus × mucronulatus
R. pulcherrimus × mucronatus ?
FIRST RECORD: 1909.
REFERENCES: **Kew Gardens.** 1909, *Rolfe & A.B. Jackson* (S10: 373): Borders of thickets in Queen's Cottage grounds. Not rare. Specimens of three different bushes were collected in August, 1909, and sent to Mr Rogers, and all were determined as above. We had referred them doubtfully to [*R. mucronulatus*], which on the whole it most resembles, and with which it was probably included by Nicholson.
CURRENT STATUS: This is probably *R. iodnephes* W.C.R.Watson (D.E.Allen, pers. comm.), but there are no recent sightings.

625. Rubus rhombifolius Weihe ex Boenn.
R. gratus var. *sciaphilus* auct.
R. platyacanthus auct. non P.J.Müll. & Lefèvre
FIRST RECORD: 1908.
REFERENCES: **Kew Gardens.** 1909, *Rolfe & A.B. Jackson* (S10: 370, 371; SFS: 268 and see Exsicc.): Queen's Cottage grounds; not common. Of specimens collected in August, 1908, Mr Rogers remarks: "Except for the stronger stem-prickles and the nearly bald stem, these two sheets seem identical with Danish specimens of *R. sciaphilus*, now in my Herbarium".
EXSICC.: **Kew Gardens.** Queen's Cottage Grounds, in one or two places; 4 viii 1908; *A.B.Jackson & Rolfe* s.n. (K*). • Queen's Cottage Grounds, Richmond end; 9 viii 1909; *A.B.Jackson & Rolfe* s.n. (K*).
CURRENT STATUS: No recent sightings.

626. Rubus sprengelii Weihe
FIRST RECORD: 1866.
NICHOLSON: **1873/4 survey**: Not recorded. • **1906 revision**: Not recorded.
EXSICC.: **Kew Gardens.** [Unlocalised]; viii 1866; *Baker* s.n. (K*).
CURRENT STATUS: No recent sightings.

627. Rubus anglocandicans A.Newton
R. macrophyllus sensu Nicholson, non Weihe & Nees
FIRST RECORD: 1878.
NICHOLSON: **1906 revision** (AS: 78 and in SFS: 270): **Q:** Abundant.
OTHER REFERENCES: **Kew Gardens.** 1909, *Rolfe & A.B. Jackson* (S10: 371): Nicholson records this [*R. macrophyllus*] as abundant in Queen's Cottage grounds, but we have not come across any bush which can be assigned to any form of that variable species, nor can we find a dried specimen of it. The plant may possibly be included here under another name.
EXSICC.: **Kew Gardens.** Corner of Queen's Cottage Grounds near Isleworth Ferry Gate; vii 1878; *Nicholson* s.n. (ABD).

CURRENT STATUS: The specimen cited above, identified by Nicholson as *R. macrophyllus*, has been redetermined by Newton as *R. anglocandicans* (DEA: 33). No recent sightings.

628. *Rubus armeniacus Focke cv. **'Himalayan Giant'**
R. procerus auct. non P.J. Müll. ex Boulay
FIRST RECORD: 1963.
EXSICC.: **Kew Gardens.** Queen's Cottage Grounds; 30 viii 1963, *Neumann* R.42 (K).
CURRENT STATUS: This cultivated bramble is rapidly spreading in the Conservation Area and elsewhere and should now be considered a serious weed.

629. Rubus armipotens W.C.Barton ex A.Newton
R. godronii Lecoq & Lamotte
FIRST RECORD: 1908.
REFERENCES: **Kew Gardens.** 1909, *Rolfe & A.B.Jackson* (S10: 371 and see Exsicc.): Collected in Queen's Cottage grounds in August, 1908, and a few bushes seen again this year. Mr Rogers has confirmed the identification. There is a specimen in [K], collected in Queen's Cottage grounds in 1899 and left unnamed, which is certainly identical. • 1909, *Rolfe & A.B. Jackson* (S10: 370; SFS: 266 and see Exsicc.): Nicholson records this as here and there in Queen's Cottage grounds. We have only succeeded in finding a single bush on the towing-path side of the ha-ha, which Mr Rogers considers to be correctly named. We have not seen a specimen of Nicholson's plant [this under discussion of a plant identified as *R. dumnoniensis*, determined later by D.E.Allen as *R. armipotens*]. • 1909, *Rolfe & A.B.Jackson* (S10: 371 and see Exsicc.): Here and there in Queen's Cottage grounds. A piece collected in August, 1909, was sent to Mr Rogers, who marked it: "Probably *R. leucostachys* × *rusticanus*, though in panicle nearer to *R. leucostachys* than is usual in this hybrid" ... and as *R. rusticanus* is absent from the Queen's Cottage grounds we incline to call it simply *R. leucostachys*.
EXSICC.: **Kew Gardens.** Queen's Cottage Grounds; 4 viii 1908; *A.B.Jackson & Rolfe* s.n. (K* sub *argentatus*). • Towing path near Isleworth Gate to Kew Gardens; a single bush; 8 viii 1908; *A.B. Jackson & Rolfe* s.n. (K*). • Queen's Cottage Grounds; 30 vii 1963; *Neumann* R42 (K). • Queen's Cottage Grounds; 9 viii 1909; *A.B.Jackson & Rolfe* s.n. (K* sub *leucostachys*).
CURRENT STATUS: Still to be found in the Conservation Area.

630. Rubus hylophilus Rip. ex Genev.
R. carpinifolius sensu Rogers, non Weihe & Nees
R. platyacanthus auct. non P.J. Mueller & Lef.
FIRST RECORD: 1906.
NICHOLSON: **1906 revision** (AS: 78; SFS: 264): **Q.** "Form with exceptionally broad panicle and coarsely-toothed leaves."
OTHER REFERENCES: **Kew Gardens.** 1909, *Rolfe & A.B. Jackson* (S10: 370 and see Exsicc.): Several bushes found in Queen's Cottage grounds, and showing a certain amount of variation.
EXSICC.: **Kew Gardens.** Entrance gate to Queen's Cottage Grounds, Brentford end; 9 viii 1909; *A.B. Jackson & Rolfe* s.n. (K*). • Queen's Cottage Grounds; 30 vii 1963; *Neumann* R44-44a
CURRENT STATUS: Specimens filed under *R. platyacanthus* at K appear to be a mixture of *R. hylophilus* (det. D.E.Allen as cf. *R. hylophilus*) and *R. rhombifolius*. The species is still present in the Conservation Area.

631. Rubus ulmifolius Schott
R. rusticanus Mercier
FIRST RECORD: 1869 [1909].
NICHOLSON: **1873/4 survey:** Not recorded. **1906 revision** (AS: 78): **Strip.** Here and there by towing-path.
OTHER REFERENCES: **Kew Gardens.** 1909, *Rolfe & A.B. Jackson* (S10: 371): Nicholson records this as occurring here and there by towing-path, where only we have found it. The double-flowered variety (var. *flore pleno*) occurs near Queen's Cottage, but only as a cultivated plant.
EXSICC.: **Kew Gardens.** [Unlocalised]; 1 x 1869; *Baker* s.n. (K* sub *rusticanus*). • Pleasure Ground; xi 1872; *anon.* s.n. (K).
CULT.: 1909, unspec. (K* sub *rusticanus*). Not currently in cult.
CURRENT STATUS: No recent sightings.

632. Rubus adscitus Genev.
FIRST RECORD: 1908.
EXSICC.: **Kew Gardens.** Queen's Cottage Grounds; 4 viii 1908; *A.B.Jackson & Rolfe* s.n. (K*). • Queen's Cottage Grounds; 30 vii 1963, *Neumann* R43 (K).
CURRENT STATUS: Still in the Conservation Area and is relatively abundant. The species may include those specimens referred to by A.B.Jackson & Rolfe as *R. hypoleucus* (DEA: 33); see under *R. micans* for details.

633. Rubus mucronulatus Boreau
R. mucronatus Bloxam
FIRST RECORD: 1869.
NICHOLSON: **1906 revision** (AS: 78): **Q:** Not uncommon along borders of plantations.
OTHER REFERENCES: **Kew Gardens.** 1909, *Rolfe & A.B. Jackson* (S10: 371 and see Exsicc.): Nicholson records this as not uncommon along borders of plantations in Queen's Cottage grounds, where we have also gathered it. A piece collected in August, 1909, and forwarded to Mr Rogers was returned marked: "I suppose certainly a *mucronatus* form (?f. *umbrosa*) with foliage less hairy than usual, and exceptionally long petiolule to terminal leaflet." **Kew Environs.** 1878, *Nicholson* (BEC18: 15; SFS: 272 and see Exsicc.): Mr G.Nicholson

sends, from the neighbourhood of Kew, specimens of a London bramble which is regarded by Babington as a variety of *mucronulatus*. The true *mucronulatus* ... we do not get anywhere in the neighbourhood of London.

EXSICC.: **Kew Gardens.** [Unlocalised]; ix 1869; *Bloxam* s.n. (K*). • Queen's Cottage Grounds, near fence by pinetum; 9 viii 1909; *A.B.Jackson & Rolfe* s.n. (K*; D.E.Allen notes "certainly not *R. mucronulatus*"). • Queen's Cottage Grounds; 4 viii 1963; *Neumann* R51 (K). **Kew Environs.** Kew; viii 1878; *Nicholson* 429 (K*).

CURRENT STATUS: No recent sightings. The Neumann specimen cited was determined by Neumann as *R. mucronatus* but is filed at Kew under *R. mucronatiformis* (Sudre) W.C.R.Watson. The species would be anomalous in Kew Gardens.

634. Rubus echinatus × erythrops

R. echinatus × rosaceus?

FIRST RECORD: 1909.

REFERENCES: **Kew Gardens.** 1909, *Rolfe & A.B. Jackson* (S10: 373 and see Exsicc.): A fine clump of a *Rubus* was found in a thicket in Queen's Cottage grounds in August, 1908, and was collected as *R. fuscoater?* Mr Rogers replied: "Looks like a hybrid. Not *R. fuscoater*." This year we collected abundant material, and sent it with the record that it grew associated with *R. rudis, R. rosaceus, R. pulcherrimus,* and *R. echinatus*. Mr Rogers replied: "In all respects looks like a hybrid. The strong development of prickles and the barrenness of all the panicles especially point to such an origin, but I cannot suggest any parents." After a comparison with its associates we think *R. echinatus × rosaceus* the most likely combination. The resemblance to *R. rosaceus* is very marked, and except in its longer spines it may be described as fairly intermediate between the two.

EXSICC.: **Kew Gardens.** Queen's Cottage Grounds; 14 viii 1909; *A.B.Jackson & Rolfe* 483C (K).

CURRENT STATUS: Probably recorded in error; its presence is very doubtful (D.E.Allen, pers. comm.). The specimen cited is filed under *R. leptadenes × rosaceus* but labelled *R. echinatus × rosaceus*; it has not recently been redetermined. No recent sightings.

635. Rubus glareosus W.M.Rogers

R. hostilis sensu Salmon, non P.J.Müll. & Wirtg. ex Focke

R. rosaceus var. *hystrix* W.M.Rogers

FIRST RECORD: 1875.

NICHOLSON: **1873/4 survey**: Not recorded. • **1906 revision** (AS: 79; SFS: 276 [as pallidus], 277 and see Exsicc.): **Q.** Very common in shady places.

OTHER REFERENCES: **Kew Gardens.** 1909, *Rolfe & A.B. Jackson* (S10: 372 and see Exsicc.): Very common in shady places. Mr Rogers writes: "I know this form well, having seen a great deal of it since I first met with it in the Witley neighbourhood in 1890. It has long seemed to me curiously intermediate between *R. rosaceus* and *R. pallidus*, and I agree with Mr Marshall in considering it to be the common sand form of *R. rosaceus* throughout S.W. Surry ..." A very common and characteristic form at Kew, frequently with very narrow leaves. • 1931, *C.E.Salmon* (SFS: 277; S33: 746): Queen's Cottage Grounds, Kew, G.Nicholson. Requires confirmation.

EXSICC.: **Kew Gardens.** Queen's Cottage Grounds; 28 ix 1875; *Nicholson* 433 (K*). • Queen's Cottage Grounds; 29 vii 1908; *A.B.Jackson & Rolfe* s.n. (K*). • Queen's Gardens Grove; 27 vii 1908; *E.G.Gilbert* s.n. (K). • Queen's Cottage Grounds; 9 viii 1909; *A.B.Jackson & Rolfe* s.n. (K*). • Queen's Cottage Grounds; same as last but more in shade; 9 viii 1909; *A.B.Jackson & Rolfe* s.n. (K*). • Queen's Cottage Grounds; 30 vii 1963; *Neumann* R37 (K).

CURRENT STATUS: Still present in the Conservation Area. A specimen in SLBI is the basis for Salmon's record; (DEA: 34).

636. Rubus lintonii Focke ex Bab.

FIRST RECORD: 1897.

EXSICC.: **Kew Gardens.** [unlocalised]; 28 vi 1897 & 4 ix 1898; *anon.* (K).

CURRENT STATUS: No recent sightings.

637. Rubus micans Godr.

R. anglosaxonicus Gelert

R. hypoleucus nom. dub.

FIRST RECORD: 1909.

REFERENCES: **Kew Gardens.** 1909, *Rolfe & A.B. Jackson* (S10: 371; SFS: 271, 273 and see Exsicc.): Found in several spots in Queen's Cottage grounds in August, 1908. Mr Rogers remarks: "I think a highly glandular and aciculate form of this." An unusually vigorous form [of *R. anglosaxonicus*] was collected in the Queen's Cottage grounds in August, 1909, of which Mr Rogers remarks: "The panicle is rather suggestive of a hybrid origin, but there seems little room for doubt that the stem pieces belong to *R. anglosaxonicus*."

EXSICC.: **Kew Gardens.** Queen's Cottage Grounds; 9 viii 1909; *A.B.Jackson & Rolfe* s.n. (K* sub *R. anglosaxonicus*).

CURRENT STATUS: Very doubtful; the reference may, in part, be applicable to *R. adscitus* Genev. (D.E. Allen, pers. comm.), which is still to be found. A specimen collected by A.B.Jackson & Rolfe, referred to in the literature as *R. hypoleucus* is filed away at Kew under *R. adscitus* (q.v.). Plants recorded may simply be luxuriant forms of *R. rudis* (DEA: 34) but there is nothing recent.

638. Rubus moylei W.C.Barton & Ridd.

R. lejeunei var. *ericetorum* auct. non Lefèvre

FIRST RECORD: 1908.

REFERENCES: **Kew Gardens.** 1909, *Rolfe & A.B. Jackson* (S10: 372 and see Exsicc.): A specimen

collected in the Queen's Cottage grounds in August, 1908, was returned by Mr Rogers with the remark: "Looks to me nearer to *R. ericetorum* than to *R. mucronatus*, but it seems rather peculiar and not in a very good condition for determination."
EXSICC.: **Kew Gardens.** Queen's Cottage Grounds; 27 vii 1908; *A.B.Jackson & Rolfe* s.n. (K*). • [Unlocalised]; 27 vii 1910; *E.G.Gilbert* s.n. (K). • [Unlocalised]; 20 vii 1914; *E.G.Gilbert* s.n. (K).
CURRENT STATUS: Gilbert's unlocalised specimen is undoubtedly this species (DEA: 34) and it has recently been found on the towpath. No specimens have been found in the conservation Area.

639. Rubus infestus Weihe ex Boenn.
FIRST RECORD: 1910.
EXSICC.: **Kew Gardens.** [Unlocalised]; 27 vii 1910; *E.G.Gilbert* s.n. (K).
CURRENT STATUS: No recent sightings.

640. Rubus leyanus W.M.Rogers
FIRST RECORD: 2000.
REFERENCES: **Kew Gardens.** 2000, *Allen & Cope* (LN85: 34).
CURRENT STATUS: There is one patch in the Conservation Area. The species is previously unrecorded and likely to be a new arrival.

641. Rubus echinatus Lindl.
R. discerptus P.J.Müll.
FIRST RECORD: 1906 [1869].
NICHOLSON: **1906 revision** (AS: 79): **Q.** "May perhaps be a shade form of this species, but if so, very much off type."
OTHER REFERENCES: **Kew Gardens.** 1909, *Rolfe & A.B. Jackson* (S10: 372 and see Exsicc.): Fairly common and quite typical in Queen's Cottage grounds. One of our most distinct brambles and locally common.
EXSICC.: **Kew Gardens.** Grove in Queen's Cottage Gardens; 27 vii 1908; *E.G.Gilbert* s.n. (K). • Queen's Cottage Grounds; 4 viii 1908; *A.B.Jackson & Rolfe* s.n. (K*).
CULT.: 1869, unspec. (K). Not currently in cult.
CURRENT STATUS: No recent sightings. The Gilbert specimen cited is filed at Kew under *R. discerptus.*

642. Rubus flexuosus P.J.Müll. & Lefèvre
Rubus foliosus Weihe & Nees
FIRST RECORD: 1906.
NICHOLSON: **1906 revision** (AS: 79 and in SFS: 276): **Q.** Common in shade.
OTHER REFERENCES: **Kew Gardens.** 1909, *Rolfe & A.B. Jackson* (S10: 372 and see Exsicc.): Nicholson records this as common in shade in Queen's Cottage grounds, and there is a specimen in [K]. We found only a few examples of it in 1908 and none in 1909.
EXSICC.: **Kew Gardens.** Queen's Cottage Grounds; 4 viii 1908; *A.B.Jackson & Rolfe* s.n. (K*).
CURRENT STATUS: No recent sightings.

643. Rubus iodnephes W.C.R.Watson
FIRST RECORD: 1908.
REFERENCES: 1909, *Rolfe & A.B.Jackson* (S10: 372 and see Exsicc.): Queen's Cottage grounds. Nicholson records a form so named by Mr Rogers, of which there is a specimen in [K sub *podophyllus*]. We found an identical bush in August, 1908, and Mr Rogers felt some doubt about the correct identification.
EXSICC.: **Kew Gardens.** Queen's Cottage Grounds; in one or two spots; 4 viii 1908; *A.B.Jackson & Rolfe* s.n. (K* sub *podophyllus*). • Queen's Cottage Grounds; 30 vii 1963; *Neumann* R46 (K).
CURRENT STATUS: Still to be found in the Conservation Area but uncommon.

644. Rubus malvernicus Edees
?*R. norrefaetus* 'M.L.' nom. nud.
FIRST RECORD: 1963.
EXSICC.: **Kew Environs.** Kew; Thames bank; 8 vii 1963; *Neumann* R3 (K).
CURRENT STATUS: The name *R. norrefaetus* cannot be traced; the specimen is in a folder at Kew with another specimen which has been determined by D.E.Allen as *R. malvernicus*. The two specimens look very much alike. No recent sightings.

645. Rubus radula Weihe ex Boenn.
FIRST RECORD: 1866.
REFERENCES: **Kew Gardens.** 1909, *Rolfe & A.B. Jackson* (S10: 372 and see Exsicc.): A few bushes in the Queen's Cottage and Kew Palace grounds. In both cases Mr Rogers agreed as to the determination, though suggesting that the latter showed an approach to *R. anglosaxonicus raduloides*, in one direction and to *R. mutabilis*, in the other.
EXSICC.: **Kew Gardens.** Palace Grounds; 27 viii 1908; *A.B.Jackson & Rolfe* s.n. (K*). • Queen's Cottage Grounds, near Ferry gate; 14 viii 1909; *A.B.Jackson & Rolfe* s.n. (K*; D.E.Allen notes "not *radula*"). **Kew Environs.** Kew, 1866, *Baker* s.n. (OXF, fide D.E.Allen) • Kew Zone, 1867, *Baker* s.n. (CGE, fide D.E.Allen).
CURRENT STATUS: No recent sightings.

646. Rubus rudis Weihe
FIRST RECORD: 1906.
NICHOLSON: **1906 revision** (AS: 79 and in SFS: 274): **Q.** In open places, not uncommon.
OTHER REFERENCES: **Kew Gardens.** 1909, *Rolfe & A.B. Jackson* (S10: 372 and see Exsicc.): Common in open places in Queen's Cottage grounds, and found also on the strip by the side of the towing-path.
EXSICC.: **Kew Gardens.** Queen's Cottage Gardens grove; 27 vii 1908; *E.G.Gilbert* s.n. (K). • Queen's Gardens; shady thicket; 27 vii 1908; *E.G.Gilbert* s.n. (K). • Queen's Cottage Grounds, abundant; 4 viii 1908; *A.B.Jackson & Rolfe* s.n. (K*). • Queen's Cottage Grounds; ix 1910; *Rolfe* s.n. (K*). • Queen's

Cottage Grounds; 30 vii 1963; *Neumann* R36 (K). **Kew Environs.** Kew; 18 viii 1909; *E.G.Gilbert* s.n. (K). • Kew; 30 viii 1909; ?*E.G.Gilbert* s.n. (K).
CURRENT STATUS: Abundant in the Conservation Area.

647. ¶Rubus sectiramus W.C.R.Watson
FIRST RECORD: 1908.
EXSICC.: **Kew Gardens.** Queen's Cottage Grounds; 29 vii 1908; *A.B.Jackson & Rolfe* s.n. (K*). • Queen's Cottage Grounds; 30 vii 1963; *Neumann* R38 (K).
CURRENT STATUS: Common in the Conservation Area.

648. Rubus radulicaulis Sudre
FIRST RECORD: 1909.
EXSICC.: **Kew Gardens.** Grown in Kew Gardens; 21 vii 1909; *E.G.Gilbert* s.n. (K).
CURRENT STATUS: Although the label reads 'grown in Kew Gardens' it is unlikely that this plant was cultivated. No recent sightings.

649. Rubus spadix W.C.R.Watson
R. podophyllus sensu Rogers pro parte et auct., non P.J. Müll.
FIRST RECORD: 1906.
NICHOLSON: **1906 revision** (AS: 79): **Q.** Or form between it and *R. oligocladus* [nom. dub.].
CURRENT STATUS: No recent sightings.

650. Rubus bercheriensis (Druce ex Rogers) Rogers
FIRST RECORD: 1910.
EXSICC.: **Kew Gardens.** Queen's Cottage Grove; 20 vii 1910; *E.G.Gilbert* s.n. (K).
CURRENT STATUS: The specimen above was initially determined as *R. rosaceus* and then redetermined as *R. rudis*, before finally being filed away under *R. bercheriensis*. No recent sightings.

651. Rubus dasyphyllus (W.M.Rogers) E.Marshall
FIRST RECORD: 1963.
EXSICC.: **Kew Gardens.** Queen's Cottage Grounds; 30 vii 1963; *Neumann* R45 (K). • Queen's Cottage Grounds; 30 vii 1963; *Neumann* R50 (K).
CURRENT STATUS: No recent sightings.

652. Rubus phaeocarpus W.C.R.Watson
R. babingtonii auct. non T.B.Salter
FIRST RECORD: 1914.
EXSICC.: **Kew Gardens.** [Unlocalised]; 14 vii 1914; *E.G.Gilbert* s.n. (K).
CURRENT STATUS: The specimen was identified as *R. babingtonii* but is filed under *R. phaeocarpus*. No recent sightings.

653. Rubus pedemontanus Pinkw.
R. bellardii auct. non Weihe & Nees
FIRST RECORD: 1866.
EXSICC.: **Kew Gardens.** [Unlocalised]; viii 1866; *Baker* s.n. (K*).
CURRENT STATUS: The specimen is filed under *R. bellardii* but this name does not apparently apply to any British species. No recent sightings.

654. Rubus britannicus W.M.Rogers
FIRST RECORD: 1963.
EXSICC.: **Kew Gardens.** Queen's Cottage Grounds; 30 vii 1963, *Neumann* R39 (K).
CURRENT STATUS: Abundant in the Conservation Area, along with a specimen that may be a hybrid with *R. adscitus* (DEA: 35).

655. Rubus nemorosus Hayne & Willd.
R. balfourianus Bloxam ex Bab.
FIRST RECORD: 1866 [1869].
NICHOLSON: **1873/4 survey**: Not recorded. • **1906 revision**: Not recorded.
OTHER REFERENCES: **Kew Environs.** 1877, *Baker* (BEC18: 6; S33: 745): Kew, Surrey [as "*R. tuberculatus*, Bab."]. Not my plant, nor much like it. It is very like *R. balfourianus*, especially resembling a plant so named by me in the Herb. Borrer from Eridge Wood, near Tonbridge Wells; indeed almost the only difference is found in the more furrowed stem of the Kew plant now issued. The lower part of the stem of typical *R. balfourianus* is not furrowed, although its upper part often is so. I have never seen it with so furrowed a stem as in this from Kew. – C.C.Babington.
EXSICC.: **Kew Gardens.** [Unlocalised]; viii 1866; ?*Baker* s.n. (K*).
CULT.: 1869, unspec. (K). Not currently in cult.
CURRENT STATUS: The plant may well have been *R. tuberculatus* Bab. after all (D.E.Allen, pers. comm.), which is present in the Gardens. Nicholson probably ignored the species because the Baker specimen may well have been from a cultivated plant.

656. Rubus pruinosus Arrh.
?*R. corylifolius* var. *degener* P.J.Müll.
FIRST RECORD: 1878.
REFERENCES: **Kew Environs.** 1878, *Baker* (BEC18: 16; S33: 745): Under this name [var. *degener*], for which I am indebted to Genevier, I have distributed a few specimens, from hedges at Kew, of a bramble that comes in between *Balfourianus* and *corylifolius* var. *intermedius*.
CURRENT STATUS: A probable misidentification (D.E. Allen, pers. comm.). Nicholson might have ignored the Baker specimen because it may well have come from a cultivated plant. No recent sightings.

[**657. Rubus corylifolius** Sm., sens. lat.
CULT.: 1811, Nat. of Britain (HK2, 3: 268); 1814, Nat. of Britain (HKE: 161); 1869, unspec. (K*).
Note. The species is not accepted in the strict sense for the British Isles. The sheet in question has not been redetermined recently and could be one of a number of segregates from the complex; the specimens cited in *Hortus Kewensis* are of unknown identity.]

658. Rubus tuberculatus Bab.
R. dumetorum var. *diversifolius* Warren
FIRST RECORD: 1869.
NICHOLSON: **1873/4 survey**: Not recorded. • **1906 revision**: Not recorded.
EXSICC.: **Kew Gardens.** Hort. Kew; ix 1869; *Baker* s.n. (K*).
CURRENT STATUS: One clump by the Larch Pond in the Conservation Area. The Baker specimen was probably ignored by Nicholson because it might well have been from a cultivated plant.

659. Rubus oigocladus P.J.Müll. & Lefèvre
FIRST RECORD: 1908.
EXSICC.: **Kew Gardens.** Queen's Cottage Grounds; at margin of wood; 27 vii 1908; *E.G.Gilbert* s.n. (K).
CURRENT STATUS: The name *R. oigocladus* is placed at the end of the bramble monograph by Edees & Newton amongst names of unknown application.

660. Rubus scabrosus P.J.Müll.
R. dumetorum var. *ferox* auct.
FIRST RECORD: 1866.
NICHOLSON: **1873/4 survey**: Not recorded. • **1906 revision**: Not recorded.
EXSICC.: **Kew Gardens.** Hort. Kew; viii 1866; *Baker* s.n. (K*). • Hort. Kew; ix 1869; *Baker* s.n. (K*).
CURRENT STATUS: The name *R. scabrosus* is placed at the end of the bramble monograph by Edees & Newton amongst names of unknown application.

661. Rubus wahlbergii Arrh.
FIRST RECORD: 1869.
NICHOLSON: **1873/4 survey**: Not recorded. • **1906 revision**: Not recorded.
EXSICC.: **Kew Gardens.** [Unlocalised]; ix 1869; *Baker* s.n. (K*).
CURRENT STATUS: The name *R. wahlbergii* is placed at the end of the bramble monograph by Edees & Newton amongst names of unknown application. The specimen itself is labelled *R. corylifolius* var. *wahlbergii*.

662. Rubus caesius L.
dewberry
FIRST RECORD: 1906 [1768].
NICHOLSON: **1906 revision** (AS: 79): **Strip:** On side of towing-path.
OTHER REFERENCES: **Kew Gardens.** 1909, *Rolfe & A.B. Jackson* (S10: 372): Nicholson records this as found on the strip by the side of the towing path, where only we have found it. • 1999, *Stones* (EAK: 2): By Lake.
CULT.: 1768 (JH: 216, 454); 1789, Nat. of Britain (HK1, 2: 209); 1811, Nat. of Britain (HK2, 3: 268); 1814, Nat. of Britain (HKE: 161); 1869, unspec. (K*). Not currently in cult.
CURRENT STATUS: No recent sightings.

663. Potentilla fruticosa L.
shrubby cinquefoil
CULT.: 1768 (JH: 185); 1789, Nat. of England (HK1, 2: 212); 1811, Nat. of England (HK2, 3: 273); 1814, Nat. of England (HKE: 162); 1882, unspec. (K); 1891, unspec. (K); 1914, Arboretum (K); 1915, Arboretum (K); 1929, Arboretum (K); 1933, Arboretum (K); 1940, unspec. (K); 1995, plot 154 (K). In cult.

664. Potentilla palustris (L.) Scop.
Comarum palustre L.
marsh cinquefoil
CULT.: 1768 (JH: 186); 1789, Nat. of Britain (HK1, 2: 219); 1811, Nat. of Britain (HK2, 3: 282); 1814, Nat. of Britain (HKE: 163); 1954, Water Garden (K). In cult.

665. Potentilla anserina L.
silverweed
FIRST RECORD: 1873/4 [1768].
NICHOLSON: **1873/4 survey** (JB: 44): **B.** A weed in shrubberies. **Strip.** Common. • **1906 revision** (AS: 79): **B.** A weed in shrubberies. **Strip.** Common.
EXSICC.: **Kew Environs.** Tow-path, Kew to Richmond; 22 vii 1915; *Britton* s.n. (K). • Towpath between Kew & Richmond; in moist positions, creeping extensively; 23 v 1944; *Souster* 71A (K).
CULT.: 1768 (JH: 185); 1789, Nat. of Britain (HK1, 2: 213); 1811, Nat. of Britain (HK2, 3: 273); 1814, Nat. of Britain (HKE: 162); 1933, Arboretum (K); 1936, Herb Gdn. (K); 1977, unspec. (K). Not currently in cult.
CURRENT STATUS: Invading the lawns around the Order Beds in which it has been cultivated, otherwise it has not recently been seen as a truly wild plant.

Silverweed (*Potentilla anserina*). Once common in the area but most wild plants have disappeared. Those now seen within the Gardens have escaped from the Order Beds.

666. Potentilla rupestris L.
rock cinquefoil
CULT.: 1768 (JH: 185); 1789, Nat. of England (HK1, 2: 213); 1811, Nat. of England (HK2, 3: 274); 1814, Nat. of England (HKE: 162); 1930, unspec. (K); 1940, Herbaceous Ground (K); 1954, Chalk Garden (K); 1977, unspec. (K). In cult.

667. Potentilla argentea L.
hoary cinquefoil
FIRST RECORD: 1873/4 [1768].
NICHOLSON: **1873/4 survey** (JB: 44): **P.** Two plants in middle of Pagoda Avenue, about 50 yards from the gate leading into Bot. Gard. One plant near north wing of Winter Garden. **Strip.** Two or three about 100 yards north of Isleworth Gate. • **1906 revision** (AS: 79 and in SFS: 285): Pagoda vista, near temperate house. **Strip.**
CULT.: 1768 (JH: 185); 1789, Nat. of Britain (HK1, 2: 215); 1811, Nat. of Britain (HK2, 3: 276); 1814, Nat. of Britain (HKE: 162); 1895, unspec. (K); 1899, unspec. (K); 1923, Herb. Expt. Ground (K); 1925, Herb. Expt. Ground (K); 1926, Herb. Expt. Ground (K); 1927, Herb. Expt. Ground (K); 1928, Herb. Expt. Ground (K); 1939, Herbarium Ground (K). In cult.
CURRENT STATUS: Rare; there have been a couple of records from the northeast part of the Gardens but otherwise it seems to have declined. Plants were once regularly seen in mown grass on the slope up to the roof of the Banks Building, but once the mowing stopped because of building works they disappeared.

668. *Potentilla inclinata Vill.
grey cinquefoil
FIRST RECORD: 1836 [1931].
EXSICC.: **Kew Environs.** Kew Green; 1836; *anon.* s.n. (K).
CULT.: 1931, unspec. (K). Not currently in cult.
CURRENT STATUS: No recent sightings. It was presumably a garden escape on Kew Green though not necessarily from Kew Gardens.

669. *Potentilla recta L.
P. hirta auct. non L.
sulphur cinquefoil
FIRST RECORD: 1866 [1768].
REFERENCES: **Kew Environs.** 1866, *Baker* (LBEC9: 15; BSEC9: 687; SFS: 286; S33: 746, and see Exsicc.): Kew Green, subspontaneous.
EXSICC.: **Kew Environs.** Subspontaneous on Kew Green; viii 1866, *Baker* in *Herb. Watson* (K).
CULT.: 1768 (JH: 185); 1789, Alien (HK1, 2: 214); 1811, Alien (HK2, 3: 275); 1814, Alien (HKE: 162); 1883, unspec. (K); 1895, unspec. (K); 1924, unspec. (K); 1925, Expt. Grd. (K); 1930, unspec. (K); 1931, unspec. (K); 1976, unspec. (K). Not currently in cult.
CURRENT STATUS: Extinct on Kew Green, where it must have been an escape from nearby gardens, but recently seen in the Director's Garden where it would have been been cultivated. It was planted in the Robinsonian Meadow in front of 'Climbers & Creepers' but seems to have gone from there now. A single plant is currently known from a meadow near the Pagoda

[670. Potentilla intermedia L.
P. opaca auct. non L.
Russian cinquefoil
CULT.: 1789, Nat. of Scotland (HK1, 3: 493); 1811, Alien (HK2, 3: 276); 1814, Alien (HKE: 162).
Note. Originally listed as native, but this was corrected in later editions.]

671. Potentilla crantzii (Crantz) Beck & Fritsch
P. aurea auct. non L.
alpine cinquefoil
CULT.: 1789, Alien (HK1, 2: 215); 1811, Nat. of Scotland (HK2, 3: 277); 1814, Nat. of Scotland (HKE: 162); 1969, Herbaceous Dept. (K). In cult.

672. Potentilla neumanniana Rchb.
P. verna auct. non L.
spring cinquefoil
CULT.: 1768 (JH: 185); 1789, Nat. of Britain (HK1, 2: 215); 1811, Nat. of Britain (HK2, 3: 277); 1814, Nat. of Britain (HKE: 162). In cult. (2006). Not currently in cult.

673. Potentilla erecta (L.) Raeusch.
Tormentilla erecta L.
tormentil
CULT.: 1768 (JH: 172/6); 1789, Nat. of Britain (HK1, 2: 217); 1811, Nat. of Britain (HK2, 3: 279); 1814, Nat. of Britain (HKE: 162); 1933, unspec. (K); 1976, unspec. (K). Not currently in cult.

674. Potentilla anglica Laichard.
P. procumbens Sibth., nom. illeg.
Tormentilla reptans auct. non L.
trailing tormentil
FIRST RECORD: 1873/4 [1768].
NICHOLSON: **1873/4 survey** (JB: 43 and in SFS: 283): Common, except in division **B.** • **1906 revision** (AS: 79): Common, except in division **B.**
CULT.: 1768 (JH: 172/6); 1811, Nat. of Britain (HK2, 3: 279); 1814, Nat. of Britain (HKE: 162); c. 1887, unspec. (K). Not currently in cult.
CURRENT STATUS: No recent sightings.

675. Potentilla reptans L.
creeping cinquefoil
FIRST RECORD: 1873/4 [1768].
NICHOLSON: **1873/4 survey** (JB: 43): Everywhere. Very frequent in the open turf, particularly about wood and lake in Pleasure Grounds. • **1906 revision**

(AS: 79): Everywhere. Frequent in the open turf, particularly about wood and lake.
EXSICC.: **Kew Gardens.** Herbarium Experimental Ground; in thick stand of *Aegopodium*, drawn up & without runners; 5 vi 1930; *Summerhayes* 469 (K).
CULT.: 1789, Nat. of Britain (HK1, 2: 216); 1811, Nat. of Britain (HK2, 3: 278); 1814, Nat. of Britain (HKE: 162). In cult.
CURRENT STATUS: A significant component of short grass throughout the Gardens.

676. Potentilla sterilis (L.) Garcke
P. fragariastrum Pers.
Fragaria sterilis L.
barren strawberry
FIRST RECORD: 1873/4 [1768].
NICHOLSON: **1873/4 survey** (JB: 43): **Pal.** Common in turf in shady places near Palace. • **1906 revision** (AS: 79): Common in turf near palace.
CULT.: 1768 (JH: 216); 1789, Nat. of Britain (HK1, 2: 212); 1811, Nat. of Britain (HK2, 3: 273); 1814, Nat. of Britain (HKE: 162); c. 1887, unspec. (K). Not currently in cult.
CURRENT STATUS: No recent sightings.

[**677. Potentilla alba** L.
white cinquefoil
CULT.: 1789, Nat. of Wales (HK1, 2: 215); 1811, Nat. of Wales (HK2, 3: 277); 1814, Nat. of Wales (HKE: 162).
Note. Listed in error as native.]

[**678. Potentilla tridentata** Aiton
three-toothed cinquefoil
CULT.: 1789, Alien (HK1, 2: 216); 1811, Alien (HK2, 3: 279); 1814, Nat. of Scotland (HKE: 162).
Note. Listed in error as native.]

679. Sibbaldia procumbens L.
Sibbaldia
CULT.: 1768 (JH: 184); 1789, Nat. of Britain (HK1, 1: 398); 1811, Nat. of Britain (HK2, 2: 199); 1814, Nat. of Britain (HKE: 85); 1981, unspec. (K). Not currently in cult.

680. Fragaria vesca L.
wild strawberry
FIRST RECORD: 1998 [1768].
REFERENCES: **Kew Gardens.** 1998, *Cope* (S33: 746): North Arb.: behind Banks Building (108); Herbaceous: Duke's Garden (135); North Arb.: behind Water Lily House (163).
CULT.: 1768 (JH: 216); 1789, Nat. of Britain (HK1, 2: 211); 1811, Nat. of Britain (HK2, 3: 271); 1814, Nat. of Britain (HKE: 161); 1932, Arboretum (K); 1934, unspec. (K); 1940, Herbaceous Ground (K). In cult.
CURRENT STATUS: Rare and declining; recently found only behind the Banks Building.

[**681. Fragaria moschata** (Duchesne) Duchesne
F. elatior Ehrh., nom. illegit.
hautbois strawberry
CULT.: 1811, Nat. of England (HK2, 3: 271); 1814, Nat. of England (HKE: 161).
Note. Listed in error as native.]

682. *Fragaria ananassa (Duchesne) Duchesne
garden strawberry
FIRST RECORD: 1983/4.
REFERENCES: **Kew Gardens.** 1983/4, *Cope* (S31: 181): New student vegetable plots behind Hanover House.
CURRENT STATUS: Has not reappeared behind Hanover House since 1984.

683. *Duchesnea indica (Jacks.) Focke
yellow-flowered strawberry
FIRST RECORD: 2001 [1919].
REFERENCES: **Kew Gardens.** 2001, *Cope* (S35: 647): Naturalised in short grass by the Lake (262). Subsequently found in numerous other places throughout the Gardens; easily overlooked as *Potentilla reptans*.
CULT.: 1919, unspec. (K); 1929, unspec. (K); 1931, unspec. (K); 1940, Herbaceous Ground (K); 1951, Herbarium Ground (K); 1953, Herb. Expt. Grnd. (K). In cult. under glass.
CURRENT STATUS: Established in grass in several places in the Gardens, but it is easily mistaken for a *Potentilla* when not in flower and easily overlooked.

684. Geum rivale L.
water avens
CULT.: 1768 (JH: 185); 1789, Nat. of Britain (HK1, 2: 218); 1811, Nat. of Britain (HK2, 3: 280); 1814, Nat. of Britain (HKE: 163); pre-1920, unspec. (K). In cult.

685. Geum × intermedium Ehrh. (rivale × urbanum)
CULT.: 1912, unspec. (K); 1916, unspec. (K). Not currently in cult.

686. Geum urbanum L.
wood avens
FIRST RECORD: 1873/4 [1768].
NICHOLSON: **1873/4 survey** (JB: 44): **B.** A flower-bed weed. **P.** Here and there in shrubberies. **Strip.** By towing path. • **1906 revision** (AS: 79): **B.** A flower-bed weed. **A.** Here and there in shrubberies. **Strip.** By towing path.
CULT.: 1768 (JH: 185); 1789, Nat. of Britain (HK1, 2: 218); 1811, Nat. of Britain (HK2, 3: 280); 1814, Nat. of Britain (HKE: 163); 1933, Arboretum (K); 1936, Herbaceous Ground (K); 1936, Herb Gdn. (K). In cult.
CURRENT STATUS: Widespread but not very common. Easily distributed unwittingly because of the hooked achenes.

687. Dryas octopetala L.

mountain avens

Cult.: 1768 (JH: 226); 1789, Nat. of Scotland (HK1, 2: 219); 1811, Nat. of Britain (HK2, 3: 281); 1814, Nat. of Britain (HKE: 163). In cult.

688. Agrimonia eupatoria L.

agrimony

First record: 1873/4 [1768].
Nicholson: **1873/4 survey** (JB: 43): Uncommon. A very few plants exist in each division. • **1906 revision** (AS: 79): Uncommon. A few plants in each division.
Cult.: 1768 (JH: 183); 1789, Nat. of Britain (HK1, 2: 129); 1811, Nat. of Britain (HK2, 3: 151); 1814, Nat. of Britain (HKE: 141); 1933, unspec. (K); 1936, Herb Gdn. (K); 1937, Herb Garden (K). In cult.
Current status: Rare as a wild plant and only in the Conservation Area; cultivated elsewhere.

689. Agrimonia procera Wallr.
A. odorata auct. non (L.) Mill.

fragrant agrimony

Cult.: 1789, Alien (HK1, 2: 130); 1811, Alien (HK2, 3: 152); 1814, Alien (HKE: 141). Not currently in cult.
Note. Not discovered in Britain until 1853.

690. *Aremonia agrimonioides (L.) DC.
Agrimonia agrimonioides L.

bastard agrimony

First record: 2001 [1768].
References: **Kew Environs.** 2001, *Sheahan* (LN86: 179): St. Anne's churchyard. • 2006, *Joint LNHS/BSBI meeting* (LN86: 179): Thames path outside Kew Gardens.
Cult.: 1768 (JH: 184); 1789, Nat. of Italy (HK1, 2: 131); 1811, Nat. of Italy (HK2, 3: 152); 1814, Nat. of Italy (HKE: 141). No currently in cult.
Current status: Rare; just the two sightings listed above.

691. Sanguisorba officinalis L.

great burnet

Cult.: 1789, Nat. of Britain (HK1, 1: 155); 1810, Nat. of Britain (HK2, 1: 257); 1814, Nat. of Britain (HKE: 35); 1933, Arboretum (K); 1934, Arboretum (K); 1936, Herb Gdn. (K); 1954, Herb. Expt. Grnd (K); 1954, Order Beds (K). Not currently in cult.

692. Sanguisorba minor Scop.
Poterium sanguisorba L.
a. subsp. **minor**

salad burnet

First record: 1873/4 [1789].
Nicholson: **1873/4 survey** (JB: 43 and in SFS: 289): **Strip.** Some few score tufts in the turf about midway between Brentford and Isleworth Gates. • **1906 revision** (AS: 79): **Strip.** A few score tufts in the turf midway between Brentford and Isleworth Gates.
Exsicc.: **Kew Gardens.** Queens Cottage Grounds; [c. 1914]; *Flippance* s.n. (K).
Cult.: 1789, Nat. of England (HK1, 3: 353); 1813, Nat. of England (HK2, 5: 286); 1814, Nat. of England (HKE: 294); 1935, unspec. (K); 1936, Herbarium Ground (K). In cult.
Current status: One recent sighting in a meadow near the Pagoda.

b. subsp. **muricata** (Gremli) Briq.

fodder burnet

Cult.: 1915, unspec. (K). Not currently in cult.

693. *Acaena novae-zelandiae Kirk

pirri-pirri-bur

First record: 1999 [1897].
References: **Kew Gardens.** Bare patch in a lawn near the Temperate House (414). Also naturalised around the Princess of Wales Conservatory, by the Lake and around the Administration Block, 1999, *Cope* (S35: 647).
Cult.: 1897, unspec. (K); c. 1972, unspec. (K). In cult.
Current status: Naturalised in a number of places, especially by the Princess of Wales Conservatory and the Rockery. It appeared in the Paddock in 2004 from seed in topsoil stored there. Several other species of *Acaena* are cultivated but this seems to be the only one that has escaped.

694. Alchemilla alpina L.

alpine lady's-mantle

Cult.: 1768 (JH: 373, 388); 1789, Nat. of Britain (HK1, 1: 167); 1810, Nat. of Britain (HK2, 1: 274); 1814, Nat. of Britain (HKE: 37). In cult.

695–701. Alchemilla vulgaris L. agg.

lady's-mantle

Indet exsicc.: **Kew Environs.** Bowling Green, Kew; 4 vi 1944, *Hutchinson* s.n. (K).
Indet cult.: 1768 (JH: 373, 388); 1789, Nat. of Britain (HK1, 1: 166); 1810, Nat. of Britain (HK2, 1: 274); 1814, Nat. of Britain (HKE: 37); 1933, Herb. Ground (K); 1936, Herb Garden (K). In cult.

695. Alchemilla glaucescens Wallr.
A. hybrida (L.) Mill.
A. vulgaris var. *hybrida* auct. dub.
Cult.: 1768 (JH: 373); 1789, Nat. of Britain (HK1, 1: 166). Not currently in cult.

696. Alchemilla acutiloba Buser
Cult.: 1931, unspec. (K). Not currently in cult.

697. Alchemilla micans Buser
CULT.: In cult.

698. Alchemilla xanthochlora Rothm.
CULT.: In cult.

699. Alchemilla filicaulis subsp. **vestita** (Buser) M.E.Bradshaw
FIRST RECORD: 1946.
REFERENCES: **Kew Environs.** 1946, *Welch* (LN31: S103; S33: 746): By towpath, Kew Gardens, one plant, adventive.
CURRENT STATUS: No recent sightings. The towpath has been much modified since the 1940s and the plant has almost certainly been lost.

700. Alchemilla wichurae (Buser) Stefánsson
CULT.: In cult.

701. *Alchemilla mollis (Buser) Rothm.
FIRST RECORD: 2001.
REFERENCES: **Kew Gardens.** 2001, *Cope* (S35: 647): Naturalised in grass near the Rhododendron Dell whence it has probably escaped.
CULT.: In cult.
CURRENT STATUS: Naturalised in several places in and around the Rhododendron Dell where it is cultivated.

702. Aphanes arvensis L.
Alchemilla arvensis (L.) Scop.
A. aphanes Leers
parsley-piert
FIRST RECORD: 1873/4 [1768].
NICHOLSON: **1873/4 survey** (JB: 43): **P.** Plentiful along top of wall facing river and as a weed in beds containing newly-planted oak collections. Common in bad places in the turf. **Strip.** Abundant by side of towing-path. • **1906 revision** (AS: 79): **P, A, B.** Common where turf is poor. **Strip.** Abundant on towing-path.
EXSICC.: **Kew Gardens.** W. of Temperate house; small weed in poor lawn; 25 iv 1944; *Souster* 37 (K). **Kew Environs.** Riverside, Kew; 5 v 1894; *Hosking* s.n. (K).
CULT.: 1768 (JH: 374); 1789, Nat. of Britain (HK1, 1: 167); 1810, Nat. of Britain (HK2, 1: 274); 1814, Nat. of Britain (HKE: 37). Not currently in cult.
CURRENT STATUS: Not recently recorded; all occurrences of the genus now seem to be the following species. Either the identity of Nicholson's plants was incorrect or there has been a change in soil chemistry since the end of flooding by the river that now favours the more calcifuge of the two species. Since Nicholson left no vouchers it is impossible to check his records.

703. Aphanes australis Rydb.
slender parsely-piert
FIRST RECORD: 2001.
REFERENCES: **Kew Gardens.** 2001, *Cope* (S35: 647): In short grass between the Broadwalk and the Princess of Wales Conservatory. Subsequently found near the Evolution House (417).
CURRENT STATUS: Widespread and often in abundance where it does occur. It grows in quantity on Kew Green, opposite the Herbarium entrance, intermixed with *Ornithopus* and *Montia*.

704. Rosa arvensis Huds.
field-rose
CULT.: 1789, Nat. of Britain (HK1, 2: 202); 1811, Nat. of Britain (HK2, 3: 259); 1814, Nat. of Britain (HKE: 158). In cult.

705. Rosa × verticillacantha Mérat (arvensis × canina)
R. × kosinskiana Besser
FIRST RECORD: 1906.
NICHOLSON: **1906 revision** (AS: 79): In the Queen's Cottage grounds in a wild state.
CURRENT STATUS: A single bush is known from the towpath south of Brentford Gate.

Slender parsely-piert (*Aphanes australis*). Characteristic of acid grassland but small and easily overlooked. Readily seen on Kew Green opposite the Herbarium.

706. *Rosa × alba L. (arvensis × gallica?)
R. × collina Jacq. non Woods
white rose of York
FIRST RECORD: 1881 [1768].
NICHOLSON: **1906 revision**: Not recorded.
OTHER REFERENCES: **Kew Gardens.** 1881, *Nicholson* (BECB1: 71; S33: 746): This rose occurs along the edges of shrubberies in the Queen's Cottage grounds at Kew, in company with many forms of *R. canina* ... At present it is impossible to say whether [it] is really a truly wild plant or not. — G.Nicholson. This is close upon what I understand is the true *R. collina*. — J.G.Baker. Dr A.P.Winslow, a Swedish authority on *Rosa*, also looks upon this as good *R. collina*. — G.Nicholson.
CULT.: 1768 (JH: 454); 1811, Nat. of Britain (HK2, 3: 266); 1814, Nat. of Britain (HKE: 160). Not currently in cult.
CURRENT STATUS: No recent sightings.

707. Rosa pimpinellifolia L.
R. rubella Sm.
R. spinosissima L.
burnet rose
FIRST RECORD: 1866 [1768].
NICHOLSON: **1873/4 survey**: Not recorded. • **1906 revision**: Not recorded.
EXSICC.: **Kew Gardens.** Hort. Kew; vi 1866; *Baker* s.n. (K). • Hort. Kew; 28 viii 1866; *Baker* s.n. (K).
CULT.: 1768 (JH: 453, 454); 1789, Nat. of Britain (HK1, 2: 203); 1811, Nat. of Britain (HK2, 3: 259); 1814, Nat. of Britain (HKE: 158), Nat. of England (ibid.: Add. as *rubella*). In cult.
CURRENT STATUS: The name used on one of the herbarium specimens, *R. rubella*, applies to a variety with red fruits which may, in fact, be a hybrid between *R. pimpinellifolia* and another species. Plants still growing on Mount Pleasant have the habit of burnet rose, but are clearly not this species; they are remnant suckers of former cultivated roses but may be the source of records of burnet rose.

708. Rosa × involuta Sm. (pimpinellifolia × sherardii)
CULT.: 1811, Nat. of the Hebrides (HK2, 3: 260); 1814, Nat. of the Hebrides (HKE: 158). Not currently in cult.

709. Rosa canina L.
R. lutetiana Lem.
R. sphaerica Gren.
R. surculosa Woods
R. canina var. *lutetiana* Baker
R. canina var. *sphaerica* (Gren.) Dumort.
R. canina var. *surculosa* (Woods) Hook.
dog-rose
FIRST RECORD: 1906 [1768].
NICHOLSON: **1906 revision** (AS: 79 and in SFS: 292, 293, 296): In the Queen's Cottage grounds in a wild state.
CULT.: 1768 (JH: 454); 1789, Nat. of Britain (HK1, 2: 208); 1811, Nat. of Britain (HK2, 3: 265); 1814, Nat. of Britain (HKE: 160). In cult.
CURRENT STATUS: A single bush on the towpath south of Brentford Gate; all those within the Gardens are cultivated.

710. Rosa × dumetorum Thuill. (canina × obtusifolia)
CULT.: 1814, Nat. of England (HKE: Add.). Not currently in cult.

711. Rosa × hibernica Templeton (canina × pimpinellifolia)
FIRST RECORD: 1866 [1811].
NICHOLSON: **1873/4 survey**: Not recorded. • **1906 revision**: Not recorded.
EXSICC.: **Kew Gardens.** Hort. Kew; vi 1866; *Baker* s.n. (K). • Hort. Kew; viii 1866; *Baker* s.n. (K). • Cult. Hort. Kew; viii 1869, *Baker* in *Herb. Watson* (K).
CULT.: 1811, Nat. of Ireland (HK2, 3: 261); 1814, Nat. of Ireland (HKE: 159). Not currently in cult.
CURRENT STATUS: No recent sightings; it probably did not occur as a wild plant when collected by Baker.

712. Rosa × andegavensis Bastard (canina × stylosa)
FIRST RECORD: 2003.
CURRENT STATUS: A single bush is known from the Conservation Area.

713. Rosa caesia Sm.
R. coriifolia Fries
hairy dog-rose
FIRST RECORD: 1906 [1814].
NICHOLSON: **1906 revision** (AS: 79): In the Queen's Cottage grounds in a wild state.
EXSICC.: **Kew Gardens.** Cult. Hort. Kew [as *R. caesia*]; 2 viii 1867; *Baker* in *Herb. Watson* (K). • Cult. Hort. Kew [as *R. coriifolia*]; 2 viii 1867; *Baker* in *Herb. Watson* (K).
CULT.: 1814, Nat. of Scotland (HKE: Add.). Not currently in cult.
CURRENT STATUS: No recent sightings; the plants recorded by Nicholson may have escaped from cultivation.

714. Rosa × dumalis Bechst. (caesia × canina)
R. canina L. var. *dumalis* Dumort.
FIRST RECORD: 1906.
NICHOLSON: **1906 revision** (AS: 79 and in SFS: 294): In the Queen's Cottage grounds in a wild state.
CURRENT STATUS: No recent sightings.

715. Rosa obtusifolia Desv.
round-leaved dog-rose
FIRST RECORD: 1906.
NICHOLSON: **1906 revision** (AS: 79): In the Queen's Cottage grounds in a wild state.

EXSICC.: **Kew Gardens.** Thames bank, Kew – Richmond; highest zone amongst shrubs, near Isleworth Gate; 16 viii 1928; *Summerhayes & Turrill* s.n. (K).
CURRENT STATUS: A single bush is known from the Conservation Area.

716. Rosa tomentosa Sm.
R. scabriuscula auct. non Sm.
harsh downy-rose
CULT.: 1811, Nat. of England (HK2, 3: 266); 1814, Nat. of England (HKE: 160 & Add.). In cult.

717. Rosa sherardii Davies
Sherard's downy-rose
Cult.: In cult.

718. Rosa mollis Sm.
R. villosa auct. non L.
soft downy-rose
CULT.: 1768 (JH: 454); 1789, Nat. of Britain (HK1, 2: 203); 1811, Nat. of Britain (HK2, 3: 260); 1814, Nat. of Britain (HKE: 159 & Add.). Not currently in cult.

719. Rosa rubiginosa L.
R. eglanteria Mill.
sweet-briar
CULT.: 1768 (JH: 454); 1789, Nat. of Britain (HK1, 2: 206); 1811, Nat. of Britain (HK2, 3: 264); 1814, Nat. of Britain (HKE: 160). In cult.

720. Rosa micrantha Borrer ex Sm.
small-flowered sweet-briar
CULT.: 1814, Nat. of Britain (HKE: Add.). Not currently in cult.

721. Rosa agrestis Savi
small-leaved sweet-briar
CULT.: In cult.

722. Prunus spinosa L.
blackthorn
CULT.: 1789, Nat. of Britain (HK1, 2: 166); 1811, Nat. of Britain (HK2, 3: 199); 1814, Nat. of Britain (HKE: 149); 1880, Arboretum (K); 1900, Arboretum (K); 1905, Arboretum (K); 1907, Arboretum (K); 1917, Arboretum (K); 1931, Arboretum (K); 1935, unspec. (K); 1936, Herbarium Ground (K). In cult.

723. Prunus domestica L.
a. subsp. **domestica**
wild plum
CULT.: 1768 (JH: 451); 1789, Nat. of England (HK1, 2: 165); 1811, Nat. of England (HK2, 3: 199); 1814, Nat. of England (HKE: 148); 1880, Arboretum (K); 1881, unspec. (K); 1885, unspec. (K); 1892, Arboretum (K); 1894, Canal Beds (K); 1897, unspec. (K); 1910, Arboretum (K); 1927, unspec. (K); 1931, Arboretum (K); 1935, unspec. (K); 1946, Arboretum (K); 1954, unspec. (K); 1956, Arboretum (K). In cult.

b. subsp. **insititia** (L.) Bonnier & Layens
P. insititia L.
bullace
CULT.: 1768 (JH: 451); 1789, Nat. of Britain (HK1, 2: 165); 1811, Nat. of Britain (HK2, 3: 199); 1814, Nat. of Britain (HKE: 149). In cult.
Note. Neither subspecies is native; both are considered to be archaeophytes.

724. Prunus avium (L.) L.
wild cherry
CULT.: 1768 (JH: 451); 1789, Nat. of Britain (HK1, 2: 165); 1880, Arboretum (K); 1881, unspec. (K); 1884, Arboretum (K); 1887, Arboretum (K); 1890, unspec. (K); 1891, Arboretum (K); 1897, Arboretum (K); 1905, Arboretum (K); 1927, unspec. (K); 1931, Arboretum (K); 1960, Arboretum South (K); 1982, Kew Green, bridge end of W. Kew Green opposite King's Arms (K). In cult.

725. Prunus cerasus L.
dwarf cherry
CULT.: 1768 (JH: 451); 1789, Nat. of England (HK1, 2: 164); 1811, Nat. of England (HK2, 3: 198); 1814, Nat. of England (HKE: 148); 1880, Arboretum (K); 1881, Arboretum (K); 1884, Arboretum (K); 1886, unspec. (K); 1890, Nursery (K); 1894, Arboretum (K); 1895, Arboretum (K); 1900, Arboretum (K); 1901, Arboretum (K); 1903, Arboretum (K); 1905, unspec. (K); 1920, Arboretum (K); 1930, Arboretum (K); 1931, Arboretum (K); 1939, nr. Cumberland Gate (K). In cult.

726. Prunus padus L.
P. rubra Willd.
P. padus var. *rubra* (Willd.) auct. dub.
bird cherry
CULT.: 1768 (JH: 451); 1789, Nat. of Britain (HK1, 2: 162); 1811, Nat. of Britain (HK2, 3: 196); 1814, Nat. of Britain (HKE: 147); 1880, Arboretum (K); 1884, Arboretum (K); 1924, Arboretum (K); 1927, Arboretum (K); 1931, Arboretum (K); 1954, Arboretum (K); 1960, Arboretum South (K). In cult.
Note. Originally grown as a native but now considered to be an archaeophyte.

727. Pyrus cordata Desv.
Plymouth pear
CULT.: 1881, Arboretum (K); 1882, unspec. (K); 1907, Arboretum (K); 1955, by cart road west of Seven Sisters lawn (K); 1956, unspec. (K). In cult.

728. Pyrus communis L.
pear
FIRST RECORD: 1918 [1768].
REFERENCES: **Kew Environs.** 1918, *Bishop* (LN31: S111; S33: 746): Thames bank between Kew and Richmond. • 1932, *anon.* (LN11: S45; S33: 746): Between Kew and Richmond, by the river.
CULT.: 1768 (JH: 451); 1789, Nat. of England (HK1, 2: 174); 1811, Nat. of England (HK2, 3: 208); 1814, Nat. of England (HKE: 150); 1880, Arboretum (K); 1881, Arboretum (K); 1894, Arboretum (K); 1898, unspec. (K); 1905, Arboretum (K); 1927, unspec. (K); 1955, in collection left of cross path at entrance from Pagoda side (K); 1955, in border on right of grassy walk leading to Temp. house, West end (K). In cult.
CURRENT STATUS: No recent sightings. It is unlikely to have been anything but a relict of cultivation.

729. Malus sylvestris (L.) Mill.
Pyrus malus L.
crab apple
CULT.: 1789, Nat. of Britain (HK1, 2: 175); 1811, Nat. of Britain (HK2, 3: 208); 1814, Nat. of Britain (HKE: 151). In cult.

730. Sorbus domestica L.
service-tree
CULT.: 1768 (JH: 455); 1789, Nat. of England (HK1, 2: 171); 1811, Nat. of England (HK2, 3: 204); 1814, Nat. of England (HKE: 150); 1843-1853, Arboretum (K); 1880, Arboretum (K); 1881, Arboretum (K); 1881, Old Arboretum (K); 1882, Arboretum (K); 1943, near Main Gate (K). In cult.

731. Sorbus aucuparia L.
rowan
FIRST RECORD: 2001 [1768].
REFERENCES: **Kew Gardens.** 2001, *Cope* (S35: 647): Conservation Area (310). Hitherto unrecorded but seems well naturalised in woodland.
CULT.: 1768 (JH: 455); 1789, Nat. of Britain (HK1, 2: 171); 1811, Nat. of Britain (HK2, 3: 204); 1814, Nat. of Britain (HKE: 150); 1880, Arboretum (K); 1881, Arboretum (K); 1882, Arboretum (K); 1905, Arboretum (K); 1906, Arboretum (K); 1924, Arboretum (K); 1931, Arboretum (K); 1934, Arboretum (K); 1934, unspec. (K); 1942, Arboretum (K); 1943, Arboretum (K); 1946, Arboretum (K); 1959, South end of Temperate House in *Cotoneaster* bed, 1st canal bed on left (K); 1962, Arboretum (K); 1971, unspec. (K). In cult.
CURRENT STATUS: In the Conservation Area; self-sown from trees that were originally planted.

732. Sorbus pseudofennica E.F.Warb.
CULT.: 1994, area 339 (K). In cult.

[**733. Sorbus hybrida** L.
Swedish service-tree
CULT.: 1789, Alien (HK1, 2: 171); 1811, Nat. of England (HK2, 3: 204); 1814, Nat. of England (HKE: 150).
Note. Listed in error as native.]

734. Sorbus arranensis Hedl.
CULT.: In cult.

735. Sorbus minima (Ley) Hedl.
CULT.: In cult.

736. Sorbus anglica Hedl.
CULT.: In cult.

737. Sorbus aria (L.) Crantz
Crataegus aria L.
Pyrus aria (L.) Ehrh.
common whitebeam
CULT.: 1768 (JH: 438); 1789, Nat. of Britain (HK1, 2: 167); 1811, Nat. of Britain (HK2, 3: 210); 1814, Nat. of Britain (HKE: 152); 1880, Arboretum (K); 1881, Arboretum (K); 1882, unspec. (K); 1885, Arboretum (K); 1894, Arboretum (K); 1905, Arboretum (K); 1905, at N. end of Pagoda Vista near big Beech (K); 1907, unspec. (K); 1919, Arboretum (K); 1931, Arboretum (K); 1932, Arboretum (K); 1942, unspec. (K). In cult.

738. Sorbus leptophylla E.F.Warb.
CULT.: In cult.

739. Sorbus eminens E.F.Warb.
CULT.: In cult.

740. Sorbus porrigentiformis E.F.Warb.
CULT.: In cult.

741. Sorbus lancastriensis E.F.Warb.
CULT.: In cult.

742. Sorbus rupicola (Syme) Hedl.
CULT.: In cult.

743. Sorbus vexans E.F.Warb.
CULT.: 1984, area 339 (K). In cult.

744. Sorbus devoniensis E.F.Warb.
CULT.: 1963, Arboretum (K); 1994, area 339 (K). In cult.

745. Sorbus bristoliensis Wilmott
CULT.: In cult.

746. *Sorbus latifolia (Lam.) Pers.
broad-leaved whitebeam
FIRST RECORD: 1938 [1843–53].
REFERENCES: **Kew Gardens.** 1938 (WAT17: 396; S33: 746, and see Exsicc.).
EXSICC.: **Kew Gardens.** [Unlocalised]; 28 x 1938, *Ellis* s.n. (LANC).
CULT.: 1843–1853, Arboretum (K); 1881, end plant (group of three) nearest Winter Garden (K); 1903, Arboretum (K); 1905, Arboretum (K); 1950, Arboretum (K); 1959, Hydrangea collection near Oxenhouse Gate (K); 1976, Herbarium Ground (K). In cult.
CURRENT STATUS: No recent sightings.

747. Sorbus torminalis (L.) Crantz
Crataegus torminalis L.
Pyrus torminalis (L.) Ehrh.
wild service-tree
CULT.: 1768 (JH: 438); 1789, Nat. of England (HK1, 2: 167); 1811, Nat. of England (HK2, 3: 210); 1814, Nat. of England (HKE: 152).; 1908, Arboretum (K); 1954, Arboretum (K). In cult.

748. *Cotoneaster obtusus Wall. ex Lindl.
Dartford cotoneaster
FIRST RECORD: 2003.
CULT.: In cult.
CURRENT STATUS: Cultivated in various parts of the Gardens but naturalised on the towpath opposite the Conservation Area and around the Herbarium car park. Both localitites are known to J.Fryer (pers. comm.) and this casts some doubt on the identity of the seedlings of *C. bacillaris* below.

749. *Cotoneaster bacillaris Wall. ex Lindl.
open-fruited cotoneaster
FIRST RECORD: 1983/4 [1856].
REFERENCES: **Kew Gardens.** 1983/4, *Cope* (S31: 181): Seedling in new student vegetable plots behind Hanover House.
CULT.: 1856, Pleasure Grounds (K); 1869, in bed facing new Temp. House (K); 1880, Arboretum (K); 1881, Arboretum (K); 1882, unspec. (K); 1894, Arboretum (K); 1905, Arboretum (K); 1907, Arboretum (K); 1923, Arboretum (K); 1924, Arboretum (K); 1939, Arboretum (K); 1957, Arboretum (K). In cult.
CURRENT STATUS: Has not persisted behind Hanover House. In view of the records of the previous species, there may be doubt about the correct identity of the seedlings in question.

750. Mespilus germanica L.
medlar
CULT.: 1789, Nat. of England (HK1, 2: 172); 1811, Nat. of England (HK2, 3: 205); 1814, Nat. of England (HKE: 150); 1880, Arboretum (K); 1955, Arboretum (K). In cult.

751. Crataegus monogyna Jacq.
C. monogyna var. *pteridifolia* (Loudon) Druce
C. oxyacantha L. pro parte
hawthorn
FIRST RECORD: 1942 [1789].
REFERENCES: **Kew Environs.** 1942, *Batko* (LN32: S113; S33: 746): One small tree on the Thames riverside between Richmond and Kew.
CULT.: 1789, Nat. of Britain (HK1, 2: 170); 1811, Nat. of Britain (HK2, 3: 203); 1814, Nat. of Britain (HKE: 149); 1880, Arboretum (K); 1881, Arboretum (K); 1882, Arboretum (K); 1885, Arboretum (K); 1887, unspec. (K); 1890, unspec. (K); 1892, Arboretum (K); 1893, Arboretum (K); 1895, Arboretum (K); 1900, Arboretum (K); 1903, unspec. (K); 1905, Arboretum (K); 1908, Arboretum (K); 1927, unspec. (K); 1930, Palace Ground nursery (K); 1960, lawn near old *Ginkgo* (K); 1963, Arboretum (K); 1968, unspec. (K); 1987, plot 435 (K). In cult.
CURRENT STATUS: Still on the towpath but presumably originally planted. Occasional seedlings occur within the Gardens derived from cultivated trees, some of these being highly ornamental varieties.

752. Crataegus laevigata (Poir.) DC.
C. oxyacantha L.
Midland hawthorn
CULT.: 1768 (JH: 439); 1880, Arboretum (K); 1881, Arboetum (K); 1881, top plant nr. Pagoda Old Collection (K); 1892, Arboretum (K); 1895, Arboretum (K); 1898, near Museum No. III (K); 1900, Arboretum (K); 1905, Arboretum (K); 1942, unspec. (K); 1989, back of staff houses between Admin. and Wood Museum (K). In cult.

753. Crataegus × macrocarpa Hegetschw. (laevigata × monogyna)
C. × *media* auct. non Bechst
C. monogyna Jacq. × *oxyacanthoides* Thuill.
FIRST RECORD: 1942.
REFERENCES: **Kew Environs.** 1942, *Batko* (LN32: S113; S33: 746): Thames riverside between Kew and Richmond.
CURRENT STATUS: No recent sightings. The tree, or trees, in question have not be positively relocated.

Fabaceae

754. *Robinia pseudoacacia L.
false-acacia
First record: 1983/4 [1768].
References: **Kew Gardens.** 1983/4, *Cope* (S31: 181): Seedling in new student vegetable plots behind Hanover House.
Cult.: 1768 (JH: 291, 453); 1880, Arboretum (K); 1881, Arboretum (K); 1882, Arboretum (K); 1883, Arboretum (K); 1885, Arboretum (K); 1896, unspec. (K); 1898, unspec. (K); 1905, Arboretum (K); 1933, Arboretum (K); 1987, zone 476 (K); 1987, zone 477 (K). In cult.
Current status: Has not persisted behind Hanover House; occasional seedlings occur near the Pagoda and mature trees can be found on the golf course in the Old Deer Park. Many named varieties have been cultivated in the Gardens, any of which could be the parents of seedlings.

755. *Galega officinalis L.
goat's-rue
First record: 2004 [1969].
Cult.: In cult.
Current status: A single record from the golf course in the Old Deer Park, and another from rough ground outside the Stable Yard.

756. Astragalus danicus Retz.
A. hyptoglottis auct. non L.
purple milk-vetch
Cult.: 1789, Nat. of Britain (HK1, 3: 76); 1812, Nat. of Britain (HK2, 4: 367); 1814, Nat. of Britain (HKE: 235); 1916, unspec. (K). Not currently in cult.

757. Astragalus alpinus L.
alpine milk-vetch
Cult.: 1897, unspec. (K). Not currently in cult.

758. Astragalus glycyphyllos L.
wild liquorice
First record: 2001 [1768].
References: **Kew Gardens.** 2001, *Cope* (S35: 646): In rough grass near the Stable Yard (344). Subsequently in a few other places (187, 261, 343). Originally noted by P.J.Edwards who drew my attention to it.
Cult.: 1768 (JH: 301); 1789, Nat. of Britain (HK1, 3: 74); 1812, Nat. of Britain (HK2, 4: 364); 1814, Nat. of Britain (HKE: 235); 1933, unspec. (K); 1940, Herbaceous Ground (K); 1976, Order Beds (K). In cult.
Current status: In several places in rough grass far from any cultivated specimens but doubtless originally escaping from cultivation.

759. Oxytropis halleri Bunge ex W.D.J.Koch
Astragalus uralensis L.
purple oxytropis
Cult.: 1789, Nat. of Scotland (HK1, 3: 77); 1812, Nat. of Scotland (HK2, 4: 370); 1814, Nat. of Scotland (HKE: 236). Not currently in cult.

760. Oxytropis campestris (L.) DC.
yellow oxytropis
Cult.: 1928, Herb. Dept. (K); 1951, Herbarium Ground (K); 1953, Herbarium Expt. Gd. (K). In cult.

761. Onobrychis viciifolia Scop.
Hedysarum onobrychis L.
sainfoin
First record: 2006 [1768].
Cult.: 1768 (JH: 295); 1789, Nat. of Britain (HK1, 3: 66); 1812, Nat. of Britain (HK2, 4: 348); 1814, Nat. of Britain (HKE: 233); 1953, Herb. Expt. Grnd. (K); 1997, unspec. (K). Not currently in cult.
Current status: A single plant was found in the Conservation Area, far removed from any obvious cultivated source.

Wild liquorice (*Astragalus glycyphyllos*). Rather rare in rough grass and doubtless an escape from cultivation.

762. Anthyllis vulneraria L.
kidney vetch
CULT.: 1768 (JH: 306); 1789, Nat. of Britain (HK1, 3: 25); 1812, Nat. of Britain (HK2, 4: 282); 1814, Nat. of Britain (HKE: 225); 1970s, plot 157 (K); 1979, Order Beds (K); 1997, unspec. (K). In cult.

763. Lotus glaber Mill.
narrow-leaved bird's-foot-trefoil
CULT.: 1983, bed 156-39 (K). Not currently in cult.

764. Lotus corniculatus L.
common bird's-foot-trefoil
FIRST RECORD: 1873/4 [1768].
NICHOLSON: **1873/4 survey** (JB: 43): Abundant in all the divisions. • **1906 revision** (AS: 78): Abundant in all the divisions.
EXSICC.: **Kew Gardens.** Queen's Cottage; [c. 1914]; *Flippance* s.n. (K). **Kew Environs.** Golf Course, Kew; growing on sandy soil with *Galium verum* & *Thymus serpyllum*; 10 v 1948; *Ward* 77 (K).
CULT.: 1768 (JH: 297); 1789, Nat. of Britain (HK1, 3: 93); 1812, Nat. of Britain (HK2, 4: 396); 1814, Nat. of Britain (HKE: 238); 1934, unspec. (K); 1949, Herbarium Ground (K); 1955, Herbarium Ground (K); 1970, in glasshouse (K); 1973, Herb. & Alpine Dept, S. Arboretum (K). In cult.
CURRENT STATUS: Widespread and sometimes abundant in short grass.

765. Lotus pedunculatus Cav.
L. major auct.
L. uliginosus Schkuhr
greater bird's-foot-trefoil
FIRST RECORD: 1873/4 [1812].
NICHOLSON: **1873/4 survey** (JB: 43): Common round edge of lake, growing among the *Juncus*. • **1906 revision** (AS: 78): **A.** Common round edge of lake, amongst *Juncus*, etc.
OTHER REFERENCES: **Kew Gardens.** 1999, *Stones* (EAK: 2): By Lake.
EXSICC.: **Kew Gardens.** [Unlocalised]; erect plant in rough grass; 24 vii 1944; *Souster* 124 (K).
CULT.: 1812, Nat. of Britain (HK2, 4: 395); 1814, Nat. of Britain (HKE: 238); 1940, Herbaceous Ground (K). Not currently in cult.
CURRENT STATUS: Occasional in and around the Conservation Area and amongst tall vegetation by the Lake.

766. Lotus subbiflorus Lag.
hairy bird's-foot-trefoil
FIRST RECORD: 1990 [1940].
REFERENCES: **Kew Gardens.** 1990, *Cope* (S31: 182): Paddock, after restoration in 1990.
CULT.: 1940, Herbaceous Ground (K); 1976, Order Beds (K); 1979, Order Beds (K). In cult.
CURRENT STATUS: No recent sightings; it has not reappeared in the Paddock since 1990.

Bird's-foot (*Ornithopus perpusillus*). A tiny herb known principally from Kew Green, opposite the Herbarium, and in a patch of *Festuca filiformis* grassland north of the Rhododendron Dell. Both sites are vulnerable.

767. Lotus angustissimus L.
L. diffusus Sm.
slender bird's-foot-trefoil
FIRST RECORD: 1990 [1812].
REFERENCES: **Kew Gardens.** 1990, *Cope* (S31: 182): Paddock, after restoration in 1990.
CULT.: 1812, Nat. of England (HK2, 4: 393); 1814, Nat. of England (HKE: 238). Not currently in cult.
CURRENT STATUS: No recent sightings; it has not reappeared in the Paddock since 1990.

768. Ornithopus perpusillus L.
bird's-foot
FIRST RECORD: 1873/4 [1768].
NICHOLSON: **1873/4 survey** (JB: 43 and in SFS: 246): Everywhere. Large pieces of turf between King William's Temple and Winter Garden were composed almost entirely of this plant in 1873–4. • **1906 revision** (AS: 78): Everywhere. The turf behind King William's temple and near winter garden was almost entirely composed of this plant in 1873-4.
EXSICC.: **Kew Gardens.** Near Temperate House; weed in poor lawn; 2 vii 1944; *Souster* 110 (K).
CULT.: 1768 (JH: 300); 1789, Nat. of Britain (HK1, 3: 59); 1812, Nat. of Britain (HK2, 4: 333); 1814, Nat. of Britain (HKE: 232). Not currently in cult.
CURRENT STATUS: Widespread but rather rare. It is confined to the more acidic grassland, especially on Kew Green and in one small patch north of the Rhododendron Dell where it occurs amongst *Festuca filiformis*.

769. Hippocrepis comosa L

horseshoe vetch

CULT.: 1768 (JH: 294); 1789, Nat. of England (HK1, 3: 61); 1812, Nat. of England (HK2, 4: 335); 1814, Nat. of England (HKE: 232); 1896, unspec. (K). Not currently in cult.

770. Vicia orobus DC.

Orobus sylvaticus L.

wood bitter-vetch

CULT.: 1768 (JH: 298); 1789, Nat. of Britain (HK1, 3: 39); 1812, Nat. of Britain (HK2, 4: 305); 1814, Nat. of Britain (HKE: 228); 1897, unspec. (K); 1922, unspec. (K). In cult.

771. Vicia cracca L.

tufted vetch

FIRST RECORD: 1873/4 [1768].

NICHOLSON: **1873/4 survey** (JB: 43): **Pal, P** and **Q:** Very much less common than [*V. hirsuta*]. • **1906 revision** (AS: 78): **A, P, Q:** Much less common than [*V. hirsuta*].

EXSICC.: **Kew Gardens.** [Unlocalised]; climbing to 2' in rough places; 24 vii 1944; *Souster* 126 (K).

CULT.: 1768 (JH: 300); 1789, Nat. of Britain (HK1, 3: 44); 1812, Nat. of Britain (HK2, 4: 311); 1814, Nat. of Britain (HKE: 229); 1882, unspec. (K); 1933, unspec. (K); 1934, unspec. (K); 1961, Order Beds (K). In cult.

CURRENT STATUS: It once regularly appeared at the edge of the Paddock, but was usually mown before it fruited and has now probably been destroyed by the creation of a temporary car park on the site. Other sites are near King William's Temple and on the golf course in the Old Deer Park.

772. Vicia sylvatica L.

wood vetch

CULT.: 1768 (JH: 299); 1789, Nat. of Britain (HK1, 3: 44); 1812, Nat. of Britain (HK2, 4: 311); 1814, Nat. of Britain (HKE: 229); 1940, Herbaceous Ground (K). Not currently in cult.

773. Vicia hirsuta (L.) Gray

Ervum hirsutum L.

hairy tare

FIRST RECORD: 1873/4 [1789].

NICHOLSON: **1873/4 survey** (JB: 43): **Pal, P** and **Q:** Common both in open turf and as a weed in beds and shrubberies. • **1906 revision** (AS: 78): Common both in open turf and as a weed in beds and shrubberies.

CULT.: 1789, Nat. of Britain (HK1, 3: 47); 1812, Nat. of Britain (HK2, 4: 317); 1814, Nat. of Britain (HKE: 229); 1922, unspec. (K); 1932, unspec. (K); 1946, Herbarium Ground (K); 1948, unspec. (K); 1950, unspec. (K); 1971, Herb. & Alpine Dept. (K). Not currently in cult.

CURRENT STATUS: Sporadic in the longer grass and abundant in the Ancient Meadow.

774. Vicia tetrasperma (L.) Schreb.

Ervum tetraspermum L.

smooth tare

FIRST RECORD: 1990 [1768].

REFERENCES: **Kew Gardens.** 1990, *Cope* (S31: 182): Paddock, after restoration in 1990.

CULT.: 1768 (JH: 294); 1789, Nat. of Britain (HK1, 3: 47); 1812, Nat. of Britain (HK2, 4: 317); 1814, Nat. of Britain (HKE: 229). Not currently in cult.

CURRENT STATUS: Occasional in longer grass and usually with the preceding species.

775. Vicia sepium L.

bush vetch

CULT.: 1768 (JH: 300); 1789, Nat. of Britain (HK1, 3: 46); 1812, Nat. of Britain (HK2, 4: 315); 1814, Nat. of Britain (HKE: 229); 1940, Herbaceous Ground (K). Not currently in cult.

776. Vicia sativa L.

common vetch

a. subsp. **nigra** (L.) Ehrh.

V. angustifolia L. var. *angustifolia*

V. angustifolia var. *bobartii* Koch

V. sativa auct. non L. s. str.

FIRST RECORD: c. 1779–1805 [1768].

NICHOLSON: **1873/4 survey** (JB: 43): **Pal, P** and **Q.** Abundant. The var. *bobartii* occurs sparingly, and by intermediate gradations merges into *V. angustifolia*. • **1906 revision** (AS: 78 and in SFS: 251): **A, P, Q.** Abundant. Var. *bobartii*: Sparingly, with the type.

EXSICC.: **Kew Environs.** Richmond Gardens, c. 1779–1805, *Goodenough* s.n. (K).

CULT.: 1768 (JH: 300); 1789, Nat. of Britain (HK1, 3: 45); 1812, Nat. of Britain (HK2, 4: 314); 1814, Nat. of Britain (HKE: 229); 1940, under glass (K). Not currently in cult.

CURRENT STATUS: Widespread; mostly in the longer grass.

b. subsp. ***sativa**

FIRST RECORD: 1998 [1940].

REFERENCES: **Kew Gardens.** 1998, *Cope* (S33: 746): North Arb.: Paddock (104); lawn opposite the Orangery, in a wild patch (127).

CULT.: 1940, Herbaceous Ground (K). Not currently in cult.

CURRENT STATUS: No recent sightings; it has not persisted in the Paddock.

777. Vicia lathyroides L.

Ervum soloniense L.

spring vetch

CULT.: 1768 (JH: 294, 300); 1789, Nat. of Britain (HK1, 3: 45); 1812, Nat. of Britain (HK2, 4: 314); 1814, Nat. of Britain (HKE: 229). Not currently in cult.

778. Vicia lutea L.
V. laevigata Sm.
yellow-vetch
First record: 1990 [1768].
References: **Kew Gardens.** 1990, *Cope* (S31: 182): Paddock, after restoration in 1990.
Cult.: 1768 (JH: 300); 1789, Nat. of England (HK1, 3: 45); 1812, Nat. of Britain (HK2, 4: 314), Nat. of England (ibid.: as *laevigata*); 1814, Nat. of Britain (HKE: 229), Nat. of England (ibid.: as *laevigata*); 1898, unspec. (K); 1921, unspec. (K); 1940, Herbaceous Ground (K). Not currently in cult.
Current status: No recent sightings; it has not reappeared in the Paddock since 1990.

779. Vicia bithynica (L.) L.
bithynian vetch
Cult.: 1768 (JH: 300); 1812, Nat. of England (HK2, 4: 315); 1814, Nat. of England (HKE: 229); 1940, Herbaceous Ground (K). Not currently in cult.

[**780. Vicia hybrida** L.
hairy yellow-vetch
Cult.: 1789, Nat. of England (HK1, 3: 46); 1812, Nat. of England (HK2, 4: 314); 1814, Nat. of England (HKE: 229).
Note. Listed in error as native.]

781. Lathyrus japonicus Willd.
Pisum maritimum L.
sea pea
Cult.: 1768 (JH: 293); 1789, Nat. of England (HK1, 3: 37); 1812, Nat. of England (HK2, 4: 302); 1814, Nat. of England (HKE: 228); 1940, Herbaceous Ground (K); 1976, Order Beds (K). In cult.

782. Lathyrus linifolius (Reichard) Bässler
Orobus tuberosus L.
O. pyrenaicus L.
bitter-vetch
Cult.: 1768 (JH: 298); 1789, Nat. of Britain (HK1, 3: 38); 1812, Nat. of Britain (HK2, 4: 304); 1814, Nat. of Britain (HKE: 228); 1940, Herbaceous Ground (K); 1986, unspec. (K). Not currently in cult.

783. Lathyrus pratensis L.
meadow vetchling
First record: 1873/4 [1768].
Nicholson: **1873/4 survey** (JB: 43): **Pal** and **Q.** Uncommon. **Strip.** A large patch some two or three yards long by towing path near Isleworth Gate. • **1906 revision** (AS: 78): **P, Q, Strip.** Uncommon.
Cult.: 1768 (JH: 304); 1789, Nat. of Britain (HK1, 3: 42); 1812, Nat. of Britain (HK2, 4: 309); 1814, Nat. of Britain (HKE: 228); 1934, unspec. (K); 1940, Herbaceous Ground (K); 1991, unspec. (K). Not currently in cult.
Current status: Widespread in the butterfly conservation areas and in other long grass.

784. Lathyrus palustris L.
marsh pea
Cult.: 1768 (JH: 304); 1789, Nat. of Britain (HK1, 3: 43); 1812, Nat. of Britain (HK2, 4: 310); 1814, Nat. of Britain (HKE: 228); 1953, Bog Garden (K). Not currently in cult.

785. Lathyrus sylvestris L.
narrow-leaved everlasting-pea
Cult.: 1768 (JH: 304); 1789, Nat. of Britain (HK1, 3: 43); 1812, Nat. of Britain (HK2, 4: 309); 1814, Nat. of Britain (HKE: 228); 18th cent., unspec. (K); 1940, Herbaceous Ground (K). In cult.

786. *Lathyrus latifolius L.
broad-leaved everlasting-pea
First record: 2000 [1768].
References: **Kew Gardens.** 2000, *Cope* (S35: 647): In rough grass near Syon Vista (255).
Cult.: 1768 (JH: 304); 1789, Nat. of England (HK1, 3: 43); 1812, Nat. of England (HK2, 4: 309); 1814, Nat. of England (HKE: 228). In cult.
Current status: Naturalised in long grass near the Lake. Plants that were found behind Kew Palace may have escaped from the bee garden.

787. *Lathyrus hirsutus L.
hairy vetchling
First record: 1990 [1768].
References: **Kew Gardens.** 1990, *Cope* (S31: 182): Paddock, after restoration in 1990.
Cult.: 1768 (JH: 304); 1789, Nat. of England (HK1, 3: 42); 1812, Nat. of England (HK2, 4: 308); 1814, Nat. of England (HKE: 228); 1940, Herbaceous Ground (K); 1991, Scientific Support Glasshouse (K). In cult.
Current status: No recent sightings; it has not reappeared in the Paddock since 1990.

788. Lathyrus nissolia L.
grass vetchling
First record: 2002 [1768].
References: **Kew Gardens.** 2002, *Cope* (S35: 647): In rough grassy area near the Bamboo Garden (252).
Cult.: 1768 (JH: 304); 1789, Nat. of England (HK1, 3: 39); 1812, Nat. of England (HK2, 4: 305); 1814, Nat. of England (HKE: 228); 1940, Herbaceous Ground (K). Not currently in cult.
Current status: A single record from near the Bamboo Garden a very long way from the last known record of its cultivation, in the Order Beds.

789. Lathyrus aphaca L.
yellow vetchling
FIRST RECORD: 1927 [1768].
EXSICC.: **Kew Gardens.** On rubbish heap in lower nursery; 3 vi 1927; *Hubbard* s.n. (K).
CULT.: 1768 (JH: 304); 1789, Nat. of England (HK1, 3: 39); 1812, Nat. of England (HK2, 4: 305); 1814, Nat. of England (HKE: 228); 1921, unspec. (K); 1940, Herbaceous Ground (K); 1976, Order Beds (K); 1979, Order Beds (K); 1991, unspec. (K). In cult.
CURRENT STATUS: No recent sightings.

790. Ononis reclinata L.
small restharrow
CULT.: 1919, unspec. (K). Not currently in cult.

791. Ononis spinosa L.
spiny restharrow
CULT.: 1768 (JH: 292); 1789, Nat. of Britain (HK1, 3: 21); 1812, Nat. of Britain (HK2, 4: 276); 1814, Nat. of Britain (HKE: 224); 1940, Herbaceous Ground (K). In cult.

792. Ononis repens L.
O. arvensis auct.
common restharrow
FIRST RECORD: 1873/4 [1768].
NICHOLSON: **1873/4 survey** (JB: 42; SFS: 227): **Strip.** Two plants by side of towing-path midway between Brentford and Isleworth Gates. • **1906 revision** (AS: 77): **Strip.** Two plants by side of towing-path between Brentford and Isleworth Gates.
CULT.: 1768 (JH: 292); 1812, Nat. of Britain (HK2, 4: 276); 1814, Nat. of Britain (HKE: 224); 1978, unspec. (K). In cult.
CURRENT STATUS: Known only from the golf course in the Old Deer Park.

793. Melilotus altissimus Thuill.
tall melilot
FIRST RECORD: 1946 [1940].
REFERENCES: **Kew Environs.** 1946, *Welch* (LN31: S59; S33: 746): River wall, Kew.
CULT.: 1940, Herbaceous Ground (K). Not currently in cult.
CURRENT STATUS: No recent sightings.

794. *Melilotus albus Medik.
Trifolium officinale L. var. b
white melilot
FIRST RECORD: 1873/4 [1789].
NICHOLSON: **1873/4 survey** (JB: 42): **P.** A couple of large plants in the hollow between "Unicorn Gate" and "Douglas Spar." • **1906 revision** (AS: 78): **A.** A couple of large plants in the hollow between unicorn gate and flagstaff.
EXSICC.: **Kew Environs.** Thames banks, Kew; 15 ix 1883; *Fraser* s.n. (K). • Kew; waste ground; 9 vii 1902; *Worsdell* s.n. (K).
CULT.: 1789, Nat. of Britain (HK1, 3: 84); 1812, Nat. of Britain (HK2, 4: 380); 1814, Nat. of Britain (HKE: 237); 1886, unspec. (K); 1933, Herb Garden (K). In cult.
CURRENT STATUS: Found in 2008, after a long absence, in rough ground along the boundary wall adjacent to the Order Beds.

795. *Melilotus officinalis (L.) Pall.
Trifolium officinale L. var. a
ribbed melilot
FIRST RECORD: 1929 [1768].
EXSICC.: **Kew Environs.** Thames bank at Kew; common; vii 1929; *Hubbard* s.n. (K).
CULT.: 1768 (JH: 302); 1789, Nat. of Britain (HK1, 3: 84); 1812, Nat. of Britain (HK2, 4: 380); 1814, Nat. of Britain (HKE: 237); 1887, unspec. (K); 1895, unspec. (K); 1933, Herb Garden (K); 1934, unspec. (K); 1936, unspec. (K); 1937, Herbaceous Ground (K); 1953, Herbarium Ground (K). In cult.
CURRENT STATUS: No recent sightings.

796. *Melilotus indicus (L.) All.
Trifolium indicum L.
small melilot
FIRST RECORD: 1990 [1768].
REFERENCES: **Kew Gardens.** 1990, *Cope* (S31: 182): Paddock, after restoration in 1990.
CULT.: 1768 (JH: 302); 1921, unspec. (K); 1936, Herbaceous Ground (K). Not currently in cult.
CURRENT STATUS: After a long absence this plant reappeared in 2004 in topsoil stored in the Paddock.

797. Medicago lupulina L.
black medick
FIRST RECORD: 1873/4 [1768].
NICHOLSON: **1873/4 survey** (JB: 42): **Pal, P** and **Strip.** Most common in the two first-named divisions. • **1906 revision** (AS: 77): Common.
EXSICC.: **Kew Gardens.** Queen's Cottage; [c. 1914]; *Flippance* s.n. (K). • Pasture by Herbarium; 10 ix 1942; *Melville* s.n. (K). **Kew Environs.** Towpath, Kew; 5 viii 1944; *Souster* 139 (K).
CULT.: 1768 (JH: 305); 1789, Nat. of Britain (HK1, 3: 97); 1812, Nat. of Britain (HK2, 4: 400); 1814, Nat. of Britain (HKE: 239); 1899, unspec. (K); 1934, unspec. (K); 1940, Herbaceous Ground (K). Not currently in cult.
CURRENT STATUS: Widespread in short grass though noticeably absent from the southeast part of the Gardens.

798. Medicago sativa L.
sickle medick
a. subsp. **falcata** (L.) Arcang.
M. falcata L.
M. falcata var. *tenuifoliata* auct. dub.
FIRST RECORD: 1880 [1768].
NICHOLSON: **1906 revision:** Not recorded.

Other references: **Kew Gardens.** 1880, *Baker* (BECB1: 40; S33: 746): Thames bank opposite Sion House, midway between Kew and Richmond.
Exsicc.: **Kew Environs.** Tow path by Thames, above Kew; one plant; 21 vii 1912; *Britton* 767 (K). • Towing path by Old Deer Park, Richmond; 20 vii 1914; Fraser s.n. (K). • On bank of R. Thames between Kew & Richmond; 9 viii 1929; *Hubbard* s.n. (K). • Thames banks, east end of Kew; 21 viii 1932; *Fraser* s.n. (K).
Cult.: 1768 (JH: 305); 1789, Nat. of England (HK1, 3: 97); 1812, Nat. of England (HK2, 4: 400); 1814, Nat. of England (HKE: 240); 19th cent., unspec. (K); 1897, unspec. (K); 1908, unspec. (K); 1922, unspec. (K); 1940, Herbaceous Ground (K). Not currently in cult.
Current status: No recent sightings.

b. subsp. ***sativa**

Lucerne

First record: 1873/4 [1768].
Nicholson: **1873/4 survey** (JB: 42): **Strip.** A single plant in the turf about 15 yards from Brentford Ferry. • **1906 revision** (AS: 77): **Strip.** In turf near Brentford Ferry.
Cult.: 1768 (JH: 305); 1789, Nat. of England (HK1, 3: 96); 1812, Nat. of England (HK2, 4: 400); 1814, Nat. of England (HKE: 240); 1984, bed 156-37 (K). In cult. (2006). In cult.
Current status: A single plant has persisted in the island bed on the edge of the Paddock beneath one of the plane trees near the boundary wall.

799. Medicago minima (L.) Bartal.

bur medick

Cult.: 1812, Nat. of England (HK2, 4: 406); 1814, Nat. of England (HKE: 239). Not currently in cult.

800. Medicago polymorpha L.
M. polymorpha var. *muricata* L.
M. muricata (L.) With.

toothed medick

First record: 1878 [1768].
Exsicc.: **Kew Environs.** Kew; waste ground; vii 1878; *G.Nicholson* s.n. (K).
Cult.: 1768 (JH: 305); 1789, Alien (HK1, 3: 98); 1812, Nat. of England (HK2, 4: 405); 1814, Nat. of England (HKE: 239); 1855, unspec. (K); 1940, Herbaceous Ground (K). In cult.
Current status: Occurs in patches in short grass on the plinth in front of the Palm House.

801. Medicago arabica (L.) Huds.
M. polymorpha var. *arabica* L.
M. maculata Sibth.

spotted medick

First record: 1873/4 [1789].
Nicholson: **1873/4 survey** (JB: 42): Common everywhere within our limits. • **1906 revision** (AS: 77): Common everywhere within Kew limits.
Exsicc.: **Kew Environs.** River side, Kew; 13 v 1894; *anon.* (K).
Cult.: 1789, Alien (HK1, 3: 98); 1812, Nat. of England (HK2, 4: 404); 1814, Nat. of England (HKE: 239); 1979, Order Beds (K). Not currently in cult.
Current status: Widespread in short grass; once a common feature of the Herbarium Lawn and Paddock, both of which have been damaged by current building works.

802. Trifolium ornithopodioides L.

bird's-foot clover

Cult.: 1768 (JH: 302); 1789, Nat. of Britain (HK1, 3: 84); 1812, Nat. of Britain (HK2, 4: 380); 1814, Nat. of Britain (HKE: 237). Not currently in cult.

803. Trifolium repens L.

white clover

First record: 1873/4 [1768].
Nicholson: **1873/4 survey** (JB: 43): A common component of the turf everywhere. • **1906 revision** (AS: 78): A common component of the turf everywhere.
Exsicc.: **Kew Environs.** Kew; 1883; *Fraser* s.n. (K). • Towpath, Kew; 20 vi 1944; *Souster* 100 (K).
Cult.: 1768 (JH: 302); 1789, Nat. of Britain (HK1, 3: 85); 1812, Nat. of Britain (HK2, 4: 382); 1814, Nat. of Britain (HKE: 238); 1856, unspec. (K); 1894, unspec. (K); 1911, unspec. (K); 1933, unspec. (K); 1933, Herbarium Ground (K); 1935, unspec. (K); 1936, Herbaceous Ground (K); 1976, Order Beds (K); 2000, plot 156-45 (K). In cult.
Current status: Ubiquitous in grass.

804. *Trifolium hybridum L.

alsike clover

First record: 1873/4 [1768].
Nicholson: **1873/4 survey** (JB: 43): **P.** A few plants in turf near Juniper Collection. • **1906 revision** (AS: 78): **A.** Near juniper collection. **Q.**
Exsicc.: **Kew Gardens.** Wild in beds near 'Seven Sisters Lawn'; 5 viii 1915; *Turrill* s.n. (K). • In Herbarium field; 22 viii 1933; *Hubbard* s.n. (K).
Cult.: 1768 (JH: 302); 1812, Alien (HK2, 4: 382); 1814, Alien (HKE: 238); 1897, unspec. (K); 1934, unspec. (K); 1976, Order Beds (K). Not currently in cult.
Current status: Appeared near King William's Temple in 2004. It also made a brief appearance in the Paddock but has not returned. Currently, it occurs in rough ground outside the Stable Yard but the area is due for redevelopment.

805. Trifolium glomeratum L.
clustered clover
FIRST RECORD: 1805 [1768].
REFERENCES: **Kew Environs.** 1805, *Goodenough* (SFS: 239): Kew Green. • 1863, *Brewer* (BFS: 61): Kew Green. • 1931, *Fraser* (SFS: 239): Still there in 1920. • 1977, *Meikle* (S30: 451): In some quantity on part of Kew Green near the Herbarium (see *Herbarium and Library News* No. 1016). It was found on the Green in 1805 by Bishop Goodenough (first record for Surrey) and subsequently in 1920, by Mr John Fraser. N.Y. Sandwith repeatedly searched in vain for it and J.E.Lousley, *Flora of Surrey* (1976) records it as 'very rare and perhaps extinct.' [in Surrey] Its reappearance in 1977 might be attributable to the exceptional summer in 1976. • 1983, *Burton* (FLA: 56): ... Kew Green, where it was rediscovered in 1977 having been previously known from 1805 to 1920. • 1987, *Leslie* (FSSC: 109): Refound in good quantity on Kew Green in 1977 and seen there each year since; it was first noted there in 1805, but had not been seen since 1920. • 1993, *Hastings* (S31: 182): Has been seen on Kew Green every year up to 1991 since R.D.Meikle found it including an abundance in 1987.
EXSICC.: **Kew Environs.** Kew Green; 11 ix 1920; *Fraser* s.n. (K). • Kew Green near Herbarium on right hand side of path to Church; in considerable abundance; 8 vii 1977; *Meikle* s.n. (K). • ibid.; 1 vi 1978; *Meikle* s.n. (K).
CULT.: 1768 (JH: 302); 1789, Nat. of England (HK1, 3: 88); 1812, Nat. of England (HK2, 4: 388); 1814, Nat. of England (HKE: 237); 1919, unspec. (K); 1922, unspec. (K). Not currently in cult.
CURRENT STATUS: It was first recorded on Kew Green in 1805 and was still present in 1920, but thereafter it was not seen again until the spring following the drought of 1976 which severely affected the grass. It is now firmly re-established and can be found most years near the cricket nets.

806. Trifolium suffocatum L.
suffocated clover
CULT.: 1812, Nat. of England (HK2, 4: 388); 1814, Nat. of England (HKE: 237); 19th cent., unspec. (K); 1923, unspec. (K). Not currently in cult.
Note. Discovered in the wild in Britain in 1794.

807. Trifolium fragiferum L.
strawberry clover
CULT.: 1768 (JH: 303); 1789, Nat. of England (HK1, 3: 89); 1812, Nat. of England (HK2, 4: 389); 1814, Nat. of England (HKE: 238); 1898, unspec. (K); 1940, Herbaceous Ground (K); 1976, unspec. (K). Not currently in cult.

808. *Trifolium aureum Pollich
large trefoil
FIRST RECORD: 1881 [1897].
REFERENCES: **Kew Environs.** 1881, *Baker* (BECB1: 47; S33: 746): Surrey side of Thames, between Kew and Richmond; 'I would name this a very luxuriant form of *T. aureum*' – J.T.Boswell.
CULT.: 1897, unspec. (K); 1973, unspec. (K). Not currently in cult.
CURRENT STATUS: No recent sightings.

809. Trifolium campestre Schreb.
T. procumbens auct. non L.
T. agrarium auct. non L.
hop trefoil
FIRST RECORD: 1873/4 [1768].
NICHOLSON: **1873/4 survey** (JB: 43): **B.** Here and there in the turf behind Herbaceous ground wall. **P.** Plentiful in the dry gravelly soil near Pagoda. • **1906 revision** (AS: 78): **A.** Abundant in the dry gravelly soil near pagoda, and elsewhere.
CULT.: 1768 (JH: 303); 1789, Nat. of Britain (HK1, 3: 89, 90); 1812, Nat. of Britain (HK2, 4: 390); 1814, Nat. of Britain (HKE: 237); 1921, unspec. (K). Not currently in cult.
CURRENT STATUS: Rare and very sporadic.

810. Trifolium dubium Sibth.
T. minus Sm.
lesser trefoil
FIRST RECORD: 1873/4 [1812].
NICHOLSON: **1873/4 survey** (JB: 43): Generally diffused over the turf, also common as a weed in flower-beds and on walks. • **1906 revision** (AS: 78): Generally diffused, common as a weed in flower-beds and on walks.
EXSICC.: **Kew Gardens.** Near herbarium & in lawns; in short grass; 16 vi 1928; *Hubbard* s.n. (K). • Near herbarium; in sandy soil among short grasses; 20 vi 1928; *Hubbard* s.n. (K). • [Unlocalised]; found in dry, gravell soil in association with *Senecio jacobaea* & *Impatiens parviflora*; 3 vii 1948; *Parker* s.n. (K).
CULT.: 1812, Nat. of Britain (HK2, 4: 391); 1814, Nat. of Britain (HKE: 237); 1933, unspec. (K); 1940, Herbaceous Ground (K). Not currently in cult.
CURRENT STATUS: Abundant in grass throughout the Gardens.

811. Trifolium micranthum Viv.
T. filiforme L.
slender trefoil
FIRST RECORD: 1866 [1768].
NICHOLSON: **1873/4 survey** (JB: 43): **B.** On most of the lawns. Plentiful on the one near House No. 1, also behind Herbaceous ground wall. Grows to a large size in open places such as edges of shrubberies etc. **P.** Here and there about lake. • **1906 revision** (AS: 78): On most of the lawns.

Exsicc.: **Kew Gardens.** In sandy soil among short grass & in lawn near herbarium; 20 vi 1928; *Hubbard* s.n. (K). **Kew Environs.** Kew Green; 22 vi 1866; *Thiselton Dyer* in *Herb. Watson* (K).
Cult.: 1768 (JH: 303); 1789, Nat. of Britain (HK1, 3: 90); 1812, Nat. of Britain (HK2, 4: 391); 1814, Nat. of Britain (HKE: 237). Not currently in cult.
Current status: Uncommon but widespread; it grows with the preceding species in many places and is possibly still under-recorded.

812. Trifolium pratense L.
red clover
a. var. **pratense**
First record: 1873/4 [1768].
Nicholson: **1873/4 survey** (JB: 43): Everywhere, though not so abundant as [*T. subterraneum*]. • **1906 revision** (AS: 78): Everywhere.
Exsicc.: **Kew Environs.** Towpath, Kew; 23 v 1944; *Souster* 70 (K).
Cult.: 1768 (JH: 302); 1789, Nat. of Britain (HK1, 3: 86); 1812, Nat. of Britain (HK2, 4: 384); 1814, Nat. of Britain (HKE: 238); 1933, unspec. (K); 1936, Herbaceous Ground (K). Not currently in cult.
Current status: Mostly confined to the longer grass and not very common; it has suffered a steep decline since Nicholson's first survey.

b. *var. **sativum** Schreb.
T. pratense var. *americanum* Harz
First record: 1907 [1933].
References: **Kew Gardens.** 1908, *A.B.Jackson* (S7: 125): Rough grassy and shady places in the Arboretum and elsewhere.
Exsicc.: **Kew Gardens.** [Unlocalised]; 24 viii 1907; *Domin* s.n. (K). • Queen's Cottage; [c. 1914]; *Flippance* s.n. (K). • In herbarium field; 22 viii 1933; *Hubbard* s.n. (K).
Cult.: 1933, Herbarium Ground (K); 1935, unspec. (K). Not currently in cult.
Current status: Recently, but fleetingly, appeared in the Paddock, otherwise it appears to have been lost.

813. Trifolium medium L.
T. alpestre auct. non L.
zigzag clover
First record: 1873/4 [1789].
Nicholson: **1873/4 survey** (JB: 43): **P.** A single plant near Winter Garden. Two in wood near "Engine House." • **1906 revision** (AS: 78; SFS: 236): **A:** Near temperate house. In wood near pumping station.
Cult.: 1789, Nat. of Scotland (HK1, 3: 86 as *alpestre*); 1812, Nat. of England (HK2, 4: 384), Alien (ibid.: as *alpestre*); 1814, Nat. of England (HKE: 238), Alien (ibid.: as *alpestre*); 1922, unspec. (K); 1940, Herbaceous Ground (K); 1967, Herbaceous Ground (K); 1976, Order Beds (K). In cult.
Current status: No recent sightings.

814. Trifolium ochroleucon Huds.
sulphur cover
Cult.: 1789, Nat. of England (HK1, 3: 87); 1812, Nat. of England (HK2, 4: 386); 1814, Nat. of England (HKE: 238); 1976, Order Beds (K). Not currently in cult.

[815. Trifolium stellatum L.
starry clover
Cult.: 1789, Nat. of England (HK1, 3: 87); 1812, Nat. of England (HK2, 4: 387); 1814, Nat. of England (HKE: 237).
Note. Listed in error as native, but naturalised since 1804.**]**

816. *Trifolium incarnatum L. subsp. **incarnatum**
crimson clover
First record: 2003 [1768].
References: **Kew Gardens.** 2003, *Cope* (S35: 647): Sown into the new 'native' wheatfield at the southern end of the Lake (261). An unfortunate addition to the mix as it is not native to the British Isles.
Cult.: 1768 (JH: 302); 1789, Alien (HK1, 3: 87); 1812, Alien (HK2, 4: 386); 1814, Alien (HKE: 237); 1936, unspec. (K); 1976, unspec. (K). Not currently in cult.
Current status: Introduced as a component of the Wheatfield at the south end of Syon Vista in 2001 and may persist for a while.

817. Trifolium striatum L.
knotted clover
First record: 1837 [1768].
Nicholson: **1906 revision** (AS: 78): On wall of ha-ha. Amongst turf in dry spots.
Other references: **Kew Environs.** 1837, *E.Kent* (SFS: 239): River side between Kew Bridge and Richmond, near latter. • c. 1875, *Bennett* (SFS: 239): Kew Green.
Exsicc.: **Kew Environs.** Thames Banks, Kew; vii 1883; *Fraser* s.n. (K). • Kew; (early 19th cent.) *Prior* s.n. (K). • Kew Green, near Herbarium by side of path leading to the church, with *T. glomeratum* & *T. subterraneum*; 1 vi 1978; *Meikle* s.n. (K).
Cult.: 1768 (JH: 302); 1789, Nat. of Britain (HK1, 3: 88); 1812, Nat. of Britain (HK2, 4: 388); 1814, Nat. of Britain (HKE: 237); 1953, Herb. Expt. Grnd. (K). Not currently in cult.
Current status: Still on Kew Green where it occurs with *T. glomeratum*; recently found in the grass in front of Kew Palace and in one or two other places.

818. Trifolium scabrum L.
rough clover
Cult.: 1768 (JH: 302); 1789, Nat. of Britain (HK1, 3: 88); 1812, Nat. of Britain (HK2, 4: 388); 1814, Nat. of Britain (HKE: 237). Not currently in cult.

819. Trifolium arvense L.
hare's-foot clover
FIRST RECORD: 1873/4 [1768].
NICHOLSON: **1873/4 survey** (JB: 43): **Pal.** Common in the dry open parts. **P.** Some hundreds of plants about where the *Melilotus* grew. Frequent near Winter Garden. • **1906 revision** (AS: 78 and in SFS: 238): **P.** Common in dry gravelly spots.
CULT.: 1768 (JH: 302); 1789, Nat. of Britain (HK1, 3: 87); 1812, Nat. of Britain (HK2, 4: 387); 1814, Nat. of Britain (HKE: 237); 1922, unspec. (K); 1976, Order Beds (K); 1985, unspec. (K). Not currently in cult.
CURRENT STATUS: Rare; it is very sporadic in appearance and has clearly suffered a steep decline since Nicholson's first survey.

820. Trifolium squamosum L.
T. maritimum Huds.
sea clover
CULT.: 1812, Nat. of England (HK2, 4: 385); 1814, Nat. of England (HKE: 237); 19th cent., unspec. (K). Not currently in cult.

821. *Trifolium constantinopolitanum Ser.
T. constantinopolitanum var. *phleoides* Boiss.
FIRST RECORD: 1855.
REFERENCES: **Kew Environs.** 1855, *Blake* (BSEC4: 407; BSEC9: 686; S33: 746): Between Kew and Richmond.
CURRENT STATUS: No recent sightings.

Subterranean clover (*Trifolium subterraneum*). Once as common as white clover but now very rare and still declining. It prefers barish ground and worn-out grassland, both habitats subject to 'improvement'. Still in reasonable numbers on Kew Green.

822. *Trifolium echinatum M.Bieb.
T. supinum Savi
hedgehog clover
FIRST RECORD: 1878.
REFERENCES: **Kew Environs.** 1878, *Nicholson* (BEC18: 14; S33: 746): Waste ground on the Surrey side of the Thames, near Kew.
CURRENT STATUS: No recent sightings.

823. Trifolium subterraneum L.
subterranean clover
FIRST RECORD: 1834 [1768].
NICHOLSON: **1873/4 survey** (JB: 43): Very frequent. Nearly as common as *T. repens.* • **1906 revision** (AS: 78): Very frequent.
OTHER REFERENCES: **Kew Gardens.** 1863, *J.D.Salmon* (BFS: 59; SFS: 235): Kew Green. • 1931, *B.D. Jackson* (SFS: 235): Turf in Kew Gardens near the lake, abundant. • 1945, *Welch* (LN31: S62): Turf in Kew Gardens. • 1976, *Lousley* (LFS: 173): A persistent plant still, for example, on Kew Green and in Kew Gardens. • 1987 & 1991, *Hastings* (S31: 182): Kew Green (R.B.G. side).
EXSICC.: **Kew Gardens.** Kew Green; 27 ix 1918; *Fraser* s.n. (K). • Kew Green; 26 x 1919; *Fraser* s.n. (K). • Kew Green (triangle opposite Herbarium House); 14 v 1976; *Brenan & Meikle* s.n. (K). • Plot 261, about 30m from W end of the Lake towards the river and Sion House, abundant over an area of about 50 m; 8 v 2000; *Brummitt* 20368 (K). **Kew Environs.** Kew; 1834; *Prior* s.n. (K).
CULT.: 1768 (JH: 302); 1789, Nat. of England (HK1, 3: 85); 1812, Nat. of England (HK2, 4: 382); 1814, Nat. of England (HKE: 237); 1952, Herbarium Ground (K). Not currently in cult.
CURRENT STATUS: Still to be found near the Lake although the creation of the Wheatfield in 2001 may have eradicated most of it. It has persisted on Kew Green, particularly in the oval bed immediately in front of the Main Gate, but in May 2008 the turf in this bed was removed for a sowing of wildflowers. As an autumn-germinating annual it is dependant on careful management for its survival. There are one or two scattered localities elsewhere, especially near the Temperate House, and it recently turned up in a deliberately disturbed piece of ground in the Ancient Meadow. However, it is not nearly as common as intimated by Nicholson and is probably still declining.

824. *Trifolium spumosum L.
FIRST RECORD: 1878 [1789].
REFERENCES: **Kew Environs.** 1878, *Nicholson* (BEC18: 14; S33: 746): Waste ground on the Surrey side of the Thames, near Kew.
CULT.: 1789, Alien (HK1, 3: 87); 1812, Alien (HK2, 4: 389); 1814, Alien (HKE: 237). Not currently in cult.
CURRENT STATUS: No recent sightings.

825. *Lupinus polyphyllus Lindl.
garden lupin
FIRST RECORD: 2001.
CURRENT STATUS: Naturalised on the golf course in the Old Deer Park and sown into the oval bed in front of the Main Gate.

826. Cytisus scoparius (L.) Link
broom
a. subsp. **scoparius**
Spartium scoparium L.
FIRST RECORD: 1873/4 [1768].
NICHOLSON: [**1873/4 survey** (JB: 42): Before the lake was made, its present site was covered with this and *Ulex europaeus* (q.v.).]. • **1906 revision** (AS: 77): Along border of wood, etc. [also see under *Ulex europaeus*].
CULT.: 1768 (JH: 307, 455); 1789, Nat. of Britain (HK1, 3: 12); 1812, Nat. of Britain (HK2, 4: 257); 1814, Nat. of Britain (HKE: 221); 1880, Arboretum (K); 1882, Arboretum (K); 1890, Arboretum (K); 1890, unspec. (K); 1934, Herb Garden (K); 1976, St Anne's Graveyard (K); 1985, zone 412 (K); 1987, unspec. (K); 1999, bed 412-07 (K). In cult.
CURRENT STATUS: Rare; only on Mount Pleasant near its original locality, intermixed with cultivated varieties of assorted colours. Present in the Old Deer Park, on the golf course, where it is probably native. When the soil profile was experimentally reversed near the Pagoda it came up in abundance and now needs careful management.

b. subsp. **maritimus** (Rouy) Heywood
CULT.: 1932, Herbarium Ground (K); 1936, Herbarium Ground (K); 1937, Herbarium Ground (K); 1938, Herbarium Ground (K); 1939, Herbarium Ground (K). In cult.

827. Genista tinctoria L.
Dyer's greenweed
CULT.: 1768 (JH: 285, 442); 1789, Nat. of England (HK1, 3: 15); 1812, Nat. of Britain (HK2, 4: 259); 1814, Nat. of Britain (HKE: 222); 1843-1853, Arboretum (K); 1880, Arboretum (K); 1881, Arboretum (K); 1885, unspec. (K); 1889, unspec. (K); 1892, Arboretum (K); 1901, Arboretum Nursery (K); 1901, Arboretum (K); 1906, Arboretum (K); 1926, Herb. Exp. Ground (K); 1927, Herb. Exp. Ground (K); 1927, Arboretum (K); 1928, Herb. Exp. Ground (K); 1936, Herbarium Ground (K); 1956, unspec. (K); 1998, unspec. (K); 1998, zone 412 (K). In cult.

828. Genista pilosa L.
hairy greenweed
CULT.: 1789, Nat. of England (HK1, 3: 15); 1812, Nat. of England (HK2, 4: 260); 1814, Nat. of England (HKE: 222); 1905, Arboretum (K). In cult.

829. Genista anglica L.
petty whin
FIRST RECORD: 1894 [1768].
NICHOLSON: **1906 revision:** Not recorded
EXSICC.: **Kew Gardens.** Arboretum; 25 v 1894; *anon.* (K).
CULT.: 1768 (JH: 285, 442); 1789, Nat. of Britain (HK1, 3: 15); 1812, Nat. of Britain (HK2, 4: 260); 1814, Nat. of Britain (HKE: 222); 1985, zone 412 (K). Not currently in cult.
CURRENT STATUS: No recent sightings.

830. Ulex europaeus L.
gorse
FIRST RECORD: 1873/4 [1768].
NICHOLSON: **1873/4 survey** (JB: 42): I believe this to be a bona fide native of Kew. Many young plants may easily be found in the turf in Pleasure Grounds, although they get continually cut down by the scythes. Before the lake was made, its present site was covered with this and [*Cytisus scoparius*]. • **1906 revision** (AS: 77): Before the lake was made its site was covered with this and *Cytisus scoparius*.
EXSICC.: **Kew Gardens.** Self-sown seedlings; 1914-1948; *Lowne* s.n. (K).

Gorse (*Ulex europaeus*). Once native in the Gardens where the Lake now stands and seedlings can occasionally be found. Still in the Old Deer Park.

CULT.: 1768 (JH: 308, 458); 1789, Nat. of Britain (HK1, 3: 17); 1812, Nat. of Britain (HK2, 4: 265); 1814, Nat. of Britain (HKE: 223); 1905, Arboretum (K); 1959, Experimental Ground (K). In cult.
CURRENT STATUS: Uncommon; seedlings occurred abundantly in the kidney-shaped beds along the Broad Walk in 2001 but were weeded out the following year. It was once native in the Gardens, especially in the area now occupied by the Lake, but it is likely that all recent seedlings have been derived from cultivars. It is still on the golf course in the Old Deer Park.

831. Ulex europaeus × gallii
CULT.: 1959, Experimental Ground (K). Not currently in cult.

832. Ulex gallii Planch
western gorse
CULT.: 1911, unspec. (K); 1959, Experimental Ground (K). Not currently in cult.

833. Ulex minor Roth
U. nanus Forst.
dwarf gorse
CULT.: 1789, Nat. of Britain (HK1, 3: 17); 1812, Nat. of Britain (HK2, 4: 265); 1814, Nat. of Britain (HKE: 223). In cult.

ELAEAGNACEAE

834. Hippophae rhamnoides L.
sea-buckthorn
CULT.: 1768 (JH: 443); 1789, Nat. of England (HK1, 3: 396); 1813, Nat. of England (HK2, 5: 372); 1814, Nat. of England (HKE: 307). In cult.

HALORAGACEAE

835. Myriophyllum verticillatum L.
whorled water-milfoil
FIRST RECORD: 1838 [1768].
NICHOLSON: **1873/4 survey**: Not recorded. • **1906 revision**: Not recorded.
OTHER REFERENCES: **Kew Gardens.** 1838, *Anderson* (SFS: 320; S33: 746): Ditch round Kew Gardens.
CULT.: 1768 (JH: 379); 1789, Nat. of England (HK1, 3: 352); 1813, Nat. of England (HK2, 5: 282); 1814, Nat. of England (HKE: 294). Not currently in cult.
CURRENT STATUS: No recent sightings. The ditch around the gardens is no longer suitable.

836. Myriophyllum spicatum L.
spiked water-milfoil
FIRST RECORD: 1873/4 [1768].

Purple-loosestrife (*Lythrum salicaria*). Native on the towpath outside the Gardens but planted by the Lake.

NICHOLSON: **1873/4 survey** (JB: 44): **Strip.** Common nearly the whole length of moat. • **1906 revision** (AS: 79 and in SFS: 320): **A.** Lake. **Strip.** Common the whole length of ha-ha.
EXSICC.: **Kew Environs.** Ha-ha, Richmond; vi 1927; *Findlay* s.n. (K).
CULT.: 1768 (JH: 379); 1789, Nat. of Britain (HK1, 3: 351); 1813, Nat. of Britain (HK2, 5: 282); 1814, Nat. of Britain (HKE: 294). Not currently in cult.
CURRENT STATUS: No recent sightings. The Moat (see the preceding species where it is referred to as the ditch) is no longer suitable.

LYTHRACEAE

837. Lythrum salicaria L.
purple-loosestrife
FIRST RECORD: 1873/4 [1768].
NICHOLSON: **1873/4 survey** (JB: 44): **Strip.** Common by moat. The plants in division **B** and **P** were planted in 1873. • **1906 revision** (AS: 80): **Strip.** Common along the ha-ha.
OTHER REFERENCES: **Kew Gardens.** 1999, Stones (EAK: 1): By Lake.

Exsicc.: **Kew Environs.** Thames bank, Kew; growing in top zone on mud of high bank; 7 viii 1928; *Summerhayes & Turrill* s.n. (K). • Thames Embankment, Kew; waterside plant; ass. plants:- *Lysimachia vulgaris, Carex riparia*; 23 viii 1950; *Naylor* 179 (K). • Banks of Ha-Ha, River Thames, Richmond; 14 viii 1956; *Sealy & Ross-Craig* s.n. (K).
Cult.: 1768 (JH: 221); 1789, Nat. of Britain (HK1, 2: 128); 1811, Nat. of Britain (HK2, 3: 149); 1814, Nat. of Britain (HKE: 140); pre-1867, unspec. (K); 1881, unspec. (K). In cult.
Current status: Around the Lake, where it was originally planted, and on the towpath, and sometimes in unexpectedly dry places.

838. Lythrum hyssopifolium L.
grass-poly
First record: 1877 [1768].
References: **Kew Environs.** 1877, *Baker* (BEC18: 16; S33: 746, and see Exsicc.): This I gathered last year in small quantity on the Surrey side of the Thames above Kew Bridge.
Exsicc.: **Kew Environs.** Just above Kew Bridge on Surrey side of Thames; vii 1878; *Baker* s.n. (K).
Cult.: 1768 (JH: 221); 1789, Nat. of England (HK1, 2: 128); 1811, Nat. of England (HK2, 3: 150); 1814, Nat. of England (HKE: 140). Not currently in cult.
Current status: No recent sightings.

839. Lythrum portula (L.) D.A.Webb
Peplis portula L.
water-purslane
First record: 1839 [1768].
Nicholson: **1873/4 survey** (JB: 44 and in SFS: 324): **P.** Frequent near the water's edge at Palm House end of lake. • **1906 revision** (AS: 80): **A.** Frequent along edge of lake.
Other references: **Kew Environs.** 1839, *Francis* (SFS: 324): Riverside between Kew Bridge and Richmond.
Cult.: 1768 (JH: 220); 1789, Nat. of Britain (HK1, 1: 480); 1811, Nat. of Britain (HK2, 2: 316); 1814, Nat. of Britain (HKE: 103). Not currently in cult.
Current status: No recent sightings.

Thymelaeaceae

840. Daphne mezereum L.
mezereon
Cult.: 1768 (JH: 329, 440); 1789, Nat. of England (HK1, 2: 25); 1811, Nat. of England (HK2, 2: 409); 1814, Nat. of England (HKE: 115). In cult.

841. Daphne laureola L.
spurge-laurel
Cult.: 1768 (JH: 329, 440); 1789, Nat. of Britain (HK1, 2: 26); 1811, Nat. of Britain (HK2, 2: 410); 1814, Nat. of Britain (HKE: 116). In cult.

Great willowherb (*Epilobium hirsutum*). A weed in rough and damp places, but a feature of the Lake margins.

Onagraceae

842. Epilobium hirsutum L.
great willowherb
First record: 1873/4 [1768].
Nicholson: **1873/4 survey** (JB: 44): **P.** About lake. **Strip.** Very common. All about pond in **B** were planted in 1873. • **1906 revision** (AS: 80): **A.** About lake. **Strip.** Very common.
Other references: **Kew Gardens.** 1999, *Stones* (EAK: 1 & 5): By Lake and Palm House Pond.
Exsicc.: **Kew Gardens.** By the tow-path just outside Kew Gardens; growing on eroded area with *Urtica*; 19 iii 1928; *Summerhayes & Turrill* s.n. (K). **Kew Environs.** Thames bank, Kew; growing in top zone on mud of high bank; 7 viii 1928; *Summerhayes & Turrill* s.n. (K). • Tow Path Richmond to Kew; 1 viii 1948; *O.J.Ward* 98 (K). • River Thames bank, at Kew; growing in cracks between granite blocks which have been used to face the bank amid a now luxuriant vegetation; 11 vii 1957; *Ross-Craig & Sealy* 1868 (K).
Cult.: 1768 (JH: 172/8); 1789, Nat. of Britain (HK1, 2: 5); 1811, Nat. of Britain (HK2, 2: 345); 1814, Nat. of Britain (HKE: 108). In cult.
Current status: In damp and rough ground throughout the Gardens.

843. Epilobium parviflorum Schreb.
E. villosum Curt.
hoary willowherb
First record: 1873/4 [1789].
Nicholson: **1873/4 survey** (JB: 44): **P.** About 100 plants in an open place in wood near the lake end of the "Hollow Walk." • **1906 revision** (AS: 80): **A.** Here and there. Abundant years ago on site now occupied by bamboo garden.
Cult.: 1789, Nat. of Britain (HK1, 2: 5); 1811, Nat. of Britain (HK2, 2: 345); 1814, Nat. of Britain (HKE: 108). Not currently in cult.
Current status: Widespread, but never common.

844. Epilobium montanum L.
broad-leaved willowherb
First record: 1873/4 [1768]
Nicholson: **1873/4 survey** (JB: 44): Rather common everywhere. • **1906 revision** (AS: 80): Common everywhere.
Exsicc.: **Kew Gardens.** Near Main Gate; in bed of Vinca; 29 vi 1932; *Airy Shaw* s.n. (K).
Cult.: 1768 (JH: 172/8); 1789, Nat. of Britain (HK1, 2: 6); 1811, Nat. of Britain (HK2, 2: 346); 1814, Nat. of Britain (HKE: 108). Not currently in cult.
Current status: A common weed throughout the Gardens.

845. Epilobium lanceolatum Sebast. & Mauri
spear-leaved willowherb
First record: 1964.
References: **Kew Gardens.** 1964, *Raven* (WAT6: 36; LFS: 208; S33: 746; and see Exsicc.): By Herbarium.
Exsicc.: **Kew Gardens.** Weed by the herbarium; *Raven* 16087 (BM).
Current status: No recent sightings.

846. Epilobium × neogradense Borbás (lanceolatum × montanum)
First record: 1947.
References: **Kew Gardens.** 1947, *Sandwith* (WAT1: 44; S33: 746): Weed in experimental ground, the herbarium.
Current status: No recent sightings.

847. Epilobium tetragonum L.
E. lamyi F.W.Schultz
square-stalked willowherb
First record: 1873/4 [1768].
Nicholson: **1873/4 survey** (JB: 44): **Pal.** Common in the kitchen garden ground. **P.** Here and there at edges of beds about Syon Vista, also about lake. • **1906 revision** (AS: 80 and in SFS: 329): Common as weed in shrubberies, etc.
Other references: **Kew Gardens.** 1958, *Airy Shaw* (S24: 188): A single plant on an overgrown allotment behind the Herbarium.
Exsicc.: **Kew Gardens.** Near the Herbarium; among grass; 21 viii 1958; *Airy Shaw* s.n. (K).
Cult.: 1768 (JH: 172/8); 1789, Nat. of Britain (HK1, 2: 6); 1811, Nat. of Britain (HK2, 2: 346); 1814, Nat. of Britain (HKE: 108); 1930, unspec. (K); 1931, unspec. (K); 1932, unspec. (K); 1934, unspec. (K); 1958, Order Beds (K). Not currently in cult.
Current status: Widely scattered but not common.

848. Epilobium obscurum Schreb.
short-fruited willowherb
First record: 1873/4.
Nicholson: **1873/4 survey** (JB: 44 and in SFS: 330): Here and there with [*E. tetragonum*], though not so frequent. • **1906 revision** (AS: 80): Here and there with [*E. tetragonum*].
Current status: No recent sightings.

849. Epilobium roseum Schreb.
pale willowherb
First record: 1902 [1811].
Nicholson: **1906 revision:** Not recorded.
Other references: **Kew Gardens.** 1902, *Beeby* (SFS: 328): Kew Gardens Moat wall. • 1949, *Airy Shaw* (S22: 287): In flower bed by the drive inside main gates, with *E. montanum*. It is curious that this species has not previously been recorded from the Gardens. • 1964, *Raven* (WAT6: 36, and see Exsicc.): By Herbarium.
Exsicc.: **Kew Gardens.** Weed by the herbarium; *Raven* 16193 (K).
Cult.: 1811, Nat. of England (HK2, 2: 346); 1814, Nat. of England (HKE: 108). Not currently in cult.
Current status: An occasional weed in the northern end of the Gardens.

850. *Epilobium ciliatum Raf.
American Willowherb
First record: 1953.
References: **Kew Gardens.** 1998, *Cope* (S35: 647): Originally noted in the Riverside Walk (211), but subsequently found throughout the Gardens. The commonest willowherb in the Gardens and increasing.
Exsicc.: **Kew Gardens.** Garden of Temperate House Lodge; 17 viii 1953; *Souster* s.n. (K).
Current status: The commonest willowherb in the Gardens and still increasing.

851. Epilobium palustre L.
marsh willowherb
Cult.: 1768 (JH: 172/8); 1789, Nat. of Britain (HK1, 2: 6); 1811, Nat. of Britain (HK2, 2: 346); 1814, Nat. of Britain (HKE: 108). Not currently in cult.

852. Epilobium anagallidifolium Lam.
alpine willowherb
Cult.: 1789, Nat. of Britain (HK1, 2: 6); 1811, Nat. of Britain (HK2, 2: 347); 1814, Nat. of Britain (HKE: 108). Not currently in cult.

853. Epilobium alsinifolium Vill.
chickweed willowherb
CULT.: 1811, Nat. of Britain (HK2, 2: 346); 1814, Nat. of Britain (HKE: 108). Not currently in cult.

854. Chamerion angustifolium (L.) Holub
Epilobium angustifolium L.
rosebay willowherb
FIRST RECORD: 1873/4 [1768].
NICHOLSON: **1873/4 survey** (JB: 44 and in SFS: 325): **Q.** A great many plants in a clump of trees opposite Syon House. • **1906 revision** (AS: 80): **Q.** Elsewhere planted.
EXSICC.: **Kew Gardens.** Herbarium Experimental Ground; ii 1927; *Summerhayes* s.n. (K). • Herbarium Experimental Ground; 21 ii 1928, *Summerhayes* s.n. (K). • On Experimental Ground; 28 iii 1928; *Summerhayes* s.n. (K). • Herbarium Exper. Ground; 23 iv 1928; *Summerhayes* s.n. (K). • Near Office of Works; in shade under trees & at base of wall; 6 v 1928; *Summerhayes & Turrill* s.n. (K). • Herbarium Exper. Ground; 4 vii 1928; *Summerhayes* s.n. (K).
CULT.: 1768 (JH: 172/8); 1789, Nat. of Britain (HK1, 2: 4); 1811, Nat. of Britain (HK2, 2: 344); 1814, Nat. of Britain (HKE: 108). Not currently in cult.
CURRENT STATUS: Widespread as an occasional weed but abundant in the Conservation Area.

855. Ludwigia palustris (L.) Elliott
Isnardia palustris L.
Hampshire-purslane
CULT.: 1789, Alien (HK1, 1: 164); 1810, Alien (HK2, 1: 266); 1814, Alien (HKE: 36). Not currently in cult.
Note. Listed as an alien, but known as a native since 1666.

856. *Oenothera glazioviana P.Micheli ex Mart.
O. erythrosepala Borbás
large-flowered evening-primrose
FIRST RECORD: 1920 [1970].
REFERENCES: **Kew Gardens.** 1920, *Turrill* (WAT14: 18; S33: 746 and see Exsicc.): Spontaneous in Kew Gardens (cult. in 1911). • 1931, *Sprague* (WAT14: 18): ibid. • 1948, *Souster* (WAT14: 18): ibid. • 1952, *Stearn* (WAT14: 18): ibid.
EXSICC.: **Kew Gardens.** Near Tennis Courts; escape from cultivation, on rubbish heaps; 1920; *Turrill* s.n. (K). • Near Kew Palace; waste ground; 29 viii 1931; *Sprague* s.n. (K). • [Unlocalised]; found growing spontaneously among trees; 12 viii 1948; *Souster* 874 (K).
CULT.: 1970, Decorative Dept. (K). Not currently in cult.
CURRENT STATUS: A rare escape.

Rosebay willowherb (*Chamerion angustifolium*). An occasional weed of flower-beds but forms large stands in the Conservation Area.

857. *Oenothera biennis L.
common evening-primrose
FIRST RECORD: 1873/4 [1768].
NICHOLSON: **1873/4 survey** (JB: 44): **B.** Two plants near pond. **P.** Waste ground near Winter Garden. • **1906 revision** (AS: 80 and in SFS: 332): This formerly occurred in waste-places in arboretum.
CULT.: 1768 (JH: 172/4); 1789, Alien (HK1, 2: 2); 1811, Alien (HK2, 2: 341); 1814, Alien (HKE: 108); 1907, unspec. (K); 1933, Herb Garden (K); 1936, Herb. Gdn. (K). In cult.
CURRENT STATUS: Occasional as an escape.

858. *Oenothera × fallax Renner (biennis × glazioviana)
intermediate evening-primrose
FIRST RECORD: 2002.
REFERENCES: **Kew Gardens.** 2002, *Cope* (S35: 647): Neglected flowerbed in the Lower Nursery (120). Subsequently found near the Stable Yard (353). The plant may be an escape from cultivation, or it may have arisen on site by the hybridisation of *O. glazioviana* and *O. biennis*.
CURRENT STATUS: A handful of widely scattered records and first noted outside a greenhouse in the Lower Nursery.

Dogwood (*Cornus sanguinea*). As a wild plant this has not been seen since the 1930s, but it is still in cultivation.

859. *Oenothera cambrica Rostanski
small-flowered evening-primrose
FIRST RECORD: 2007.
CURRENT STATUS: One or two plants appear in 2007 from seed in topsoil stored in the Paddock. It is a relatively recently recognised taxon in Britain and there are no previous records of it in the Gardens either wild or in cultivation. It could conceivably have arrived by wind-blown seed from outside but more probably the plants in the Paddock were cultivated in the Gardens under a different name.

860. *Oenothera stricta Ledeb. ex Link
fragrant evening-primrose
FIRST RECORD: 1856 [1879].
NICHOLSON: **1873/4 survey:** Not recorded. • **1906 revision:** Not recorded.
EXSICC.: **Kew Gardens.** [Unlocalised]; 1856; *anon* s.n. (K).
CULT.: 1879, unspec. (K); 1936, unspec. (K); 1999, plot 156.44 (K). Not currently in cult.
CURRENT STATUS: No recent sightings.

861. Circaea lutetiana L.
enchanter's-nightshade
FIRST RECORD: 1873/4 [1768].
NICHOLSON: **1873/4 survey** (JB: 44): **P.** "Merlin's Cave." • **1906 revision** (AS: 80): **A.** Formerly in wood about "Merlin's Cave."
EXSICC.: **Kew Gardens.** [Unlocalised]; growing in shady places under trees; 5 viii 1944; *Souster* 138 (K).
CULT.: 1768 (JH: 156); 1789, Nat. of Britain (HK1, 1: 17); 1810, Nat. of Britain (HK2, 1: 26); 1814, Nat. of Britain (HKE: 5). Not currently in cult.
CURRENT STATUS: A widespread weed that threatens to become a serious problem.

862. Circaea alpina L.
alpine enchanter's-nightshade
CULT.: 1768 (JH: 156); 1789, Nat. of Britain (HK1, 1: 17); 1810, Nat. of Britain (HK2, 1: 26); 1814, Nat. of Britain (HKE: 5); 1934, unspec. (K). Not currently in cult.

CORNACEAE

863. Cornus sanguinea L.
dogwood
FIRST RECORD: 1931 [1768].
EXSICC.: **Kew Gardens.** Shrubbery adjoining bowling green, Kew Palace; 29 viii 1931; *Sprague* s.n. (K).
CULT.: 1768 (JH: 438); 1789, Nat. of Britain (HK1, 1: 158); 1810, Nat. of Britain (HK2, 1: 261); 1814, Nat. of Britain (HKE: 35); 1880, Arboretum (K); 1890, unspec. (K); 1894, Arboretum (K); 1934, Arboretum (K); 1970 & 1971, *Cornus* beds at King William's Temple (both K); 1972, Arboretum (K). In cult.
CURRENT STATUS: No recent sightings of truly wild plants.

864. Cornus suecica L.
Dwarf Cornel
CULT.: 1768 (JH: 172/5, 438); 1789, Nat. of Britain (HK1, 1: 157); 1810, Nat. of Britain (HK2, 1: 260); 1814, Nat. of Britain (HKE: 35). Not currently in cult.

SANTALACEAE

865. Thesium humifusum DC.
T. linophyllon auct. non L.
bastard-toadflax
CULT.: 1789, Nat. of England (HK1, 1: 292); 1811, Nat. of England (HK2, 2: 63); 1814, Nat. of England (HKE: 65). Not currently in cult.

LORANTHACEAE

866. Loranthus europaeus Jacq.
FIRST RECORD: 1873 [2005].
OTHER REFERENCES: See below and E.J.Clement in BSN 108: 44 (2008).
CULT.: 2005, Area 233 Arboretum, on *Quercus velutina* (K). In cult.
CURRENT STATUS: My attention was first drawn to this species by Tony Hall of the arboretum staff who had spotted it growing on a specimen of the American species *Quercus velutina* Lam. in zone 233. Subsequent to confirmation of its identity it was accessioned as a cultivated plant and placed on the Living Collections Database. The article by Clement (see above) suggests that this species

***Loranthus europaeus*. An enigmatic species that had been growing unnoticed on an oak tree until finally spotted in 2005. Its origin is entirely unknown as it has never been cultivated nor recorded in the wild in this country.**

was spotted in the Gardens in 1873 by D.Moore of Dublin, but this is a mis-reading of the letter written by F.W.Moore to W.J.Bean at Kew in 1911; the context clearly indicates that the plants were growing in Glasnevin. An article by D.Moore, first delivered at a meeting of the Royal Dublin Society in 1873, recounts how the plants in question were grown from seed by Moore, but later died. The two bushes on the current host plant at Kew are of considerable size and age but since their host was grown from seed (and not imported with the mistletoe already attached) there is a mystery concerning their origin. It remains to be seen whether the species spontaneously spreads to other nearby oaks, and indeed whether it exists as a wild plant in southern England.

VISCACEAE

867. Viscum album L.

mistletoe

FIRST RECORD: 1823 [1768].

NICHOLSON: **1873/4 survey** (JB: 45): This occurs on the thorn, poplar, and lime. Some large plants are growing on a tall poplar in **P**. All the rest occur in **B**. On two large limes about 158 yards in a northern direction from north door of Palm House, on thorns on both sides of Broad Walk and on a large white lime and a black Italian poplar in Old Arboretum. • **1906 revision** (AS: 87 and in SFS: 574): This occurs at present on poplar and lime in **P** and **A.**

OTHER REFERENCES: **Kew Gardens.** 1940, *Cooke* (S32: 657; LN34: S247): On *Salix purpurea* and *Crataegus*.

EXSICC.: **Kew Gardens.** Arboretum; growing on Apple tree in "Canal Bed"; 3 x 1905; *Bean* s.n. (K). • Arboretum; 16 vii 1934; *Dallimore* 3 (K).

CULT.: 1768 (JH: 420, 458); 1789, Nat. of England (HK1, 3: 395); 1813, Nat. of England (HK2, 5: 371); 1814, Nat. of England (HKE: 307). Said to be in cult. but it is not known how successful the experimental introduction has been.

CURRENT STATUS: Extinct. It has not been seen for many years and the host of the last known colony had to be felled in the late 1970s or early 1980s. Specimens planted on apples near the Pagoda have not survived. The earliest record appears to be that noted by Ray Desmond in *Kew: The History of the Royal Botanic Gardens* (1995): 'When John Smith ... took charge of the small arboretum in 1823 ... *Populus alba*, wreathed in mistletoe, soared to over 60 feet'. Recent attempts at reintroduction, after initial failure, are beginning to look promising.

Mistletoe (*Viscum album*). Once abundant but now extinct. Last seen by the author in the 1970s on a Poplar near Brentford Gate, but the tree had to be felled. A reintroduction programme is under way.

Spindle (*Euonymus europaeus*). Naturalised in the Conservation Area where it has been cultivated for many years.

Celastraceae

868. Euonymus europaeus L.

spindle

First record: 2001 [1768].

References: **Kew Gardens.** 2001, *Cope* (S35: 646): Conservation Area (310). Probably introduced-naturalised; unrecorded by Nicholson and no subsequent records.

Cult.: 1768 (JH: 441); 1789, Nat. of Britain (HK1, 1: 273); 1811, Nat. of Britain (HK2, 2: 28); 1814, Nat. of Britain (HKE: 60); 1880, Arboretum (K); 1881, Arboretum (K); 1892, Arboretum (K); 1907, unspec. (K); 1908, unspec. (K); 1929, unspec. (K); 1944, unspec. (K); 1966, Arboretum South (K); 1968, Arobretum South (K); 1975, unspec. (K); 1997, Arboretum Nursery (K). In cult.

Current status: Naturalised in the Conservation Area and probably a relict of former cultivation.

Aquifoliaceae

869. Ilex aquifolium L.

holly

First record: 1998 [1768].

References: **Kew Gardens.** 1998, *Cope* (S33: 746): Self-sown seedlings in following:- North Arb. (113, 116, 132, 162, 183); West Arb. (121, 211—214, 216, 223); Herbaceous (135).

Cult.: 1768 (JH: 444); 1789, Nat. of Britain (HK1, 1: 168); 1810, Nat. of Britain (HK2, 1: 277); 1814, Nat. of Britain (HKE: 38); pre-1867, unspec. (K); 1881, Arboretum (K); 1882, Arboretum (K); 1883, Arboretum (K); 1905, Arboretum (K); 1906, unspec. (K); 1921, Arboretum (Herb. Arnold Arboretum, photo K); 1930, Arboretum (K); 1932, Arboretum (K); 1935, Arboretum (K); 1936, Arboretum (K); 1950s, unspec. (K); 1969, unspec. (K); 1980, bed 413.01 (K); 1980, zone 354 (K); 1981, zone 354 (K); 1981, zone 147 (K); 1981, zone 417 (K); 1983, Holly Walk (K); 1985, zone 344 (K); 1985, zone 354 (K); 1985, bed 413.02 (K); 1985, zone 417 (K); 1990, Arboretum Nursery (K); 1994, zone 321 (K). In cult.

Current status: Seedlings are abundant throughout the Gardens wherever the tree is cultivated, but are seldom allowed to exceed a few cm before being removed. The cultivated specimens represent innumerable varieties and cultivars.

Holly (*Ilex aquifolium*). Countless cultivars are grown throughout the Gardens and seedlings are abundant.

BUXACEAE

870. Buxus sempervirens L.
box
CULT.: 1768 (JH: 435); 1789, Nat. of England (HK1, 3: 339); 1813, Nat. of England (HK2, 5: 260); 1814, Nat. of England (HKE: 291); 1880, Arboretum (K); 1882, Arboretum (K); 1883, unspec. (K); 1884, Arboretum (K); 1887, unspec. (K); 1897, unspec. (K); 1920, Arboretum (K); 1935, Arboretum (K); 1940, Herbarium Ground (K); 1944, near Tennis Courts (K); 1947, Herbarium Ground (K); 1962, Arboretum (K); 1979, Temperate Nurs. (K). In cult.

EUPHORBIACEAE

871. Mercurialis perennis L.
dog's mercury
FIRST RECORD: 1873/4 [1768].
NICHOLSON: **1873/4 survey** (JB: 73): **P.** Among stones near Merlin's Cave. A few patches in wood. • **1906 revision** (AS: 87): **A.** Here and there in wood.
CULT.: 1768 (JH: 422); 1789, Nat. of Britain (HK1, 3: 408); 1813, Nat. of Britain (HK2, 5: 398); 1814, Nat. of Britain (HKE: 311); 1888, unspec. (K). Not currently in cult.
CURRENT STATUS: Rare; the only current record is from the Conservation Area.

872. Mercurialis annua L.
annual mercury
FIRST RECORD: 1873/4 [1768].
NICHOLSON: **1873/4 survey** (JB: 73): **P.** Waste ground near Temperate House. • **1906 revision** (AS: 87): Common as weed in cultivated ground.
EXSICC.: **Kew Gardens.** Mound of earth from excavation for foundations of new wing; 14 x 1965; *Verdcourt* 4245 (K). • Stable Yard; soil heaps; viii 1976; *Butler* 54 (K). **Kew Environs.** Thames bank at Kew; vii 1929; *Hubbard* s.n. (K). • Kew; weed in allotment; 12 x 1942; *Hutchinson* s.n. (K). • Kew Green, St Anne's Church wall; 11 vi 2000; *Verdcourt* 5520A (K).
CULT.: 1768 (JH: 422); 1789, Nat. of Britain (HK1, 3: 408); 1813, Nat. of Britain (HK2, 5: 398); 1814, Nat. of Britain (HKE: 311). Not currently in cult.
CURRENT STATUS: Widely scattered and sometimes abundant on disturbed ground.

873. Euphorbia peplis L.
purple spurge
CULT.: 1768 (JH: 172/3); 1789, Nat. of England (HK1, 2: 140); 1811, Nat. of England (HK2, 3: 163); 1814, Nat. of England (HKE: 141). Not currently in cult.

874. *Euphorbia maculata L.
spotted spurge
FIRST RECORD: 1917 [1768].
REFERENCES: **Kew Gardens.** 1917, *anon.* fide *Clement* (FSSC: 44; BSN15: 12; S33: 747): Weed in rock garden.
EXSICC.: **Kew Gardens.** Rock Garden; weed; vii 1917; *anon.* (K).
CULT.: 1768 (JH: 172/3); 1789, Alien (HK1, 2: 139); 1811, Alien (HK2, 3: 162); 1814, Alien (HKE: 141); 1964, unspec. (K); 1988, plot 154-06 (K). Not currently in cult.
CURRENT STATUS: Grew from seed in topsoil stored in the Paddock and behind the Princess of Wales Conservatory. It was absent for more than 80 years before returning.

875. *Euphorbia corallioides L.
coral spurge
FIRST RECORD: 1988 [1962].
EXSICC.: **Kew Gardens.** Behind the herbarium; at least 20 plants found scattered over a wide area of disturbed or dumped ground; 22 iv 1988; *Milne* 3 (K). • Behind herbarium; recently introduced but scattered and very well established; 22 iv 1988; *Milne* 4 (K).
CULT.: 1962, unspec. (K); 1963, Experimental Ground (K); 1980, plot 154-07 (K). In cult.
CURRENT STATUS: Has not reappeared since the original collections. It is likely to have escaped from experimental material grown in the Herbarium grounds.

876. Euphorbia hyberna L.
Tithymalus hybernus (L.) Hill
Irish spurge
CULT.: 1789, Nat. of Ireland & England (HK1, 2: 145); 1811, Nat. of Britain (HK2, 3: 170); 1814, Nat. of Britain (HKE: 142); 1903, unspec. (K); 1987, plot 152-03 (K). Not currently in cult.

877. Euphorbia platyphyllos L.
Tithymalus platyphyllos (L.) Hill
E. verrucosa L.
broad-leaved spurge
FIRST RECORD: 1988 [1768].
EXSICC.: **Kew Gardens.** Waste ground behind Herbarium; scattered individuals mainly on more compact soil; this plant is apparently a larger more robust form induced by competition. Height over a metre (other specimens under 60 cm); vi 1988; *Milne* 31 (K).
CULT.: 1768 (JH: 172/3); 1789, Nat. of England (HK1, 2: 144, and 143 as *verrucosa*); 1811, Nat. of England (HK2, 3: 168); 1814, Nat. of England (HKE: 141); 1947, Herbarium Ground (K). Not currently in cult.

CURRENT STATUS: Recently reappeared after a long absence, with several plants amongst long grass in disturbed ground at the edge of the Paddock. Very sporadic, but mostly near the Herbarium where it may once have been cultivated in experimental plots.

878. Euphorbia serrulata Thuill.
upright spurge
FIRST RECORD: 1962 [1940].
REFERENCES: **Kew Gardens.** 1990, *Cope* (S31: 182): Paddock, after restoration in 1990. **Kew Environs.** 1990, *MacPherson* (LN71: 179): Thames path outside the Royal Botanic Gardens, from which the seed had perhaps sprung.
EXSICC.: **Kew Gardens.** Experimental Ground; naturalised on lime heap; vii 1962, *Radcliffe-Smith* s.n. (K). • Experimental Ground; on sandy soil; vii 1962; *Radcliffe-smith* s.n. (K). • Behind herbarium; established in quantity — area was staff allotments in the 60s and apparently *Euphorbia* seeds were dormant in the soil until it was dug up 18 months ago; 20 vi 1988; *Milne* 20 (K). • Behind herbarium; waste ground; 20 vi 1988; *Milne* 21 (K).
CULT.: 1940, Herbarium Ground (K). Not currently in cult.
CURRENT STATUS: Has recently reappeared in the Paddock and another specimen was found behind the Admin. Building. It occasionally turned up on the edge of the vegetable plots behind Hanover House but the site is now being built on. It is doubtless a relict of former cultivation in the Herbarium experimental plots.

879. Euphorbia helioscopia L.
Tithymalus helioscopius (L.) Hill
sun spurge
FIRST RECORD: 1873/4 [1768].
NICHOLSON: **1873/4 survey** (JB: 73): **P.** In Rose-bed near Pagoda. **Strip.** In company with *Lamium amplexicaule*. Mr Lynch has known of it in this spot for several years, but the present season, 1874, I have only been able to find there a single starved plant. In 1873 not one was seen. • **1906 revision** (AS: 87): Here and there in cultivated ground.
EXSICC.: **Kew Gardens.** Herbarium Experimental Ground; subspontaneous; vi 1964; *Radcliffe-Smith* s.n. (K).
CULT.: 1768 (JH: 172/3); 1789, Nat. of Britain (HK1, 2: 142); 1811, Nat. of Britain (HK2, 3: 167); 1814, Nat. of Britain (HKE: 141). Not currently in cult.
CURRENT STATUS: A weed of disturbed ground, mostly around the Herbarium and Banks Building.

880. Euphorbia lathyris L.
Tithymalus lathyris (L.) Hill
caper spurge
FIRST RECORD: 1858 [1768].
REFERENCES: **Kew Gardens.** 1983, *Burton* (FLA: 93): Only in the spinney, a part of Kew Gardens with no public access, does look ... permanently established. • 1983/4, *Cope* (S31: 181): New student vegetable plots behind Hanover House.
EXSICC.: **Kew Gardens.** [Unlocalised]; waste places and cult. ground; [?1936]; *Hutchinson* s.n. (K). **Kew Environs.** Kew; 1858; *Hooker* s.n. (K).
CULT.: 1768 (JH: 172/3); 1789, Alien (HK1, 2: 140); 1811, Alien (HK2, 3: 164); 1814, Alien (HKE: 141); 1918, unspec. (K); 1962, unspec. (K); 1965, Herbarium Experimental Plot (K). In cult.
CURRENT STATUS: Occasional; most common around the Herbarium but otherwise widely scattered.

Dwarf spurge (*Euphorbia exigua*). Rather rare and declining, but often a feature of disturbed ground in the Paddock and in the lawn behind the Herbarium.

881. Euphorbia exigua L.
Tithymalus exiguus (L.) Lam.
dwarf spurge

FIRST RECORD: 1999 [1768].
REFERENCES: **Kew Gardens.** 1999, *Cope* (S35: 648): Herbarium lawn.
CULT.: 1768 (JH: 172/3); 1789, Nat. of Britain (HK1, 2: 140); 1811, Nat. of Britain (HK2, 3: 164); 1814, Nat. of Britain (HKE: 141); 1959, Experimental Ground (K); 1987, unspec. (K); 1999, plot 157-03 (K). Not currently in cult.
CURRENT STATUS: Known for many years in the Paddock and more recently found in the Herbarium lawn, but sporadic and declining in both places. It appeared in quantity in the Paddock in 2004 in topsoil being stored there.

882. Euphorbia peplus L.
Tithymalus peplus (L.) Gaertn.
petty spurge

FIRST RECORD: 1873/4 [1768].
NICHOLSON: **1873/4 survey** (JB: 73): Very common everywhere. • **1906 revision** (AS: 87): Very common everywhere.
OTHER REFERENCES: **Kew Gardens.** 1999, *Stones* (EAK: 5): By Palm House Pond.
EXSICC.: **Kew Gardens.** Herbarium enclosure; 28 viii 1930; *Bullock* 7 (K). • Arboretum; 6 vii 1934; *Dallimore* 2 (K). • Herb Garden; 1 vii 1936; [?*Hutchinson*] s.n. (K). • Allotments by Herbarium; weed; 10 ix 1942; *Melville* s.n. (K). • Mound of earth from excavation for foundations of new wing; 14 x 1965; *Verdcourt* 4244 (K). • Rhododendron Dell, small nursery; vii 1979, *McNamara* 14 (K). **Kew Environs.** Kew; 13 vi 1931; *Fraser* s.n. (K). • Kew; 17 vi 1931; *Fraser* s.n. (K). • Kew; garden weed; 20 x 1942; *Hutchinson* s.n. (K).
CULT.: 1768 (JH: 172/3); 1789, Nat. of Britain (HK1, 2: 140); 1811, Nat. of Britain (HK2, 3: 163); 1814, Nat. of Britain (HKE: 141); 1934, Herbarium Ground (K); 1935, Herbarium Ground (K); 1962, unspec. (K). Not currently in cult.
CURRENT STATUS: A widespread and common weed.

883. Euphorbia portlandica L.
Tithymalus portlandicus (L.) Hill
Portland spurge

FIRST RECORD: 2004 [1768].
CULT.: 1768 (JH: 172/3); 1789, Nat. of England (HK1, 2: 142); 1811, Nat. of England (HK2, 3: 166); 1814, Nat. of England (HKE: 141); 1900, unspec. (K); 1962, unspec. (K); 1963, Experimental Ground (K); 1980, Alpine Dept. (K). In cult.
CURRENT STATUS: Several plants grew from seed in topsoil stored in the Paddock and by the Princess of Wales Conservatory.

884. Euphorbia paralias L.
Tithymalus paralias (L.) Hill
sea spurge

CULT.: 1768 (JH: 172/3); 1789, Nat. of England (HK1, 2: 142); 1811, Nat. of England (HK2, 3: 166); 1814, Nat. of England (HKE: 142); 1962, unspec. (K); 1991, plot 157-03 (K). Not currently in cult.

[885. Euphorbia esula L.
leafy spurge

CULT.: 1789, Alien (HK1, 2: 144); 1811, Nat. of Britain (HK2, 3: 169); 1814, Nat. of Britain (HKE: 142).
Note. Listed in error as native; a difficult complex including the next species.**]**

[886. Euphorbia cyparissias L.
cypress spurge

CULT.: 1789, Alien (HK1, 2: 145); 1811, Nat. of England (HK2, 3: 169); 1814, Nat. of England HKE: 142).
Note. Listed in error as native; part of the difficult *E. esula* complex (see above).**]**

887. Euphorbia amygdaloides L.
Tithymalus amygdaloides (L.) Hill
T. sylvaticus (L.) Hill
wood spurge

CULT.: 1768 (JH: 172/4); 1789, Nat. of England (HK1, 2: 146); 1811, Nat. of England (HK2, 3: 170); 1814, Nat. of England (HKE: 142); 1920, unspec. (K); 1938, Herbaceous Dept. (K); 1977, Duke's Garden (K). In cult.

[888. Euphorbia characias L.
Mediterranean spurge

CULT.: 1789, Nat. of England (HK1, 2: 146); 1811, Nat. of England (HK2, 3: 170); 1814, Nat. of England (HKE: 142).
Note. Listed in error as native.**]**

RHAMNACEAE

889. Rhamnus cathartica L.
buckthorn

FIRST RECORD: 1918 [1768].
REFERENCES: **Kew Environs.** 1918, *Bishop* (LN31: S53; S33: 747): Thames bank between Kew and Richmond, one or two bushes.
CULT.: 1768 (JH: 452); 1789, Nat. of England (HK1, 1: 263); 1811, Nat. of England (HK2, 2: 14); 1814, Nat. of England (HKE: 58); 1880, Arboretum (K); 1897, unspec. (K); 1898, unspec. (K); 1934, unspec. (K). In cult.

Buckthorn (*Rhamnus cathartica*). Along the towpath and in the Conservation Area. It may have been originally planted in one location and spread to the other, most likely from the Gardens to the towpath.

CURRENT STATUS: Several specimens can be seen in the Conservation Area adjacent to the Moat; these will almost certainly have been planted although there is a possibility that at least some of them may have arrived as seed brought in from trees on the towpath. However, it is just as likely that those on the towpath escaped from the Gardens.

890. Frangula alnus Mill.
Rhamnus frangula L.
alder buckthorn
FIRST RECORD: 1856 [1768].
NICHOLSON: **1873/4 survey:** Not recorded. • **1906 revision:** Not recorded.
EXSICC.: **Kew Gardens.** Pleasure Grounds; 1856; *Hooker* (K).
CULT.: 1768 (JH: 452); 1789, Nat. of England (HK1, 1: 264); 1811, Nat. of England (HK2, 2: 16); 1814, Nat. of England (HKE: 58); 1880, Arboretum (K); 1881, Arboretum (K); 1891, Arboretum (K); 1933, Arboretum (K); 1934, unspec. (K); 1936, Arboretum (K); 1949, Arboretum (K). In cult.
CURRENT STATUS: No recent sightings; the specimen collected by Hooker was probably cultivated.

VITACEAE

891. *Vitis vinifera L.
grape-vine
FIRST RECORD: 1947 [1768].
REFERENCES: **Kew Gardens.** 1983/4, *Cope* (S31: 181): Seedling in new student vegetable plots behind Hanover House. **Kew Environs.** 1947, *Lousley* (LN31: S54): Thames bank, Kew Bridge, known here for many years.
EXSICC.: **Kew Environs.** Towpath, West side, Kew Bridge; 9 vii 1964; *D.L.Smith* s.n. (K).
CULT.: 1768 (JH: 458); 1789, Cult. (HK1, 1: 282); 1811, Cult. (HK2, 2: 51); 1814, Cult. (HKE: 63). In cult.
CURRENT STATUS: No recent sightings.

LINACEAE

892. Linum bienne Mill.
L. angustifolium Huds.
L. tenuifolium auct. non L.
pale flax
FIRST RECORD: 1990 [1789].
REFERENCES: **Kew Gardens.** 1990, *Cope* (S31: 182): Paddock, after restoration in 1990.
CULT.: 1789, Nat. of England (HK1, 1: 387); 1811, Nat. of England (HK2, 2: 186); 1814, Nat. of England (HKE: 83); 1934, Herbarium Ground (K); 1935, Herbarium Ground (K). Not currently in cult.
CURRENT STATUS: No recent sightings; it has not reappeared in the Paddock since 1990.

893. *Linum usitatissimum L.
flax
FIRST RECORD: 1999 [1768].
REFERENCES: **Kew Gardens.** 1999, *Cope* (S35: 646): Under trees by Pagoda Vista (184). Subsequently near the Flagstaff (482).
CULT.: 1768 (JH: 200); 1789, Nat. of Britain (HK1, 1: 386); 1811, Nat. of Britain (HK2, 2: 184); 1814, Nat. of Britain (HKE: 83); 1918, unspec. (K); 1932, unspec. (K); 1933, unspec. (K); 1934, unspec. (K); 1934, Herb Garden (K); 1936, Herbaceous Ground (K). In cult.
CURRENT STATUS: Casual; there are just two widely scattered recent records.

Pale flax (*Linum bienne*). Appeared in the Paddock after its restoration in 1990, but has not returned.

894. Linum perenne L.

perennial flax

FIRST RECORD: 19th cent. [1768]

EXSICC.: **Kew Gardens.** Hort. Bot. Kew; [19th cent. label, no other information].

CULT.: 1768 (JH: 200); 1789, Nat. of England (HK1, 1: 386); 1811, Nat. of England (HK2, 2: 185); 1814, Nat. of England (HKE: 83); 19th cent., unspec. (K); 1885, unspec. (K); 1932, unspec. (K); 1951, Chalk Garden (K); 1985, bed 154-04 (K). In cult.

CURRENT STATUS: No recent sightings.

895. Linum catharticum L.

fairy flax

CULT.: 1768 (JH: 200); 1789, Nat. of Britain (HK1, 1: 389); 1811, Nat. of Britain (HK2, 2: 188); 1814, Nat. of Britain (HKE: 83). In cult.

896. *Linum grandiflorum Desf.

crimson flax

FIRST RECORD: 1880.

NICHOLSON: **1906 revision:** Not recorded.

OTHER REFERENCES: **Kew Gardens.** 1880, *Baker* (BECB1: 40; S33: 747): Thames bank opposite Sion House, midway between Kew and Richmond.

CULT.: In cult.

CURRENT STATUS: No recent sightings.

897. Radiola linoides Roth

Linum radiola L.

R. millegrana Sm.

allseed

CULT.: 1768 (JH: 200); 1789, Nat. of Britain (HK1, 1: 389); 1810, Nat. of Britain (HK2, 1: 282); 1814, Nat. of Britain (HKE: 39). Not currently in cult.

POLYGALACEAE

898. Polygala vulgaris L.

common milkwort

FIRST RECORD: 1873/4 [1768].

NICHOLSON: **1873/4 survey** (JB: 11 and in SFS: 165): **P.** Here and there in turf on both sides of "Syon Vista." • **1906 revision** (AS: 76): **A.** Here and there in turf on both sides of "Syon Vista."

CULT.: 1768 (JH: 309); 1789, Nat. of Britain (HK1, 3: 4); 1812, Nat. of Britain (HK2, 4: 243); 1814, Nat. of Britain (HKE: 220). Not currently in cult.

CURRENT STATUS: No recent sightings.

899. Polygala amarella Crantz

P. amara auct. non L.

dwarf milkwort

CULT.: 1789, Alien (HK1, 3: 4); 1812, Alien (HK2, 4: 242); 1814, Alien (HKE: 220). Not currently in cult.

Note. Not discovered in Britain until 1853.

Common milkwort (*Polygala vulgaris*). Occurred along Syon Vista in the late 19th and very early 20th centuries, but has not been seen since.

Horse-chestnut (*Aesculus hippocastanum*). Widely cultivated in the Gardens and seedlings can be found in the vicinity of most mature trees.

STAPHYLEACEAE

[900. Staphylea pinnata L.
bladdernut
CULT.: 1789, Nat. of England (HK1, 1: 375); 1811, Nat. of England (HK2, 2: 171); 1814, Nat. of England (HKE: 81).
Note. Listed in error as native.]

HIPPOCASTANACEAE

901. *Aesculus hippocastanum L.
horse-chestnut
FIRST RECORD: 1949 [1768].
REFERENCES: **Kew Gardens.** 1998, *Cope* (S33: 747): Self-sown seedlings in following:- North Arb.: near Banks Building (108); on the RBG side of Kew Green (118); wild area by Main Gate (132).
EXSICC.: **Kew Gardens.** Herbarium Experimental Ground [seedling]; vi 1949, *Sealy* s.n. (K).
CULT.: 1768 (JH: 434); 1789, Alien (HK1, 1: 493); 1811, Alien (HK2, 2: 335); 1814, Alien (HKE: 106); 1869, Pleasure Ground (K); 1880, Arboretum (K); 1883, opposite end of *Crataegus* Avenue (K); 1884, unspec. (K); 1904, inside Main Entrance (K); 1907, Arboretum (K); 1911, unspec. (K); 1912, nr. the Herbarium (K); 1927, unspec. (K); 1951, south of lake (K); 1962, Arboretum (K); 1978, zone 471 (K); 1978, zone 473 (K); 1978, zone 475 (K); 1996, zone 352 (K). In cult.
CURRENT STATUS: Seedlings occur wherever the species is cultivated, but are seldom allowed to exceed a few cm before being removed. Mature trees have all been planted.

ACERACEAE

902. Acer campestre L.
field maple
CULT.: 1768 (JH: 433); 1789, Nat. of Britain (HK1, 3: 435); 1813, Nat. of Britain (HK2, 5: 448); 1814, Nat. of Britain (HKE: 318); 1843–1853, Arboretum (K); 1880, Arboretum (K); 1881, Arboretum (K); 1907, Arboretum (K); 1941, unspec. (K); 1971, Arboretum North (K); 1971, Arboretum South (K). In cult.

903. *Acer pseudoplatanus L.
sycamore
FIRST RECORD: 1928 [1768].
REFERENCES: **Kew Gardens.** 1983/4, *Cope* (S31: 181): Seedling in new student vegetable plots behind Hanover House.
EXSICC.: **Kew Gardens.** Thames bank, Kew; in cleft at top of wall, near Isleworth Gate; 13 viii 1928; *Summerhayes & Turrill* s.n. (K). • Queen's Cottage Grounds, common; 15 v 1979; *Casey* 7 (K). **Kew Environs.** • River-bank, Kew; v 1929; *Pearce* s.n. (K). • Tow Path, Kew Bridge; 14 viii 1951; *Ross-Craig & Sealy* s.n. (K). • R. Thames, bank near Kew Bridge; in stone faced embankment; 1 v 1952; *Sealy* 1675 (K).
CULT.: 1768 (JH: 433); 1789, Nat. of Britain (HK1, 3: 434); 1813, Nat. of Britain (HK2, 5: 446); 1814, Nat. of Britain (HKE: 318); 1880, Arboretum (K); 1881, Arboretum (K); 1882, Arboretum (K); 1885, Arboretum (K); 1891, Arboretum (K); 1903, Arboretum (K); 1904, Arboretum (K); 1905, Arboretum (K); 1907, Arboretum (K); 1927, unspec. (K); 1961, border opposite tennis courts at back of Ladies' Cloakroom (K); 1971, Arboretum South (K); 1979, zone 465 (K). In cult.
CURRENT STATUS: Cultivated in a number of loacations within the Gardens but widespread as a self-sown weed; it is seldom allowed to persist for long.

Simaroubaceae

904. *Ailanthus altissima* (Mill.) Swingle
tree-of-heaven
First record: 1948.
References: **Kew Gardens.** 1983/4, *Cope* (S31: 181): Seedling in new student vegetable plots behind Hanover House. **Kew Environs.** 1948, *Welch* (LN31: S52): Bombed site, Kew Road, Richmond.
Cult.: In cult.
Current status: No recent sightings. The seedling was removed soon after its discovery and no others have been seen.

Oxalidaceae

905. *Oxalis corniculata* L.
procumbent yellow-sorrel
First record: 1873/4 [1768].
Nicholson: **1873/4 survey** (JB: 42; SFS: 218): A common flower-bed weed. • **1906 revision** (AS: 77): A common flower-bed weed.
Cult.: 1768 (JH: 132); 1789, Alien (HK1, 2: 114); 1811, Nat. of England (HK2, 3: 130); 1814, Nat. of England (HKE: 136). Not currently in cult.
Current status: An increasing and persistent weed of flower-beds in both its green- and bronze-leaved forms.

Procumbent yellow-sorrel (*Oxalis corniculata*). A persistent and increasing weed in the Gardens often seen in this easily overlooked bronze-leaved form.

Sycamore (*Acer pseudoplatanus*). Widely cultivated in the Gardens and freely seeding into lawns and flower-beds.

906. *Oxalis stricta* L.
O. europaea Jord.
upright yellow-sorrel
First record: 1873/4 [1768].
Nicholson: **1873/4 survey** (JB: 42; SFS: 218): **B.** Here and there in shrubberies, etc., with [*O. corniculata*]. • **1906 revision** (AS: 77): Here and there in shrubberies with [*O. corniculata*].
Other references: **Kew Gardens.** 1945, *Welch* (LN31: S50): Kew Gardens, weed.
Cult.: 1768 (JH: 132); 1789, Alien (HK1, 2: 115); 1811, Alien (HK2, 3: 130); 1814, Alien (HKE: 136). Not currently in cult.
Current status: Known only from the Director's Garden and as a weed in the Order Beds.

907. Oxalis acetosella L.
wood-sorrel
First record: 1873/4 [1768].
Nicholson: **1873/4 survey** (JB: 42): [**P.** "Merlin's Cave" and "Old Ruined Arch."]. • **1906 revision** (AS: 77): [**A.** Merlin's Cave, old ruined arch.]
Cult.: 1768 (JH: 132); 1789, Nat. of Britain (HK1, 2: 113); 1811, Nat. of Britain (HK2, 3: 121); 1814, Nat. of Britain (HKE: 135). Not currently in cult.
Current status: No recent sightings. Merlin's Cave had long disappeared even by the time Nicholson was born; it is interesting to speculate on where he got the record of this plant.

908. *Oxalis debilis Kunth
O. corymbosa DC.
large-flowered pink-sorrel
FIRST RECORD: 1951.
REFERENCES: **Kew Gardens.** 1951, *Young* (WAT4: 63): A plant traditionally known as this species [*O. debilis*] grows as a weed in Kew Gardens, with *O. corymbosa*. There is a specimen (K) from there dated 1879 ... The Kew plant is a clone differing from the usual form of *O. corymbosa* ...1976, *Lousley* (LFS: 161): ... Dr Young hesitated in treating it as specifically distinct [from *O. corymbosa*]. He noted it at Kew Gardens as a weed in the Rose Garden and other flower-beds mixed with *O. corymbosa*. • 1983/4, *Cope* (S31: 181): New student vegetable plots behind Hanover House.
EXSICC.: The specimen referred to above has not been found.
CURRENT STATUS: A common weed of flower-beds throughout the Gardens.

909. *Oxalis latifolia Kunth
O. vespertilionis Zucc.
garden pink-sorrel
FIRST RECORD: 1935.
EXSICC.: **Kew Gardens.** Garden of Herbarium House; 1935; *Turrill & Meikle* s.n. (K).
CULT.: In cult. under glass.
CURRENT STATUS: No recent sightings.

GERANIACEAE

910. *Geranium versicolor L.
G. striatum L.
pencilled crane's-bill
FIRST RECORD: 1998 [1768].
REFERENCES: **Kew Gardens.** 1998, *Cope* (S33: 747): North Arb.: neglected area behind Aiton House (120).
CULT.: 1768 (JH: 208); 1885, unspec. (K); 1933, unspec. (K); 1940, Herbaceous Ground (K). In cult.
CURRENT STATUS: Not seen elsewhere since the first record and certainly an escape from cultivation there.

[**911. Geranium nodosum** L.
knotted crane's-bill
CULT.: 1789, Nat. of England (HK1, 2: 434); 1812, Nat. of England (HK2, 4: 186); 1814, Nat. of England (HKE: 212).
Note. Listed in error as native.]

912. Geranium rotundifolium L.
round-leaved crane's-bill
FIRST RECORD: 1891 [1768].
NICHOLSON: **1906 revision:** Not recorded.
OTHER REFERENCES: **Kew Gardens.** 1950, *Welch* (LN30: 6; LN31: S48): Tow-path near Isleworth Gate. • 1983, *Burton* (FLA: 46): ... scruffier parts of Kew Gardens. • 1990, *Cope* (S31: 182): Paddock, after restoration in 1990.
EXSICC.: **Kew Gardens.** [Unlocalised]; weed; 21 ix 1891; *Clarke* 47060B (K). • Lower nursery; on large heap of leaf soil, with *Solanum nigrum*, *Chenopodium polyspermum* etc.; 27 ix 1929; *Hubbard* s.n. (K).
CULT.: 1768 (JH: 209); 1789, Nat. of England (HK1, 2: 436); 1812, Nat. of England (HK2, 4: 190); 1814, Nat. of England (HKE: 212); 1929, unspec. (K); 1951, Order Beds (K); 1975, unspec. (K). In cult.
CURRENT STATUS: Widely scattered but quite rare; it is most abundant beside the path between the Herbarium and Kew Palace.

913. Geranium sylvaticum L.
wood crane's-bill
FIRST RECORD: Probably between 1779 and 1805 [1768].
EXSICC.: **Kew Environs.** Kew; probably between 1779 and 1805; *Goodenough* (K).
CULT.: 1768 (JH: 208); 1789, Nat. of England (HK1, 2: 435); 1812, Nat. of Britain (HK2, 4: 187); 1814, Nat. of Britain (HKE: 212); 1856, unspec. (K); 1923, unspec. (K); 1947, Herbarium Ground (K); 1951, Herb. Exp. Grd. (K); 1972, unspec. (K); 1975, unspec. (K); 1985, zone W751 (K). In cult.
CURRENT STATUS: Goodenough's specimen was probably an escape from cultivation. The species has never been recorded as a wild plant this far south and there is no evidence that Goodenough himself ever took specimens from within the Botanic Garden.

914. Geranium pratense L.
meadow crane's-bill
FIRST RECORD: 1909 [1768].
REFERENCES: **Kew Environs.** 1952, *anon.* (LN31: S46; S33: 747): Frequent, but decreasing, by the Thames from Weybridge to Kew.
EXSICC.: **Kew Environs.** By the ditch outside the Old Deer Park, Richmond; 13 vii 1909; *Britton* s.n. (K).
CULT.: 1768 (JH: 208); 1789, Nat. of Britain (HK1, 2: 435); 1812, Nat. of Britain (HK2, 4: 187); 1814, Nat. of Britain (HKE: 212); 1934, unspec. (K). In cult.
CURRENT STATUS: No recent sightings in the Gardens as a wild plant. Naturalised in the Robinsonian Meadow in front of 'Climbers & Creepers', in the bee garden behind Kew Palace and rarely elsewhere. In quantity in the Old Deer Park where it is native.

915. Geranium sanguineum L.
bloody crane's-bill
CULT.: 1768 (JH: 209); 1789, Nat. of Britain (HK1, 2: 432); 1812, Nat. of Britain (HK2, 4: 184); 1814, Nat. of Britain (HKE: 212); 1940, Herb Garden (K); 1951, Order Beds (K); 1977, unspec. (K). In cult.

916. Geranium columbinum L.
long-stalked crane's-bill
CULT.: 1768 (JH: 209); 1789, Nat. of Britain (HK1, 2: 437); 1812, Nat. of Britain (HK2, 4: 190); 1814, Nat. of Britain (HKE: 212); 1918, unspec. (K); 1934, unspec. (K). Not currently in cult.

917. Geranium dissectum L.
cut-leaved crane's-bill
FIRST RECORD: 1873/4 [1768].
NICHOLSON: **1873/4 survey** (JB: 12): **P.** Three or four plants with [*G. pusillum*]. **Q.** Abundant the whole length of the hedge-row facing "Old Deer Park." • **1906 revision** (AS: 77): **A.** A few plants. **Q.** Abundant, the whole length of the hedge-row facing Old Deer Park.
CULT.: 1768 (JH: 208); 1789, Nat. of Britain (HK1, 2: 437); 1812, Nat. of Britain (HK2, 4: 190); 1814, Nat. of Britain (HKE: 212); 1918, unspec. (K). Not currently in cult.
CURRENT STATUS: Common in long grass throughout the Gardens.

918. *Geranium pyrenaicum Burm.f.
hedgerow crane's-bill
FIRST RECORD: 1852 [1789].
NICHOLSON: **1873/4 survey** (JB: 12): **Pal.** Many plants in company with *Malva moschata*. **Strip.** Common. • **1906 revision** (AS: 77): **P, Strip.** Common.
OTHER REFERENCES: **Kew Environs.** 1944, *Welch* (LN31: S47): By the Thames, Kew.
EXSICC.: **Kew Gardens.** Field near the Herbarium; x 1915; *Turrill* s.n. (K). • In meadow at back of the Herbarium; vi 1928; *Ballard* s.n. (K). • Tennis Courts; rubbish heap; vii 1931; *Bullock* s.n. (K). • In field near Herbarium; amongst grasses; 25 v 1933; *Hubbard* s.n. (K). • Herbarium field; 20 vi 1935; *Turrill* s.n. (K). **Kew Environs.** Kew; v 1852; *Stevens* s.n. (K).
CULT.: 1789, Nat. of Britain (HK1, 2: 436); 1812, Nat. of Britain (HK2, 4: 189); 1814, Nat. of Britain (HKE: 212); 1881, unspec. (K); 1999, Alpine & Herbaceous (K). In cult.
CURRENT STATUS: Common near the Herbarium; scattered and rather rare elsewhere.

919. Geranium pusillum L.
small-flowered crane's-bill
FIRST RECORD: 1873/4 [1768].
NICHOLSON: **1873/4 survey** (JB: 12; SFS: 212): **P.** A few plants here and there about lake. • **1906 revision** (AS: 77): **A.** A few plants here and there about lake.
OTHER REFERENCES: **Kew Gardens.** 1972, *Airy Shaw* (S29: 406): Herbarium quadrangle, and edge of lawn to E. of Wing C. Only noted near the Lake in 1906. **Kew Environs.** 1943–49, *Welch* (LN31: S48): Frequent by the Thames from Kingston to Kew.
EXSICC.: **Kew Environs.** Kew; 10 vii 1892; *Clarke* 47199D (K).

Cut-leaved crane's-bill (*Geranium dissectum*). An inconspicuous component of much of Kew's grassland.

CULT.: 1768 (JH: 209); 1812, Nat. of England (HK2, 4: 191); 1814, Nat. of England (HKE: 212). Not currently in cult.
CURRENT STATUS: Widespread in similar places to *G. molle*.

920. Geranium molle L.
dove's-foot crane's-bill
FIRST RECORD: 1873/4 [1768].
NICHOLSON: **1873/4 survey** (JB: 12): Everywhere, both in beds and turf. • **1906 revision** (AS: 77): Everywhere, both in beds and turf.
EXSICC.: **Kew Gardens.** Near herbarium; in short grass; 16 vi 1928; *Hubbard* s.n. (K). • Herbarium Expl. Ground, on path; 5 vii 1930; *Summerhayes* 544 (K). • Herbarium field; 21 v 1935; *Turrill* s.n. (K). • [Unlocalised]; summer annual of waste land; 16 v 1944; *Souster* 62 (K).
CULT.: 1768 (JH: 208); 1789, Nat. of Britain (HK1, 2: 426); 1812, Nat. of Britain (HK2, 4: 190); 1814, Nat. of Britain (HKE: 212); 1976, unspec. (K). Not currently in cult.
CURRENT STATUS: Widespread and often abundant in short grass.

921. Geranium lucidum L.
shining crane's-bill
FIRST RECORD: 2008 [1768].
CULT.: 1768 (JH: 208); 1789, Nat. of Britain (HK1, 2: 436); 1812, Nat. of Britain (HK2, 4: 189); 1814, Nat. of Britain (HKE: 212); 1973, Herb. & Alpine Dept. (K). In cult.
CURRENT STATUS: Very rare. A single record under a tree near Brentford Gate.

922. Geranium robertianum L.
herb-Robert
FIRST RECORD: 1873/4 [1768].
NICHOLSON: **1873/4 survey** (JB: 12): The rarest species of *Geranium* in our Flora. **P.** "Merlin's Cave" and "Old Ruined Arch." A plant or two in each place. • **1906 revision** (AS: 77): The rarest species of geranium in the Kew flora. **A.** Merlin's Cave (now destroyed), old ruined arch.
EXSICC.: **Kew Gardens.** Near Tennis Courts; on rubbish heap; 4 vi 1944; *Hutchinson* s.n. (K).
CULT.: 1768 (JH: 208); 1789, Nat. of Britain (HK1, 2: 437); 1812, Nat. of Britain (HK2, 4: 191); 1814, Nat. of Britain (HKE: 212); 1923, unspec. (K); 1933, Herb. Ground (K). In cult.
CURRENT STATUS: Widespread in shady places but uncommon.

Common stork's-bill (*Erodium cicutarium*). Rather rare; the best population is adjacent to the Evolution House.

923. Geranium purpureum Vill.
little-robin
CULT.: 1949, Herbarium Ground (K). Not currently in cult.

[924. Geranium phaeum L.
dusky crane's-bill
CULT.: 1789, Nat. of England (HK1, 2: 434); 1812, Nat. of England (HK2, 4: 185); 1814, Nat. of England (HKE: 212).
Note. Listed in error as native.**]**

925. *Geranium sessiliflorum subsp. **novae-zelandiae** Carolin
FIRST RECORD: 2004 [1969].
CULT.: 1969, Alpine & Herbaceous Dept. (K). In cult.
CURRENT STATUS: Many plants grew from seed in topsoil stored in the Paddock and behind the Princess of Wales Conservatory. Both the green- and bronze-leaved forms were present, the latter more common.

926. Erodium maritimum (L.) L'Hér.
Geranium maritimum L.
sea stork's-bill
CULT.: 1768 (JH: 208); 1789, Nat. of England (HK1, 2: 416); 1812, Nat. of England (HK2, 4: 158); 1814, Nat. of England (HKE: 209). Not currently in cult.

927. Erodium moschatum (L.) L.Hér.
Geranium moschatum L.
musk stork's-bill
CULT.: 1768 (JH: 208); 1789, Nat. of England (HK1, 2: 414); 1812, Nat. of England (HK2, 4: 156); 1814, Nat. of England (HKE: 209); 1951, Order Beds (K). Not currently in cult.
Note. Originally grown as a native but now considered an archaeophyte.

928. Erodium cicutarium (L.) L'Hér.
Geranium cicutarium L.
common stork's-bill
FIRST RECORD: 1866 [1768].
NICHOLSON: **1873/4 survey** (JB: 12): **B.** In the lawn between No. 5 and Museum No. 2. **P.** On the turf slopes and waste ground near Winter Garden. • **1906 revision** (AS: 77): **B.** In the lawn between No. 5 and museum No. 2. **A.** On the turf slopes near temperate house.
EXSICC.: **Kew Environs.** Kew; vi 1866; *Baker* in *Herb. Watson* (K). • Sandy ground, Kew; v & vii 1871; *Baker* in *Herb. Watson* (K).
CULT.: 1768 (JH: 208); 1789, Nat. of Britain (HK1, 2: 414); 1812, Nat. of Britain (HK2, 4: 156); 1814, Nat. of Britain (HKE: 209); 1933, Herb. Ground (K). Not currently in cult.
CURRENT STATUS: Still to be found in reasonable quantity near the Temperate House (adjacent to the Evolution House) and rarely in one or two other places.

Limnanthaceae

929. *Limnanthes douglasii R.Br.
meadow-foam
First record: 2004 [1966].
Cult.: In cult.
Current status: Naturalised in long grass near King William's Temple and growing from seed in topsoil stored in the Paddock.

Balsaminaceae

930. Impatiens noli-tangere L.
touch-me-not balsam
Cult.: 1789, Nat. of England (HK1, 3: 293); 1811, Nat. of England (HK2, 2: 50); 1814, Nat. of England (HKE: 63); 1951, border alongside Museum II (K). Not currently in cult.

931. *Impatiens capensis Meerb.
orange balsam
First record: 1929.
References: **Kew Gardens.** • 1999, *Stones* (EAK: 2): By Lake. **Kew Environs.** 1951, *Lousley* (LN31: S51; S33: 747): Frequent by the Thames from Weybridge to Kew.
Exsicc.: **Kew Environs.** River-bank, Kew; viii 1929; *Pearce* s.n. (K).
Current status: In patches around the Lake and along the towpath.

932. *Impatiens parviflora DC.
small balsam
First record: 1855.
Nicholson: **1873/4 survey** (JB: 42): **B.** Very troublesome about "Rockwork." **P.** Frequent in beds and shrubberies. • **1906 revision** (AS: 77): A troublesome weed.
Exsicc.: **Kew Gardens.** Weed in Kew Gardens; vii 1871; *Baker* in *Herb. Watson* (K). • [Unlocalised]; found in gravel soil in association with *Ranunculus repens* & *Trifolium dubium*; 3 vii 1940; *Parker* s.n. (K). • [Unlocalised]; summer annual with yellow flowers, a frequent weed on cultivated land; 1 vii 1944; *Souster* 109 (K). • From shrubbery outside Herbarium; 31 vii 1951; *Ross-Craig* s.n. (K). **Kew Environs.** Thames side, Kew; 1855, *Hooker* s.n. (K). • Thames bank, Kew; viii 1888; *Gamble* 20218 (K). • Waste ground near the river, Kew; 31 vii 1895; *Worsdell* s.n. (K).
Cult.: In cult. under glass.
Current status: Widepsread as a weed and particularly common in the Conservation Area.

933. *Impatiens glandulifera Royle
Indian balsam
First record: 1932 [1951].

Small balsam (*Impatiens parviflora*). Abundant in the Conservation Area and widespread elsewhere. A noted feature of Kew Gardens.

References: **Kew Gardens.** 1998, *Cope* (S33: 747): West Arb.: Riverside Walk (213, 215). **Kew Environs.** 1932, *Lousley* (LN31: S52; S33: 747): Thames side between Richmond and Kew.
Exsicc.: **Kew Environs.** River Bank between Kew and Richmond; in association with *Epilobium* & *Caltha palustris*, also *Scirpus maritimus*; soil – sandy gravell over silty deposits; 1 viii 1948; *O.J. Ward* 8 (K). • Kew, tow-path; 23 ix 1965; *Johnson* 65.1232 (K).
Cult.: 1951, Order Beds (K); 1961, Order Beds (K); 1966, unspec. (K); 1967, unspec. (K). In cult.
Current status: Widespread; it is still rather rare but seems to be increasing. First noted inside the Gardens in 1980.

Araliaceae

934. Hedera helix L.
Incl. *H. hibernica* Kirschner
ivy
First record: 1873/4 [1768].
Nicholson: **1873/4 survey** (JB: 45): Frequent. An undoubted native of the Kew Flora. • **1906 revision** (AS: 80): Frequent. An undoubted member of the Kew flora.
Exsicc.: **Kew Gardens.** Herbarium House; 7 i 1918; *Stapf* s.n. (K). • Queen's Cottage Grounds; growing on trees; 10 ix 1978; *T.Casey* 5 (K). • Lower Nursery; shaded brick wall; 20 x 1981; *M.Looker* 14a (K).

Ivy (*Hedera helix*). Throughout the Gardens in both wild and cultivated forms, but mainly noted for being the host of Ivy Broomrape.

CULT.: 1768 (JH: 215, 443); 1789, Nat. of Britain (HK1, 1: 281); 1811, Nat. of Britain (HK2, 2: 51); 1814, Nat. of Britain (HKE: 63); 1880, Arboretum (K); 1881, Arboretum (K); 1906, King William's Temple (K); 1920, unspec. (K). In cult.
CURRENT STATUS: Throughout the Gardens in shrubberies and shady places but in many places it will undoubtedly have been planted. A number of populations may be *H. hibernica* but no attempt has yet been made to distinguish them. In many places it is host to *Orobanche hederae*.

APIACEAE

935. Hydrocotyle vulgaris L.

marsh pennywort

FIRST RECORD: 1873/4 [1768].
NICHOLSON: **1873/4 survey** (JB: 44): **P.** Truly wild near water's edge at Palm House end of lake. Brought with soil to each of the newly-planted Magnolias on both sides of walk from Temple of Minden to end of Pagoda Avenue. • **1906 revision** (AS: 80 and in SFS: 334): **A.** Round edges of lake.
EXSICC.: **Kew Gardens.** [Unlocalised]; found as weed in potting soil when very small, grown on to identify; 3 viii 1944; *Souster* 137 (K).
CULT.: 1768 (JH: 76); 1789, Nat. of Britain (HK1, 1: 327); 1811, Nat. of Britain (HK2, 2: 117); 1814, Nat. of Britain (HKE: 73). Not currently in cult.
CURRENT STATUS: One small patch may still exist on the margin of the southwest part of the Lake but it has not been seen for several years.

936. *Hydrocotyle ranunculoides L.f.

floating pennywort

FIRST RECORD: 2003.
CURRENT STATUS: A recent invader of the Moat south of Brentford Gate.

937. Sanicula europaea L.

sanicle

CULT.: 1768 (JH: 90); 1789, Nat. of Britain (HK1, 1: 328); 1811, Nat. of Britain (HK2, 2: 118); 1814, Nat. of Britain (HKE: 73); 1934, unspec. (K). In cult.

938. Eryngium maritimum L.

sea-holly

CULT.: 1768 (JH: 82); 1789, Nat. of Britain (HK1, 1: 326); 1811, Nat. of Britain (HK2, 2: 116); 1814, Nat. of Britain (HKE: 73). In cult.

Marsh pennywort (*Hydrocotyle vulgaris*). Accidentally introduced to odd corners of the Gardens but once native in a single small patch by the Lake; it has not been seen in recent years and may have been lost.

939. Eryngium campestre L.
field eryngo
CULT.: 1768 (JH: 82); 1789, Nat. of England (HK1, 1: 326); 1811, Nat. of England (HK2, 2: 116); 1814, Nat. of Britain (HKE: 73); 1933, unspec. (K); 1934, unspec. (K). In cult.

[940. Echinophora spinosa L.
CULT.: 1811, Nat. of England (HK2, 2: 123); 1814, Nat. of England (HKE: 74).
Note. Listed in error as native.]

[941. Chaerophyllum aureum L.
golden chervil
CULT.: 1789, Alien (HK1, 1: 358); 1811, Nat. of Scotland (HK2, 2: 153); 1814, Nat. of Scotland (HKE: 78).
Note. Listed in error as native.]

942. Chaerophyllum temulum L.
C. temulentum L.
rough chervil
FIRST RECORD: 1873/4 [1768].
NICHOLSON: **1873/4 survey** (JB: 45): **Q.** One or two plants only. • **1906 revision** (AS: 80): A few plants in Queen's Cottage grounds.
CULT.: 1768 (JH: 100); 1789, Nat. of Britain (HK1, 1: 357); 1811, Nat. of Britain (HK2, 2: 153); 1814, Nat. of Britain (HKE: 78). Not currently in cult.
CURRENT STATUS: No recent sightings; always rare and now seems to have been lost.

943. Anthriscus sylvestris (L.) Hoffm.
Chaerophyllum sylvestre L.
cow parsley
FIRST RECORD: 1873/4 [1768].
NICHOLSON: **1873/4 survey** (JB: 45): **Pal, P** and **Q.** Abundant. • **1906 revision** (AS: 80): Abundant in all the divisions.
EXSICC.: **Kew Gardens.** Queen's Cottage Grounds; 15 vi 1914; *Flippance* s.n. (K). • Between Herbarium and Tennis Courts; 11 v 1928; *Sprague* s.n. (K). • [Unlocalised]; sandy soil; 27 v 1928; *Hubbard* s.n. (K). • In the Herbarium field; v 1928; *C.A.Smith* 6020 (K). • Herbarium field; amongst grasses; v 1933; *Hubbard* s.n. (K). • Near Herbarium; grass verge beneath pine trees; 5 v 1950; *Sealy* s.n. (K). • Paddock behind Herbarium; 27 v 1958; *Sealy & Ross-Craig* 1882, 1882A (both K). • Paddock behind Herbarium; 9 vi 1958; *Sealy & Ross-Craig* 1882B (K). **Kew Environs.** Riverside, Kew; 17 v 1894; *anon.* s.n. (K).
CULT.: 1768 (JH: 100); 1789, Nat. of Britain (HK1, 1: 357); 1811, Nat. of Britain (HK2, 2: 152); 1814, Nat. of Britain (HKE: 78). In cult.
CURRENT STATUS: Abundant in shady places throughout the Gardens.

[944. Anthriscus cerefolium (L.) Hoffm.
Scandix cerefolium L.
garden chervil
CULT.: 1789, Alien (HK1, 1: 356); 1811, Nat. of England (HK2, 2: 151); 1814, Nat. of England (HKE: 78).
Note. Listed in error as native.]

945. Anthriscus caucalis M.Bieb.
Scandix anthriscus L.
bur chervil
FIRST RECORD: 2004 [1768].
CULT.: 1768 (JH: 99); 1789, Nat. of Britain (HK1, 1: 356); 1811, Nat. of Britain (HK2, 2: 152); 1814, Nat. of Britain (HKE: 78); 1962, unspec. (K). Not currently in cult.
CURRENT STATUS: Appeared briefly near King William's Temple in 2004.

946. Scandix pecten-veneris L.
S. pecten L.
shepherd's-needle
FIRST RECORD: 1863 [1768].
EXSICC.: **Kew Environs.** Kew; waste ground; vii 1863; *Brocas* s.n. (K).
CULT.: 1768 (JH: 99); 1789, Nat. of Britain (HK1, 1: 356); 1811, Nat. of Britain (HK2, 2: 151); 1814, Nat. of Britain (HKE: 78); 1971, Herbarium Experimental Ground (K). Not currently in cult.
CURRENT STATUS: No recent sightings. Seed is very short-lived and the plant is unlikely to reappear from the soil seed-bank after disturbance.

Shepherd's-needle (*Scandix pecten-veneris*). Recorded just once, in 1868. Its seed is short-lived so any remaining in the soil from early populations will no longer be viable.

947. *Myrrhis odorata (L.) Scop.
Scandix odorata L.
sweet cicely
FIRST RECORD: 1920 [1768].
REFERENCES: **Kew Gardens.** 1920, *Turrill* (S16: 216): In quantity and spreading near the Herbarium.
EXSICC.: **Kew Gardens.** Herbarium field; amongst grasses; 20 v 1933; *Hubbard* s.n. (K).
CULT.: 1768 (JH: 99); 1789, Nat. of Britain (HK1, 1: 356); 1811, Nat. of Britain (HK2, 2: 151); 1814, Nat. of Britain (HKE: 78); 1936, Herb Gdn. (K); 1958, Order Beds (K). In cult.
CURRENT STATUS: No recent sightings.

948. *Coriandrum sativum L.
coriander
FIRST RECORD: 1944 [1768].
REFERENCES: **Kew Environs.** 1944, *Davies* (LN32: S135; S33: 747): Riverside just below Kew Gardens, near Ferry Lane.
CULT.: 1768 (JH: 98); 1789, Nat. of England (HK1, 1: 355); 1811, Nat. of England (HK2, 2: 151), 1814, Nat. of England (HKE: 78); 1919, unspec. (K); 1934, Arboretum (K); 1936, Herb Gdn. (K); 1939, unspec. (K); 1949, unspec. (K); 1950, Herbarium Ground (K). In cult.
CURRENT STATUS: An occasional escape around vegetable plots; the largest and most persistent population, around the plots behind Hanover House, was destroyed during building works.

Perfoliate Alexanders (*Smyrnium perfoliatum*). An aggressive and persistent weed of the Gardens attempting to spread to all corners from its stronghold in the Conservation Area, but seldom allowed to get far.

949. Smyrnium olusatrum L.
Alexanders
FIRST RECORD: 1981 [1768].
REFERENCES: **Kew Environs.** 1981, *Norman* (LN61: 103; S33: 747): Kew Green.
CULT.: 1768 (JH: 102); 1789, Nat. of Britain (HK1, 1: 362); 1811, Nat. of Britain (HK2, 2: 158); 1814, Nat. of Britain (HKE: 79); 1927, unspec. (K); 1930, Herb. Expt. Ground (K); 1932, Herbarium Expt. Ground (K). In cult.
CURRENT STATUS: Recently found in shade just north of the Rhododendron Dell and near the Herbarium, but the former of these two specimens was removed soon after the observation. It has long gone from Kew Green.

950. *Smyrnium perfoliatum L.
perfoliate Alexanders
FIRST RECORD: 1948 [1768].
REFERENCES: **Kew Gardens.** 1973, *Airy Shaw* (S29: 407): Naturalised in grass near Queen's Cottage. • 1981, *Hastings* (FSSC: 103): Recorded as a weed in many places in Kew Gardens. • 1983, *Burton* (FLA: 87): It is ... well naturalised within the walls of the Royal Botanic Gardens at Kew, but no longer, as it was for a few years around 1950, outside them under bushes in Kew Road. **Kew Environs.** 1948, *Welch* (LN29: 10; LN32: S127): Under bushes, Kew Road, Richmond, opposite Stanmore Road. • 1948–53, *Welch* (LFS: 216): Kew Road, escape from Kew Gardens.
CULT.: 1768 (JH: 102); 1789, Alien (HK1, 1: 361); 1811, Alien (HK2, 2: 157); 1814, Alien (HKE: 79); 1897, unspec. (K); 1916, unspec. (K); 1925, unspec. (K); 1936, Herbarium Ground (K). Not currently in cult.
CURRENT STATUS: Abundant in the Conservation Area; it is becoming a weed throughout the Gardens and was recently seen (briefly) on Kew Green. Steps have been taken to limit its range within the Conservation Area and large numbers of plants are now regularly pulled.

951. Bunium bulbocastanum L.
great pignut
CULT.: 1768 (JH: 93); 1789, Nat. of Britain (HK1, 1: 337); 1811, Nat. of Britain (HK2, 2: 130); 1814, Nat. of Britain (HKE: 75); 1895, unspec. (K). Not currently in cult.

952. Conopodium majus (Gouan) Loret
Bunium flexuosum Stokes
C. denudatum W.D.J.Koch
pignut
FIRST RECORD: 1873/4 [1811].
NICHOLSON: **1873/4 survey** (JB: 44): **P.** Common in wood near "Princess's Gate." • **1906 revision** (AS: 80): **A.** Common in the woods.

Pignut (*Conopodium majus*). Rare but easily overlooked. As likely to be found in uncut grassland as in woodland shade.

CULT.: 1811, Nat. of England (HK2, 2: 130); 1814, Nat. of England (HKE: 75). Not currently in cult.
CURRENT STATUS: Rare; widely scattered in the rougher grassy areas, often in the light shade of trees.

953. Pimpinella major (L.) Huds.
P. magna L.
greater burnet-saxifrage
FIRST RECORD: 1909 [1789].
REFERENCES: **Kew Gardens.** 1909, *Turrill* in *Rolfe & A.B.Jackson* (S10: 373): In quantity near the Herbarium. **Kew Environs.** 1942, *D.H.Kent* (LN32: S130): Kew.
CULT.: 1789, Nat. of England (HK1, 1: 363); 1811, Nat. of England (HK2, 2: 160); 1814, Nat. of England (HKE: 79). In cult.
CURRENT STATUS: No recent sightings.

954. Pimpinella saxifraga L.
burnet-saxifrage
FIRST RECORD: 1873/4 [1768].
NICHOLSON: **1873/4 survey** (JB: 44): **Strip.** Plentiful between the third and fourth seats counting from Brentford Ferry. • **1906 revision** (AS: 80): **Strip.** Plentiful in turf bordering towing-path.
CULT.: 1768 (JH: 103); 1789, Nat. of Britain (HK1, 1: 363); 1811, Nat. of Britain (HK2, 2: 159); 1814, Nat. of Britain (HKE: 79); 1973, Herb. & Alpine Dept. (K). Not currently in cult.
CURRENT STATUS: Known only from the golf course in the Old Deer Park.

955. Aegopodium podagraria L.
ground-elder
FIRST RECORD: 1873/4 [1768].
NICHOLSON: **1873/4 survey** (JB: 44): In all the divisions, though not frequent. • **1906 revision** (AS: 80): In all the divisions.
EXSICC.: **Kew Gardens.** Under Horse Chestnut near Office of Works; soil a medium loam; 9 ii 1928; *Turrill & Summerhayes* s.n. (K). • On dryish mound by Office of Works; 9 ii 1928; *Turrill & Summerhayes* s.n. (K). • Herbarium Experimental Ground; soil a rather light loam; 10 ii 1928; *Summerhayes & Turrill* s.n. (K). • Herbarium Exper. Ground; in fallow plot cult. last year; 24 iv 1928; *Summerhayes* s.n. (K). • On waste ground near Herbarium; 16 vi 1928; *Hubbard* s.n. (K). • In cult. ground east of the Herbarium; 17 viii 1928; *Summerhayes & Turrill* s.n. (K). • Experimental Ground; in loose soil; 19 iv 1929; *Summerhayes* s.n. (K). • Herbaceous Ground; 15 vi 1937; *Howes* s.n. (K).
CULT.: 1768 (JH: 103); 1789, Nat. of Britain (HK1, 1: 365); 1811, Nat. of Britain (HK2, 2: 161); 1814, Nat. of Britain (HKE: 80); 1905, unspec. (K). Not currently in cult.
CURRENT STATUS: A widespread and persistent weed of flower-beds and grassy areas and particularly abundant near the Main Gate.

Ground-elder (*Aegopodium podagraria*). A widespread and aggressive weed, difficult to remove and, in places, beginning to become dominant.

956. Sium latifolium L.
greater water-parsnip
CULT.: 1768 (JH: 96); 1789, Nat. of England (HK1, 1: 349); 1811, Nat. of England (HK2, 2: 144); 1814, Nat. of England (HKE: 77); 1905, unspec. (K); 1948, Aquatic Garden (K). Not currently in cult.

957. Berula erecta (Huds.) Coville
Sium angustifolium L.
lesser water-parsnip
CULT.: 1789, Nat. of Britain (HK1, 1: 349); 1811, Nat. of Britain (HK2, 2: 144); 1814, Nat. of Britain (HKE: 77); 1897, unspec. (K). Not currently in cult.

958. Crithmum maritimum L.
rock samphire
CULT.: 1768 (JH: 95); 1789, Nat. of Britain (HK1, 1: 342); 1811, Nat. of Britain (HK2, 2: 135); 1814, Nat. of Britain (HKE: 76). In cult.

959. Seseli libanotis (L.) W.D.J.Koch
Athamanta libanotis L.
A. oreoselinum L.
moon carrot
CULT.: 1768 (JH: 94); 1789, Nat. of England (HK1, 1: 339, and: 340 as *oreoselinum*); 1811, Nat. of England (HK2, 2: 132), Alien (ibid.: 133 as *oreoselinum*); 1814, Nat. of England (HKE: 75), Alien (ibid.: as *oreoselinum*). Not currently in cult.

960. Oenanthe fistulosa L.
tubular water-dropwort
CULT.: 1768 (JH: 97); 1789, Nat. of Britain (HK1, 1: 353); 1811, Nat. of Britain (HK2, 2: 148); 1814, Nat. of Britain (HKE: 77); 1933, unspec. (K). Not currently in cult.

961. Oenanthe silaifolia M.Bieb.
narrow-leaved water-dropwort
O. peucedanifolia auct. non Pollich
CULT.: 1811, Nat. of England (HK2, 2: 148); 1814, Nat. of England (HKE: 78); 1899, unspec. (K). Not currently in cult.
Note. Not known to be native until 1794.

962. Oenanthe pimpinelloides L.
corky-fruited water-dropwort
CULT.: 1768 (JH: 97); 1789, Nat. of England (HK1, 1: 353); 1811, Nat. of England (HK2, 2: 148); 1814, Nat. of England (HKE: 78); 1940, Bog Garden (K). Not currently in cult.

963. Oenanthe lachenalii C.C.Gmel.
parsley water-dropwort
CULT.: 1931, unspec. (K). Not currently in cult.

964. Oenanthe crocata L.
hemlock water-dropwort
FIRST RECORD: 1873/4 [1768].
NICHOLSON: **1873/4 survey** (JB: 45; SFS: 347): **P.** Several plants round lake. **Strip.** Abundant. • **1906 revision** (AS: 80): **A.** Round lake. **Strip.** Abundant the whole length of the river boundary.
OTHER REFERENCES: **Kew Gardens.** 1999, *Stones* (EAK: 2 & 5): By Lake and Palm House Pond. **Kew Environs.** 1953, *Welch* (LN32: S133): Frequent by the Thames from opposite Staines to Barnes
EXSICC.: **Kew Gardens.** Thames-bank; in *Petasites* zone near Isleworth Gate; 25 vi 1929; *Summerhayes & Turrill* 2478 (K). **Kew Environs.** Thames bank, Kew; 12 vii 1888; *Gamble* 20206 (K). • River bank, Kew; growing under *Salix* in damp stony ground; 24 iv 1928; *Turrill* s.n. (K). • Kew, Thames bank; middle zone with *Phalaris* & *Senecio*; 13 vii 1928; *Turrill* s.n. (K). • Thames bank, Kew; growing under willow bushes; 30 vii 1928; *Summerhayes & Turrill* 2477 (K). • Thames bank, Kew; 5 vi 1942; *Hutchinson & Howes* s.n. (K). • Bank of river Thames, at Kew; growing in between granite blocks of retaining wall of bank; 11 vi 1958; *Ross-Craig & Sealy* 1888 (K).
CULT.: 1768 (JH: 97); 1789, Nat. of Britain (HK1, 1: 353); 1811, Nat. of Britain (HK2, 2: 148); 1814, Nat. of Britain (HKE: 77); 1933, Herb. Ground (K). In cult.
CURRENT STATUS: Still occurs abundantly around the Lake and along the towpath; it can also be found in one or two other places on unexpectedly dry soil.

965. Oenanthe aquatica (L.) Poir.
Phellandrium aquaticum L.
fine-leaved water-dropwort
CULT.: 1768 (JH: 100); 1789, Nat. of Britain (HK1, 1: 354); 1811, Nat. of Britain (HK2, 2: 149); 1814, Nat. of Britain (HKE: 78). Not currently in cult.

966. Aethusa cynapium L.
fool's parsley
FIRST RECORD: 1873/4 [1768].
NICHOLSON: **1873/4 survey** (JB: 45): **B.** A flower-bed weed. **Q.** Kitchen garden ground. • **1906 revision** (AS: 80): A flower-bed weed. Everywhere in cultivated ground.
EXSICC.: **Kew Gardens.** Experimental Ground; spontaneous; 23 x 1958; *Ross-Craig* s.n. (K). • Experimental Ground; 23 x 1968; *Ross-Craig* s.n. (K). **Kew Environs.** Kew; in hedges; 8 viii 1886; *Fraser* s.n. (K). • Towing path between Richmond & Kew; 25 vi 1905; *Sprague* s.n. (K).
CULT.: 1768 (JH: 99); 1789, Nat. of Britain (HK1, 1: 354); 1811, Nat. of Britain (HK2, 2: 150); 1814, Nat. of Britain (HKE: 78); c. 1900, unspec. (K). In cult.
CURRENT STATUS: Occasional in neglected areas and abundant at times near the rear entrance to the Banks Building.

967. Foeniculum vulgare Mill.
Anethum foeniculum L.
fennel
FIRST RECORD: 2001 [1768].
REFERENCES: **Kew Gardens.** 2001, *Cope* (S35: 647): Rough ground in the Lower Nursery (120).
CULT.: 1768 (JH: 103); 1789, Nat. of England (HK1, 1: 362); 1811, Nat. of England (HK2, 2: 159); 1814, Nat. of England (HKE: 79). In cult.
CURRENT STATUS: A rare casual of rough and neglected ground.

968. Silaum silaus (L.) Schinz & Thell.
Peucedanum silaus L.
pepper-saxifrage
CULT.: 1789, Nat. of England (HK1, 1: 341); 1811, Nat. of England (HK2, 2: 134); 1814, Nat. of England (HKE: 76). Not currently in cult.

969. Meum athamanticum Jacq.
Aethusa meum Murr.
Athamanta meum L.
spignel
CULT.: 1768 (JH: 94); 1789, Nat. of Britain (HK1, 1: 355); 1811, Nat. of Britain (HK2, 2: 150); 1814, Nat. of Britain (HKE: 78); 1900, unspec. (K). Not currently in cult.

970. Physospermum cornubiense (L.) DC.
Ligusticum cornubiense L.
bladderseed
CULT.: 1811, Nat. of England (HK2, 2: 142); 1814, Nat. of England (HKE: 77). Not currently in cult.

971. Conium maculatum L.
hemlock
FIRST RECORD: 1983/4 [1768].
REFERENCES: **Kew Gardens.** 1983/4, *Cope* (S31: 181): New student vegetable plots behind Hanover House.
CULT.: 1768 (JH: 94); 1789, Nat. of Britain (HK1, 1: 338); 1811, Nat. of Britain (HK2, 2: 130); 1814, Nat. of Britain (HKE: 75); 1934, unspec. (K); 1937, Herbaceous Ground (K). Not currently in cult.
CURRENT STATUS: Occasional; a few recent sightings, but being biennial it does not often manage to reach flowering before being weeded out.

972. Bupleurum tenuissimum L.
slender hare's-ear
CULT.: 1768 (JH: 93); 1789, Nat. of England (HK1, 1: 330); 1811, Nat. of England (HK2, 2: 122); 1814, Nat. of England (HKE: 74). Not currently in cult.

973. Bupleurum rotundifolium L.
thorow-wax
CULT.: 1768 (JH: 93); 1789, Nat. of England (HK1, 1: 329); 1811, Nat. of England (HK2, 2: 120); 1814, Nat. of England (HKE: 74); 1968, plot 157-77 (K); 1968, plot 157-78 (K). Not currently in cult.

974. Trinia glauca (L.) Dumort.
Pimpinella dioica Huds.
P. glauca L.
honewort
CULT.: 1768 (JH: 103); 1789, Nat. of England (HK1, 1: 364); 1811, Nat. of England (HK2, 2: 160); 1814, Nat. of England (HKE: 79). Not currently in cult.

975. Apium graveolens L.
wild celery
CULT.: 1768 (JH: 89); 1789, Nat. of Britain (HK1, 1: 364); 1811, Nat. of Britain (HK2, 2: 161); 1814, Nat. of Britain (HKE: 80); 1933, Herb Garden (K); 1975, unspec. (K). In cult.

976. Apium nodiflorum (L.) Lag.
Sium nodiflorum L.
Helosciadium nodiflorum (L.) Koch
fool's-water-cress
FIRST RECORD: 1873/4 [1768].
NICHOLSON: **1873/4 survey** (JB: 44): **Strip.** Abundant near river. • **1906 revision** (AS: 80): **Strip.** Abundant near river.
EXSICC.: **Kew Environs.** By the Thames bet. Kew & Richmond; viii 1907; *A.B.Jackson & Domin* s.n. (K). • Thames bank, Kew; growing on mud in lower zone; 7 viii 1928; *Summerhayes & Turrill* s.n. (K).
CULT.: 1768 (JH: 96); 1789, Nat. of Britain (HK1, 1: 350); 1811, Nat. of Britain (HK2, 2: 144); 1814, Nat. of Britain (HKE: 77). In cult.
CURRENT STATUS: Still present in the Moat and near the Banks Building.

977. Apium repens (Jacq.) Lag.
Sium repens Jacq.
creeping marshwort
CULT.: 1811, Nat. of Britain (HK2, 2: 144); 1814, Nat. of Britain (HKE: 77). In cult.

978. Apium inundatum (L.) Rchb.f.
Sison inundatum L.
lesser marshwort
CULT.: 1768 (JH: 91); 1789, Nat. of Britain (HK1, 1: 351); 1811, Nat. of Britain (HK2, 2: 146); 1814, Nat. of Britain (HKE: 77). Not currently in cult.

979. Petroselinum crispum (Mill.) Nyman ex A.W.Hill
garden parsley
Apium petroselinum L.
CULT.: 1768 (JH: 89); 1918, unspec. (K). Not currently in cult.

980. Petroselinum segetum (L.) W.D.J.Koch
Sison segetum L.
corn parsley
CULT.: 1768 (JH: 91); 1789, Nat. of England (HK1, 1: 351); 1811, Nat. of England (HK2, 2: 145); 1814, Nat. of England (HKE: 77). Not currently in cult.

981. Sison amomum L.
stone parsley
CULT.: 1768 (JH: 91); 1789, Nat. of England (HK1, 1: 350); 1811, Nat. of England (HK2, 2: 145); 1814, Nat. of England (HKE: 77). Not currently in cult.

982. Cicuta virosa L.
Cowbane
CULT.: 1768 (JH: 99); 1789, Nat. of Britain (HK1, 1: 354); 1811, Nat. of Britain (HK2, 2: 149); 1814, Nat. of Britain (HKE: 78); 1919, unspec. (K). Not currently in cult.

983. *Ammi majus L.
bullwort
FIRST RECORD: 1966 [1768].
REFERENCES: **Kew Gardens.** 1966, *J.L.Gilbert* (S29: 407): Near Herbarium door.
CULT.: 1768 (JH: 88); 1789, Alien (HK1, 1: 337); 1811, Alien (HK2, 2: 129); 1814, Alien (HKE: 75); 1898, unspec. (K); 1899, unspec. (K); 1937, Herbarium Ground (K). Not currently in cult.
CURRENT STATUS: No recent sightings.

984. *Ammi visnaga (L.) Lam.
Daucus visnaga L.
toothpick-plant
FIRST RECORD: 1969 [1768].
REFERENCES: **Kew Gardens.** 1969, *Brenan* (S29: 407): In garden of Herbarium House, 55 Kew Green. Bird-seed alien.
CULT.: 1768 (JH: 89); 1789, Alien (HK1, 1: 336); 1811, Alien (HK2, 2: 128); 1814, Alien (HKE: 75). Not currently in cult.
CURRENT STATUS: No recent sightings.

985. Carum carvi L.
caraway
CULT.: 1768 (JH: 101); 1789, Nat. of Britain (HK1, 1362); 1811, Nat. of Britain (HK2, 2: 159); 1814, Nat. of Britain (HKE: 79); 1887, unspec. (K); 1918, Herbaceous Ground (K); 1934, unspec. (K). In cult.
Note. Originally grown as a native but now considered to be an archaeophyte.

986. Carum verticillatum (L.) W.D.J.Koch
Sison verticillatum L.
whorled caraway
CULT.: 1789, Nat. of Britain (HK1, 1: 351); 1811, Nat. of Britain (HK2, 2146); 1814, Nat. of Britain (HKE: 77). Not currently in cult.

987. Selinum carvifolia (L.) L.
Cambridge milk-parsley
CULT.: 1768 (JH: 97); 1789, Alien (HK1, 1: 338); 1811, Alien (HK2, 2: 131); 1814, Alien (HKE: 75); 1972, plot 157-78 (K). Not currently in cult.
Note. Not known as a British plant until 1881.

988. Ligusticum scoticum L.
Scots lovage
CULT.: 1768 (JH: 95); 1789, Nat. of Britain (HK1, 1: 347); 1811, Nat. of Britain (HK2, 2: 141); 1814, Nat. of Britain (HKE: 77); 1900, unspec. (K); 1936, Herbaceous Ground (K). Not currently in cult.

989. Angelica sylvestris L.
A. sylvestris var. *decurrens* Fisch.
wild angelica
FIRST RECORD: 1873/4 [1768].
NICHOLSON: **1873/4 survey** (JB: 45): **Strip.** Rather frequent. • **1906 revision** (AS: 80): **Strip.** Rather frequent.
OTHER REFERENCES: **Kew Environs.** 1920, *Britton* (BSEC6: 23): By the Thames above London – Mortlake, Kew, etc. [refers to var. *decurrens*].
EXSICC.: **Kew Environs.** Thames bank, Kew; 15 ix 1893; *Fraser* s.n. (K). • Thames bank, Kew; growing in top zone on mud of high bank; 7 viii 1928; *Summerhayes & Turrill* s.n. (K). • Towpath between Kew & Richmond; on river bank, subject to flooding; 18 vii 1948; *Souster* 885 (K).
CULT.: 1768 (JH: 92); 1789, Nat. of Britain (HK1, 1: 348); 1811, Nat. of Britain (HK2, 2: 143); 1814, Nat. of Britain (HKE: 77); 1926, Herb. Expt. Ground (K); 1927, Herb. Exp. Ground (K). Not currently in cult.
CURRENT STATUS: Not recently found along the river since strengthening of the banks has destroyed the zoning of the vegetation, but present in the Old Deer Park, on the golf course.

990. *Angelica archangelica L.
garden angelica
FIRST RECORD: 1929 [1768].
REFERENCES: **Kew Environs.** 1931–32, *Cooke* (LN32: S134; S33: 747): Scattered along the Thames bank from Kingston to Kew. • 1933, *anon.* (LN12: S52; S33: 747): Along the banks of the Thames between Barnes and Kingston.
EXSICC.: **Kew Environs.** On bank of R. Thames between Kew & Richmond; 9 vii 1929; *Hubbard* s.n. (K). • River-wall, near Kew Bridge; 8 vii 1934; *Turrill* s.n. (K). • River Thames at Kew; growing between granite blocks retaining the banks, in quantity; 11 vi 1958; *Ross-Craig & Sealy* 1889 (K).
CULT.: 1768 (JH: 92); 1789, Alien (HK1, 1: 348); 1811, Nat. of England (HK2, 2: 143); 1814, Nat. of England (HKE: 77); 1862, unspec. (K); 1908, unspec. (K); 1922, unspec. (K); 1933, Arboretum (K); 1934, unspec. (K); 1937, Herbaceous Ground (K). In cult.

Current status: Only as a cultivated plant within the Gardens but still present along the river, particularly near the Bridge.

991. Peucedanum officinale L.
hog's fennel
Cult.: 1768 (JH: 94); 1789, Nat. of England (HK1, 1: 341); 1811, Nat. of England (HK2, 2: 134); 1814, Nat. of England (HKE: 76); 1863, unspec. (K). Not currently in cult.

992. Peucedanum palustre (L.) Moench
Selinum palustre L.
milk-parsley
Cult.: 1789, Nat. of England (HK1, 1: 338); 1811, Nat. of England (HK2, 2: 131); 1814, Nat. of England (HKE: 75). Not currently in cult.

993. Peucedanum ostruthium (L.) W.D.J.Koch
Imperatoria ostruthium L.
masterwort
Cult.: 1789, Nat. of Scotland (HK1, 1: 358); 1811, Nat. of Scotland (HK2, 2: 154); 1814, Nat. of Scotland (HKE: 78). Not currently in cult.
Note. Originally grown as a native but now considered to be an archaeophyte.

994. Pastinaca sativa L.
parsnip
First record: 1977 [1789]
References: **Kew Environs.** 1977, *Cope* (unpubl.): towpath, near Kew Bridge.
Cult.: 1789, Nat. of England (HK1, 1: 361); 1811, Nat. of England (HK2, 2: 157); 1814, Nat. of England (HKE: 79). Not currently in cult.
Current status: No recent sightings.

995. Heracleum sphondylium L.
H. angustifolium auct. non Jacq.
hogweed
First record: 1873/4 [1768].
Nicholson: **1873/4 survey** (JB: 45): Fairly common in all divisions except **B.** • **1906 revision** (AS: 80): Fairly common in all divisions except **B**.
Cult.: 1768 (JH: 97); 1789, Nat. of Britain (HK1, 1: 345), Nat. of England (ibid.: 346 as *angustifolium*); 1811, Nat. of Britain (HK2, 2: 140); 1814, Nat. of Britain (HKE: 76); 1935, Herbarium Ground (K). Not currently in cult.
Current status: Widespread but no longer common.

996. *Heracleum mantegazzianum Sommier & Levier
H. sibiricum sphalm.
giant hogweed
First record: 1988 [1917].
References: **Kew Gardens.** 1988, *Cope* (S31: 182): Behind Sir Joseph Banks Building, before restoration of the 'Paddock.'

Giant hogweed (*Heracleum mantegazzianum*). Formerly near the Herbarium but now more or less confined to a conspicuous stand on Cumberland Mound.

Cult.: 1917, unspec. (K). Not currently in cult.
Current status: Rare; well established on Cumberland Mound. It is no longer in the Paddock where it was previously noted in 1977. It occasionally turns up in other places but is soon weeded out.

[**997. Tordylium maximum** L.
T. officinale auct. non L.
hartwort
Cult.: 1789, Alien (HK1, 1: 333), Nat. of England (ibid.: as *officinale*); 1811, Nat. of England (HK2, 2: 125); 1814, Nat. of England (HKE: 74).
Note. Listed in error as native.]

998. Torilis japonica (Houtt.) DC.
T. anthriscus (L.) Gmel.
Caucalis anthriscus (L.) Huds.
upright hedge-parsley
First record: 1873/4 [1789].
Nicholson: **1873/4 survey** (JB: 45): **P.** A couple of plants near the "Temperate House." **Strip.** Plentiful by side of towing path between Isleworth Gate and beginning of "Old Deer Park." • **1906 revision** (AS: 80): **Strip.** Plentiful along towing-path. Here and there in Queen's Cottage grounds, etc.
Cult.: 1789, Nat. of Britain (HK1, 1: 335); 1811, Nat. of Britain (HK2, 2: 127); 1814, Nat. of Britain (HKE: 74); 1934, unspec. (K). Not currently in cult.
Current status: No recent sightings.

999. Torilis arvensis (Huds.) Link
Caucalis arvensis Huds.
C. infesta (L.) Curtis
spreading hedge-parsley
CULT.: 1789, Nat. of Britain (HK1, 1: 334); 1811, Nat. of Britain (HK2, 2: 127); 1814, Nat. of Britain (HKE: 74); 1972, unspec. (K). Not currently in cult.
Note. Originally grown as a native but now considered to be an archaeophyte.

1000. Torilis nodosa (L.) Gaertn.
Tordyluim nodosum L.
Caucalis nodosa (L.) Scop.
knotted hedge-parsley
CULT.: 1768 (JH: 96); 1789, Nat. of Britain (HK1, 1: 335); 1811, Nat. of Britain (HK2, 2: 127); 1814, Nat. of Britain (HKE: 74). Not currently in cult.

[1001. Torilis leptophylla (L.) Rchb.f.
CULT.: 1789, Nat. of England (HK1, 1: 334); 1811, Alien (HK2, 2: 127); 1814, Alien (HKE: 74).
Note. Originally listed as native but this corrected in later editions.**]**

[1002. Caucalis platycarpos L.
C. daucoides L.
small bur-parsley
CULT.: 1811, Nat. of England (HK2, 2: 126); 1814, Nat. of England (HKE: 74).
Note. Listed in error as native.**]**

[1003. Turgenia latifolia (L.) Hoffm.
Caucalis latifolia L.
greater bur-parsley
CULT.: 1789, Nat. of England (HK1, 1: 334); 1811, Nat. of England (HK2, 2: 126); 1814, Nat. of England (HKE: 74).
Note. Listed in error as native.**]**

1004. Daucus carota L.
a. subsp. **carota**
wild carrot
FIRST RECORD: 1873/4 [1768].
NICHOLSON: **1873/4 survey** (JB: 45): **B.** A few plants on the slope on north side of lake. • **1906 revision** (AS: 80): **A.** A few plants on the north and south slopes near lake.
CULT.: 1768 (JH: 89); 1789, Nat. of Britain (HK1, 1: 336); 1811, Nat. of Britain (HK2, 2: 128); 1814, Nat. of Britain (HKE: 75); 1934, Arboretum (K); 1938, Herb. Exper. Ground (K); 1939, Herb. Exper. Ground (K). Not currently in cult.
CURRENT STATUS: Found on one occasion along the Broad Walk but has not recently been seen near the Lake.

Common centaury (*Centaurium erythraea*). Extinct. Last recorded near the Lake and in Kew Palace Grounds around a hundred years ago.

b. subsp. **gummifer** (Syme) Hook.f.
D. maritimus Lam.
sea carrot
CULT.: 1814, Nat. of England (HKE: Add.). Not currently in cult.

GENTIANACEAE

1005. Cicendia filiformis (L.) Delarbre
Gentiana filiformis L.
yellow centaury
CULT.: 1768 (JH: 133). Not currently in cult.

1006. Centaurium scilloides (L.f.) Samp.
perennial centaury
CULT.: 1925, unspec. (K). In cult. under glass.

1007. Centaurium erythraea Rafn
Erythraea centaurium auct. non (L.) Pers.
Gentiana centaurium L.
Chironia centaurium (L.) Schmidt
common centaury
FIRST RECORD: 1873/4 [1768].
NICHOLSON: **1873/4 survey** (JB: 48): **P.** Two or three plants near lake, 1873–4. **Pal.** Perhaps a score plants. • **1906 revision** (AS: 84): **A.** Two or three plants near lake, 1873-4. **P.** About a score plants.

CULT.: 1768 (JH: 133); 1789, Nat. of Britain (HK1, 1: 323); 1811, Nat. of Britain (HK2, 2: 6); 1814, Nat. of Britain (HKE: 57); 1895, unspec. (K); 1901, unspec. (K); 1907, unspec. (K); 1915, unspec. (K); 1919, unspec. (K); 1923, Herbaceous Pits (K). Not currently in cult.
CURRENT STATUS: No recent sightings and probably lost.

1008. Centaurium littorale (Turner ex Sm.) Gilmour
Chironia littoralis Turner ex Sm.
seaside centaury
CULT.: 1814, Nat. of Britain (HKE: Add.); 1919, unspec. (K). Not currently in cult.
Note. Not found in Britain until 1805.

1009. Centaurium pulchellum (Sw.) Druce
Chironia pulchella Sw.
lesser centaury
CULT.: 1811, Nat. of England (HK2, 2: 6); 1814, Nat. of England (HKE: 57); 1882, unspec. (K). Not currently in cult.
Note. Not found in Britain until 1796.

1010. Blackstonia perfoliata (L.) Huds.
Gentiana perfoliata L.
Chlora perfoliata (L.) L.
yellow-wort
CULT.: 1768 (JH: 133); 1789, Nat. of Britain (HK1, 2: 8); 1811, Nat. of Britain (HK2, 2: 352); 1814, Nat. of Britain (HKE: 109); Undated, unspec. (K). Not currently in cult.

1011. Gentianella campestris (L.) Börner
Gentiana campestris L.
field gentian
CULT.: 1768 (JH: 133); 1789, Nat. of England (HK1, 1: 324); 1811, Nat. of Britain (HK2, 2: 113); 1814, Nat. of Britain (HKE: 72); 1926, unspec. (K). In cult. under glass.

1012. Gentianella amarella (L.) Börner
Gentiana amarella L.
autumn gentian
CULT.: 1768 (JH: 133); 1789, Nat. of Britain (HK1, 1: 324); 1811, Nat. of Britain (HK2, 2: 113); 1814, Nat. of Britain (HKE: 72). In cult.

1013. Gentiana pneumonanthe L.
marsh gentian
CULT.: 1768 (JH: 133); 1789, Nat. of England (HK1, 1: 322); 1811, Nat. of England (HK2, 2: 111); 1814, Nat. of England (HKE: 72); 1904, unspec. (K); 1922, unspec. (K); 1936, unspec. (K); 1987, Alpine #10 (K). In cult.

1014. Gentiana verna L.
spring gentian
CULT.: 1768 (JH: 133); 1811, Nat. of England & Ireland (HK2, 2: 112); 1814, Nat. of Britain (HKE: 73). Not currently in cult.

1015. Gentiana nivalis L.
alpine gentian
CULT.: 1811, Nat. of Scotland (HK2, 2: 113); 1814, Nat. of Scotland (HKE: 72). Not currently in cult.

[**1016. Gentiana acaulis** L. sens. lat.
CULT.: 1789, Alien (HK1, 1: 323); 1811, Nat. of Wales (HK2, 2: 112); 1814, Nat. of Wales (HKE: 72).
Note. Listed in error as native.]

[**1017. Swertia perennis** L.
CULT.: 1789, Nat. of England (HK1, 1: 321); 1811, Nat. of England (HK2, 2: 109); 1814, Nat. of England (HKE: 72).
Note. Listed in error as native.]

APOCYNACEAE

1018. Vinca minor L.
lesser periwinkle
CULT.: 1768 (JH: 149); 1789, Nat. of Britain (HK1, 1: 295); 1811, Nat. of Britain (HK2, 2: 66); 1814, Nat. of Britain (HKE: 66); 1940, Herb Garden (K); 1968, unspec. (K); 1974, unspec. (K). In cult.
Note. Originally grown as a native but now considered to be an archaeophyte.

1019. *Vinca major L.
greater periwinkle
FIRST RECORD: 2001 [1768].
REFERENCES: **Kew Gardens.** 2001, *Cope* (S35: 648): Originally planted but now thoroughly naturalised in flower beds in the Lower Nursery (120).
CULT.: 1768 (JH: 149); 1789, Nat. of England (HK1, 1: 295); 1811, Nat. of England (HK2, 2: 66); 1814, Nat. of England (HKE: 66); 1880, Arboretum (K); 1918, unspec. (K); 1930, unspec. (K); 1940, Herb Garden (K); 1944, unspec. (K); 1946, shrubbery by entrance to nursery (K); 1970, Jodrell Greenhouse (K); 1973, Rock Garden (K); 2000, new hedge by Aiton House (K). In cult.
CURRENT STATUS: Thoroughly naturalised in beds by Aiton House and has spread to the bank at the back of the Brentford Gate car park.

Greater periwinkle (*Vinca major*). Thoroughly naturalised by Aiton House and at the back of the Brentford Gate car park.

SOLANACEAE

1020. *Nicandra physalodes (L.) Gaertn.
Atropa physalodes L.
apple-of-Peru

FIRST RECORD: 1951 [1768].
REFERENCES: **Kew Gardens.** 1951, *Welch* (LN33: S199): Towpath between Richmond and Kew, on old potting soil from Kew Gardens. • 1983/4, *Cope* (S31: 181): New student vegetable plots behind Hanover House. • 1985, *London Natural History Society* (LN65: 196): [Around what is now the Princess of Wales Conservatory], 1 plant.
CULT.: 1768 (JH: 147); 1882, unspec. (K); 1954, unspec. (K). In cult.
CURRENT STATUS: Occasional on heavily disturbed ground.

1021. *Lycium barbarum L.
Duke of Argyll's teaplant

FIRST RECORD: 2004 [1968].
CULT.: In cult.
CURRENT STATUS: Naturalised in a bed between Aiton House and the Brentford Gate car park though it probably originated from cultivated plants.

1022. Atropa belladonna L.
deadly nightshade

FIRST RECORD: 1945 [1768].
REFERENCES: **Kew Gardens.** 1984, *Hastings* (LN64: 117; S33: 747): A new mound of earth at the herbarium allotments, 2 plants. **Kew Environs.** 1945, *Welch* (LN33: S199; S33: 747): By River Thames, Kew. • 1950, *Bull* (LN33: S199): Crevices of old tombs, St. Anne's churchyard, Kew Green. •
CULT.: 1768 (JH: 147); 1789, Nat. of Britain (HK1, 1: 243); 1810, Nat. of Britain (HK2, 1: 392); 1814, Nat. of Britain (HKE: 54); 1918, Herbaceous Ground (K); 1933, Herb Garden (K); 1936, Herb. Ground (K); 1942, Trial Plot (K). In cult.
CURRENT STATUS: Widespread and probably increasing in shady places.

1023. Hyoscyamus niger L.
henbane

FIRST RECORD: 1877 [1768].
REFERENCES: **Kew Gardens.** 1998, *Cope* (S33: 747): North Arb.: lawn behind Herbarium, after disturbance during building works. **Kew Environs.** 1940, *Cooke* (LN33: S200; S33: 747): Thames bank, Kew.
EXSICC.: **Kew Environs.** Rubbish heap at Kew; vii 1877; *Baker* s.n. (K).
CULT.: 1768 (JH: 137); 1789, Nat. of Britain (HK1, 1: 240); 1810, Nat. of Britain (HK2, 1: 388); 1814, Nat. of Britain (HKE: 54); 1910, unspec. (K); 1918, Herbaceous Ground (K); 1933, Arboretum (K); 1934, Arboretum (K); 1936, Herbaceous Ground (K); 1951, Herbarium Ground (K); 1953, Herbarium Ground (K). In cult.
CURRENT STATUS: It made a brief appearance behind the Herbarium in 1998, but had not returned by the time the new building works had again damaged this lawn.

Apple-of-Peru (*Nicandra physalodes*). An exotic-looking species that seems to appear whenever there are major soil disturbances.

Deadly nightshade (*Atropa belladonna*). Widely scattered; rare but possibly on the increase.

1024. *Physalis philadelphica Lam.
large-flowered tomatillo
FIRST RECORD: 1967.
EXSICC.: **Kew Gardens.** Waste ground by Kew Palace; 25 viii 1967; *Reid* s.n. (K).
CURRENT STATUS: Grew from seed in topsoil stored in the Paddock. It has not been seen growing wild for forty years and is apparently not in cultivation in the Gardens; it is therefore likely to have originated from a vegetable plot, either one of the originals in the Paddock lost when the Banks Building was constructed, or one of those created behind the Herbarium to replace them.

1025. *Jaltomata procumbens (Cav.) J.L.Gentry
FIRST RECORD: 2005 [1917].
CULT.: 1917, unspec. (K). Not currently in cult.
CURRENT STATUS: Grew from seed in topsoil stored in the Paddock.

1026. *Lycopersicon esculentum Mill.
Solanum lycopersicum L.
tomato
FIRST RECORD: 1983/4 [1768].
REFERENCES: **Kew Gardens.** 1983/4, *Cope* (S31: 181): New student vegetable plots behind Hanover House.
CULT.: 1768 (JH: 148); 1946, unspec. (K). In cult.
CURRENT STATUS: Occasional; probably germinating from discarded sandwiches or bird droppings.

1027. Solanum nigrum L.
black nightshade
FIRST RECORD: 1873/4 [1768].
NICHOLSON: **1873/4 survey** (JB: 48): Plentiful. Often appears where turf has become bare. • **1906 revision** (AS: 84): Abundant. Often appears where turf has become bare.
EXSICC.: **Kew Gardens.** Lower Nursery; on heap of "leaf soil"; 9 ix 1929; *Hubbard & Nelmes* s.n. (K). • [Unlocalised]; annual weed; 24 vii 1944; *Souster* 122 (K). • Mound of earth from excavation for foundations of new wing; 14 x 1965; *Verdcourt* 4236 (K). • Students' vegetable plots; dry cultivated ground; ix 1980; *Tasker* 8 (K). **Kew Environs.** Kew; 12 viii 1900; *Clarke* 97B (K). • Kew; 1904; *Gamble* 29207 (K).
CULT.: 1768 (JH: 149); 1789, Nat. of Britain (HK1, 1: 249); 1810, Nat. of Britain (HK2, 1: 399); 1814, Nat. of Britain (HKE: 55); 1907, unspec. (K); 1911, unspec. (K); 1933, unspec. (K); 1934, Arboretum (K); 1936, Herbaceous Ground (K). Not currently in cult.
CURRENT STATUS: A fairly common flower-bed weed throughout the Gardens.

1028. *Solanum villosum Mill.
S. luteum Mill.
red nightshade
FIRST RECORD: 1877 [1789].
REFERENCES: **Kew Gardens.** 1984, *Hastings* (FSSC: 103; S33: 747): Waste ground, Kew Gardens. • 1984, *Hastings* (LN64: 117; S33: 747): A new mound of earth at the herbarium allotments, 4 plants.
EXSICC.: **Kew Environs.** Kew; 1877; *Baker* 874 (K).
CULT.: 1789, Alien (HK1, 1: 249); 1810, Alien (HK2, 1: 399); 1814, Alien (HKE: 55); 1907, unspec. (K); 1930, unspec. (K); 1971, unspec. (K). Not currently in cult.

Henbane (*Hyoscyamus niger*). Very rare; seen only once in recent years, in disturbed soil behind the Herbarium. The plant was mown before it could set seed.

CURRENT STATUS: A couple of plants occurred near the Temporary Cycad House, but were lost through redevelopment of the site in 2004 as 'Climbers & Creepers'. It reappeared in quantity in 2007 from seed in topsoil stored in the Paddock.

1029. Solanum dulcamara L.

bittersweet

FIRST RECORD: 1873/4 [1768].
NICHOLSON: **1873/4 survey** (JB: 48): **Q.** Common in the shrubberies skirting river. A very pubescent form occurs opposite the large elms on Strip. • **1906 revision** (AS: 84): **Q.** Common in shrubberies and wood. A very pubescent form occurs on **Strip**.
OTHER REFERENCES: **Kew Gardens.** 1999, *Stones* (EAK: 1): By Lake.
EXSICC.: **Kew Gardens.** Arboretum; 9 viii 1880, *G. Nicholson* 2078 (K). • Queen's Cottage Grounds; 15 vi 1914; *Flippance* s.n. (K). • Arboretum; 17 viii 1916; *Dallimore* s.n. (K). • [Unlocalised]; viii 1942; *Hutchinson* s.n. (K). **Kew Environs.** Thames bank, Kew; from crevices of new stone wall, half-way down; 2 viii 1928; *Summerhayes & Turrill* s.n. (K). • Towpath, Kew; 23 v 1944; *Souster* 69 (K).
CULT.: 1768 (JH: 149); 1789, Nat. of Britain (HK1, 1: 247); 1810, Nat. of Britain (HK2, 1: 397); 1814, Nat. of Britain (HKE: 55); 1905, Arboretum (K); 1907, unspec. (K); 1924, unspec. (K); 1933, unspec. (K); 1936, Herbaceous Ground (K). In cult.
CURRENT STATUS: Common throughout the Gardens.

1030. *Solanum sisymbriifolium Lam.

red buffalo-bur

FIRST RECORD: 1963 [1897].
EXSICC.: **Kew Gardens.** Waste ground by tennis courts (between Kew Palace and Herbarium); 15 x 1963; *P. Taylor & Polhill* s.n. (K).
CULT.: 1897, unspec. (K); 1912, unspec. (K); 1932, unspec. (K); 1984, unspec. (K). In cult.
CURRENT STATUS: Grew from seed in topsoil stored in the Paddock.

1031. *Solanum capsicoides All.

cockroach-berry

FIRST RECORD: 1985 [1969].
REFERENCES: **Kew Gardens.** 1985, *London Natural History Society* (LN65: 196; S33: 747): [Around what is now the Princess of Wales Conservatory], 1 plant.
CULT.: 1969, Tropical Dept. (K). In cult. under glass.
CURRENT STATUS: No recent sightings.

1032. *Nicotiana paniculata L.

FIRST RECORD: 2005 [1768].
CULT.: 1768 (JH: 138); 1789, Nat. of Peru (HK1, 1: 242); 1810, Nat. of Peru (HK2, 1: 391); 1814, Peru (HKE: 54); 1917, unspec. (K); 1940, Herbaceous Ground (K); 1999, plot 157-31 (K). Not currently in cult.
CURRENT STATUS: Grew from seed in topsoil stored in the Paddock.

1033. *Datura stramonium L.

thorn-apple

FIRST RECORD: 1965 [1768].
REFERENCES: **Kew Gardens.** 1983, *Hastings* (LN63: 143): Many disturbed places in Kew Gardens. • 1985, *London Natural History Society* (LN65: 196): [Around what is now the Princess of Wales Conservatory]. • 1985, *Cope* (S31: 183): Site of demolished T-Range.
EXSICC.: **Kew Gardens.** Mound of earth from excavation for foundations of new wing; 14 x 1965; *Verdcourt* 4238 (K).
CULT.: 1768 (JH: 136); 1789, Nat. of England (HK1, 1: 239); 1810, Nat. of England (HK2, 1: 387); 1814, Nat. of England (HKE: 53); 1906, unspec. (K); 1922, unspec. (K); 1933, unspec. (K); 1936, Herbaceous Ground (K); 1948, Herbarium Ground (K); 1966, Lower Nursery, Decorative Dept. (K); 1971, unspec. (K). In cult.
CURRENT STATUS: Sporadic in disturbed ground. It can be found somewhere in most years but is very unpredictable.

1034. *Petunia linearis (Hoek.) Paxton

FIRST RECORD: 2005.
EXSICC.: **Kew Gardens.** In unkempt ground below security fence between edge of Grass Beds and Jodrell Laboratory Extension building site; 5 ix 2005; *Cope* 705 (K).
CURRENT STATUS: A single plant was found growing as a weed but without any record of ever having been cultivated.

CONVOLVULACEAE

1035. Convolvulus arvensis L.

field bindweed

FIRST RECORD: 1873/4 [1768].
NICHOLSON: **1873/4 survey** (JB: 48): **P.** Common on the slopes and elsewhere about Temperate House. • **1906 revision** (AS: 84): **A.** Common on slopes and elsewhere about temperate house.
EXSICC.: **Kew Gardens.** Herbarium Experimental Ground; in cultivated ground; 15 ii 1928; *Summerhayes* s.n. (K). • Herbarium Experimental Ground; in loose loam; 10 iv 1928; *Summerhayes* s.n. (K). • Herbarium Expermtal. Ground; growing in grass turf on fallow ground; 26 iv 1928; *Summerhayes* s.n. (K). • Herbarium Exper. Ground; in loose garden soil; 27 vi 1928; *Summerhayes* s.n. (K). • Ground east of the Herbarium; 17 viii 1928; *Summerhayes & Turrill* s.n. (K). • Herbarium Expermtl. Ground; in almost closed turf mown once during summer; 13 ix 1928; *Summerhayes* s.n. (K). • Herbarium Experimental Ground; in cultivated earth, rather loose; 19 iv 1929; *Summerhayes* s.n. (K). • Kew Gardens – riverside; in meadowland on rather poor soil; 13 vi

1948; *Parker* s.n. (K). **Kew Environs.** Towpath between Kew & Richmond; dry position in rough grass; 18 vii 1948; *Souster* 886 (K).
CULT.: 1768 (JH: 116); 1789, Nat. of Britain (HK1, 1: 207); 1810, Nat. of Britain (HK2, 1: 327); 1814, Nat. of Britain (HKE: 46). Not currently in cult.
CURRENT STATUS: A common and persistent weed.

1036. Calystegia soldanella (L.) R.Br.
Convolvulus soldanella L.
sea bindweed
CULT.: 1768 (JH: 116); 1789, Nat. of Britain (HK1, 1: 214); 1810, Nat. of Britain (HK2, 1: 337); 1814, Nat. of Britain (HKE: 46); 1917, unspec. (K). Not currently in cult.

1037. Calystegia sepium (L.) R.Br.
Convolvulus sepium L.
hedge bindweed
FIRST RECORD: 1873/4 [1768].
NICHOLSON: **1873/4 survey** (JB: 48): **B.** Plentiful in the Rhododendron beds behind Palm House. **P.** Not common, but found here and there in several shrubberies. • **1906 revision** (AS: 84): Here and there as a weed in shrubberies.
OTHER REFERENCES: **Kew Gardens.** 1999, *Stones* (EAK: 2 & 5): By Lake and Palm House Pond.
EXSICC.: **Kew Gardens.** [Unlocalised]; 2 vii 1933; *Dallimore* s.n. (K). • [Unlocalised]; on hedges; 1943; *Hutchinson* s.n. (K). **Kew Environs.** Thames bank, Kew; flooded at high tide, climbing up *Oenanthe* in top zone; 13 viii 1928; *Summerhayes & Turrill* 2480 (K).
CULT.: 1768 (JH: 116); 1789, Nat. of Britain (HK1, 1: 207); 1810, Nat. of Britain (HK2, 1: 327); 1814, Nat. of Britain (HKE: 46); 1934, unspec. (K); 1935, unspec. (K). In cult.
CURRENT STATUS: A common weed throughout the Gardens.

1038. *Calystegia pulchra Brummitt & Heywood
hairy bindweed
FIRST RECORD: 1958 [1977].
REFERENCES: **Kew Environs.** 1958–1981 (S33: 747, and see Exsicc).
EXSICC.: **Kew Environs.** Kew, along the Towing Path climbing over brambles near the C.M.I.; 17 vii 1958; *J.L.Gilbert* 316 (K). • Climbing on wire mesh fence between Kew Bridge & back of Herbarium grounds; 1960; *Townsend* s.n. (K). • Hedge facing river, Kew side of Kew Bridge; 13 vii 1961; *Turrill* s.n. (K). • Overgrown waste place by tow path at end of garden of 69 Kew Green; 3 vii 1981; *Verdcourt* 5394 (K).
CULT.: 1977, unspec. (K). Not currently in cult.
CURRENT STATUS: No recent sightings; it was last seen on the riverbank in 1982.

Hedge bindweed (*Calystegia sepium*). A common and persistent weed scrambling over trees, shrubs and flower-beds throughout the Gardens.

1039. *Calystegia silvatica (Kit.) Griseb.
large bindweed
FIRST RECORD: 1998.
REFERENCES: **Kew Gardens.** 1998, *Cope* (S33: 748): North Arb.: behind the Banks Building (108); grassy bank between Banks Building and Kew Palace entrance.
CURRENT STATUS: Less common in the Gardens than *C. sepium*, but more common on the towpath.

1040. *Ipomoea purpurea Roth
Convolvulus purpureus L., nom. illeg.
C. hederaceus L.
common morning-glory
FIRST RECORD: 1959 [1768].
REFERENCES: **Kew Gardens.** 1959, *Hepper* (S24: 188; and see Exsicc.).
EXSICC.: **Kew Gardens.** By Wing A of Herbarium; beneath bird table; vi 1959; *Hepper* 2900 (K).
CULT.: 1768 (JH: 116); 1898, unspec. (K); 1934, unspec. (K); 1980s, unspec. (K). In cult.
CURRENT STATUS: No recent sightings. Despite being in cultivation for many years the one recorded occurrence as a wild plant undoubtedly resulted from spilled birdseed.

1041. *Cuscuta campestris Yunck.
yellow dodder
FIRST RECORD: 1950–1956.
REFERENCES: **Kew Gardens.** 1950–1956, *Milne-Redhead, D.H.Kent & Lousley* (LFS: 264; LN36: S354): Herbarium grounds, apparently parasitic on *Bromus carinatus*. • 1952, *Milne-Redhead* (S24: 188): By W. side of Wing A of Herbarium, parastising *Convolvulus arvensis*, and every year till 1959, but absent in 1960, probably killed by the dry summer of 1959.

Dodder (*Cuscuta epithymum*). Probably extinct. Last recorded for certainty as a wild plant a hundred years ago but there have been occasional unconfirmed sightings since.

Exsicc.: **Kew Gardens.** Growing under window of Herbarium Wing A; ?possibly introduced with bird food; viii 1953; *Milne-Redhead* 6453 (K).
Current status: No recent sightings. Its host is abundant, and indeed increasing, but Yellow Dodder has not been seen since the 1950s.

1042. Cuscuta europaea L.
greater dodder
First record: 1947 [1768].
References: **Kew Gardens.** 1983, *Burton* (FLA: 117): ... accidentally introduced into the natural order beds and allowed to grow on. • 1989, *comm. Rudall* (S32: 657 and see Exsicc.): Order beds, on *Lamium garganicum*; probably an imported strain, usually on nettle. **Kew Environs.** 1947, *Lousley* (LN33: S197): Between Kew and Richmond. • 1949, *Welch* (LN33: S197): Between Kew and Richmond; site destroyed by flood of March 1949.
Exsicc.: **Kew Gardens.** Tow-path by River Thames between Isleworth Gate and Isleworth Ferry; rampant over *Urtica* & *Galium*; 23 vii 1948; *Ross-Craig & Sealy* 1571 (K). • Growing on *Lamium* in Order Beds; 1989; *Rudall* s.n. (K).
Cult.: 1768 (JH: 144); 1789, Nat. of England (HK1, 1: 168); 1810, Nat. of England (HK2, 1: 275); 1814, Nat. of England (HKE: 38). Not currently in cult.
Current status: No recent sightings. There are occasional unconfirmed sightings from the Order Beds, so the plant may still be around.

1043. Cuscuta epithymum (L.) L.
dodder
First record: 1873/4 [1810].
Nicholson: **1873/4 survey** (JB: 48): **B.** In the flower-beds behind Palm House on the bordering of variegated ivy. In 1873 on Gladioli, Penstemons, etc. In 1874 on Mesembryanthemums, Alternantheras, etc. • **1906 revision** (AS: 84 and in SFS: 472): **B.** On ivy behind palm house. In 1873 in flower borders, on *Alternanthera*, *Mesembryanthemum*, *Gladiolus*, *Penstemon*, etc.
Cult.: 1810, Nat. of Britain (HK2, 1: 275); 1814, Nat. of Britain (HKE: 38). Not currently in cult.
Current status: No recent sightings.

Menyanthaceae

1044. Menyanthes trifoliata L.
bogbean
Cult.: 1768 (JH: 130); 1789, Nat. of Britain (HK1, 1: 197); 1810, Nat. of Britain (HK2, 1: 312); 1814, Nat. of Britain (HKE: 43). In cult.

1045. Nymphoides peltata Kuntze
Menyanthes nymphoides L.
fringed water-lily
First record: 1849 [1768].
Exsicc.: **Kew Environs.** Kew; viii 1849; *Wing* s.n. (K). • Ha-ha, Old Deer Park, Richmond; 1883; *Fraser* s.n. (K). • Ha-ha, Old Deer Park, Richmond; vii 1913; *Fraser* s.n. (K). • Ha-ha near Richmond; viii 1927; *Findlay* s.n. (K).
Cult.: 1768 (JH: 130); 1789, Nat. of Britain (HK1, 1: 196); 1810, Nat. of England (HK2, 1: 312); 1814, Nat. of England (HKE: 43); 1995, Alpine & Herbaceous (K). Not currently in cult.
Current status: Not seen inside Kew Gardens. The Ha-ha mentioned above was probably not that between Kew Gardens and the Old Deer Park but rather between the Park and the River south of Kew Gardens.

Polemoniaceae

1046. Polemonium caeruleum L.
Jacob's-ladder
Cult.: 1768 (JH: 139); 1789, Nat. of Britain (HK1, 1: 218); 1810, Nat. of England (HK2, 1: 342); 1814, Nat. of England (HKE: 47); 1884, unspec. (K); 1887, unspec. (K); 1898, unspec. (K); 1929, unspec. (K); 1933, Herb Garden (K); 1937, Herbaceous Ground (K); 1940, Herb Garden (K); 1986, unspec. (K); 2000, Alpine & Herb., plot 157-40 (K). In cult.

1047. *Polemonium pauciflorum S.Watson
FIRST RECORD: 1972.
REFERENCES: **Kew Gardens.** 1972, *J.G.Gilbert* (S29: 407): A plant on north side of Wing D of Herbarium; transplanted and potted up; flowered May 1973.
EXSICC.: **Kew Gardens.** 1923, unspec. (K); 1924, unspec. (K); 1972, unspec. (K).
CULT.: In cult.
CURRENT STATUS: In 2004 this species reappeared in the Paddock, after a long absence, from seed in topsoil being stored there. It is curious that it was first found near the Herbarium and then returned to the same spot from a different source.

1048. *Gilia achilleifolia Benth.
FIRST RECORD: 2001 [1895].
REFERENCES: **Kew Gardens.** 2001, *Cope* (S35: 648): Flowerbed in front of the Herbarium. Presumably introduced in compost.
CULT.: 1895, unspec. (K). Not currently in cult.
CURRENT STATUS: Casual in flower-beds by the Herbarium; introduced with compost and persistent for a short time.

1049. *Gilia capitata Sims
blue-thimble-flower
FIRST RECORD: 2001 [1884].
REFERENCES: **Kew Gardens.** 2001, *Cope* (S35: 648): Flowerbed in front of the Herbarium. Presumably introduced in compost.
CULT.: 1884, unspec. (K); 1926, unspec. (K); 1930, unspec. (K). Not currently in cult.
CURRENT STATUS: Casual in flower-beds by the Herbarium; introduced with compost but has not persisted.

1050. *Navarretia squarrosa (Eschsch.) Hook. & Arn.
skunkweed
FIRST RECORD: 1990 [1904].
REFERENCES: **Kew Gardens.** 1990, *Cope* (S31: 182): Paddock, after restoration in 1990.
CULT.: 1904, unspec. (K). Not currently in cult.
CURRENT STATUS: No recent sightings; it has not reappeared in the Paddock since 1990.

1051. *Collomia linearis Nutt.
FIRST RECORD: 1940.
REFERENCES: **Kew Gardens.** 1940, *Marshall* (S31: 183): Herbarium Grounds.
EXSICC.: **Kew Gardens.** Herbarium Grounds; 1940; *Marshall* s.n. (K).
CURRENT STATUS: No recent sightings. There is some doubt about this record for although Marshall implied that the species was found wild in the Herbarium grounds, the voucher specimen has been annotated with the word 'Cult.'

HYDROPHYLLACEAE

1052. *Phacelia parviflora Pursh
small-flowered phacelia
FIRST RECORD: 1880.
NICHOLSON: **1906 revision:** Not recorded.
OTHER REFERENCES: **Kew Gardens.** 1880, *Baker* (BECB1: 40; S33: 748): Thames bank opposite Sion House, midway between Kew and Richmond.
CURRENT STATUS: No recent sightings.

BORAGINACEAE

1053. Lithospermum purpureocaeruleum L.
purple gromwell
CULT.: 1768 (JH: 124); 1789, Nat. of England (HK1, 1: 177); 1810, Nat. of England (HK2, 1: 288); 1814, Nat. of England (HKE: 40); 1922, Rock Garden (K); 1993; unspec. (K). Not currently in cult.

1054. Lithospermum officinale L.
common gromwell
CULT.: 1768 (JH: 123); 1789, Nat. of Britain (HK1, 1: 176); 1810, Nat. of Britain (HK2, 1: 286); 1814, Nat. of Britain (HKE: 40). Not currently in cult.

1055. Lithospermum arvense L.
field gromwell
FIRST RECORD: 1951 [1768].
EXSICC.: **Kew Gardens.** [Unlocalised]; weed on manure heap with other cornfield plants; 26 vi 1951; *Souster* 1203 (K).
CULT.: 1768 (JH: 123); 1789, Nat. of Britain (HK1, 1: 176); 1810, Nat. of Britain (HK2, 1: 287); 1814, Nat. of Britain (HKE: 40). Not currently in cult.
CURRENT STATUS: No recent sightings.

1056. Echium vulgare L.
viper's-bugloss
FIRST RECORD: 2006 [1768].
CULT.: 1768 (JH: 122); 1789, Nat. of Britain (HK1, 1: 187); 1810, Nat. of Britain (HK2, 1: 301); 1814, Nat. of Britain (HKE: 41); 1884, unspec. (K); 1933, unspec. (K); 1936, Herb Gdn. (K). In cult.
CURRENT STATUS: A single sighting near Brentford Gate; the plant was likely to have escaped from a nearby student demonstration garden. It now seems to be spreading.

1057. Echium plantagineum L.
E. violaceum L.
purple viper's-bugloss
FIRST RECORD: 1880 [1987].
NICHOLSON: **1906 revision:** Not recorded.

Viper's-bugloss (*Echium vulgare*). A small patch occurs near Brentford Gate where it persisted from a temporary demonstration garden.

OTHER REFERENCES: **Kew Gardens.** 1880, *Baker* (BECB1: 40; S33: 748): Thames bank opposite Sion House, midway between Kew and Richmond.
CULT.: 1987, plot 157-36 (K). In cult. under glass.
CURRENT STATUS: No recent sightings.

1058. *Echium pininana Webb & Berthel.
giant viper's-bugloss
FIRST RECORD: 2004 [1995].
CULT.: In cult.
CURRENT STATUS: Established for many years in beds around Aiton House. It recently appeared at the back of the Mycology Building but was destroyed before it could flower when the building was demolished in 2006. There has also been a recent sighting near the Broad Walk.

[**1059. Echium italicum** L.
pale bugloss
CULT.: 1789, Nat. of England (HK1, 1: 187); 1810, Nat. of Jersey (HK2, 1: 301); 1814, Nat. of Jersey (HKE: 41).
Note. Listed in error as native.]

[**1060. Pulmonaria officinalis** L.
lungwort
CULT.:1789, Nat. of Britain (HK1, 1: 181); 1810, Nat. of England (HK2, 1: 292); 1814, Nat. of England (HKE: 40).
Note. Listed in error as native.]

1061. Pulmonaria obscura Dumort.
unspotted lungwort
CULT.: 1927, unspec. (K); 1993, unspec. (K). In cult.

1062. Pulmonaria longifolia (Bastard) Boreau
narrow-leaved lungwort
CULT.: 1789, Alien (HK1, 1: 181); 1810, Nat. of England & Wales (HK2, 1: 292); 1814, Nat. of Britain (HKE: 40); 1934, Herbarium Ground (K). Not currently in cult.

1063. *Nonea lutea (Desr.) DC.
yellow nonea
FIRST RECORD: 1983 [1883].
REFERENCES: **Kew Gardens.** 1983, *Cope* (S31: 181): New student vegetable plots behind Hanover House.
CULT.: 1883, unspec. (K); 1894, unspec. (K). In cult.
CURRENT STATUS: No recent sightings; it has not reappeared behind Hanover House since 1983.

1064. Symphytum officinale L.
S. officinale var. *patens* Sibth.
common comfrey
FIRST RECORD: 1873/4 [1768].
NICHOLSON: **1873/4 survey** (JB: 71): This and its var. *patens* grow abundantly on the Palace side of the wooden fence stretching from the "Princess's" to Brentford Gate. Not uncommon on **Strip. • 1906 revision** (AS: 84): **Strip.** Not uncommon.
EXSICC.: **Kew Gardens.** Herbarium field; 15 vi 1932; *Turrill & A.K.Jackson* s.n. (K). • Kew, bank between towpath & Ha-ha opposite Brentford; 23 v 1947; *Burtt* 100 (K). **Kew Environs.** Towpath, Kew; 7 v 1944; *Souster* 58 (K).
CULT.: 1768 (JH: 124); 1789, Nat. of Britain (HK1, 1: 182); 1810, Nat. of Britain (HK2, 1: 294); 1814, Nat. of Britain (HKE: 40). In cult.
CURRENT STATUS: Occasional, but often abundant in isolated patches.

1065. *Symphytum × uplandicum Nyman (asperum × officinale)
Russian comfrey
FIRST RECORD: 1932 [1898].
EXSICC.: **Kew Gardens.** Herbarium field; 15 vi 1932; *A.K.Jackson & Turrill* s.n. (K). • Fresh mound of earth outside Wing A; 13 v 1966; *Verdcourt* 4249 (K).
CULT.: 1898, unspec. (K); 1940, Herbaceous Ground (K). In cult.
CURRENT STATUS: Currently known only from the towpath near Brentford Gate.

1066. Symphytum tuberosum L.
tuberous comfrey
Cult.: 1789, Alien (HK1, 1: 182); 1810, Nat. of Scotland (HK2, 1: 294); 1814, Nat. of Scotland (HKE: 41); 1933, unspec. (K); 1940, Herbaceous Ground (K); 1968, unspec. (K). Not currently in cult.

1067. *Symphytum orientale L.
white comfrey
First record: 1998 [1968].
References: **Kew Gardens.** 1998, *Cope* (S35: 648): In rough grass near the Beech Circle (242). Subsequently in one or two other spots, including the Herbarium lawn (113).
Cult.: 1968, Duke's Garden (K). In cult.
Current status: Occasional; sometimes forming extensive patches.

1068. *Symphytum caucasicum M.Bieb.
Caucasian comfrey
First record: 2004 [1969].
Cult.: In cult.
Current status: A small group grows on the bank at the back of the Brentford Gate car park.

1069. *Symphytum bulbosum K.F.Schimp.
bulbous comfrey
First record: 1998.
References: **Kew Gardens.** 1998, *Cope* (S33: 748): Paddock, by Herbarium.
Current status: Rare; known only from a handful of widely scattered localities.

[1070. Anchusa officinalis L.
alkanet
Cult.: 1789, Alien (HK1, 1: 178); 1810, Nat. of Britain (HK2, 1: 289); 1814, Nat. of Britain (HKE: 40).
Note. Listed in error as native.]

1071. Anchusa arvensis (L.) M.Bieb.
Lycopsis arvensis L.
bugloss
First record: 1873/4 [1768].
Nicholson: **1873/4 survey** (JB: 71): **P.** A couple of plants at lake end of Juniper collection. Here and there on waste ground about Temperate House. • **1906 revision** (AS: 84 and in SFS: 462): Here and there on waste ground.
Cult.: 1768 (JH: 125); 1789, Nat. of Britain (HK1, 1: 185); 1810, Nat. of Britain (HK2, 1: 298); 1814, Nat. of Britain (HKE: 41); 1883, unspec. (K). Not currently in cult.
Current status: Recently appeared in front of the Mycology Building, but this has since been demolished and the site redeveloped. A single specimen was subsequently seen on the golf course in the Old Deer Park.

1072. *Pentaglottis sempervirens (L.) Tausch ex L.H.Bailey
Anchusa sempervirens L.
green alkanet
First record: 1943 [1768].
References: **Kew Gardens.** 1991, *Cope* (S31: 183): Restored 'Paddock.' **Kew Environs.** 1943, *Welch* (LN33: S192): Thames bank between Kew and Richmond. • 1949, *Boniface* (LN33: S192): Thames bank between Kew and Richmond.
Cult.: 1768 (JH: 126); 1789, Nat. of Britain (HK1, 1: 179); 1810, Nat. of Britain (HK2, 1: 290); 1814, Nat. of Britain (HKE: 40). In cult.
Current status: A persistent and increasing weed in many parts of the Gardens.

1073. *Borago officinalis L.
borage
First record: 2000 [1768].
References: **Kew Gardens.** 2000, *Cope* (S35: 648): Weed of flowerbeds (127).
Cult.: 1768 (JH: 124); 1789, Nat. of England (HK1, 1: 184); 1810, Nat. of England (HK2, 1: 296); 1814, Nat. of England (HKE: 41); 1894, unspec. (K); 1933, Arboretum (K); 1933, Herb Garden (K); 1934, Arboretum (K); 1936, Herbaceous Ground (K); 1971, unspec. (K). In cult.
Current status: A casual weed especially around allotments.

Green alkanet (*Pentaglottis sempervirens*). A persistent and increasing weed in parts of the Gardens. Here photographed outside the now-demolished Mycology Building.

1074. *Trachystemon orientalis (L.) Don
Borago orientalis L.
Abraham-Isaac-Jacob
FIRST RECORD: 2002 [1768].
REFERENCES: **Kew Gardens.** 2002, *Cope* (S35: 648): A sizeable patch under trees between the Bamboo Garden and Mount Pleasant (252).
CULT.: 1768 (JH: 124); 1897, unspec. (K); 1921, unspec. (K); 1922, unspec. (K); 1981, Alpine (K). In cult.
CURRENT STATUS: A single patch is naturalised in woodland near the Temperate House.

1075. Mertensia maritima (L.) Gray
Pulmonaria maritima L.
Pneumaria maritima Hill
oysterplant
CULT.: 1768 (JH: 123); 1789, Nat. of Britain (HK1, 1: 182); 1810, Nat. of Britain (HK2, 1: 293); 1814, Nat. of Britain (HKE: 40). Not currently in cult.

[1076. Asperugo procumbens L.
madwort
CULT.: 1789, Nat. of Britain (HK1, 1: 185); 1810, Nat. of Britain (HK2, 1: 297); 1814, Nat. of Britain (HKE: 41).
Note. Listed in error as native.]

1077. Myosotis scorpioides L.
M. palustris L.
water forget-me-not
FIRST RECORD: 1873/4 [1768].
NICHOLSON: **1873/4 survey** (JB: 71): **P.** Here and there near edge of lake. **Strip.** Rather frequently by moat. • **1906 revision** (AS: 84): Here and there by lake. **Strip.** Rather frequent by ha-ha.
CULT.: 1768 (JH: 126); 1789, Nat. of Britain (HK1, 1: 175); 1810, Nat. of Britain (HK2, 1: 285); 1814, Nat. of Britain (HKE: 39). Not currently in cult.
CURRENT STATUS: No recent sightings as a wild plant, but introduced around the margin of the new pond in the Conservation Area.

1078. Myosotis alpestris F.W.Schmidt
M. rupicola Sm.
alpine forget-me-not
CULT.: 1814, Nat. of Scotland (HKE: Add.). Not currently in cult.
Note. Not found as a British native until 1813.

1079. Myosotis sylvatica Hoffm.
wood forget-me-not
FIRST RECORD: 1928.
EXSICC.: **Kew Gardens.** [Unlocalised]; on old rubbish heap; white flowers [det. A.E. Wade as var. *lactea*]; v 1928; *Hubbard* s.n. (K).
CURRENT STATUS: A very rare casual; there has been only one recent sighting.

1080. Myosotis arvensis (L.) Hill
field forget-me-not
FIRST RECORD: 1873/4 [1789].
NICHOLSON: **1873/4 survey** (JB: 71): **B, Pal** and **P.** Common in bare places in the turf about lake, also in beds. • **1906 revision** (AS: 84): **A, B, P.** Common in bare places in turf about lake, also in beds.
CULT.: 1789, Nat. of Britain (HK1, 1: 175); 1810, Nat. of Britain (HK2, 1: 286); 1814, Nat. of Britain (HKE: 39). In cult.
CURRENT STATUS: An occasional weed of flower-beds and short grass; widespread but rather uncommon.

1081. Myosotis ramosissima Rochel
M. collina auct.
early forget-me-not
FIRST RECORD: 1906.
NICHOLSON: **1906 revision** (AS: 84): **Strip.** On wall of ha-ha near river. In turf in dry gravelly places.
CURRENT STATUS: No recent sightings.

1082. Myosotis discolor Pers.
M. balbisiana auct.
M. versicolor Sm.
changing forget-me-not
FIRST RECORD: 1873/4.
NICHOLSON: **1873/4 survey** (JB: 71 and in SFS: 467): **P.** About lake with [*Myosotis arvensis*]. On wall skirting moat. • **1906 revision** (AS: 84 and in SFS: 467): In company with [*Myosotis ramosissima*]; also abundant in dry places near pagoda, where the smaller-flowered yellow form, *M. balbisiana*, was first recognised as a British plant.
EXSICC.: **Kew Gardens.** Near the Pagoda; perfectly wild & naturalised in the shrubbery beds; vi 1887; *anon.* s.n. (K). **Kew Environs.** Kew; v 1885; *G. Nicholson* s.n. (K).
CURRENT STATUS: No recent sightings.

1083. Cynoglossum officinale L.
hound's-tongue
FIRST RECORD: 1999 [1768].
REFERENCES: **Kew Gardens.** 1999, *Cope* (S35: 648): Under a pine tree near the Banks Building (104). Flowered, but cut before seed was set.
CULT.: 1768 (JH: 125); 1789, Nat. of Britain (HK1, 1: 179); 1810, Nat. of Britain (HK2, 1: 290); 1814, Nat. of Britain (HKE: 40); 1934, unspec. (K). In cult.
CURRENT STATUS: Made a brief appearance under trees on the edge of the Paddock, but was removed by strimming before it had set seed.

1084. Cynoglossum germanicum Jacq.
C. officinale var. *sempervirens* auct. dub.
C. sylvaticum Haenke
green hound's-tongue
CULT.: 1789, Nat. of Britain (HK1, 1: 179); 1810, Nat. of Britain (HK2, 1: 290 & 291); 1814, Nat. of Britain (HKE: 40). Not currently in cult.

Hound's-tongue (*Cynoglossum officinale*). Photographed during its single appearance on the edge of the Paddock in 1999. It escaped mowing of the grass because it was close to a tree, but later succumbed to strimming before it could set seed.

VERBENACEAE

1085. Verbena officinalis L.
vervain
FIRST RECORD: 1873/4 [1768].
NICHOLSON: **1873/4 survey** (JB: 71): Mound behind House No. 1. Two or three plants near Palace. • **1906 revision** (AS: 85): **P, Q** and **Strip.** Here and there.
OTHER REFERENCES: **Kew Gardens.** 1999, *Stones* (EAK: 2): By Lake. • 1995, *Lock* (S33: 748): several patches on lawns dried up after period of drought bordering 'The Pond' in front of Palm House.
EXSICC.: **Kew Gardens.** Arboretum; 12 ix 1934; *Dallimore* s.n. (K).
CULT.: 1768 (JH: 267); 1789, Nat. of Britain (HK1, 1: 33); 1812, Nat. of Britain (HK2, 4: 41); 1814, Nat. of Britain (HKE: 191); 1887, unspec. (K); 1936, Herbaceous Ground (K). In cult.
CURRENT STATUS: Occasional; a persistent component of some of the lawns despite attempts at eradication. It survives mowing well and soon regrows.

1086. *Verbena bonariensis L.
Argentinian vervain
FIRST RECORD: 2008 [1768].
CULT.: 1768 (JH: 267); 1789, Alien (HK1, 1: 32); 1812, Alien (HK2, 4: 39); 1814, Alien (HKE: 191); 1914, unspec. (K). In cult.
CURRENT STATUS: Naturalised on the wall of the Palm House Pond, having escaped from the nearby Hardy Display and has been noted on a heap of topsoil being stored in the Paddock.

LAMIACEAE

1087. Stachys officinalis (L.) Trevis.
Betonica officinalis L.
betony
CULT.: 1768 (JH: 255); 1789, Nat. of Britain (HK1, 2: 299); 1811, Nat. of Britain (HK2, 3: 396); 1814, Nat. of Britain (HKE: 180); 1930, unspec. (K); 1933, Herb Garden (K); 1936, Herbaceous Ground (K); 1936, Herb Gdn. (K); 1971, Alpine & Herbaceous Dept. (K); 1975, unspec. (K). In cult.

1088. Stachys germanica L.
downy woundwort
CULT.: 1768 (JH: 256); 1789, Nat. of England (HK1, 2: 301); 1811, Nat. of England (HK2, 3: 399); 1814, Nat. of England (HKE: 181); 1898, unspec. (K); 1978, unspec. (K). In cult.

Vervain (*Verbena officinalis*). An inconspicuous plant with tiny pink flowers that is a persistent weed in some of the lawns.

1089. Stachys sylvatica L.

hedge woundwort

FIRST RECORD: 1873/4 [1768].

NICHOLSON: **1873/4 survey** (JB: 71): **P.** A few plants in wood near large cedar (the one so conspicuous from behind Palm House). **Q.** A large plot containing 20 or 30 plants. • **1906 revision** (AS: 86): **A.** Here and there in wood. **Q.** Not uncommon.

CULT.: 1768 (JH: 256); 1789, Nat. of Britain (HK1, 2: 300); 1811, Nat. of Britain (HK2, 3: 398); 1814, Nat. of Britain (HKE: 181); 1908, unspec. (K); 1933, Herb Garden (K). In cult.

CURRENT STATUS: Uncommon, in shady places; it appears to be on the decline.

1090. Stachys palustris L.

marsh woundwort

FIRST RECORD: 1873/4 [1768].

NICHOLSON: **1873/4 survey** (JB: 71): **Strip.** Three or four plants by moat 100 yards north of Isleworth Gate. • **1906 revision** (AS: 86): **Strip.** A few plants by ha-ha near Isleworth Gate.

EXSICC.: **Kew Environs.** Thames bank between Kew and Richmond; growing out of stone wall; 23 vii 1946; *Sandwith* 3138 (K).

CULT.: 1768 (JH: 256); 1789, Nat. of Britain (HK1, 2: 301); 1811, Nat. of Britain (HK2, 3: 398); 1814, Nat. of Britain (HKE: 181); 1933, Arboretum (K). Not currently in cult.

CURRENT STATUS: Known within the Gardens from a single plant near the Oxenhouse Gate. It has not been refound on the towpath.

1091. Stachys × ambigua Sm. (palustris × sylvatica)

hybrid woundwort

CULT.: 1811, Nat. of Britain (HK2, 3: 398); 1814, Nat. of Britain (HKE: 181). Not currently in cult.
Note. Not known to be a hybrid when originally cultivated.

1092. Stachys arvensis (L.) L.

field woundwort

CULT.: 1789, Nat. of Britain (HK1, 2: 303); 1811, Nat. of Britain (HK2, 3: 401); 1814, Nat. of Britain (HKE: 181). Not currently in cult.
Note. Originally grown as a native but now considered to be an archaeophyte.

1093. Ballota nigra L. subsp. **meridionalis** (Bég.) Bég.

B. nigra subsp. *foetida* (Vis.) Hayek

B. nigra var. *foetida* Koch

B. alba L.

black horehound

FIRST RECORD: 1873/4 [1768].

NICHOLSON: **1873/4 survey** (JB: 71): **P.** A few plants on waste ground near "Spar." **Strip.** Three or four by towing-path. • **1906 revision** (AS: 86): **Strip.** By towing-path. **Q.** Not uncommon.

CULT.: 1768 (JH: 258); 1789, Nat. of Britain (HK1, 2: 303), Nat. of England (ibid.: 304 as *alba*); 1811, Nat. of Britain (HK2, 3: 402); 1814, Nat. of Britain (HKE: 181). Not currently in cult.

CURRENT STATUS: Widespread and sometimes forming large stands.

1094. *Leonurus cardiaca L.

L. tataricus L.

motherwort

FIRST RECORD: 1943–1945 [1768].

REFERENCES: **Kew Gardens.** 1949–50, *Boniface & Welch* (LN33: S226): Opposite Syon house, a single large plant. **Kew Environs.** 1943–45, *Welch* (LN33: S226; S33: 748): By Thames Towing-path between Kew and Richmond, near Kew Bridge.

CULT.: 1768 (JH: 250); 1789, Nat. of Britain (HK1, 2: 306); 1811, Nat. of Britain (HK2, 3: 406); 1814, Nat. of Britain (HKE: 181); 1896, unspec. (K); 1933, Herb Garden (K); 1936, Herbaceous Ground (K). In cult.

CURRENT STATUS: A single plant briefly appeared beside the Palm House Pond and several others grew from seed in topsoil stored in the Paddock; another appeared in a small bed near the Dutch House in 2008.

Motherwort (*Leonurus cardiaca*). An uncommon species that appears at irregular intervals. Here photographed near the Broad Walk in its most recent occurrence.

1095. Lamiastrum galeobdolon (L.) Ehrend. & Polatschek
Galeopsis galeobdolon L.
Galeobdolon luteum Huds.
yellow archangel
FIRST RECORD: 1934 [1768].
REFERENCES: **Kew Gardens.** 1998, *Cope* (S33: 748): North Arb.: behind Kew Palace (111).
EXSICC.: **Kew Gardens.** Arboretum; 6 vii 1934; *Dallimore* 106 (K).
CULT.: 1768 (JH: 256); 1789, Nat. of Britain (HK1, 2: 298); 1811, Nat. of Britain (HK2, 3: 396); 1814, Nat. of Britain (HKE: 180); 1921, unspec. (K). In cult.
CURRENT STATUS: Rare; mostly persisting from plants that may have been originally cultivated, though none occurs in the vicinity of currently cultivated specimens. Dallimore's plant from the Arboretum has been identified as subsp. *montanum* (Pers.) Ehrend. & Polatschek, but other subspp. may be in cultivation; they still need to be checked.

1096. Lamium album L.
white dead-nettle
FIRST RECORD: 1856 [1768].
NICHOLSON: **1873/4 survey** (JB: 71): Very frequent in **Strip**. Uncommon in the other divisions. • **1906 revision** (AS: 86): **Strip.** Very frequent. Less common in other divisions.
EXSICC.: **Kew Gardens.** Near the Tennis Courts; vi 1932; *Turrill* s.n. (K). • Herbarium grounds; 12 vi 1935; *Ross-Craig* 414 (K). • Herbarium field; 7 v 1945; *Ross-Craig & Sealy* 1051 (K). • Herbarium Ground; 21 v 1966; Ross-Craig & *Sealy* s.n. (K). **Kew Environs.** Kew; 25 vii 1856; *E.G.Western* s.n. (K).
CULT.: 1768 (JH: 255); 1789, Nat. of Britain (HK1, 2: 297); 1811, Nat. of Britain (HK2, 3: 393); 1814, Nat. of Britain (HKE: 180); 1918, Herbaceous Ground (K); 1933, Herb Garden (K); 1936, Herb. Gdn. (K); 1960, unspec. (K); 1983, unspec. (K). Not currently in cult.
CURRENT STATUS: Widespread but seldom common.

1097. *Lamium garganicum L.
FIRST RECORD: 1985 [1768].
REFERENCES: **Kew Gardens.** South facing wall by Alpine House; 3 v 1985; *Payne* (comm. *T.B.Ryves*).
CULT.: 1768 (JH: 255); 1789, Alien (HK1, 2: 297); 1811, Alien (HK2, 3: 393); 1814, Alien (HKE: 180); 1977, inspec. (K); 1979, plot 257 (K); 1983, unspec. (K). In cult.
CURRENT STATUS: No recent sightings; the original location, near the Alpine House, is now the site of the new Jodrell Laboratory extension. *Cuscuta europaea* has been known to parasitise this species in the Order Beds.

1098. *Lamium maculatum (L.) L.
spotted dead-nettle
FIRST RECORD: 1920 [1974].
REFERENCES: **Kew Environs.** 1998, *Cope* (S33: 748): Ferry Lane, roadside verge opposite the Herbarium.
EXSICC.: **Kew Gardens.** Field nr. Herbarium; 15 iv 1920; *Turrill* s.n. (K). • On rubbish heap; v 1928; *Hubbard* s.n. (K). • Nr. Herbarium; vi 1928; *C.A. Smith* 6071 (K).
CULT.: 1974, unspec. (K); 1983, unspec. (K). In cult.
CURRENT STATUS: Rare; only one recent sighting doubtless as an escape from cultivation.

1099. Lamium purpureum L.
red dead-nettle
FIRST RECORD: 1873/4 [1768].
NICHOLSON: **1873/4 survey** (JB: 71): Common in all bare places. • **1906 revision** (AS: 86): Common in all the divisions.
OTHER REFERENCES: **Kew Gardens.** 1999, *Stones* (EAK: 2): By Lake.
EXSICC.: **Kew Gardens.** Near the Tennis Courts; rubbish heap; shady place surrounded by docks; 11 v 1928; *Turrill* s.n. (K). • Cult. ground near Herbarium; 15 v 1928; *Summerhayes* s.n. (K). • In herbarium field; 25 v 1929; *Hubbard* s.n. (K). • Meadow behind herbarium; 27 iv 1930; *Bullock* 5 (K). • Grass outside the Herbarium; 9 iv 1936; *Sealy* 439 (K). • Aiton House plots; cultivated soil; 7 xii 1980; *Close* 15 (K). • [Unlocalised]; 1 v 1981; *Jellis* s.n. (K). • Herbarium car park [var. *incisum* and var. *purpureum* on separate sheets]; 6 iv 1987; *Harley* s.n. (K).
CULT.: 1768 (JH: 255); 1789, Nat. of Britain (HK1, 2: 297); 1811, Nat. of Britain (HK2, 3: 394); 1814, Nat. of Britain (HKE: 180); 1922, Herbarium Ground (K). Not currently in cult.
CURRENT STATUS: Widespread and common throughout the Gardens.

1100. Lamium hybridum Vill.
L. incisum Willd.
cut-leaved dead-nettle
FIRST RECORD: 1991 [1811].
REFERENCES: **Kew Gardens.** 1991, *Cope* (S31: 183): Restored 'Paddock.'
CULT.: 1811, Nat. of Britain (HK2, 3: 394); 1814, Nat. of Britain (HKE: 180). Not currently in cult.
CURRENT STATUS: This has not persisted in the Paddock but has recently appeared in some quantity between the Nash Conservatory and the Main Gate where it is usually mown before it can set seed. It is probably declining once more after a brief spell of increase.

Henbit dead-nettle (*Lamium amplexicaule*). Occasional in short grassland and a frequent weed of flower-beds.

1101. Lamium amplexicaule L.
henbit dead-nettle
FIRST RECORD: 1873/4 [1768].
NICHOLSON: **1873/4 survey** (JB: 71): **Strip.** In turf by side of towing-path about 150 yards south of Brentford Ferry. Not seen within the Gardens. • **1906 revision** (AS: 86): **Strip.** In turf by towing-path near Brentford Ferry.
OTHER REFERENCES: **Kew Gardens.** 1945, *Welch* (LN33: S228): Kew Gardens.
EXSICC.: **Kew Gardens.** Weed in riverside nursery; 9 iv 1945; *Melville* s.n. (K). • Behind Aiton House; cultivated ground on light soil; ix 1979; *McNamara* 6 (K).
CULT.: 1768 (JH: 255); 1789, Nat. of Britain (HK1, 2: 298); 1811, Nat. of Britain (HK2, 3: 394); 1814, Nat. of Britain (HKE: 180). Not currently in cult.
CURRENT STATUS: Fairly common in short grass and in flower-beds.

1102. Galeopsis segetum Neck.
G. villosa Huds.
downy hemp-nettle
FIRST RECORD: 2007 [1811].
CULT.: 1811, Nat. of Britain (HK2, 3: 395); 1814, Nat. of Britain (HKE: 180). Not currently in cult.
CURRENT STATUS: Grew from seed in a heap of topsoil being stored in the Paddock behind the Banks Building.

1103. Galeopsis angustifolia Ehrh. ex Hoffm.
red hemp-nettle
CULT.: 1904, unspec. (K). Not currently in cult.

1104. Galeopsis speciosa Mill.
G. versicolor Curt.
large-flowered hemp-nettle
CULT.: 1811, Nat. of Britain (HK2, 3: 395); 1814, Nat. of Britain (HKE: 180); 1984, unspec. (K). Not currently in cult.
Note. Originally grown as a native but now considered to be an archaeophyte.

1105. Galeopsis tetrahit L.
common hemp-nettle
CULT.: 1768 (JH: 256); 1789, Nat. of Britain (HK1, 2: 298); 1811, Nat. of Britain (HK2, 3: 395); 1814, Nat. of Britain (HKE: 180); 1922, unspec. (K); 1984, unspec. (K); 1986, Order Beds (K). Not currently in cult.

1106. Galeopsis bifida Boenn.
bifid hemp-nettle
FIRST RECORD: 2007.
CURRENT STATUS: Grew from seed in topsoil stored in the Paddock. Cultivated specimens are likely to have been included in *G. tetrahit*, above.

[1107. Galeopsis ladanum L.
broad-leaved hemp-nettle
CULT.: 1789, Nat. of England (HK1, 2: 298); 1811, Nat. of Britain (HK2, 3: 395); 1814, Nat. of Britain (HKE: 180).
Note. Listed in error as native.]

1108. Melittis melissophyllum L.
M. grandiflora Sm.
bastard balm
CULT.: 1768 (JH: 246); 1789, Nat. of England (HK1, 2: 320); 1811, Nat. of England (HK2, 3: 421, and: 422 as *grandiflora*); 1814, Nat. of England (HKE: 184 and as *grandiflora*); 19th cent., unspec. (K); 1976, unspec. (K). In cult.

1109. Marrubium vulgare L.
white horehound
CULT.: 1768 (JH: 258); 1789, Nat. of Britain (HK1, 2: 305); 1811, Nat. of Britain (HK2, 3: 404); 1814, Nat. of Britain (HKE: 181); 1925, Herb. Expt. Ground (K); 1927, Herb. Expt. Ground (K); 1933, Herb Garden (K); 1936, Herb Garden (K); 1936, Herbaceous Ground (K). In cult.

1110. Scutellaria galericulata L.
skullcap
FIRST RECORD: 1873/4 [1768].
NICHOLSON: **1873/4 survey** (JB: 71): **Strip.** Plentiful by moat near Isleworth Gate. • **1906 revision** (AS: 86): **Strip.** Plentiful by ha-ha near Isleworth Gate.

OTHER REFERENCES: **Kew Gardens.** 1999, *Stones* (EAK: 2): By Lake.
EXSICC.: **Kew Environs.** Thames bank, Kew; 12 vii 1888; *Gamble* 20204 (K). • Thames bank, Kew; on old wooden wall, between crevices; 2 viii 1928; *Summerhayes & Turrill* s.n. (K). • Towpath, Kew; 5 viii 1944; *Souster* 140 (K).
CULT.: 1768 (JH: 242); 1789, Nat. of Britain (HK1, 2: 324); 1811, Nat. of Britain (HK2, 3: 428); 1814, Nat. of Britain (HKE: 184); 1934, Arboretum (K); 1937, Herb Garden (K); 1990, Jodrell Glass (K). In cult. under glass.
CURRENT STATUS: No recent sightings.

1111. Scutellaria minor Huds.
lesser skullcap
CULT.: 1768 (JH: 242); 1789, Nat. of Britain (HK1, 2: 324); 1811, Nat. of Britain (HK2, 3: 428); 1814, Nat. of Britain (HKE: 184); 1949, Herbarium Ground (K); 1988, Jodrell Glass (K). In cult. under glass.

1112. Teucrium scorodonia L.
wood sage
FIRST RECORD: 1873/4 [1768].
NICHOLSON: **1873/4 survey** (JB: 71): **P.** About "Old Ruined Arch," and "Merlin's Cave." • **1906 revision** (AS: 86): **A.** About "ruined arch." Here and there in wood.
CULT.: 1768 (JH: 253); 1789, Nat. of Britain (HK1, 2: 278); 1811, Nat. of Britain (HK2, 3: 369); 1814, Nat. of Britain (HKE: 176); 1933, unspec. (K); 1936, Herbaceous Ground (K); 1971, Alpine & Herb. Dept. (K). In cult.
CURRENT STATUS: No recent sightings.

[**1113. Teucrium chamaedrys** L.
wall germander
CULT.: 1789, Nat. of England (HK1, 2: 279); 1811, Nat. of England (HK2, 3: 370); 1814, Nat. of England (HKE: 176).
Note. Listed in error as native.]

1114. Teucrium scordium L.
water germander
CULT.: 1768 (JH: 253); 1789, Nat. of England (HK1, 2: 279); 1811, Nat. of England (HK2, 3: 369); 1814, Nat. of England (HKE: 176); 1904, unspec. (K); 1922, unspec. (K). Not currently in cult.

1115. Ajuga reptans L.
bugle
FIRST RECORD: 1873/4 [1768].
NICHOLSON: **1873/4 survey** (JB: 71): Here and there on most of the lawns in **B.** Much less frequent elsewhere. • **1906 revision** (AS: 86): Here and there on most of the lawns.
CULT.: 1768 (JH: 254); 1789, Nat. of Britain (HK1, 2: 275); 1811, Nat. of Britain (HK2, 3: 364); 1814, Nat. of Britain (HKE: 176); 1940, Herb Garden (K); 1992, unspec. (K). In cult.
CURRENT STATUS: No recent sightings. It has probably been driven out by intensive close mowing of the lawns.

1116. Ajuga pyramidalis L.
pyramidal bugle
CULT.: 1768 (JH: 254); 1789, Nat. of Britain (HK1, 2: 274); c. 1779–1805, unspec. (K); 1811, Nat. of Britain (HK2, 3: 363); 1814, Nat. of Britain (HKE: 176); 1969, unspec. (K). Not currently in cult.

1117. Ajuga chamaepitys (L.) Schreb.
Teucrium chamaepitys L.
ground-pine
FIRST RECORD: 1998 [1768].
REFERENCES: **Kew Gardens.** 1998, *Cope* (S33: 748): North Arb.: flower bed at the back of the Banks Building (107).
CULT.: 1768 (JH: 252); 1789, Nat. of England (HK1, 2: 275); 1811, Nat. of England (HK2, 3: 364); 1814, Nat. of England (HKE: 176); 1887, unspec. (K); 1978, unspec. (K). In cult.
CURRENT STATUS: Extremely rare; probably now extinct since it was growing in an area under renovation but it is just possible that seed may persist and the plant reappear.

[**1118. Ajuga genevensis** L.
CULT.: 1789, Alien (HK1, 2: 274); 1811, Nat. of England (HK2, 3: 363); 1814, Nat. of England (HKE: 176).
Note. Listed in error as native.]

Ground-pine (*Ajuga chamaepitys*). Possibly extinct; its only appearance was in an area under renovation. but it may have left viable seed behind.

1119. Nepeta cataria L.
cat-mint
FIRST RECORD: 1873/4 [1768].
NICHOLSON: **1873/4 survey** (JB: 71 and in SFS: 523): A few plants near Winter Garden. • **1906 revision** (AS: 86): **A.** A few plants near temperate house.
OTHER REFERENCES: **Kew Environs.** 1943–44, *Welch* LN33: S224): Riverside waste, Kew.
CULT.: 1768 (JH: 256); 1789, Nat. of Britain (HK1, 2: 284); 1811, Nat. of Britain (HK2, 3: 377); 1814, Nat. of Britain (HKE: 178); 1936, Herbaceous Ground (K); 1983, unspec. (K); 1985, unspec. (K); 1999, Jodrell Glass (K). In cult.
CURRENT STATUS: No recent sightings.

1120. *Nepeta grandiflora M.Bieb.
FIRST RECORD: 2007 [1923].
CULT.: 1923, Herbaceous Ground (K); 1934, unspec. (K); 1962, Decorative Dept. (K); 1986, plot 19309 (K). Not currently in cult.
CURRENT STATUS: One or two plants grew from seed in topsoil stored in the Paddock.

1121. Glechoma hederacea L.
Nepeta glechoma Benth.
Glechoma hederacea var. *micrantha* Moric.
ground-ivy
FIRST RECORD: 1873/4 [1768].
NICHOLSON: **1873/4 survey** (JB: 71): Common under trees, and in shrubberies. • **1906 revision** (AS: 86): Common under trees and in shrubberies.
OTHER REFERENCES: **Kew Gardens.** 1999, *Stones* (EAK: 2): By Lake.
EXSICC.: **Kew Gardens.** Behind the Tennis Courts; growing on slack from furnaces; 9 v 1919, *Turrill* s.n. (K). • Near the Herbarium; growing in the shade of young pines; 20 v 1919; *Turrill* s.n. (K). • Meadow near Herbarium; in open grass; 22 v 1919; *Turrill* s.n. (K). • Meadow near Herbarium (near iron-fence at end of West Wing); 22 v 1919; *Turrill* s.n. (K). • Field near the Herbarium; 20 v 1920; *Turrill* s.n. (K). • Field near the Herbarium; growing amongst grass in the open; 10 v 1921; *Turrill* s.n. (K). • Near Rhododendron Dell; 29 iv 1927; *Turrill* s.n. (K). • Near tennis courts; rubbish heap; i 1928; *Ballard* s.n. (K). • Near tennis courts; on bank of rubbish heap; 5 iii 1928; *Summerhayes & Turrill* s.n. (K). • Near tennis courts; rubbish heap; iv 1928; *Ballard* s.n. (K). • In copse alongside experimental ground, near Works Dept.; 27 vi 1928; *Ballard* s.n. (K). • Near tennis courts; rubbish heap; vi 1928; *Ballard* s.n. (K). **Kew Environs.** Base of wall facing north nr. river, Kew; 22 v 1919, *Turrill* s.n. (K).
CULT.: 1768 (JH: 250); 1789, Nat. of Britain (HK1, 2: 296); 1811, Nat. of Britain (HK2, 3: 392); 1814, Nat. of Britain (HKE: 180); 1928, Herb. Expt. Ground (K). Not currently in cult.
CURRENT STATUS: Common in grass under shrubs and trees throughout the Gardens.

1122. Prunella vulgaris L.
selfheal
FIRST RECORD: 1873/4 [1768].
NICHOLSON: **1873/4 survey** (JB: 71): **P.** Two or three plants by side of Irish Yew Avenue leading from Pagoda to Arbour. • **1906 revision** (AS: 86): Here and there in turf in all the divisions.
OTHER REFERENCES: **Kew Gardens.** 1999, *Stones* (EAK: 2): By Lake.
CULT.: 1768 (JH: 244); 1789, Nat. of Britain (HK1, 2: 325); 1811, Nat. of Britain (HK2, 3: 429); 1814, Nat. of Britain (HKE: 185); 1934, Arboretum (K); 1936, Herbaceous Ground (K); 1952, Herb. Exp. Ground (K). In cult.
CURRENT STATUS: A common constituent of lawns, particularly under trees, throughout the Gardens

1123. *Prunella grandiflora (L.) Scholler
large selfheal
FIRST RECORD: 2001 [1789].
REFERENCES: **Kew Gardens.** 2001, *Cope* (S35: 648): Incorporated in the Robinsonian Meadow sown in front of the temporary Cycad House (224), and now naturalised.
CULT.: 1789, Alien (HK1, 2: 325); 1811, Alien (HK2, 3: 429); 1814, Alien (HKE: 185); 1898, unspec. (K); 1904, unspec. (K); 1974, Alpine, plot 154-04 (K). Not currently in cult.
CURRENT STATUS: Incorporated in the Robinsonian Meadow in front of 'Climbers & Creepers' but rapidly declining.

1124. *Melissa officinalis L.
balm
FIRST RECORD: 2001 [1768].
REFERENCES: **Kew Gardens.** 2001, *Cope* (S35: 648): Edge of the lawn by the Banks Building pond. May have escaped from the Bee Garden behind Kew Palace.
CULT.: 1768 (JH: 246); 1789, Alien (HK1, 2: 315); 1811, Alien (HK2, 3: 417); 1814, Alien (HKE: 183); 1933, Arboretum (K); 1936, Herbaceous Ground (K). In cult.
CURRENT STATUS: In a flower-bed beside the lawn between Kew Palace and the Banks Building and likely to have escaped from the bee garden behind the Palace.

1125. Clinopodium menthifolium (Host) Stace
Calamintha sylvatica Bromf.
wood calamint
CULT.: 1923, unspec. (K); 2001, Alpine & Herb. (K). In cult.

1126. Clinopodium ascendens (Jord.) Samp.
Calamintha ascendens Jord.
Melissa calamintha L.
common calamint
CULT.: 1768 (JH: 247); 1789, Nat. of England (HK1, 2: 316); 1811, Nat. of England (HK2, 3:

417); 1814, Nat. of England (HKE: 183); 1927, Herb. Expt. Ground (K); 1934, Herbarium Ground (K); 1937, Herbarium Ground (K); 1951, Herbarium Ground (K). Not currently in cult.

1127. Clinopodium calamintha (L.) Stace
Melissa nepeta L.
Calamintha nepeta (L.) Savi
lesser calamint
CULT.: 1768 (JH: 247); 1789, Nat. of England (HK1, 2: 316); 1811, Nat. of England (HK2, 3: 417); 1814, Nat. of England (HKE: 183); 1903, unspec. (K); 1982, unspec. (K). Not currently in cult.

1128. Clinopodium vulgare L.
Wild Basil
CULT.: 1768 (JH: 246); 1789, Nat. of Britain (HK1, 2: 310); 1811, Nat. of Britain (HK2, 3: 411); 1814, Nat. of Britain (HKE: 182); 1979, unspec. (K). Not currently in cult.

1129. Clinopodium acinos (L.) Kuntze
Thymus acinos L.
Acinos arvensis (Lam.) Dandy
basil thyme
CULT.: 1768 (JH: 243); 1789, Nat. of Britain (HK1, 2: 314); 1811, Nat. of Britain (HK2, 3: 414); 1814, Nat. of Britain (HKE: 183); 1930, unspec. (K); 1987, Order Beds (K); 1990, Order Beds, plot 157-15 (K). Not currently in cult.

1130. Origanum vulgare L.
wild marjoram
CULT.: 1769 (JH: 245); 1789, Nat. of Britain (HK1, 2: 312); 1811, Nat. of Britain (HK2, 3: 413); 1814, Nat. of Britain (HKE: 182); 1905, unspec. (K); 1923, unspec. (K); 1931, unspec. (K); 1933, Herbarium Ground (K); 1934, Arboretum (K); 1936, Herbaceous Ground (K); 1945, Herbarium Garden (K); 1945, Allotment Plot, behind Herbarium (K); 1985, Arboretum (K); 1996, Jodrell Glass (K). In cult.

1131. Thymus pulegioides L.
T. chamaedrys Fr.
T. glaber Mill.
T. ovatus Mill.
large thyme
FIRST RECORD: 1873/4 [1925].
NICHOLSON: **1873/4 survey** (JB: 71): Common in every dry piece of turf within the limits of our Flora. Frequent on wall facing river. • **1906 revision** (AS: 85): Common in every dry piece of turf within the limits of our flora. **Strip.** Frequent on wall facing river.
OTHER REFERENCES: **Kew Gardens.** 1902, *Beeby* (SFS: 518): Walls of Kew Moat. • 1931, *Baker* (SFS: 517): Kew Gardens.
EXSICC.: **Kew Gardens.** Pleasure Grounds; *anon.* s.n. (K). **Kew Environs.** Kew; 20 vii 1856; *E.G. Western* s.n. (K).
CULT.: 1925, unspec. (K); 1932, unspec. (K); 1933, unspec. (K); 1934, Herbarium Ground (K); 1962, unspec. (K); 1974, Temperate (K); 1976, unspec. (K); 1985, unspec. (K). In cult.
CURRENT STATUS: No recent sightings. A once-common species that we seem to have lost.

1132. *Thymus × oblongifolius Opiz (pulegioides × serpyllum)
FIRST RECORD: 1923.
REFERENCES: **Kew Environs.** 1923, *Druce* (BSEC7: 238 & 8: 509; S33: 748): Kew.
CURRENT STATUS: No recent sightings.

1133. Thymus polytrichus subsp. **britannicus** (Ronniger) Kerguélen
T. praecox auct. non Opiz
T. serpyllum auct. non L.
wild thyme
CULT.: 1789, Nat. of Britain (HK1, 2: 313); 1811, Nat. of Britain (HK2, 3: 414); 1814, Nat. of Britain (HKE: 183); 1940, Herbarium Ground (K); 1971, Alpine & Herb. Dept. (K); 1974, unspec. (K); 1977, unspec. (K); 1978, plot 154-04 (K); 1978, plot 154-06 (K); 1981, unspec. (K). In cult.

1134. Thymus serpyllum L.
Breckland thyme
CULT.: 1768 (JH: 243); 1925, unspec. (K); 1933, Herb Garden (K); 1934, Arboretum (K); 1936, Herb Garden (K); 1936, Herbaceous Ground (K); 1971, unspec. (K); 1999, Alpine & Herb. (K). In cult.

1135. Lycopus europaeus L.
gypsywort
FIRST RECORD: 1873/4 [1768].
NICHOLSON: **1873/4 survey** (JB: 71): **B.** Common about pond. **P.** Abundant all round lake. • **1906 revision** (AS: 85): Abundant about pond, lake, and along **Strip.**
OTHER REFERENCES: **Kew Gardens.** 1999, *Stones* (EAK: 1 & 5): By Lake and Palm House Pond.
EXSICC.: **Kew Environs.** Thames bank, Kew; 1883; *Fraser* s.n. (K). • Thames bank, Kew; growing in crevices near top of wall; 30 vii 1928; *Summerhayes & Turrill* 3336 (K). • Thames bank, Kew; growing under willow bushes; 30 vii 1928; *Summerhayes & Turrill* 3337 (K).
CULT.: 1768 (JH: 248); 1789, Nat. of Britain (HK1, 1: 34); 1810, Nat. of Britain (HK2, 1: 47); 1814, Nat. of Britain (HKE: 8); 1934, unspec. (K); 1936, Herbaceous Ground (K). In cult.
CURRENT STATUS: In several areas but mostly around the Lake, in the Conservation Area and on the towpath.

1136. Mentha arvensis L.
Mentha agrestis Sole
corn mint

FIRST RECORD: pre-1922 [1768].
EXSICC.: **Kew Environs.** Riverside, Kew; pre-1922; *anon.* s.n. (K).
CULT.: 1768 (JH: 251); 1789, Nat. of Britain (HK1, 2: 295); 1811, Nat. of Britain (HK2, 3: 389); 1814, Nat. of Britain (HKE: 179), Nat. of England (ibid.: Add. as *agrestis*); 1918, Kew Green (K); 1919, unspec. (K); 1920, Kew Green (K); 1921, unspec. (K); 1922, unspec. (K); 1923, unspec. (K); 1930, unspec. (K); 1931, unspec. (K); 1934, unspec. (K). Not currently in cult.
CURRENT STATUS: No recent sightings. One of the specimens cultivated in 1922 originated from the riverside at Kew (see Exsicc.).

1137. Mentha × gracilis Sole (arvensis × spicata)
M. gentilis auct. non L.
CULT.: 1768 (JH: 251); 1789, Nat. of England (HK1, 2: 295); 1811, Nat. of Britain (HK2, 3: 389), Nat. of England (ibid.: as *gentilis*); 1814, Nat. of Britain (HKE: 179), Nat. of England (ibid.: as *gentilis*); 1904, unspec. (K); 1926, unspec. (K); 1927, unspec. (K). Not currently in cult.

1138. Mentha aquatica L.
M. hirsuta Huds.
water mint

FIRST RECORD: 1873/4 [1768].
NICHOLSON: **1873/4 survey** (JB: 71): **P.** Several patches near edge of lake. • **1906 revision** (AS: 85): Abundant round lake and along **Strip.**
OTHER REFERENCES: **Kew Gardens.** 1999, *Stones* (EAK: 1): By Lake.
EXSICC.: **Kew Gardens.** Thames bank Kew – Richmond, Richmond side of Isleworth Gate at upper edge of *Phalaris* zone; 16 viii 1928; *Summerhayes & Turrill* s.n. (K).
CULT.: 1768 (JH: 251); 1789, Nat. of Britain (HK1, 2: 294), Nat. of England (ibid.: as *hirsuta*); 1811, Nat. of Britain (HK2, 3: 388); 1814, Nat. of Britain (HKE: 179); 1897, unspec. (K); 1923, unspec. (K); 1931, unspec. (K); 1993, Jodrell Glass (K); 1998, Jodrell Glass (K). In cult.
CURRENT STATUS: Plentiful around the margins of the Lake and by the Banks Building pond.

1139. Mentha × verticillata L. (aquatica × arvensis)
whorled mint

M. sativa L.
M. acutifolia Sm.
CULT.: 1768 (JH: 251); 1789, Nat. of Britain (HK1, 2: 294); 1811, Nat. of England (HK2, 3: 338); 1814, Nat. of England (HKE: 179 as *acutifolia*: Add. as *sativa*); 1897, unspec. (K); 1922, unspec. (K); 1924, unspec. (K); 1926, unspec. (K); 1929, unspec. (K). Not currently in cult.

1140. Mentha × piperita L. (aquatica × spicata)
M. odorata Sole
peppermint

CULT.: 1768 (JH: 251); 1789, Nat. of England (HK1, 2: 294); 1811, Native of England (HK2, 3: 388); 1814, Nat. of England (HKE: 179). In cult.

1141. Mentha spicata L.
M. viridis L.
M. crispa L.
spear mint

CULT.: 1768 (JH: 251); 1789, Nat. of England (HK1, 2: 293); 1811, Nat. of England (HK2, 3: 388); 1814, Nat. of England (HKE: 179); 1921, unspec. (K); 1927, unspec. (K); 1929, unspec. (K); 1930, unspec. (K); 1931, unspec. (K); 1935, unspec. (K); 1936, Herbaceous Ground (K); 1993, Jodrell Glass (K). In cult.

[1142. Mentha × villosa Huds. (spicata × suaveolens)
M. rubra Mill. pro parte
apple-mint

CULT.: 1811, Nat. of Britain (HK2, 3: 389); 1814, Nat. of Britain (HKE: 179).
Note. Originally grown as a native, but now considered a neophyte.**]**

1143. Mentha suaveolens Ehrh.
M. rotundifolia (L.) Huds.
round-leaved mint

CULT.: 1768 (JH: 251); 1789, Nat. of Britain (HK1, 2: 293); 1811, Nat. of England (HK2, 3: 387); 1814, Nat. of England (HKE: 179); 1975, Lower Nursery (K); 1990, plot 157-22 (K); 1993, Jodrell Glass (K). In cult.

1144. Mentha pulegium L.
M. exigua L.
pennyroyal

CULT.: 1768 (JH: 251); 1789, Nat. of Britain (HK1, 2: 295); 1811, Nat. of Britain (HK2, 3: 390); 1814, Nat. of Britain (HKE: 179); 1906, unspec. (K); 1920, unspec. (K); 1921, unspec. (K); 1923, unspec. (K); 1933, unspec. (K); 1965, Water Garden (K); 1975, Alpine & Herbaceous (K). Not currently in cult.

1145. *Mentha longifolia (L.) Huds.
M. sylvestris L.
horse mint

FIRST RECORD: 1906 [1768].

NICHOLSON: **1906 survey** (AS: 85 and in SFS: 507): **Strip.** A large patch or two near Brentford Ferry Gate.
CULT.: 1768 (JH: 251); 1789, Nat. of Britain (HK1, 2: 293); 1811, Nat. of England (HK2, 3: 387); 1814, Nat. of England (HKE: 179); 1921, unspec. (K); 1979, unspec. (K); 1990, plot 157-24 (K); 1993, Jodrell Glass (K). In cult.
CURRENT STATUS: No recent sightings; almost certainly a misidentification of *M.* × *villosonervata* (q.v.) but difficult to prove without reference to vouchers.

1146. *Mentha × villosonervata Opiz (longifolia × spicata)
sharp-toothed mint
FIRST RECORD: 1998 [1923].
REFERENCES: **Kew Gardens.** 1998, *Cope* (S33: 748): West Arb.: Riverside Walk (215), needs checking; a 1906 record from near Brentford Ferry Gate is *M. longifolia*.
CULT.: 1923, unspec. (K). Not currently in cult.
CURRENT STATUS: Occasional, but sometimes forming large patches. This plant may account for what could be erroneous records of Horse Mint; the latter has never been refound but the hybrid occurs in the same locality.

1147. Mentha × rotundifolia var. **webberi** (J.Fraser) Harley (longifolia × suaveolens)
false apple-mint
CULT.: 1934, unspec. (K). Not currently in cult.

1148. Salvia pratensis L.
meadow-clary
CULT.: 1768 (JH: 241); 1789, Nat. of England (HK1, 1: 40); 1810, Nat. of England (HK2, 1: 57); 1814, Nat. of England (HKE: 9); 1862, unspec. (K); 1926, 1927, 1928, 1929, 1930, Herb. Expt. Ground (K); 1936, Herbaceous Ground (K); 1939, Herbarium Ground (K); 1968, Palace Garden (K). In cult.

1149. Salvia verbenaca L.
S. horminoides Pourr.
wild clary
FIRST RECORD: 1920 [1768].
REFERENCES: **Kew Gardens.** 1966, *Verdcourt* (S33: 748): Fresh mound of earth outside Herbarium, Wing A. • 1983, *Burton* (FLA: 133; S33: 748): At Kew [it] has perhaps been deliberately introduced, although it was on the riverside nearby 50 years ago. There is no chance of the species increasing naturally in the area any more. **Kew Environs.** 1920, *Tremayne* (LN33: S223; S33: 748): By river Thames between Kew and Richmond. • 1934, *anon.* (LN13: S79; S33: 748): Between Kew and Richmond.
EXSICC.: **Kew Gardens.** Fresh mound of earth outside Wing A; 13 v 1966; *Verdcourt* 4250 (K).

Wild clary (*Salvia verbenaca*). The only wild population grew outside the Mycology Building and was lost during construction of a new wing to the Herbarium. Seed was saved and progeny have been planted elsewhere in the Gardens.

CULT.: 1768 (JH: 241); 1789, Nat. of Britain (HK1, 1: 41); 1810, Nat. of Britain (HK2, 1: 58); 1814, Nat. of Britain (HKE: 9); 1887, unspec. (K); 1894, unspec. (K); 1930, 1931, 1932, Herb. Expt. Ground (K); 1933, 1936, Herbarium Ground (K); 1936, Herbaceous Ground (K); 1938, 1940, Herbarium Ground (K); 1947, Herbarium Experimental Ground (K); 1949, Herbarium Ground (K); 1976, unspec. (K). Not currently in cult.
CURRENT STATUS: Known for a long time from the grass bordering the old students' vegetable plots between the Herbarium and Hanover House. Several years ago this grass was removed, the ground repaired and new grass sown. The Wild Clary plants that had been growing there were dug up and planted in the grass behind the Mycology Building where they failed to survive. The plant meantime returned to the grass around the vegetable plots, either from roots left behind or from seed, and the population eventually numbered about twenty plants. These were regularly mown, often before setting seed, and the population was considered to be under serious threat. Recently, the plants were temporarily protected from mowing and seed was taken in 2002 for the MSB. New plans calling for an extension to

the Herbarium to be built on the land occupied by the species have now been put into action and the site has effectively been destroyed, but not before the plants themselves were removed for safe-keeping. However, in 2008, a plant appeared spontaneously at the west end of the Herbarium lawn in soil heavily disturbed by building works, and this is currently the only known wild plant of this species in the Gardens. Plants derived from the original population have now been planted in the meadow near the Main Gate, and other populations will eventually be established.

1150. *Salvia viridis L.
S. horminum L.
annual clary
FIRST RECORD: 1998 [1768].
REFERENCES: **Kew Gardens.** 1998, *Cope* (S33: 748): North Arb.: Student vegetable plots beside Herbarium (113) (plant destroyed when rubbish skips replaced).
CULT.: 1768 (JH: 241); 1789, alien (HK1, 1: 39); 1810, Alien (HK2, 1: 55); 1814, Alien (HKE: 8); 1923, unspec. (K); 1924, Herb. Expt. Ground (K). In cult.
CURRENT STATUS: Has not reappeared in the area since 1998. The plants grew from seed that had presumably been brought in with compost but a rubbish skip for vegetable waste was placed right on top of them and they subsequently died.

1151. *Salvia × superba Stapf (× sylvestris [pratensis × nemorosa] × villicaulis)
FIRST RECORD: 2001.
REFERENCES: **Kew Gardens.** 2001, *Cope* (S35: 648): Incorporated in the Robinsonian Meadow sown in front of the temporary Cycad House (224), and now naturalised.
CULT.: In cult.
CURRENT STATUS: Incorporated in the Robinsonian Meadow in front of 'Climbers & Creepers' but declining.

HIPPURIDACEAE

1152. Hippuris vulgaris L.
mare's-tail
CULT.: 1768 (JH: 411); 1789, Nat. of Britain (HK1, 1: 5); 1810, Nat. of Britain (HK2, 1: 13); 1814, Nat. of Britain (HKE: 2). Not currently in cult.

CALLITRICHACEAE

1153. Callitriche hermaphroditica L.
C. autumnalis L.
autumnal water-starwort
CULT.: 1768 (JH: 157); 1789, Nat. of Britain (HK1, 1: 6). Not currently in cult.

1154. Callitriche stagnalis Scop.
common water-starwort
FIRST RECORD: 1894.
REFERENCES: **Kew Environs.** 1932, *anon.* (LN11: S47; S33: 748): By the R. Thames, near Kew.
EXSICC.: **Kew Environs.** Muddy shore of River Thames, Kew; name given me by Mr J.G.Baker on the spot; 17 v 1894; *Hosking* s.n. (K). • Ha-ha, Richmond; vi 1927; *Findlay* s.n. (K). • Riverbank, Kew; on new stone wall, in crevices, submerged at high tides; 14 vi 1928; *Summerhayes & Turrill* s.n. (K). • Riverside, Kew; vi 1929; *Pearce* s.n. (K).
CURRENT STATUS: Still occurs in the river near Kew Bridge. It has never been found within the Gardens.

1155. Callitriche platycarpa Kütz.
C. aquatica Huds.
C. verna auct. non L.
various-leaved water-starwort
FIRST RECORD: 1873/4 [1768].
NICHOLSON: **1873/4 survey** (JB: 44): Frequent in the moat, but occurs nowhere else. • **1906 revision** (AS: 80): Frequent in the ha-ha.
CULT.: 1768 (JH: 157); 1789, Nat. of Britain (HK1, 1: 6); 1810, Nat. of Britain (HK2, 1: 13); 1814, Nat. of Britain (HKE: 3). Not currently in cult.
CURRENT STATUS: No recent sightings.

1156. Callitriche hamulata Kütz. ex W.D.J.Koch
intermediate water-starwort
FIRST RECORD: 1902.
NICHOLSON: **1906 revision:** Not recorded.
OTHER REFERENCES: **Kew Gardens.** 1902, *Beeby* (SFS: 322; S33: 748): Kew Gardens Moat.
CURRENT STATUS: No recent sightings.

PLANTAGINACEAE

1157. Plantago coronopus L.
buck's-horn plantain
FIRST RECORD: 1873/4 [1768].
NICHOLSON: **1873/4 survey** (JB: 72 and in SFS: 539): **B.** Common in every dry place. Mounds near Museums 1 and 3. Slopes about Palm House. **Strip.** Very abundant on the top of ridge by side of towing-path. • **1906 revision** (AS: 85): Common in every dry place. **Strip.** Very abundant on the top of ridge by the side of towing-path.
EXSICC.: **Kew Environs.** Kew Green, abundant in places particularly near 'dog walk'; 10 vii 2001; *Verdcourt* 5526 (K).
CULT.: 1768 (JH: 121); 1789, Nat. of Britain (HK1, 1: 153); 1810, Nat. of Britain (HK2, 1: 255); 1814, Nat. of Britain (HKE: 34); c. 1887, unspec. (K); 1947, Herb. Expt. Ground (K); 1948, Herbarium Ground (K). In cult.
CURRENT STATUS: Uncommon in the Gardens but is abundant on Kew Green between Kew Road and the footpath from Ferry Lane towards the church.

1158. Plantago maritima L.
sea plantain
Cult.: 1768 (JH: 121); 1789, Nat. of Britain (HK1, 1: 153); 1810, Nat. of Britain (HK2, 1: 254); 1814, Nat. of Britain (HKE: 35). In cult.

1159. Plantago major L.
greater plantain
First record: 1873/4 [1768].
Nicholson: **1873/4 survey** (JB: 72): Common in all the divisions, both in turf and elsewhere. • **1906 revision** (AS: 84): Common in all the divisions.
Other references: **Kew Gardens.** 1999, *Stones* (EAK: 2 & 5): By Lake and Palm House Pond.
Exsicc.: **Kew Gardens.** Herbarium Experimental Ground; on paths among grass; 9 ii 1928; *Summerhayes & Turrill* s.n. (K). • [Unlocalised]; sandy soil; 20 vi 1928; *Hubbard* s.n. (K). • Herbarium Exper. Ground; growing on path between beds; 4 vii 1928; *Summerhayes* s.n. (K). • Near Tennis Courts; on loose soil in shade on rubbish heap; 2 viii 1928; *Summerhayes & Turrill* s.n. (K). • Tennis Court rubbish heap; 14 vii 1931; *Bullock* s.n. (K). • Arboretum; 5 vii 1933; *Dallimore* s.n. (K). • Herbarium Ground; weed; 13 vii 1935; *anon.* s.n. (K). • [Unlocalised]; waste ground; vii 1940; *Ridley* s.n. (K). • The Director's Garden; 26 vi 1947; *Salisbury* H918/47 (K). • Temperate House; disturbed ground around the house on light soil; ix 1979, *McNamara* 3 (K). • Student vegetable plot; along the path, between the plots; 10 ix 1980; *Iida* 5 (K). **Kew Environs.** Thames bank, Kew; from crevices of new stone wall, half-way down; 2 viii 1928; *Turrill* s.n. (K). • Kew; 10 vi 1931; *Fraser* s.n. (K).
Cult.: 1768 (JH: 121); 1789, Nat. of Britain (HK1, 1: 150); 1810, Nat. of Britain (HK2, 1: 250); 1814, Nat. of Britain (HKE: 34); 1927, unspec. (K); 1928, Herb. Expt. Ground (K); 1931, Herb. Exp. Ground (K); 1934, Herb Garden (K); 1949, Herbarium Ground [monstrous plants] (K). In cult.
Current status: Widespread and common in grassland throughout the Gardens.

1160. Plantago media L.
hoary plantain
First record: 1873/4 [1768].
Nicholson: **1873/4 survey** (JB: 72 and in SFS: 541): Almost as common as [*Plantago major*] in many places. • **1906 revision** (AS: 84): Almost as common as [*Plantago major*] in many places.
Cult.: 1768 (JH: 121); 1789, Nat. of Britain (HK1, 1: 151); 1810, Nat. of Britain (HK2, 1: 251); 1814, Nat. of Britain (HKE: 34); c. 1887, unspec. (K). In cult.
Current status: No recent sightings. It is difficult to account for its apparent catastrophic decline since 1906 unless the original determination was an error, but this seems very unlikely.

1161. Plantago lanceolata L.
ribwort plantain
First record: 1873/4 [1768].
Nicholson: **1873/4 survey** (JB: 72): Everywhere. • **1906 revision** (AS: 84): Everywhere.
Exsicc.: **Kew Gardens.** Near Herbarium; grass; 28 v 1919; *Turrill* s.n. (K). • [Unlocalised]; sandy soil; 20 vi 1928; *Hubbard* s.n. (K). • *Carpinus* collection; in unkempt grassland; 8 vii 1980; *Wallace* 5 (K). **Kew Environs.** Tow path, Kew – Richmond; on open grassland on rather poor gravel soil, associated with *Convolvulus arvensis* & *Trifolium repens*; 13 vi 1948; *Parker* s.n. (K).
Cult.: 1768 (JH: 121); 1789, Nat. of Britain (HK1, 1: 152); 1810, Nat. of Britain (HK2, 1: 252); 1814, Nat. of Britain (HKE: 34). In cult.
Current status: Widespread and abundant in grassland throughout the Gardens.

1162. *Plantago arenaria Waldst. & Kit.
P. psyllium L.
branched plantain
First record: Pre-1875 [1768].
Exsicc.: **Kew Environs.** Kew Bridge; [pre-1875]; *Mill* s.n. (K).
Cult.: 1768 (JH: 121); 1942, unspec. (K). In cult.
Current status: No recent sightings.

Hoary plantain (*Plantago media*). Extinct. Last recorded a hundred years ago when it was described as 'almost as common as *Plantago major* in many places'.

Butterfly-bush (*Buddleja davidii*). A cultivated shrub that is likely to appear from seed in any corner of the Gardens.

1163. *Plantago afra L.
P. indica auct. non L.
glandular plantain
FIRST RECORD: 1941 [1768].
REFERENCES: **Kew Environs.** 1941, *Cooke* (LN34: S229; S33: 748): River bank, Kew.
CULT.: 1768 (JH: 121); 1789, Alien (HK1, 1: 154); 1810, Alien (HK2, 1: 256); 1814, Alien (HKE: 34); 1934, Arboretum (K); 1936, Herbaceous Ground (K). In cult.
CURRENT STATUS: No recent sightings.

[**1164. Plantago loeflingii** L.
CULT.: 1789, Nat. of England (HK1, 1: 154); 1810, Nat. of England (HK2, 1: 255); 1814, Nat. of England (HKE: 34).
Note. Listed in error as native.]

1165. Littorella uniflora (L.) Asch.
Plantago uniflora L.
L. lacustris L.
shoreweed
CULT.: 1768 (JH: 121); 1789, Nat. of Britain (HK1, 3: 335); 1813, Nat. of Britain (HK2, 5: 257); 1814, Nat. of Britain (HKE: 290). Not currently in cult.

BUDDLEJACEAE

1166. *Buddleja davidii Franch.
butterfly-bush
FIRST RECORD: 1983/4 [1897].
REFERENCES: **Kew Gardens.** 1983/4, *Cope* (S31: 181): New student vegetable plots behind Hanover House. • 1999, *Stones* (EAK: 5): By Palm House Pond.
CULT.: 1897, unspec. (K); 1905, Arboretum (K); 1906, Arboretum (K); 1908, unspec. (K); 1919, Nursery (K); 1924, Arboretum (K); 1925, unspec. (K); 1926, unspec. (K); 1931, unspec. (K); 1936, Arboretum (K); 1944, unspec. (K); 1951, *Buddleja* Collection (K); 1962, Arboretum (K); 1965, Temperate House (K); 1969, Arboretum (K); 1971, Arboretum (K); 1994, unspec. (K). In cult.
CURRENT STATUS: Seedlings are often found in rough and neglected ground but mature plants, other than cultivars, are scarce since seedlings are seldom allowed to persist.

OLEACEAE

1167. Fraxinus excelsior L.
F. simplicifolia Willd.
ash
FIRST RECORD: 1910 [1768].
EXSICC.: **Kew Gardens.** Nr. Tennis Courts; 17 ix 1924; *Turrill* s.n. (K). **Kew Environs.** Towing path, Kew; self-sown; 9 ix 1910; *Fraser* s.n. (K).
CULT.: 1768 (JH: 441); 1789, Nat. of Britain (HK1, 3: 444); 1813, Nat. of Britain (HK2, 5: 475), Nat. of England (ibid.: as *simplicifolia*); 1814, Nat. of Britain (HKE: 322), Nat. of England (ibid.: 321 as *simplicifolia*); 1880, Arboretum (K); 1881, Arboretum (K); 1882, Arboretum (K); 1883, unspec. (K); 1885, nr. Railway Gate (K); 1887, Arboretum (K); 1890, unspec. (K); 1891, unspec. (K); 1893, Arboretum (K); 1907, Arboretum (K); 1908, Arboretum (K); 1911, Arboretum (K); 1930, Arboretum (K); 1965, Arboretum (K); 1969, Arboretum (K). In cult.
CURRENT STATUS: Occasional seedlings are found in the Gardens, but these are seldom allowed to reach any size before removal.

1168. *Fraxinus pallisiae Wilmott
FIRST RECORD: 1957.
REFERENCES: **Kew Gardens.** 1983/4, *Cope* (S31: 181): Seedling in new student vegetable plots behind Hanover House.
CULT.: In cult.
CURRENT STATUS: Has not persisted behind Hanover House. It was cultivated in the Herbarium experimental ground in 1957 and two fine specimen trees (both on the TROBI register) still stand between the Herbarium and the River; one of the latter was doubtless the parent of the seedling recorded above.

Ash (*Fraxinus excelsior*). Common outside the Gardens but seedlings inside are derived from our own cultivated trees. [Photo: D.Philcox]

1169. Ligustrum vulgare L.

privet

CULT.: 1768 (JH: 445); 1789, Nat. of Britain (HK1, 1: 10); 1810, Nat. of Britain (HK2, 1: 19); 1814, Nat. of Britain (HKE: 3); 1880, Arboretum (K); 1882, Arboretum (K); 1883, Arboretum (K); 1885, Arboretum (K); 1892, Arboretum (K); 1896, unspec. (K); 1969, Arboretum North (K); 1970, Arboretum North (K); 1984, Oleaceae Beds, 161-04 (K); 1989, plot 162-01 (K); 1991, plot 161.01 (K); plot 162 (K). In cult.

SCROPHULARIACEAE

1170. *Verbascum blattaria L.

moth mullein

FIRST RECORD: 1965 [1768].

REFERENCES: **Kew Gardens.** 1965, *Wurzell* (LFS: 267): Queen's Cottage Grounds. • 1986, *Hastings* (LN66: 186): In the yard by Cambridge Cottage, 1 plant. • 1990, *Cope* (S31: 182): Paddock, after restoration in 1990.

EXSICC.: **Kew Gardens.** Herbarium grounds, in front of New Wing (D block); 28 v 1968; *Townsend* s.n. (K).

CULT.: 1768 (JH: 133); 1789, Nat. of England (HK1, 1: 237); 1810, Nat. of England (HK2, 1: 385); 1814, Nat. of England (HKE: 53). In cult.

CURRENT STATUS: Sporadic in short grassland and often mown before it can set seed; it is persistent in the lawn behind the Herbarium and in the Paddock where its colour varies from whitish to pale lemon yellow. See note under *V. nigrum*.

[**1171. Verbascum virgatum** Stokes

twiggy mullein

CULT.: 1810, Nat. of England (HK2, 1: 385); 1814, Nat. of England (HKE: 53).

Note. Listed in error as native.]

1172. *Verbascum phlomoides L.

orange mullein

FIRST RECORD: 1947 [1789].

REFERENCES: **Kew Environs.** 1947, *Welch* (LN33: S201; S33: 748): Allotment, Kew Green.

EXSICC.: **Kew Gardens.** Jodrell; Summer 1961; *Jones* 61.592 (K).

CULT.: 1789, Alien (HK1, 1: 236); 1810, Alien (HK2, 1: 384); 1814, Alien (HKE: 53);1882, unspec. (K); 1894, unspec. (K); 1909, unspec. (K); 1910, unspec. (K). In cult.

CURRENT STATUS: A couple of plants came up in 2007 in topsoil stored in the Paddock. See note under *V. nigrum*.

Moth mullein (*Verbascum blattaria*). Occasional in various parts of the Gardens, but especially in the Paddock opposite the Herbarium. It varies in flower colour from bright yellow to almost white.

1173. *Verbascum densiflorum Bertol.

dense-flowered mullein

FIRST RECORD: 1998.

REFERENCES: **Kew Environs.** 1998, *Cope* (S33: 749): Ferry Lane, roadside verge opposite Herbarium.

CURRENT STATUS: Rare; it is still on the towpath and recently appeared against the boundary wall at the edge of the Hebarium car park and in topsoil being stored in the Paddock. See note under *V. nigrum*.

1174. Verbascum thapsus L.

great mullein

FIRST RECORD: 1873/4 [1768].

NICHOLSON: **1873/4 survey** (JB: 48): **Pal.** Two plants near Brentford Gate. • **1906 revision** (AS: 85 and in SFS: 477): **P.** Near Brentford Ferry Gate. **A.** Here and there on border of wood.

CULT.: 1768 (JH: 133); 1789, Nat. of Britain (HK1, 1: 235); 1810, Nat. of Britain (HK2, 1: 383); 1814, Nat. of Britain (HKE: 53). Not currently in cult.

CURRENT STATUS: Widespread but scarce. Possibly the only mullein native in the Gardens.

1175. Verbascum nigrum L.

dark mullein

FIRST RECORD: 1873/4 [1768].

NICHOLSON: **1873/4 survey** (JB: 48 and in SFS: 478): **Pal.** Abundant. **Q.** Only a few plants. • **1906 revision** (AS: 85): **P** and **Q.** Abundant. **A.** Along border of wood.

EXSICC.: **Kew Gardens.** Field near Herbarium; 7 x 1919; *Turrill* s.n. (K). • In Herbarium field; 22 viii 1933; *Hubbard* s.n. (K). • Herbarium field; in grass to north of the Herbarium; 5 vii 1949; *Turrill* s.n. (K).

CULT.: 1768 (JH: 133); 1789, Nat. of England (HK1, 1: 237); 1810, Nat. of England (HK2, 1: 385); 1814, Nat. of England (HKE: 53); 1944, Order Beds (K). In cult.

CURRENT STATUS: Rare; it is most often seen on the towpath and in front of the Herbarium, but also occurs in one or two spots elsewhere in the Gardens. It is likely that all the plants within the Gardens are derived from original cultivated material. Species of *Verbascum* were at one time cultivated in the Herbarium experimental plots and this may account for most records of the genus.

1176. Verbascum pulverulentum Vill.

hoary mullein

FIRST RECORD: 2001 [1810].

REFERENCES: **Kew Gardens.** 2001, *Cope* (S35: 648): In rough grass near the Stable Yard (342).

CULT.: 1810, Nat. of England (HK2, 1: 384); 1814, Nat. of England (HKE: 53). Not currently in cult.

CURRENT STATUS: In rough ground near the Stable Yard. See note under *V. nigrum*.

1177. Verbascum lychnitis L.

V. thapsoides Huds.

white mullein

CULT.: 1768 (JH: 133); 1789, Nat. of Britain (HK1, 1: 236), Nat. of England (ibid.: as *thapsoides*); 1810, Nat. of Britain (HK2, 1: 384), Nat. of England (ibid.: 383 as *thapsoides*); 1814, Nat. of Britain (HKE: 53), Nat. of England (ibid.: as *thapsoides*). In cult.

1178. Scrophularia nodosa L.

common figwort

FIRST RECORD: 1873/4 [1768].

NICHOLSON: **1873/4 survey** (JB: 48): **Strip.** With [*Scrophularia auriculata*]. • **1906 revision** (AS: 85): **Strip.** Common along ha-ha and by river.

CULT.: 1768 (JH: 268); 1789, Nat. of Britain (HK1, 2: 340); 1812, Nat. of Britain (HK2, 4: 22); 1814, Nat. of Britain (HKE: 189). In cult.

CURRENT STATUS: Rare; only one recent sighting at the south end of the Lake.

1179. Scrophularia auriculata L.

S. aquatica auct. non L.

S. aquatica var. *pubescens* Bréb.

S. balbisii Hornem.

water figwort

FIRST RECORD: 1873/4 [1768].

NICHOLSON: **1873/4 survey** (JB: 48): **Strip.** Here and there by river and moat. • **1906 revision** (AS: 85): **Strip.** Here and there by river and ha-ha.

OTHER REFERENCES: **Kew Gardens.** 1999, *Stones* (EAK: 2): By Lake. **Kew Environs.** 1926, *Drabble* (BSEC8: 126 & 9: 690): By the Thames at Kew.

EXSICC.: **Kew Environs.** By the Thames between Kew and Richmond; 2 viii 1927; *Sandwith* s.n. (K).

CULT.: 1768 (JH: 268); 1789, Nat. of Britain (HK1, 2: 341); 1812, Nat. of Britain (HK2, 4: 22); 1814, Nat. of Britain (HKE: 189); 1933, Arboretum (K). In cult.

CURRENT STATUS: Rare; there have been recent sightings by the Lake, in the Conservation Area and on the towpath. Records indicate that those within the Gardens have been planted.

1180. Scrophularia umbrosa Dumort.

green figwort

CULT.: 1987, plot 157-28 (K). Not currently in cult.

[**1181. Scrophularia scorodonia** L.

balm-leaved figwort

Cult.: 1789, Nat. of England (HK1, 2: 341); 1812, Nat. of England (HK2, 4: 23); 1814, Nat. of England (HKE: 189).

Note. Listed in error as native.]

1182. *Scrophularia vernalis L.

yellow figwort

FIRST RECORD: 1873/4 [1768].

Yellow figwort (*Scrophularia vernalis*). Scattered and declining in the Gardens, but still reasonably abundant in the Riverside Walk and north of the Rhododendron Dell.

NICHOLSON: **1873/4 survey** (JB: 48 and in SFS: 486): **B.** Frequent on the mound behind "Rockwork." **P.** Common among the stones at "Merlin's Cave." • **1906 revision** (AS: 85): **B.** Frequent on ice-house mound. Destroyed elsewhere owing to alterations.
OTHER REFERENCES: **Kew Gardens.** 1927, *London Natural History Society* (S32: 657, LN33: S205): Kew Gardens, weed. • 1983, *Burton* (FLA: 123): Well naturalised in two parts of Kew Gardens. • 1993, *Verdcourt* (S32: 657): Queen's Cottage Grounds, near tow path ditch.
EXSICC.: **Kew Gardens.** [Unlocalised]; grass under trees; 14 v 1918; *Worsdell* s.n. (K).
CULT.: 1768 (JH: 268); 1789, Nat. of England (HK1, 2: 342); 1812, Nat. of Britain (HK2, 4: 24); 1814, Nat. of Britain (HKE: 189); 1924, unspec. (K). Not currently in cult.
CURRENT STATUS: An uncommon weed in shady places; it has declined in recent years though it persists in quantity in the Riverside Walk and immediately north of the Rhododendron Dell.

1183. *Mimulus moschatus Douglas ex Lindl.
musk
FIRST RECORD: 2004 [1948].
CULT.: In cult.
CURRENT STATUS: One or two plants grew from seed in topsoil stored in the Paddock.

1184. *Mimulus guttatus DC.
monkeyflower
FIRST RECORD: 2001 [1979].
REFERENCES: **Kew Gardens.** 2001, *Cope* (S35: 648): Naturalised around the margins of the Lily Pond (333). Subsequently as a flowerbed weed near White Peaks (124).
CULT.: 1979, Alpine (K). Not currently in cult.
CURRENT STATUS: Casual; an occasional flower-bed weed.

1185. Limosella aquatica L.
mudwort
CULT.: 1768 (JH: 267); 1789, Nat. of Britain (HK1, 2: 359); 1812, Nat. of Britain (HK2, 4: 51); 1814, Nat. of Britain (HKE: 193). Not currently in cult.

1186. *Calceolaria chelidonioides Kunth
C. glutinosa Heer & Regel
slipperwort
FIRST RECORD: 1922 [1903].
REFERENCES: **Kew Gardens.** 1922, *Turrill & Lousley* (LFS: 272; BSP5: 340; S33: 749; and see Exsicc.): Waste ground near Herbarium. • 1981, *Palmer* (FSSC: 20; LN61: 103; S33: 749): [Still a weed in Kew Gardens]. Near the Australian House and elsewhere.
EXSICC.: **Kew Gardens.** Waste ground, near Herbarium; x 1922; *Turrill* s.n. (K).
CULT.: 1903, unspec. (K); 1926, unspec. (K); 1963, unspec. (K); 1967, Temp. Dept. (K). Not currently in cult.
CURRENT STATUS: No recent sightings.

1187. *Antirrhinum majus L.
snapdragon
FIRST RECORD: 1983/4 [1768].
REFERENCES: **Kew Gardens.** 1983/4, *Cope* (S31: 181): New student vegetable plots behind Hanover House.
CULT.: 1768 (JH: 271); 1789, Nat. of England (HK1, 2: 338); 1812, Nat. of England (HK2, 4: 17); 1814, Nat. of England (HKE: 188). In cult.
CURRENT STATUS: An occasional escape from cultivation, mostly near the Herbarium.

1188. Chaenorhinum minus (L.) Lange
Antirrhinum minus L.
Linaria minor (L.) Desf.
small toadflax
FIRST RECORD: 1873/4 [1768].
NICHOLSON: **1873/4 survey** (JB: 48): **B.** A few plants in flower-beds behind Palm House. • **1906 revision** (AS: 85 and in SFS: 483): Here and there as a weed in shrubberies and flower-borders.
OTHER REFERENCES: **Kew Gardens.** 1965, *Airy Shaw* (S29: 407): A single plant below steps from 'Marquand memorial window,' N. end of Wing C. • 1985, *London Natural History Society* (LN65: 195): [Around what is now the Princess of Wales Conservatory], 1 plant.

EXSICC.: **Kew Gardens.** Herb. Expt. Ground; weed; 29 vii 1930; *Turrill* s.n. (K).
CULT.: 1768 (JH: 271); 1789, Nat. of England (HK1, 2: 336); 1812, Nat. of England (HK2, 4: 16); 1814, Nat. of England (HKE: 187). In cult.
CURRENT STATUS: On freshly turned ground in several places between the Jodrell Laboratory and the Stable Yard.

1189. Misopates orontium (L.) Raf.
Antirrhinum orontium L.
weasel's-snout
FIRST RECORD: 2004 [1768].
CULT.: 1768 (JH: 271); 1789, Nat. of England (HK1, 2: 338); 1812, Nat. of England (HK2, 4: 18); 1814, Nat. of England (HKE: 188). Not currently in cult.
CURRENT STATUS: Rare; found as a flower-bed weed near King William's Temple and a regular weed in the Order Beds where it was originally sown.

1190. *Cymbalaria muralis G.Gaertn., B.Mey. & Schreb.
Antirrhinum cymbalaria L.
Linaria cymbalaria (L.) Mill.
ivy-leaved toadflax
FIRST RECORD: 1873/4 [1768].

Weasel's-snout (*Misopates orontium*). First found in the Order Beds growing wild after escaping from the display, and later photographed as far away as King William's Temple.

NICHOLSON: **1873/4 survey** (JB: 48): Very common on walls and in dry places. • **1906 revision** (AS: 85): Very common on walls and in dry places.
EXSICC.: **Kew Gardens.** Queen's Cottage Grounds; 15 vi 1914; *Flippance* s.n. (K). • Kew Road, on brick wall of Royal Botanic Gardens; 17 vii 1948; *Souster* 875 (K). • [Unlocalised]; a trailing perennial, found growing in the mortar of an old brick wall; 23 viii 1948; *Parker* s.n. (K). • Melon Yard, near gate from yard to Duke's Garden; 26 viii 1958; *Darbyshire* s.n. (K). • Duke's Garden; wall; 7 viii 1980; *Rees* s.n. (K).
CULT.: 1768 (JH: 271); 1789, Nat. of England (HK1, 2: 331); 1812, Nat. of England (HK2, 4: 10); 1814, Nat. of England (HKE: 187). Not currently in cult.
CURRENT STATUS: Widespread but not common except on the wall along Ferry Lane and at one time on the Rockery.

1191. Kickxia elatine (L.) Dumort
Antirrhinum elatine L.
Linaria elatine (L.) Mill.
sharp-leaved fluellen
CULT.: 1768 (JH: 271); 1789, Nat. of England (HK1, 2: 332); 1812, Nat. of England (HK2, 4: 11); 1814, Nat. of England (HKE: 187). In cult.
Note. Originally grown as a native but now considered to be an archaeophyte.

1192. Kickxia spuria (L.) Dumort.
Antirrhinum spurium L.
Linaria spuria (L.) Mill.
round-leaved fluellen
FIRST RECORD: 2001 [1789].
REFERENCES: **Kew Gardens.** 2001, *Cope* (S35: 648): Weed in the Order Beds (156) from which it presumably escaped.
CULT.: 1789, Nat. of England (HK1, 2: 332); 1812, Nat. of England (HK2, 4: 11); 1814, Nat. of England (HKE: 187). Not currently in cult.
CURRENT STATUS: An occasional weed in the Order Beds and, for one season, it grew in abundance from seed in topsoil stored in the Paddock.

1193. Linaria vulgaris Mill.
Antirrhinum linaria L.
common toadflax
FIRST RECORD: 1873/4 [1768].
NICHOLSON: **1873/4 survey** (JB: 48): **P.** Many plants in young plantation between Winter Garden and Kew Road. These have been mown down before they had a chance of flowering, for some years. • **1906 revision** (AS: 85): Here and there as a weed in shrubberies.
CULT.: 1768 (JH: 271); 1789, Nat. of Britain (HK1, 2: 337); 1812, Nat. of Britain (HK2, 4: 17); 1814, Nat. of Britain (HKE: 188); 1881, unspec. (K); 1882, unspec. (K); 1907, unspec. (K); 1934, unspec. (K); 1938, Herbarium Ground (K). In cult.
CURRENT STATUS: Very rare; a single plant was found by the wall around the Order Beds.

Round-leaved fluellen (*Kickxia spuria*). Sometimes common as a weed in the Order Beds, from which it must have escaped, but also seen in the Paddock and in flower-beds around the Banks Building.

1194. Linaria repens (L.) Mill.
Antirrhinum repens L.
pale toadflax
FIRST RECORD: 2008 [1768].
CULT.: 1768 (JH: 271); 1789, Nat. of Britain (HK1, 2: 333); 1812, Nat. of England (HK2, 4: 13); 1814, Nat. of England (HKE: 188); 1895, unspec. (K); 1923, unspec. (K); 1924, unspec. (K). In cult.
CURRENT STATUS: A single occurrence in rough ground near the generator opposite the Herbarium.

1195. *Linaria pelisseriana (L.) Mill.
Antirrhinum pelisserianum L.
Jersey toadflax
FIRST RECORD: 1880 [1768].
NICHOLSON: **1906 revision:** Not recorded.
OTHER REFERENCES: **Kew Gardens.** 1880, *Baker* (BECB1: 40; S33: 749): Thames bank opposite Sion House, midway between Kew and Richmond.
CULT.: 1768 (JH: 271); 1789, Alien (HK1, 2: 334); 1811, Alien (HK2, 3: 414); 1814, Alien (HKE: 187). Not currently in cult.
CURRENT STATUS: No recent sightings.

[**1196. Linaria arvensis** (L.) Desf.
Antirrhinum arvense L.
corn toadflax
CULT.: 1789, Nat. of England (HK1, 2: 334).
Note. Listed in error as native.]

1197. *Nuttalanthus canadensis (L.) D.A.Sutton
Linaria canadensis (L.) Dum.-Cours.
FIRST RECORD: 1880.
NICHOLSON: **1906 revision:** Not recorded.
OTHER REFERENCES: **Kew Gardens.** 1880, *Baker* (BECB1: 40; S33: 749): Thames bank opposite Sion House, midway between Kew and Richmond.
CULT.: In cult.
CURRENT STATUS: No recent sightings.

1198. *Nemesia melissifolia Benth.
FIRST RECORD: 2004
CURRENT STATUS: A few plants grew from seed in topsoil stored in the Paddock.

1199. Digitalis purpurea L.
foxglove
FIRST RECORD: 1873/4 [1768].
NICHOLSON: **1873/4 survey** (JB: 48): **P.** Several plants on a rubbish heap in wood behind Winter Garden. **Q.** A few plants near "Cottage." • **1906 revision** (AS: 85): **A, Q.** In woods and along shrubbery borders.
EXSICC.: **Kew Gardens.** Queen's Cottage Grounds; [c. 1914]; *Flippance* s.n. (K). • Arboretum; vii 1942, *Hutchinson* s.n. (K).
CULT.: 1768 (JH: 269); 1789, Nat. of Britain (HK1, 2: 344); 1812, Nat. of Britain (HK2, 4: 28); 1814, Nat. of Britain (HKE: 189); 1901, unspec. (K); 1913, unspec. (K); 1933, Herb Garden (K); 1935, Herbarium Ground (K). In cult.
CURRENT STATUS: Widespread in shrubberies, but it is hard to be sure which plants are cultivated and which are genuinely wild. It is probably native only in the Conservation Area.

1200. *Digitalis lanata Ehrh.
Grecian foxglove
FIRST RECORD: 2004 [1812].
CULT.: 1812, Alien (HK2, 4: 29); 1814, Alien (HKE: 190); 1897, unspec. (K); 1915, unspec. (K); 1946, Rock Garden (K); 1979, Alpine (K); 1980, unspec. (K). In cult.
CURRENT STATUS: Grew from seed in topsoil stored behind the Princess of Wales Conservatory, but the area has now been cleared and restored to grassland.

Thyme-leaved speedwell (*Veronica serpyllifolia*). Common in the damper grass throughout the Gardens, but easily overlooked. It is more readily seen when growing as a flower-bed weed, as here.

1201. Veronica serpyllifolia L.
thyme-leaved speedwell
FIRST RECORD: 1873/4 [1768].
NICHOLSON: **1873/4 survey** (JB: 48): Frequent on every piece of lawn. • **1906 revision** (AS: 85): Frequent on every piece of lawn.
CULT.: 1768 (JH: 262); 1789, Nat. of Britain (HK1, 1: 21); 1810, Nat. of Britain (HK2, 1: 30); 1814, Nat. of Britain (HKE: 6); 1922, unspec. (K); 1924, Herbaceous Ground (K); 1955, unspec. (K); 1981, unspec. (K). Not currently in cult.
CURRENT STATUS: Common in the damper grass throughout the Gardens.

1202. Veronica alpina L.
alpine speedwell
CULT.: 1768 (JH: 262); 1789, Nat. of Scotland (HK1, 1: 21); 1810, Nat. of Scotland (HK2, 1: 30); 1814, Nat. of Scotland (HKE: 6). Not currently in cult.

1203. Veronica fruticans Jacq.
V. saxatilis Scop.
rock speedwell
CULT.: 1789, Alien (HK1, 1: 21); 1810, Nat. of Scotland (HK2, 1: 30); 1814, Nat. of Scotland (HKE: 6); 1980, Rock Garden (K); 1987, unspec. (K). In cult.
Note. Found as a British native in 1790.

1204. Veronica officinalis L.
heath speedwell
FIRST RECORD: 1873/4 [1768].
NICHOLSON: **1873/4 survey** (JB: 48): Fairly common everywhere. • **1906 revision** (AS: 85): Fairly common everywhere.
CULT.: 1768 (JH: 262); 1789, Nat. of Britain (HK1, 1: 20); 1810, Nat. of Britain (HK2, 1: 30); 1814, Nat. of Britain (HKE: 6); 1881, unspec. (K); 1929, unspec. (K); 1933, Herb Garden (K); 1936, Herbaceous Ground (K). In cult.
CURRENT STATUS: Rare; it has clearly declined since Nicholson's survey. Apart from the Paddock, where it may have arrived in topsoil as seed, it is known only from a patch of acidic grassland north of the Rhododendron Dell.

1205. Veronica chamaedrys L.
germander speedwell
FIRST RECORD: 1873/4 [1768].
NICHOLSON: **1873/4 survey** (JB: 49): **P.** Rather frequent on the slope at Richmond side of lake. • **1906 revision** (AS: 85): **A, Q.** Not infrequent.
EXSICC.: **Kew Gardens.** Field near the Herbarium; 20 v 1920; *Turrill* s.n. (K). • Field near the Herbarium; 26 viii 1922; *Turrill* s.n. (K). • Near herbarium; in short grass, sandy soil; 10 v 1928; *Hubbard* s.n. (K). • Near the Tennis Courts; 4 vi 1944; *Hutchinson* s.n. (K). **Kew Environs.** Towpath, Kew; perennial herb growing among grass; 23 v 1944; *Souster* 74 (K).
CULT.: 1768 (JH: 262); 1789, Nat. of Britain (HK1, 1: 23); 1810, Nat. of Britain (HK2, 1: 32); 1814, Nat. of Britain (HKE: 6); 1919, unspec. (K); 1922, unspec. (K); 1929, unspec. (K). In cult.
CURRENT STATUS: Abundant in grassland throughout the Gardens.

1206. Veronica montana L.
wood speedwell
FIRST RECORD: 1922 [1768].
REFERENCES: **Kew Gardens.** 1922, *Turrill* (S18: 65): In grassfield near Herbarium.
CULT.: 1768 (JH: 262); 1789, Nat. of Britain (HK1, 1: 23); 1810, Nat. of Britain (HK2, 1: 32); 1814, Nat. of Britain (HKE: 6). Not currently in cult.
CURRENT STATUS: No recent sightings.

1207. Veronica scutellata L.
marsh speedwell
CULT.: 1768 (JH: 262); 1789, Nat. of Britain (HK1, 1: 22); 1810, Nat. of Britain (HK2, 1: 31); 1814, Nat. of Britain (HKE: 6); 1931, unspec. (K). Not currently in cult.

1208. Veronica beccabunga L.
brooklime
CULT.: 1768 (JH: 262); 1789, Nat. of Britain (HK1, 1: 22); 1810, Nat. of Britain (HK2, 1: 31); 1814, Nat. of Britain (HKE: 6); 1917, unspec. (K); 1933, Herb. Ground (K); 1965, Water Garden (K). In cult.

1209. Veronica anagallis-aquatica L.
V. anagallis auct.
blue water-speedwell
FIRST RECORD: 1873/4 [1768].

NICHOLSON: **1873/4 survey** (JB: 49): **Pal.** On rubbish thrown up when moat was cleaned. • **1906 revision** (AS: 85): **Strip.** By ha-ha and river.
OTHER REFERENCES: **Kew Environs.** 1938, *E.B. Bangerter* (LN33: S208): By Thames between Kew and Richmond.
EXSICC.: **Kew Environs.** Thames bank, Kew; from crevices of new stone wall; 2 viii 1928; *Summerhayes & Turrill* 2481 (K).
CULT.: 1768 (JH: 262); 1789, Nat. of Britain (HK1, 1: 22); 1810, Nat. of Britain (HK2, 1: 31); 1814, Nat. of Britain (HKE: 6); 1934, unspec. (K). Not currently in cult.
CURRENT STATUS: Still occurs below the towpath near Kew Bridge. Although it has been in cultivation it has never been recorded wild within the Gardens except in decaying vegetation dragged out of the Moat and presumably used as a mulch.

1210. Veronica triphyllos L.
fingered speedwell
CULT.: 1768 (JH: 262); 1789, Nat. of Britain (HK1, 1: 25); 1810, Nat. of Britain (HK2, 1: 34); 1814, Nat. of Britain (HKE: 5). Not currently in cult.

1211. Veronica arvensis L.
wall speedwell
FIRST RECORD: 1873/4 [1768].
NICHOLSON: **1873/4 survey** (JB: 48): Common in all the divisions. • **1906 revision** (AS: 85): Common in all the divisions.
EXSICC.: **Kew Gardens.** [Unlocalised]; on rubbish heap; v 1928; *Hubbard* s.n. (K).
CULT.: 1768 (JH: 262); 1789, Nat. of Britain (HK1, 1: 25); 1810, Nat. of Britain (HK2, 1: 34); 1814, Nat. of Britain (HKE: 5). Not currently in cult.
CURRENT STATUS: Common in dry areas in bare soil, short grass and on wall tops.

1212. Veronica verna L.
spring speedwell
CULT.: 1789, Nat. of England (HK1, 1: 25); 1810, Nat. of England (HK2, 1: 34); 1814, Nat. of England (HKE: 5). Not currently in cult.

1213. *Veronica peregrina L.
American speedwell
FIRST RECORD: 1918 [1918].
REFERENCES: **Kew Gardens.** 1918, *Worsdell* (LFS: 275; BSP5: 310; S33: 749, and see Exsicc.): Weed of cultivated ground. • 1983, *Burton* (FLA: 125; S33: 749): Put in a brief appearance ... in Kew Gardens in 1918.
EXSICC.: **Kew Gardens.** [Unlocalised]; cultivated ground; 26 vii 1918; *Worsdell* s.n. (K). • Herbaceous Ground; weed in beds; ix 1922; *anon.* (K).
CULT.: 1918, unspec. (K). Not currently in cult.
CURRENT STATUS: No recent sightings.

1214. Veronica agrestis L.
green field-speedwell
FIRST RECORD: 1873/4 [1768].
NICHOLSON: **1873/4 survey** (JB: 48 and in SFS: 489): **B, Pal, P** and **Q.** Common. • **1906 revision** (AS: 85): Common in all the divisions.
EXSICC.: **Kew Gardens.** Experimental Ground; growing as a weed in company with *V. didyma* [*V. polita*]; 29 x 1935; *A.K.Jackson* s.n. (K). • Herbarium allotments; 27 iv 1946; *Sandwith* s.n. (K). • Near Herbarium; cultivated ground; 11 vii 1951; *Meikle & Turrill* s.n. (K). **Kew Environs.** Kew; 19 vi 1931; *Fraser* s.n. (K). • Kew; spontaneous; 22 ix 1933; *Fraser* s.n. (K). • Kew; spontaneous; 20 x 1933; *Fraser* s.n. (K). • Kew; 22 x 1933; *Fraser* s.n. (K).
CULT.: 1768 (JH: 262); 1789, Nat. of Britain (HK1, 1: 24); 1810, Nat. of Britain (HK2, 1: 33); 1814, Nat. of Britain (HKE: 5); 1935, unspec. (K). Not currently in cult.
CURRENT STATUS: Occasional and very widespread, but inconspicuous and difficult to spot.

1215. *Veronica polita Fr.
grey field-speedwell
FIRST RECORD: 1873/4.
NICHOLSON: **1873/4 survey** (JB: 48): **B.** Flower-beds and borders of shrubberies near Palm House, in company with [*Veronica persica*], but far less common than that species. • **1906 revision** (AS: 85): Flower-beds and borders of shrubberies in company with [*Veronica persica*], but not so common as that species.
EXSICC.: **Kew Gardens.** Lower Nursery; 19 iii 1935; *Sprague* s.n. (K). • Herbarium Ground; weed; 20 vi 1935; *anon.* s.n. (K). • Experimental Ground; growing as a weed with *V. agrestis*; 29 x 1935; *A.K. Jackson* s.n. (K). • Near Herbarium; cultivated ground; 11 vii 1951; *Meikle & Turrill* s.n. (K).
CURRENT STATUS: Widespread, but inconspicuous; it is probably more common than *V. agrestis* with which it often grows.

1216. *Veronica persica Poir.
V. buxbaumii Ten.
common field-speedwell
FIRST RECORD: 1873/4 [1895].
NICHOLSON: **1873/4 survey** (JB: 48): The commonest species in our Flora. • **1906 revision** (AS: 85): The commonest species in the Kew flora.
OTHER REFERENCES: **Kew Environs.** 1878, *G.Nicholson* (BEC18: 17): A variety ... from waste ground at Kew.
CULT.: 1895, unspec. (K); 1929, unspec. (K); 1965, Order Beds (K). Not currently in cult.
CURRENT STATUS: Widespread as a weed of flower-beds but no longer the commonest of our speedwells.

1217. *Veronica filiformis Sm.
slender speedwell
FIRST RECORD: 1971 [1921].
REFERENCES: **Kew Environs.** 1971, *D.H.Kent* (LN52: 118; S33: 749): River-wall just above Kew Bridge.
CULT.: 1921, unspec. (K). Not currently in cult.
CURRENT STATUS: No recent sightings; it was last noted in 1981.

1218. Veronica hederifolia L.
ivy-leaved speedwell
FIRST RECORD: 1873/4 [1768].
NICHOLSON: **1873/4 survey** (JB: 48): **B.** Common in shrubberies. I have no notes of its occurrence in other divisions. • **1906 revision** (AS: 85): Common in shrubberies.
EXSICC.: **Kew Gardens.** Near Tennis Courts; on loose deposited soil; 5 iii 1928; *Summerhayes & Turrill* s.n. (K). • Lower Nursery; on rubbish heap; v 1928; *Hubbard* s.n. (K). • [Unlocalised]; annual weed in beds and borders; 16 viii 1944; *Souster* 65 (K). **Kew Environs.** Kew; v 1885; *Fraser* s.n. (K).
CULT.: 1768 (JH: 262); 1789, Nat. of Britain (HK1, 1: 25); 1810, Nat. of Britain (HK2, 1: 34); 1814, Nat. of Britain (HKE: 5); 1934, Herbarium Ground (K); 1935, Herbarium Ground (K). Not currently in cult.
CURRENT STATUS: Abundant in the few places where it occurs.

1219. Veronica spicata L.
spiked speedwell

a. subsp. **spicata**
CULT.: 1768 (JH: 262); 1789, Nat. of England (HK1, 1: 19); 1810, Nat. of England (HK2, 1: 27); 1814, Nat. of England (HKE: 5); 1881, unspec. (K); 1926, Herb. Exp. Ground (K); 1947, Herbarium Ground (K). Not currently in cult.

b. subsp. **hybrida** (L.) Gaudin
V. hybrida L.
CULT.: 1768 (JH: 262); 1789, Nat. of Wales (HK1, 1: 19); 1810, Nat. of Wales (HK2, 1: 27); 1814, Nat. of Wales (HKE: 5). Not currently in cult.

[1220. Veronica fruticulosa L.
CULT.: 1789, Alien (HK1, 1: 21); 1810, Nat. of Scotland (HK2, 1: 30); 1814, Nat. of Scotland (HKE: 6).
Note. Listed in error as native.]

1221. Sibthorpia europaea L.
Cornish moneywort
CULT.: 1768 (JH: 268); 1789, Nat. of England (HK1, 2: 359); 1812, Nat. of England (HK2, 4: 51); 1814, Nat. of England (HKE: 193); no date, unspec. (K). Not currently in cult.
Note. The Kew specimen probably dates from between 1910 and 1925.

1222. Melampyrum cristatum L.
crested cow-wheat
CULT.: 1768 (JH: 264); 1789, Nat. of Britain (HK1, 2: 328); 1812, Nat. of England (HK2, 4: 3); 1814, Nat. of England (HKE: 186). Not currently in cult.

[1223. Melampyrum arvense L.
field cow-wheat
CULT.: 1789, Nat. of England (HK1, 3: 494); 1812, Nat. of England (HK2, 4: 3); 1814, Nat. of England (HKE: 186).
Note. Listed in error as native.]

1224. Melampyrum pratense L.
common cow-wheat
CULT.: 1768 (JH: 264); 1789, Nat. of England (HK1, 2: 328); 1812, Nat. of Britain (HK2, 4: 3); 1814, Nat. of Britain (HKE: 186). Not currently in cult.

1225. Melampyrum sylvaticum L.
small cow-wheat
CULT.: 1768 (JH: 264); 1812, Nat. of Britain (HK2, 4: 3); 1814, Nat. of Britain (HKE: 186); 1921, unspec. (K). Not currently in cult.

1226. Euphrasia officinalis L. sensu lato
eyebright
FIRST RECORD: 1873/4 [1768].
NICHOLSON: **1873/4 survey** (JB: 49): **Q.** A couple of small plants in turf near wall 50 yards south of Isleworth Gate, 1873. Not seen since. • **1906 revision** (AS: 85): **Q.** A few plants in turf near Isleworth Gate. Not noted since 1873.
CULT.: 1768 (JH: 264); 1789, Nat. of Britain (HK1, 2: 327); 1812, Nat. of Britain (HK2, 4: 2); 1814, Nat. of Britain (HKE: 185). Not currently in cult.
CURRENT STATUS: Extinct since 1873. Since there are no extant vouchers it is impossible to determine the microspecies.

1227. Odontites vernus (Bellardi) Dumort.
Euphrasia odontites L.
Bartsia odontites (L.) Huds.
red bartsia
CULT.: 1768 (JH: 264); 1789, Nat. of Britain (HK1, 2: 328); 1812, Nat. of Britain (HK2, 4: 2); 1814, Nat. of Britain (HKE: 185); 1922, unspec. (K). Not currently in cult.

1228. Bartsia alpina L.
alpine bartsia
CULT.: 1768 (JH: 263); 1812, Nat. of Britain (HK2, 4: 2); 1814, Nat. of Britain (HKE: 185). Not currently in cult.

Yellow bartsia (*Parentucellia viscosa*). A rare hemi-parasite photographed in the Paddock after its restoration in 1990 but it has not been seen since.

1229. Parentucellia viscosa (L.) Caruel
Bartsia viscosa L.
yellow bartsia

First record: 1990 [1812].
References: **Kew Gardens.** 1990, *Cope* (S31: 182): Paddock, after restoration in 1990.
Cult.: 1812, Nat. of Britain (HK2, 4: 1); 1814, Nat. of Britain (HKE: 185); 1921, unspec. (K). Not currently in cult.
Current status: No recent sightings; it has not reappeared in the Paddock since 1990.

1230. Rhinanthus minor L.
R. crista-galli L.
yellow-rattle

First record: 2002 [1768].
References: **Kew Gardens.** 2002, *Cope* (S35: 648): South end of the Riverside Walk (216). Sown into the area in order to control the growth of grass; the sowing was successful and the species seems to have established.
Cult.: 1768 (JH: 264); 1789, Nat. of Britain (HK1, 2: 327); 1812, Nat. of Britain (HK2, 4: 2); 1814, Nat. of Britain (HKE: 185). Not currently in cult.
Current status: Sown into grass in 2002 at the south end of the Riverside Walk and now well established. Its purpose was to control the grass and it seems to have performed well in this respect. It has not yet spread beyond the area of its original sowing but there is ample opportunity for it to do so.

1231. Pedicularis palustris L.
marsh lousewort

Cult.: 1768 (JH: 272); 1789, Nat. of Britain (HK1, 2: 328); 1812, Nat. of Britain (HK2, 4: 3); 1814, Nat. of Britain (HKE: 186); 1982, unspec. (K). Not currently in cult.

1232. Pedicularis sylvatica L.
lousewort

Cult.: 1768 (JH: 272); 1789, Nat. of Britain (HK1, 2: 329); 1812, Nat. of Britain (HK2, 4: 4); 1814, Nat. of Britain (HKE: 186); 1881, unspec. (K). Not currently in cult.

Yellow-rattle (*Rhinanthus minor*). Unknown as a wild plant in the Gardens. A hemi-parasite that was sown in 2002 at the southern end of the Riverside Walk as an experiment in control of grass growth and is now well established.

1233. Lathraea squamaria L.

toothwort

FIRST RECORD: 1873/4; introduced in 1834.
NICHOLSON: **1873/4 survey** (JB: 49 and in SFS: 504): **B.** On roots of Thorn near House No. 2. Very abundant by side of "Broad Walk," in the Rhododendron bed nearest Museum No. 3. This bed is full of elm roots, and I believe the *Lathraea* to be parasitic on them and not on the roots of the Rhododendron. Introduced from Dorking in 1834 by Mr A. Choules, then a foreman in the Royal Gardens under Mr Aiton (see *J. Bot.*, 1872, p. 173) [see also *Gard. Chron.*, 1872, p. 466; BFS: 169; SFS: 504]. • **1906 revision** (AS: 85; SFS: 504): **B.** In Rhododendron bed on mound near Cumberland Walk, and elsewhere. Introduced from Dorking about 1834. (*See J. Bot.*, 1872, 173).
OTHER REFERENCES: **Kew Gardens.** 1954, *Cooke* (LN33: S214): Kew Gardens.
EXSICC.: **Kew Gardens.** [Certainly?] the specimen from Kew Gardens where it is said to be native; 1848; *Watson* in *Herb. Watson* (K). • [Unlocalised]; on Rhododendron; v 1887; *Gamble* 18761 (K). • Near Trop. Fern House; on *Populus alba* v. *pyramidalis*; 14 iv 1923; *Turrill* s.n. (K). • [Unlocalised]; on roots of laurel; 10 v 1932; *Hubbard* s.n. (K). • Poplar collection; under *Prunus laurocerasus*; lost plant?; 9 iv 1949; *Souster* 977 (K).
CULT.: No date, unspec. (K). In cult.
CURRENT STATUS: Known from under the same Black Walnut that hosts *L. clandestina*; also sometimes beneath a nearby *Davidia*; another population can be seen behind the Orangery, also on Black Walnut. It appears to have been introduced to the Gardens in 1834.

1234. *Lathraea clandestina L.

purple toothwort

FIRST RECORD: 1996; introduced in 1888.
NICHOLSON: **1906 revision:** Not recorded.
OTHER REFERENCES: **Kew Gardens.** 1996, *Atkinson* (WAT21: 123; S33: 749): Presented in 1888 by Dr Schumann of the Berlin Herbarium (see Hooker in *Curtis's Bot. Mag.*, ser. 3, 46: t.7106 (1890)) and grown on a 'willow adjacent to the ornamental water in front of the Palm House' (see *Gard. Chron.*, ser. 3, 5: 652 (1889)). Seen c. 1990(!) near the Princess of Wales House.
CULT.: 1919, unspec. (K); 1923, unspec. (K). In cult.
CURRENT STATUS: Persists on Black Walnut in the Woodland Garden near the Order Beds, where it was originally introduced and now appears to be spreading; reported as spontaneous in the Conservation Area; still by the Palm House Pond where it was introduced in 1888.

1235. Orobanche purpurea Jacq.

yarrow broomrape

FIRST RECORD: 1992.
REFERENCES: **Kew Gardens.** 1992, *Cope & Spooner* (LN72: 116; S33: 749): Of natural appearance in the Natural Order beds. • 1992, *Cope & Spooner* (WAT19: 290; S33: 749): On *Achillea millefolium*. Not deliberately introduced.
CURRENT STATUS: Persisted until recently on Yarrow in the Order Beds and was often abundant; it was probably eradicated when many of the Asteraceae beds were converted into temporary student allotments. Its origin is unknown, but it was apparently not deliberately introduced.

Purple toothwort (*Lathraea clandestina*). An introduction in the Gardens that seems to be spreading. It is currently known from three locations, the most notable at the base of a Black Walnut in the Woodland Garden where a feature is made of it each spring.

1236. Orobanche caryophyllacea Sm.
bedstraw broomrape
FIRST RECORD: 1960s; see Milne-Redhead (1985).
REFERENCES: **Kew Gardens.** 1999, *Cope* (S35: 648): In the Order Beds. The species was originally introduced by Milne-Redhead to plants of *Galium verum* in the Herbarium experimental plots in the 1960s. Subsequent to 1971, the host was moved to the Order Beds on their present site and the *Orobanche* survived, not only on *G. verum*, but had also parasitised *G. odoratum* (see *BSBI News* 41: 29, 1985).
CURRENT STATUS: Introduced on bedstraw in the Order Beds and has persisted for many years although it has not been accessioned as a cultivated plant. For a history of its introduction see Milne-Redhead in *BSBI News* **41**: 29 (1985).

1237. Orobanche elatior Sutton
O. major L.
knapweed broomrape
FIRST RECORD: 1992 [1768].
REFERENCES: **Kew Gardens.** 1992, *Cope & Spooner* (unpubl.).
CULT.: 1768 (JH: 260); 1903, unspec. (K). Not currently in cult.
CURRENT STATUS: Of unknown origin, but doubtless introduced in the Order Beds. No recent sightings.

1238. Orobanche hederae Duby
ivy broomrape
FIRST RECORD: 1924 [1903].
REFERENCES: **Kew Gardens.** 1931, *Nicholson* (SFS: 502): In several places, Kew Gardens; not sown! • 1933–1953, *D.H.Kent* (LFS: 281; S32: 657; LN33: S213): Long known from various places in the Gardens on ivy, as, for example, nr the Order Beds. • 1983, *Burton* (FLA: 129): Kew Gardens. **Kew Environs.** 1948, *Spreadbury* (BSP1: 59; LN28: 28; LN33: S213; LFS: 502): Parasitic on Ivy on the tow-path between Richmond and Kew.
EXSICC.: **Kew Gardens.** Nr. Museum II; 4 ii 1924; *Turrill* s.n. (K). • By Ministry of Work's yard, Kew Palace; on ivy under Holly Oak; 7 viii 1958; *Reid* s.n. (K). • Queen's Garden mound; 22 vi 1974; *Hepper* 4832 (K). **Kew Environs.** Kew; v 1930; ?*Pearce* s.n. (K).
CULT.: 1903, unspec. (K); 1965, Border along Museum II (K). In cult.
CURRENT STATUS: Common on Ivy around the periphery of the Gardens in shady places. Introduced around the School of Horticulture but otherwise certainly native in one of its few inland localities. It survived, presumably as seed, when the Ivy was incompletely eradicated along the wall bordering Ferry Lane. It is only occasionally seen away from the boundary wall despite the quantites of ivy that are grown in the Gardens.

1239. Orobanche minor Sm.
common broomrape
a. var. **minor**
FIRST RECORD: 1885 [1878].
NICHOLSON: **1906 revision:** Not recorded.
OTHER REFERENCES: **Kew Gardens.** 1968, *Edwards* (S31: 183): Old frame just N of Mounters' Hut, ten plants.
EXSICC.: **Kew Gardens.** Herbarium grounds; vi 1885; *Hooker* s.n. (K).
CULT.: 1878, on *Wisteria* (K); 1967, unspec. (K). In cult.
CURRENT STATUS: Rare; there is one recent sighting in the Order Beds where it was probably accidentally introduced. No other plants have been found although early records suggest it may be native.

b. var. **maritima** (Pugsley) Rumsey & Jury
CULT.: 1933, unspec. (K). Not currently in cult.

Ivy broomrape (*Orobanche hederae*). A coastal species in Britain in one of its few inland sites, though since the River Thames is tidal at Kew the locality may be considered coastal. Widespread and often abundant in the Gardens, especially around the periphery.

Lentibulariaceae

1240. Pinguicula lusitanica L.
pale butterwort
Cult.: 1768 (JH: 259); 1810, Nat. of Britain (HK2, 1: 44); 1814, Nat. of Britain (HKE: 7). In cult. under glass.

1241. Pinguicula alpina L.
alpine butterwort
Cult.: 1810, Alien (HK2, 1: 45); 1814, Alien (HKE: 7). Not currently in cult.
Note. Not discovered in Britain until 1832 and now believed extinct.

1242. Pinguicula vulgaris L.
common butterwort
Cult.: 1768 (JH: 259); 1789, Nat. of Britain (HK1, 1: 30); 1810, Nat. of Britain (HK2, 1: 44); 1814, Nat. of Britain (HKE: 7). Not currently in cult.

1243. Utricularia vulgaris L.
greater bladderwort
First record: 1837 [1768].
Nicholson: **1873/4 survey:** Not recorded. • **1906 revision:** Not recorded.
Other references: **Kew Gardens.** 1837, *Francis* (SFS: 504; S33: 749): Once in the ditch around Kew Gardens.
Exsicc.: **Kew Gardens.** Moat bounding Kew Park by the Thames; vi 1845; *Stevens* s.n. (K).
Cult.: 1768 (JH: 275); 1789, Nat. of Britain (HK1, 1: 31); 1810, Nat. of Britain (HK2, 1: 45); 1814, Nat. of Britain (HKE: 7). Not currently in cult.
Current status: No recent sightings; there is now no suitable habitat for it.

Greater bladderwort (*Utricularia vulgaris*). Known as a wild plant in the Moat between 1837 and 1845 but sadly long absent from the area.

1244. Utricularia intermedia Hayne
intermediate bladderwort
Cult.: 1814, Native of Ireland (HKE: Add.). Not currently in cult.
Note. Scattered throughout the British Isles, not just in Ireland.

1245. Utricularia minor L.
lesser bladderwort
Cult.: 1768 (JH: 275); 1789, Nat. of Britain (HK1, 1: 31); 1810, Nat. of Britain (HK2, 1: 45); 1814, Nat. of Britain (HKE: 7). Not currently in cult.

Campanulaceae

1246. Campanula patula L.
spreading bellflower
Cult.: 1768 (JH: 131); 1789, Nat. of England (HK1, 1: 219); 1810, Nat. of England (HK2, 1: 345); 1814, Nat. of England (HKE: 48); 1908, unspec. (K). Not currently in cult.

1247. Campanula rapunculus L.
rampion bellflower
Cult.: 1768 (JH: 131); 1789, Nat. of England (HK1, 1: 220); 1810, Nat. of England (HK2, 1: 346); 1814, Nat. of England (HKE: 48); 1914, unspec. (K); 1970s, plot 154-06 (K). Not currently in cult.
Note. Originally grown as a native but now considered to be an archaeophyte.

1248. Campanula glomerata L.
clustered bellflower
First record: 1873/4 [1768].
Nicholson: **1873/4 survey** (JB: 48 and in SFS: 434): **Pal.** A few plants in hay-grass near Palace. • **1906 revision** (AS: 84): **P.** In hay grass near palace.
Cult.: 1768 (JH: 131); 1789, Nat. of Britain (HK1, 1: 222); 1810, Nat. of Britain (HK2, 1: 349); 1814, Nat. of Britain (HKE: 48); 1937, Herbarium Ground (K). In cult.
Current status: No recent sightings; grass is no longer grown for hay.

1249. *Campanula poscharskyana Degen
trailing bellflower
FIRST RECORD: 2004 [1932].
CULT.: 1932, Herbaceous Dept. (K); 1939, Herbarium Ground (K); 1947, Herbarium Ground (K); 1951, Mr Turrill's Garden (K); 1951, Rock Garden (K). Not currently in cult.
CURRENT STATUS: Grew from seed in topsoil stored in the Paddock and behind the Princess of Wales Conservatory. The latter area has now been restored but the Paddock remains in a highly disturbed state.

1250. Campanula latifolia L.
giant bellflower
CULT.: 1768 (JH: 131); 1789, Nat. of Britain (HK1, 1: 221); 1810, Nat. of Britain (HK2, 1: 347); 1814, Nat. of Britain (HKE: 48); 1883, unspec. (K); 1924, unspec. (K); 1971, Alpine Dept. (K). In cult.

Clustered bellflower (*Campanula glomerata*). Extinct. Last recorded as a wild plant near Kew Palace about a hundred years ago.

Nettle-leaved bellflower (*Campanula trachelium*). Although a British native it is unlikely to be native in the Gardens. Occasionally seen in shady places.

1251. Campanula trachelium L.
nettle-leaved bellflower
FIRST RECORD: 1998 [1768].
REFERENCES: **Kew Gardens.** 1998, *Cope* (S33: 749): North Arb.: behind Mycology Building (113); near Director's garden (131).
CULT.: 1768 (JH: 131); 1789, Nat. of Britain (HK1, 1: 222); 1810, Nat. of Britain (HK2, 1: 348); 1814, Nat. of Britain (HKE: 48); 1861, unspec. (K); 1868, unspec. (K); 1898, unspec. (K); 1930, unspec. (K); 1971, Alpine Dept. (K); 1971, plot 154-09 (K). In cult.
CURRENT STATUS: Rare; several recent sightings in addition to those above. Although a British native it is undoubtedly an escape from cultivation within the Gardens.

1252. *Campanula rapunculoides L.
creeping bellflower
FIRST RECORD: 1873/4 [1768].
NICHOLSON: **1873/4 survey** (JB: 48 and in SFS: 435): **Pal.** Here and there in turf among young trees near Brentford Gate. • **1906 revision** (AS: 84): **P.** Here and there in turf.
OTHER REFERENCES: **Kew Gardens.** 1944, *D.H.Kent* (LN33: S180): Kew Gardens, weed in shrubberies.
EXSICC.: **Kew Environs.** Towing Path, Kew; 25 vi 1930 & 18 vii 1931; *Fraser* s.n. (K).

CULT.: 1768 (JH: 131); 1789, Alien (HK1, 1: 222); 1810, Nat. of England (HK2, 1: 347); 1814, Nat. of England (HKE: 48); 1898, unspec. (K); 1933, Herb Garden (K). Not currently in cult.
CURRENT STATUS: Rare; naturalised on the flagpole mound and in woodland elsewhere. Doubtless both records are relicts of former cultivation.

1253. Campanula rotundifolia L.
harebell
FIRST RECORD: 1873/4 [1768].
NICHOLSON: **1873/4 survey** (JB: 48): **P.** Not uncommon about lake. • **1906 revision** (AS: 83): **A.** Not uncommon in turf near Ash collections and elsewhere.
EXSICC.: **Kew Gardens.** [Unlocalised]; in rough grass among trees; 24 vii 1944; *Souster* 121 (K).
CULT.: 1768 (JH: 131); 1789, Nat. of Britain (HK1, 1: 219); 1810, Nat. of Britain (HK2, 1: 345); 1814, Nat. of Britain (HKE: 48); 1854, unspec. (K); 1880s, unspec. (K); 1882, unspec. (K); 1908, unspec. (K); 1970s, unspec. (K); 1973, plot 154-09 (K). In cult.
CURRENT STATUS: Only a handful of recent sightings and the species is now very rare within the Gardens.

1254. *Campanula carpatica Jacq.
tussock bellflower
FIRST RECORD: 2004 [1881].
CULT.: 1881, unspec. (K); 1907, unspec. (K); 1924, unspec. (K); 1970s, unspec. (K); 1974, unspec. (K); 1984, unspec. (K). Not currently in cult.
CURRENT STATUS: Grew from seed in topsoil stored in the Paddock.

1255. *Campanula erinus L.
FIRST RECORD: 1961 [1768].
REFERENCES: **Kew Gardens.** 1961, *Milne-Redhead* (S29: 407 and see Exsicc.): One fine plant flowering and setting fruit, just outside Succulent Pits in the Melon Yard.
EXSICC.: **Kew Gardens.** Melon Yard; weed; 28 vii 1961; *Milne-Redhead* s.n. (K).
CULT.: 1768 (JH: 131); 1789, Alien (HK1, 1: 225); 1810, Alien (HK2, 1: 353); 1814, Alien (HKE: 48); 1884, unspec. (K). In cult. under glass.
CURRENT STATUS: No recent sightings.

1256. Legousia hybrida (L.) Delarbre
Campanula hybrida L.
Venus's-looking-glass
CULT.: 1768 (JH: 131); 1789, Nat. of England (HK1, 1: 224); 1810, Nat. of England (HK2, 1: 352); 1814, Nat. of England (HKE: 48). Not currently in cult.

1257. *Legousia speculum-veneris (L.) Chaix
Campanula speculum L.
large Venus's-looking-glass
FIRST RECORD: 1999 [1768].
REFERENCES: **Kew Gardens.** 1999, *Cope* (S35: 648): A small patch in long grass on a slope near the Lake (261).
CULT.: 1768 (JH: 131); 1789, Alien (HK1, 1: 224); 1810, Alien (HK2, 1: 352); 1814, Alien (HKE: 48). In cult. under glass.
CURRENT STATUS: Appeared in long grass near the south end of the Lake in 1999 and came up beds near King William's Temple in 2004. It has not persisted in either location.

1258. Wahlenbergia hederacea (L.) Rchb.
Campanula hederacea L.
ivy-leaved bellflower
CULT.: 1789, Nat. of England (HK1, 1: 225); 1810, Nat. of England (HK2, 1: 353); 1814, Nat. of England (HKE: 48). In cult. under glass.

1259. Phyteuma spicatum L.
spiked rampion
FIRST RECORD: 1969 [1789].
REFERENCES: **Kew Gardens.** 1969, *London Natural History Society* (FSSC: 74; S33: 749): Wild part of Kew Gardens. • 1983, *Burton* (FLA: 139; S33: 749): ... naturally increasing introduction in a wild part of Kew Gardens.
CULT.: 1789, Alien (HK1, 1: 226); 1810, Alien (HK2, 1: 355); 1814, Alien (HKE: 49); 1971, Alpine & Herbaceous Dept. (K). In cult.
CURRENT STATUS: No recent sightings; it has not been refound in the Conservation Area.

1260. Phyteuma orbiculare L.
round-headed rampion
CULT.: 1768 (JH: 128); 1789, Nat. of England (HK1, 1: 226); 1810, Nat. of England (HK2, 1: 354); 1814, Nat. of England (HKE: 49); 1882, unspec. (K); 1971, Alpine Dept. (K); 1987, plot 154.102 (K); 1987, plot 154.314 (K). In cult. under glass.

1261. Jasione montana L.
sheep's-bit
FIRST RECORD: 1873/4 [1768].
NICHOLSON: **1873/4 survey** (JB: 47 and in SFS: 432): **P.** Common in turf and shrubberies from Winter Garden to Pagoda. A white flowered variety appeared in 1873 by side of walk between Acer collections. • **1906 revision** (AS: 83): **A:** Common in turf and shrubberies from temperate house to pagoda. In 1873 albino form noted.
CULT.: 1768 (JH: 37, 79); 1789, Nat. of Britain (HK1, 3: 282); 1810, Nat. of Britain (HK2, 1: 343); 1814, Nat. of Britain (HKE: 47); 1971, Alpine Dept. (K). Not currently in cult.
CURRENT STATUS: No recent sightings; because of current mowing regimes there is no longer a suitable habitat for it within the Gardens.

Sheep's-bit (*Jasione montana*). Once common in turf but has been extinct for about a hundred years.

1262. Lobelia urens L.

heath lobelia

CULT.: 1789, Nat. of England (HK1, 3: 285); 1810, Nat. of England (HK2, 1: 359); 1814, Nat. of England (HKE: 49). Not currently in cult.

1263. Lobelia dortmanna L.

water lobelia

CULT.: 1768 (JH: 275); 1789, Nat. of Britain (HK1, 3: 283); 1810, Nat. of Britain (HK2, 1: 356); 1814, Nat. of Britain (HKE: 49); 1905, unspec. (K). Not currently in cult.

1263.1 Lobelia siphilitica L.

great lobelia

FIRST RECORD: 2008 [1768].
CULT.: 1768, (JH: 275); 1789, Alien (HK1, 3: 284); 1810, Alien (HK2, 1: 359); 1814, Alien (HKE: 49); 1881, unspec. (K); 1910, unspec. (K); 1934, unspec. (K). In cult.
CURRENT STATUS: Found as a weed in the Rhododendron Dell although it has never been cultivated there. The plant is probably referable to cv. 'Nana'.

1264. *Pratia angulata (G.Forst.) Hook.f.

lawn lobelia

FIRST RECORD: 1969 [1904].
REFERENCES: **Kew Gardens.** 1981, *Hastings* (FSSC: 81; LN61: 103; S33: 749): Lawn weed, Cambridge Cottage Garden. • 1981, *Hastings* (FSSC: 81): Lawn weed, Duke's Garden.
EXSICC.: **Kew Gardens.** Director's lawn; 14 vii 1969; *Lee* s.n. (K).
CULT.: 1904, unspec. (K); 1971, unspec. (K); 1976, unspec. (K); 1988, unspec. (K). In cult.
CURRENT STATUS: No recent sightings. The Director's lawn was recently renewed and the *Lobelia* was destroyed along with the old turf; it has not reappeared in the new turf but may yet do so.

1265. *Pratia pedunculata (R.Br.) Benth.

matted pratia

FIRST RECORD: 2004.
CURRENT STATUS: Several plants grew from seed in topsoil stored in the Paddock.

RUBIACEAE

1266. Sherardia arvensis L.

field madder

FIRST RECORD: 1873/4 [1768].
NICHOLSON: **1873/4 survey** (JB: 45): **B.** Here and there on most of the lawns, frequent on the one between porch of Orchid House and the Herb. Ground wall. **P.** Common about lake and elsewhere. • **1906 revision** (AS: 81): Here and there on most of the lawns. **A.** Common about lake and elsewhere.

Field madder (*Sherardia arvensis*). An easily overlooked species occasionally found in short turf, especially around the Herbarium.

CULT.: 1768 (JH: 120); 1789, Nat. of Britain (HK1, 1: 140); 1810, Nat. of Britain (HK2, 1: 234); 1814, Nat. of Britain (HKE: 32); 1907, unspec. (K). Not currently in cult.
CURRENT STATUS: Occasional in short grass, especially in the lawns around the Herbarium.

1267. Asperula cynanchica L.
squinancywort

CULT.: 1768 (JH: 120); 1789, Nat. of England (HK1, 1: 141); 1810, Nat. of England (HK2, 1: 235); 1814, Nat. of England (HKE: 32); 1905, unspec. (K); 1914, unspec. (K); 1919, unspec. (K); 1922, unspec. (K); 1929, unspec. (K). In cult.

1268. Galium boreale L.
Aparine borealis (L.) Hill
northern bedstraw

CULT.: 1768 (JH: 118); 1789, Nat. of Britain (HK1, 1: 145); 1810, Nat. of Britain (HK2, 1: 240); 1814, Nat. of Britain (HKE: 32). In cult.

Lady's bedstraw (*Galium verum*). A feature of much of Kew's grassland, but especially noticeable in the 'Ancient Meadow' where it is less frequently cut.

1269. Galium odoratum (L.) Scop.
Asperula odorata L.
woodruff

CULT.: 1768 (JH: 119); 1789, Nat. of Britain (HK1, 1: 140); 1810, Nat. of Britain (HK2, 1: 234); 1814, Nat. of Britain (HKE: 32); 1918, Herbaceous Ground (K); 1933, Arboretum (K); 1940, Herbaceous Ground (K). In cult.

1270. Galium uliginosum L.
fen bedstraw

CULT.: 1768 (JH: 118); 1789, Nat. of Britain (HK1, 1: 142); 1810, Nat. of Britain (HK2, 1: 239); 1814, Nat. of Britain (HKE: 32). Not currently in cult.

1271. Galium palustre L.
G. elongatum C.Presl
G. witheringii Sm.
common marsh-bedstraw

FIRST RECORD: Mid-19th cent. [1768].
NICHOLSON: **1873/4 survey** (JB: 45): **Strip.** A few plants of the typical form occur here and there by moat. • **1906 revision** (AS: 81): **Strip.** A few plants of the typical form occur here and there by ha-ha.
EXSICC.: **Kew Environs.** Near Kew, by the chain or ditch on the inner side of the towing path; *Newbould* in *Herb. Watson* (K).
CULT.: 1768 (JH: 118); 1789, Nat. of Britain (HK1, 1: 142); 1810, Nat. of Britain (HK2, 1: 236); 1814, Nat. of Britain (HKE: 32), Nat. of England (ibid.: Add. as *witheringii*). Not currently in cult.
CURRENT STATUS: No recent sightings. It is likely that Nicholson's 'typical form' was in fact subsp. *elongatum* (C.Presl) Arcang. which is generally commoner than subsp. *palustre*. Newbould's specimen, of unknown date but almost certainly prior to Nicholson's survey, is certainly subsp. *elongatum*.

1272. Galium verum L.
lady's bedstraw

FIRST RECORD: 1873/4 [1768].
NICHOLSON: **1873/4 survey** (JB: 45): Common in the dry open turf. • **1906 revision** (AS: 81): Common in the dry open turf.
EXSICC.: **Kew Gardens.** [Unlocalised]; erect herb in rough grass; 14 vii 1944; *Souster* 116 (K). **Kew Environs.** Ha-ha between Kew Road and Old Deer Park; in coarse grass, much dried; 1 vii 1959; *Ross-Craig & Sealy* s.n. (K). • Ditch by rugger ground, Lion Gate; *anon.* s.n. (K).
CULT.: 1768 (JH: 118); 1789, Nat. of Britain (HK1, 1: 143); 1810, Nat. of Britain (HK2, 1: 237); 1814, Nat. of Britain (HKE: 32); 1933, unspec. (K). In cult.
CURRENT STATUS: Widely scattered in short grass and once a conspicuous component of the lawn behind the Herbarium.

1273. Galium mollugo L.
G. erectum Huds.
hedge bedstraw
FIRST RECORD: 1873/4 [1768].
NICHOLSON: **1873/4 survey** (JB: 45): **Pal.** Several plants along top of wall near Brentford Gate. **P.** A few patches on the grassy slope facing Palace Grounds. **Q:** Here and there in the part slirting river. • **1906 revision** (AS: 81): **P.** Several plants along top of wall by Brentford Gate. **Q.** Here and there on portion nearest river.
CULT.: 1768 (JH: 118); 1789, Nat. of Britain (HK1, 1: 143); 1810, Nat. of Britain (HK2, 1: 238, and 237 as *erectum*); 1814, Nat. of Britain (HKE: 32, also as *erectum*); 1919, unspec. (K); 1971, Herb. & Alpine Dept. (K). Not currently in cult.
CURRENT STATUS: No recent sightings; it was last noted in 1983. No attempt has been made to distinguish between subspecies *mollugo* and *erectum* although they were listed separately in *Hortus Kewensis* ed. 2 and in the *Epitome*.

1274. Galium pumilum Murray
G. pusillum auct. non L.
slender bedstraw
CULT.: 1768 (JH: 118); 1789, Nat. of England (HK1, 1: 143); 1810, Nat. of England (HK2, 1: 237); 1814, Nat. of England (HKE: 32). Not currently in cult.

1275. Galium sterneri Ehrend.
limestone bedstraw
CULT.: In cult.

1276. Galium saxatile L.
G. montanum Huds.
heath bedstraw
FIRST RECORD: 1873/4 [1768].
NICHOLSON: **1873/4 survey** (JB: 45): Common. On the level piece of grass on either side of Syon Vista the flowers of this species are produced in such abundance as to give quite a colour to the turf, particularly at the end nearest the "Railway Gate." • **1906 revision** (AS: 81): Very common in the turf almost everywhere.
EXSICC.: **Kew Gardens.** Queen's Cottage Grounds; 15 vi 1914; *Flippance* s.n. (K).
CULT.: 1768 (JH: 118); 1789, Nat. of England (HK1, 1: 142); 1810, Nat. of Britain (HK2, 1: 237); 1814, Nat. of Britain (HKE: 32); 1933, unspec. (K). Not currently in cult.
CURRENT STATUS: Occasional; mostly near the Conservation Area and in the adjacent Old Deer Park.

1277. Galium aparine L.
Valantia aparine L.
cleavers
FIRST RECORD: 1873/4 [1789].
NICHOLSON: **1873/4 survey** (JB: 45): **B.** Several plants in the hedge separating Herbaceous from Private Grounds. **P** & **Q.** Not uncommon in open shrubberies. • **1906 revision** (AS: 81): **P** and **Q.** Not uncommon in open shrubberies.
EXSICC.: **Kew Gardens.** Thames-bank, Kew – Richmond; growing in *Petasites* zone near Isleworth Gate; 25 vi 1929, *Summerhayes & Turrill* s.n. (K). **Kew Environs.** Tow-path between Kew and Richmond; 18 vii 1948; *Souster* 879 (K).
CULT.: 1789, Nat. of Britain (HK1, 1: 145), Alien (ibid.: 3: 428 as *V. aparine*); 1810, Nat. of Britain (HK2, 1: 240); 1813, Nat. of Britain (HK2, 5: 435 as *V. aparine*); 1814, Nat. of Britain (HKE: 32, and 317 as *V. aparine*); 1934, Arboretum (K). Not currently in cult.
CURRENT STATUS: A common weed throughout the Gardens.

[**1278. Galium spurium** L.
false cleavers
CULT.: 1789, Nat. of England (HK1, 1: 143); 1814, Nat. of Scotland (HKE: Add.).
Note. Listed in error as native.]

1279. Galium tricornutum Dandy
G. tricorne Stokes
corn cleavers
CULT.: 1810, Nat. of England (HK2, 1: 237); 1814, Nat. of England (HKE: 32). Not currently in cult.
Note. Originally grown as a native but now considered to be an archaeophyte.

1280. Galium parisiense L.
G. anglicum Huds.
Aparine parisiensis (L.) Hill
wall bedstraw
CULT.: 1768 (JH: 118); 1789, Nat. of England (HK1, 1: 145); 1810, Nat. of England (HK2, 1: 240); 1814, Nat. of England (HKE: 32). Not currently in cult.

1281. Cruciata laevipes Opiz
Valantia cruciata L.
Galium cruciata (L.) Scop.
crosswort
FIRST RECORD: 1873/4 [1768].
NICHOLSON: **1873/4 survey** (JB: 45; SFS: 359): **Strip.** A patch more than a yard long by towing path, 30 or 40 yards on the Richmond side of Isleworth Gate. • **1906 revision** (AS: 81): **Strip.** A large patch on towing-path on the Richmond side of Isleworth Gate.
CULT.: 1768 (JH: 119); 1789, Nat. of England (HK1, 3: 428); 1810, Nat. of Britain (HK2, 1: 236); 1814, Nat. of Britain (HKE: 32). Not currently in cult.
CURRENT STATUS: No recent sightings.

1282. Rubia peregrina L.
wild madder
CULT.: 1768 (JH: 117); 1789, Nat. of England (HK1, 1: 146); 1810, Nat. of England (HK2, 1: 242); 1814, Nat. of England (HKE: 33). In cult.

CAPRIFOLIACEAE

1283. Sambucus nigra L.
S. laciniata Mill.
elder
FIRST RECORD: 1928 [1768].
REFERENCES: **Kew Gardens.** 1983/4, *Cope* (S31: 181): Seedling in new student vegetable plots behind Hanover House.
EXSICC.: **Kew Gardens.** Near the Tennis Courts; vi 1932; *Turrill* s.n. (K). • Isleworth Gate Path; situation of a dryish nature, soil gravelly; *Platanus acarifolia* & coarse grass growing in area; 9 vi 1948; *O.J.Ward* 28 (K). • Garden of the C.M.I. adjoining the Herbarium grounds; 11 vi 1958; *Ross-Craig* 1891 (K). **Kew Environs.** Bank of the Thames betw. Richmond & Kew Bridges; vii 1928; *C.A.Smith* 6079 (K). • Thames bank, Kew; top of bank, not normally flooded; 13 viii 1928; *Summerhayes & Turrill* s.n. (K). • Towpath, Kew; 20 vi 1944; *Souster* 102 (K).
CULT.: 1768 (JH: 455); 1789, Nat. of Britain (HK1, 1: 374); 1811, Nat. of Britain (HK2, 2: 170); 1814, Nat. of Britain (HKE: 81); 1880, Arboretum (K); 1883, unspec. (K); 1884, Arboretum (K); 1886, Nursery (K); 1897, unspec. (K); 1906, Arboretum (K); 1907, Arboretum (K); 1926, Arboretum (K); 1944, unspec. (K); 1959, Experimental Ground (K); 1985, unspec. (K). In cult.
CURRENT STATUS: Self-sown seedlings occur throughout the Gardens, but these are seldom allowed to mature.

1284. Sambucus ebulus L.
dwarf elder
CULT.: 1768 (JH: 455); 1789, Nat. of Britain (HK1, 1: 373); 1811, Nat. of Britain (HK2, 2: 170); 1814, Nat. of Britain (HKE: 81); 1909, unspec. (K); 1933, unspec. (K); 1975, unspec. (K); 1995, Arboretum (K). In cult.
Note. Originally grown as a native but now considered to be an archaeophyte.

1285. Viburnum opulus L.
guelder-rose
CULT.: 1768 (JH: 457); 1789, Nat. of Britain (HK1, 1: 372); 1811, Nat. of Britain (HK2, 2: 168); 1814, Nat. of Britain (HKE: 81); 1843–1853, Arboretum (K); 1880, Arboretum (K); 1882, Arboretum (K); 1883, Arboretum (K); 1900, Arboretum (K); 1904, Arboretum (K); 1905, Arboretum (K); 1907, Arboretum (K); 1978, plot 126 (K); ?1978, plot 414-02 (K); 1979, unspec. (K); 1992, unspec. (K); 1997, plot 109 (K); 1997, plot 181-01 (K); 1997, plot 182-02 (K); 1997, plot 182-03 (K); 1997, plot 182-07 (K); 1997, plot 182-08 (K); 1997, plot 310-02 (K); 1997, plot 333 (K); 1998, Arb. Nursery (K); 1999, plot 228-01 (K). In cult.

1286. Viburnum lantana L.
wayfaring-tree
CULT.: 1768 (JH: 457); 1789, Nat. of Britain (HK1, 1: 372); 1811, Nat. of Britain (HK2, 2: 168); 1814, Nat. of Britain (HKE: 81); 1843-1853, Arboretum (K); 1880, Arboretum (K); 1881, Arboretum (K); 1890, Arboretum (K); 1894, Arboretum (K); 1905, Arboretum (K); 1911, Arboretum (K); 1917, unspec. (K); 1937, unspec. (K); 1973, plot 116 (K); 1978, plot 126 (K); 1990, unspec. (K); 1991, unspec. (K); 1997, plot 110-04 (K); 1997, plot 122 (K); 1997, plot 126 (K); 1997, plot 181-01 (K); 1997, plot 182-05 (K); 1997, plot 182-07 (K); 1997, plot 310-03 (K). In cult.

1287. *Symphoricarpos albus (L.) S.F.Blake
snowberry
FIRST RECORD: 2002.
REFERENCES: **Kew Gardens.** 2002, *Cope* (S35: 647): Naturalised in the Conservation Area (310).
CULT.: In cult.
CURRENT STATUS: Naturalised in the Conservation Area.

1288. Linnaea borealis L.
twinflower
CULT.: 1789, Alien (HK1, 2: 358); 1812, Nat. of Scotland (HK2, 4: 51); 1814, Nat. of Scotland (HKE: 193). Not currently in cult.
Note. Not known as a British native until 1795.

1289. *Leycesteria formosa Wall.
Himalayan honeysuckle
FIRST RECORD: 2003 [1910].
REFERENCES: **Kew Environs.** 2003, *Kitchling* (LN83: 236): Thames path outside Kew Gardens.
CULT.: 1910, Arboretum (K); 1934, Arboretum (K); 1952, Arboretum (K). In cult.
CURRENT STATUS: Self-seeded along the river bank adjacent to Ferry Lane. The size of the plant suggests that it was considerably earlier than 2003 that it became established, though it doubtless escaped from within the Gardens.

[**1290. Lonicera xylosteum** L.
fly honeysuckle
CULT.: 1789, Alien (HK1, 1: 232); 1810, Nat. of England (HK2, 1: 379); 1814, Nat. of England (HKE: 52).
Note. Listed in error as native.]

Honeysuckle (*Lonicera periclymenum*). Probably native in the extreme southern end of the Gardens near Oxenhouse Gate although it has been widely cultivated.

1291. Lonicera periclymenum L.
honeysuckle
FIRST RECORD: 1999 [1768].
REFERENCES: **Kew Gardens.** 1999, *Cope* (S35: 647): Clambering over shrubs near the Oxenhouse Gate (327).
CULT.: 1768 (JH: 446); 1789, Nat. of Britain (HK1, 1: 231); 1810, Nat. of Britain (HK2, 1: 378); 1814, Nat. of Britain (HKE: 52); 1880, Arboretum (K); 1885, Arboretum (K); 1901, unspec. (K); 1957, Arboretum (K); 1968, Arboretum South (K). In cult.
CURRENT STATUS: This appears to be native between the Conservation Area and Oxenhouse Gate and has surprisingly been overlooked.

[**1292. Lonicera caprifolium** L.
perfoliate honeysuckle
CULT.: 1789, Alien (HK1, 1: 230); 1810, Nat. of England (HK2, 1: 377); 1814, Nat. of Britain (HKE: 52).
Note. Listed in error as native.]

ADOXACEAE

1293. Adoxa moschatellina L.
moschatel
CULT.: 1768 (JH: 119); 1789, Nat. of Britain (HK1, 2: 37); 1811, Nat. of Britain (HK2, 2: 425); 1814, Nat. of Britain (HKE: 118); 1921, unspec. (K). In cult.
Note. In the Woodland Garden, well established and spreading.

VALERIANACEAE

1294. Valerianella locusta (L.) Laterr.
Valeriana locusta L.
Valerianella olitoria (L.) Pollich
common cornsalad
FIRST RECORD: 1873/4 [1768].
NICHOLSON: **1873/4 survey** (JB: 45): **P.** Wall facing river. **Strip.** Near towing path. • **1906 revision** (AS: 81): **P.** Wall facing river. **Strip.** Near towing-path.
CULT.: 1768 (JH: 130); 1789, Nat. of Britain (HK1, 1: 53); 1810, Nat. of Britain (HK2, 1: 76); 1814, Nat. of Britain (HKE: 11). Not currently in cult.
CURRENT STATUS: Rather rare; it was recently found in the vegetable plots outside the Herbarium and again on the towpath, but these are the only sightings since 1981.

1295. Valerianella carinata Loisel.
keel-fruited cornsalad
FIRST RECORD: 1873/4.
NICHOLSON: **1873/4 survey** (JB: 45; SFS: 369): Here and there with [*V. locusta*], but not nearly so common. **Q.** A few plants near wall about 50 yards from Isleworth Gate. • **1906 revision** (AS: 81): Here and there with [*V. locusta*], but not so common.
CURRENT STATUS: No recent sightings.

1296. Valerianella rimosa Bastard
broad-fruited cornsalad
CULT.: 1934, unspec. (K). Not currently in cult.

1297. Valerianella dentata (L.) Pollich
Valeriana dentata L.
V. mixta L.
narrow-fruited cornsalad
CULT.: 1768 (JH: 130); 1810, Nat. of Britain (HK2, 1: 76); 1814, Nat. of Britain (HKE: 11). Not currently in cult.
Note. Not found in Britain until 1804, and considered to be an archaeophyte.

1298. Valeriana officinalis L.
V. officinalis var. *sambucifolia* H.C.Watson
common valerian
FIRST RECORD: 1873/4 [1768].
NICHOLSON: **1873/4 survey** (JB: 45): The typical form of this I have not seen within our present limits. The var. *sambucifolia* is common in the **Strip**. • **1906 revision** (AS: 81): [Var. *sambucifolia*] common in the **Strip** — the type form not noted as seen within our present limits.
EXSICC.: **Kew Environs.** Towing path above Kew; 22 vi 1899; *Fraser* s.n. (K).
CULT.: 1768 (JH: 130); 1789, Nat. of Britain (HK1, 1: 52); 1810, Nat. of Britain (HK2, 1: 74); 1814, Nat. of Britain (HKE: 11); 1933, Herb Garden (K); 1936, Herbaceous Ground (K); 1947, Herbarium Ground (K); 1951, Herbarium Ground (K); 1959, Order Beds (K); 1959, Aquatic Garden (K); 1986, plot 156-47 (K). In cult.
CURRENT STATUS: No recent sightings.

[1299. Valeriana pyrenaica L.
Pyrenean valerian
CULT.: 1789, Alien (HK1, 1: 53); 1810, Nat. of Scotland (HK2, 1: 75); 1814, Nat. of Scotland (HKE: 11).
Note. Listed in error as native.]

1300. Valeriana dioica L.
marsh valerian
CULT.: 1768 (JH: 130); 1789, Nat. of Britain (HK1, 1: 51); 1810, Nat. of Britain (HK2, 1: 73); 1814, Nat. of Britain (HKE: 11). Not currently in cult.

1301. *Centranthus ruber (L.) DC.
Valeriana rubra L.
red valerian
FIRST RECORD: 1873/4 [1768].
NICHOLSON: **1873/4 survey** (JB: 45; SFS:368): **Pal.** On wall between Brentford Gate and entrance to Kew Palace. • **1906 revision** (AS: 81): **P.** On wall between Brentford Gate and entrance to Kew Palace.
CULT.: 1768 (JH: 130); 1789, Alien (HK1, 1: 51); 1810, Alien (HK2, 1: 73); 1814, Nat. of England (HKE: 11); 1921, unspec. (K); 1940, Herbaceous Ground (K); 1959, Order Beds (K). In cult.
CURRENT STATUS: Naturalised in the Robinsonian Meadow in front of 'Climbers & Creepers' but has susbsequently rapidly declined. It has not been seen anywhere recently as a wild plant.

1302. *Centranthus calcitrapae (L.) Dufr.
annual valerian
Valeriana calcitrapae L.
FIRST RECORD: 1982 [1768].
REFERENCES: **Kew Gardens.** 1982, *Latham* (LN62: 107; S33: 749): Kew Gardens. **Kew Environs.** 1982, *Norman*; 1982, *Latham* (LN62: 107; S33: 749): Churchyard of St. Ann's, Kew. • 1985, *Latham* (BSN40: 14; S33: 749): It was well established on a wall of Kew churchyard in 1982 from wind-blown seed, an order bed in Kew Gardens being the obvious point of origin. But after weeding in that year it did not reappear at the churchyard until June 1985 when it was found in good quantity in an *Iris germanica* bed.
CULT.: 1768 (JH: 130); 1789, Alien (HK1, 1: 51); 1810, Alien (HK2, 1: 73); 1814, Alien (HKE: 11). Not currently in cult.
CURRENT STATUS: No recent sightings.

Red valerian (*Centranthus ruber*). Has persisted near 'Climbers & Creepers' where it was planted in a now-discontinued 'Robinsonian Meadow'.

Dipsacaceae

1303. Dipsacus fullonum L.
wild teasel
D. sylvestris Huds.
First record: 1863 [1768].
Nicholson: **1873/4 survey** (JB: 46): **Strip.** Two or three plants by towing path near Isleworth Gate. • **1906 revision** (AS: 81): **Strip.** By towing-path near Isleworth Gate.
Exsicc.: **Kew Environs.** Kew; waste ground, viii 1863; *Brocas* s.n. (K).
Cult.: 1768 (JH: 75); 1789, Nat. of Britain (HK1, 1: 132); 1810, Nat. of Britain (HK2, 1: 223); 1814, Nat. of Britain (HKE: 30); 1901, unspec. (K); 1933, unspec. (K). In cult.
Current status: Uncommon; mostly in the Riverside Walk and the Conservation Area. It is likely to have originally been planted in the bee garden behind Kew Palace but has now naturalised there. There was a sizeable population outside the Banks Building in 2008 but these were weeded out before setting seed.

1304. Dipsacus sativus (L.) Honck.
D. fullonum var. *sativus* L.
Fuller's teasel
Cult.: 1789, Nat. of Britain (HK1, 1: 132). Not currently in cult.

1305. *Dipsacus laciniatus L.
cut-leaved teasel
First record: 2003 [1768].
Cult.: 1768 (JH: 75); 1789, Alien (HK1, 1: 132); 1810, Alien (HK2, 1: 224); 1814, Alien (HKE: 30). Not currently in cult.
Current status: A single plant appeared in long grass (plot 161) in 2003. It was left to set seed when the grass around it was mown but has not yet returned.

1306. Dipsacus pilosus L.
small teasel
Virga pilosa (L.) Hill
Cult.: 1768 (JH: 75); 1789, Nat. of Britain (HK1, 1: 133); 1810, Nat. of Britain (HK2, 1: 224); 1814, Nat. of Britain (HKE: 30); 1922, unspec. (K). Not currently in cult.

1307. *Dipsacus strigosus Willd.
yellow-flowered teasel
First record: 1855.
References: **Kew Environs.** 1855, *Black* (BSP5: 123; S33: 749, and see Exsicc.). • 1872, *Trimen* (JB10: 268; BSEC4: 415; BSP5: 123; S33: 749): Two very large plants ... by the side of a ditch close to the Thames at Kew.
Exsicc.: **Kew Environs.** Riverside, Kew; viii 1855; *Black* s.n. (K).
Cult.: In cult. under glass.
Current status: No recent sightings.

1308. Knautia arvensis (L.) Coult.
field scabious
Scabiosa arvensis L.
First record: 2001 [1768].
References: **Kew Gardens.** 2001, *Cope* (S35: 647): Incorporated in the Robinsonian Meadow sown in front of the temporary Cycad House (224), and now naturalised.
Cult.: 1768 (JH: 80); 1789, Nat. of Britain (HK1, 1: 135); 1810, Nat. of Britain (HK2, 1: 226); 1814, Nat. of Britain (HKE: 31); 1935, Herbarium Ground (K); 1945, unspec. (K). In cult.
Current status: Incorporated in the Robinsonian Meadow in front of 'Climbers & Creepers' but unlikely to persist.

1309. Succisa pratensis Moench
Scabiosa succisa L.
devil's-bit scabious
Cult.: 1768 (JH: 80); 1789, Nat. of Britain (HK1, 1: 134); 1810, Nat. of Britain (HK2, 1: 226); 1814, Nat. of Britain (HKE: 31); 1945, unspec. (K); 1960, Order Beds (K). In cult.

Cut-leaved teasel (*Dipsacus laciniatus*). A species not currently in cultivation that appeared spontaneously in long grass in 2003. The plant was left to set seed when the grass around it was cut, but it has not yet returned.

1310. Scabiosa columbaria L.
small scabious
FIRST RECORD: 1999 [1768].
REFERENCES: **Kew Gardens.** 1999, *Cope* (S35: 647): In rough grass near the Oxenhouse Gate (327).
CULT.: 1768 (JH: 80); 1789, Nat. of Britain (HK1, 1: 136); 1810, Nat. of Britain (HK2, 1: 227); 1814, Nat. of Britain (HKE: 31); 1883, unspec. (K); 1930, unspec. (K); 1931, unspec. (K); 1932, unspec. (K); 1945, unspec. (K); 1972, plot 156-48 (K); 1977, unspec. (K); 1984, unspec. (K). In cult.
CURRENT STATUS: Apparently native in rough grass near the Oxenhouse Gate.

ASTERACEAE

1311. Carlina vulgaris L.
carline thistle
CULT.: 1768 (JH: 58); 1789, Nat. of Britain (HK1, 3: 149); 1812, Nat. of Britain (HK2, 4: 489); 1814, Nat. of Britain (HKE: 252); 1900, unspec. (K); 1917, unspec. (K); 1923, unspec. (K). Not currently in cult.

1312. Arctium lappa L.
A. majus Bernh.
Bardana arctium Hill
greater burdock
FIRST RECORD: 1864 [1768].
NICHOLSON: **1873/4 survey** (JB: 46 and in SFS: 397): **B.** A plant in shrubbery near Orchid House. **P.** Two or three in wood near "Hollow Walk." • **1906 revision** (AS: 82): **P, A.** A few plants in wood near hollow walk.
EXSICC.: **Kew Environs.** Kew; vii 1864; *Syme* in *Herb. Watson* (K). • Bank of Thames, Kew; 16 x 1879; *G.Nicholson* s.n. (K). • Kew; 5 viii 1898; *G. Nicholson* 672 (K).
CULT.: 1768 (JH: 64); 1789, Nat. of Britain (HK1, 3: 136); 1812, Nat. of Britain (HK2, 4: 471); 1814, Nat. of Britain (HKE: 250); 1898, unspec. (K); 1933, Arboretum (K); 1936, Herbaceous Ground 1956 (K); Herbarium Experimental Ground (K); 1971, plot 157-52 (K). In cult.
CURRENT STATUS: Rare in the Gardens but much more common on the towpath.

1313. Arctium minus (Hill) Bernh.
A. bardana Willd.
Bardana minor Hill
lesser burdock
FIRST RECORD: 1906 [1768].
NICHOLSON: **1906 revision** (AS: 82): **Q.** Common. **Strip.** Common by towing-path.
EXSICC.: **Kew Environs.** Thames bank, Kew; top of bank, not normally flooded; 13 viii 1928; *Summerhayes & Turrill* s.n. (K). • Towpath, Kew; 9 viii 1956; *Meikle* s.n. (K).
CULT.: 1768 (JH: 64); 1812, Nat. of Britain (HK2, 4: 472); 1814, Nat. of Britain (HKE: 250). In cult.
CURRENT STATUS: Widely scattered but not common; plentiful on the towpath.

1314. Saussurea alpina (L.) DC.
Serratula alpina L.
alpine saw-wort
CULT.: 1789, Nat. of Britain (HK1, 3: 137); 1812, Nat. of Britain (HK2, 4: 472); 1814, Nat. of Britain (HKE: 250). Not currently in cult.

1315. Carduus tenuiflorus Curtis
slender thistle
FIRST RECORD: 1976 [1812].
REFERENCES: **Kew Gardens.** 1983, *Burton* (FLA: 156; S33: 749): ... reported ... near the natural order beds in Kew Gardens, from which it has no doubt spread by seed, in 1976.
CULT.: 1812, Nat. of Britain (HK2, 4: 475); 1814, Nat. of Britain (HKE: 250); 1891, unspec. (K). Not currently in cult.
CURRENT STATUS: No recent sightings.

1316. Carduus crispus L.
C. acanthoides auct. non L.
Polycantha crispus (L.) Hill
welted thistle
FIRST RECORD: 1873/4 [1768].
NICHOLSON: **1873/4 survey** (JB: 46): **Pal.** A few plants growing with *C. nutans*. **P.** Here and there on border of wood on the Kew side of Syon Vista. • **1906 revision** (AS: 83): **P, A.** Here and there on borders of wood. **Q.**
EXSICC.: **Kew Gardens.** River-bank, Kew – Richmond; Richmond side of Isleworth Gate, top zone with *Petasites*; 16 viii 1928; *Summerhayes & Turrill* s.n. (K).
CULT.: 1768 (JH: 59); 1789, Nat. of Britain (HK1, 3: 140, and: 139 as acanthoides); 1812, Nat. of Britain (HK2, 4: 474); 1814, Nat. of Britain (HKE: 250);1932, unspec. (K); 1987, plot 157.52 (K). Not currently in cult.
CURRENT STATUS: Rare; there are only a few recent sightings, in the Paddock, in the Ancient Meadow (where it increased rapidly in 2004 after removal of some shrubs) and on the edge of the golf course in the Old Deer Park.

1317. Carduus nutans L.
Ascalea nutans (L.) Hill
musk thistle
FIRST RECORD: 1873/4 [1768].
NICHOLSON: **1873/4 survey** (JB: 46 and in SFS: 400): **Pal.** About 20 plants in young plantation facing river. • **1906 revision** (AS: 83): **P.** In young plantations facing river. **Strip.**
CULT.: 1768 (JH: 60); 1789, Nat. of Britain (HK1, 3: 139); 1812, Nat. of Britain (HK2, 4: 474); 1814, Nat.

of Britain (HKE: 250); 1949, Herbarium Ground (K). Not currently in cult.
CURRENT STATUS: Rare; only one recent sighting along the boundary adjacent to the Order Beds from which it probably escaped. It has gone from all its other locations.

1318. Cirsium eriophorum (L.) Scop.
woolly thistle
Cnicus eriophorus (L.) Roth
CULT.: 1789, Nat. of Britain (HK1, 3: 143); 1812, Nat. of Britain (HK2, 4: 480); 1814, Nat. of Britain (HKE: 251); 1900, unspec. (K); 1935, unspec. (K). Not currently in cult.

1319. Cirsium vulgare (Savi) Ten.
Carduus lanceolatus L.
Cnicus lanceolatus (L.) Willd.
Ascalea lanceata Hill
spear thistle
FIRST RECORD: 1873/4 [1768].
NICHOLSON: **1873/4 survey** (JB: 46): **Pal.** Not common. **P.** About 100 plants [on border of wood on the Kew side of Syon Vista]. • **1906 revision** (AS: 83): Not common. A few plants in each division.
EXSICC.: **Kew Gardens.** Near Kew Palace; waste ground; 29 iii 1931; *Sprague* s.n. (K). **Kew Environs.** Kew; 9 vi 1931; *Fraser* s.n. (K). • Kew; spontaneous; 4 x 1933; *Fraser* s.n. (K).

Spear thistle (*Cirsium vulgare*). Widespread and common in grassland and flower-beds.

CULT.: 1768 (JH: 60); 1789, Nat. of Britain (HK1, 3: 139); 1812, Nat. of Britain (HK2, 4: 479); 1814, Nat. of Britain (HKE: 251); 1922, unspec. (K); 1927, unspec. (K); 1929, Herb. Expt. Ground (K); 1933, unspec. (K); 1934, unspec. (K). Not currently in cult.
CURRENT STATUS: Widespread and common in grassland and as a weed in flower-beds.

1320. Cirsium dissectum (L.) Hill
Carduus dissectus L.
Cnicus pratensis (Huds.) Willd.
meadow thistle
CULT.: 1768 (JH: 63); 1789, Nat. of England (HK1, 3: 140); 1812, Nat. of Britain (HK2, 4: 481); 1814, Nat. of Britain (HKE: 251). Not currently in cult.

1321. Cirsium tuberosum (L.) All.
Cnicus tuberosus (L.) Roth
Ixine tuberosa (L.) Hill
tuberous thistle
CULT.: 1768 (JH: 61); 1812, Nat. of England (HK2, 4: 483); 1814, Nat. of England (HKE: 251). Not currently in cult.

1322. Cirsium heterophyllum (L.) Hill
Carduus helenioides L.
Cnicus heterophyllus (L.) Retz.
melancholy thistle
CULT.: 1768 (JH: 64); 1789, Nat. of Britain (HK1, 3: 143); 1812, Nat. of Britain (HK2, 4: 481); 1814, Nat. of Britain (HKE: 251); 1920, unspec. (K); 1923, unspec. (K); 1925, unspec. (K). Not currently in cult.

1323. Cirsium acaule (L.) Scop.
Carduus acaulis L.
Cnicus acaulis (L.) Willd.
dwarf thistle
CULT.: 1789, Nat. of Britain (HK1, 3: 144); 1812, Nat. of Britain (HK2, 4: 483); 1814, Nat. of Britain (HKE: 252); 1934, unspec. (K). Not currently in cult.

1324. Cirsium palustre (L.) Scop.
Carduus palustris L.
Cnicus palustris (L.) Willd.
Polycantha palustris (L.) Hill
marsh thistle
FIRST RECORD: 1873/4 [1768].
NICHOLSON: **1873/4 survey** (JB: 46): **Q.** Two large plants [in a clump of trees opposite Syon house]. • **1906 revision** (AS: 83): Here and there along borders of shrubberies in Queen's Cottage Grounds.
CULT.: 1768 (JH: 59); 1789, Nat. of Britain (HK1, 3: 140); 1812, Nat. of Britain (HK2, 4: 478); 1814, Nat. of Britain (HKE: 251); 1922, unspec. (K). Not currently in cult.
CURRENT STATUS: No recent sightings; it was last seen on the towpath in 1979.

1325. Cirsium arvense (L.) Scop.
Serratula arvensis L.
Carduus arvensis (L.) Hill
Cnicus arvensis (L.) Roth
creeping thistle
FIRST RECORD: 1873/4 [1789].
NICHOLSON: **1873/4 survey** (JB: 46): **P.** A few plants near lake. **Q.** Common in the turf of the open "ride" nearest river. • **1906 revision** (AS: 83): **Q.** Common in the turf of the open vista nearest river.
EXSICC.: **Kew Gardens.** Stable Yard; 24 ix 1978; *Roberts* 13 (K). **Kew Environs.** Thames-bank, Kew; viii 1888; *Gamble* 20216 (K). • Kew; waste ground; 1900; *Hulme* s.n. (K).
CULT.: 1789, Nat. of Britain (HK1, 3: 139); 1812, Nat. of Britain (HK2, 4: 473); 1814, Nat. of Britain (HKE: 250). Not currently in cult.
CURRENT STATUS: A widespread and persistent weed of both grassland and flower-beds.

1326. *Galactites tomentosa Moench
Centaurea galactites L.
Cirsium galactites (L.) Hill
FIRST RECORD: 1992 [1768].
REFERENCES: **Kew Environs.** 1992, *Mullin* (LN72: 116; S33: 749): Abundant in Kew Churchyard. Disappeared by July (*fide* E.Norman).
CULT.: 1768 (JH: 64); 1789, Alien (HK1, 3: 266); 1813, Alien (HK2, 5: 161); 1814, Alien (HKE: 276); 1882, unspec. (K); 1987, Herb. Beds (K). Not currently in cult.
CURRENT STATUS: No recent sightings.

1327. Onopordum acanthium L.
cotton thistle
FIRST RECORD: 1990 [1768].
REFERENCES: **Kew Gardens.** 1990, *Cope* (S31: 182): Paddock, after restoration in 1990. **Kew Environs.** pre-2000, *Bartlett* (LN76: 196): Richmond Golf Course.
CULT.: 1768 (JH: 60); 1789, Nat. of Britain (HK1, 3: 146); 1812, Nat. of Britain (HK2, 4: 485); 1814, Nat. of Britain (HKE: 252); 1895, unspec. (K); 1933, Herb Garden (K); 1948, Herbarium Ground (K); 1962, unspec. (K); 1971, unspec. (K). In cult.
CURRENT STATUS: Sporadic and very unpredictable.

1328. Silybum marianum (L.) Gaertn.
Carduus marianus L.
Mariana mariana (L.) Hill, nom. illegit.
milk thistle
FIRST RECORD: 1927 [1768].
REFERENCES: **Kew Gardens.** 1990, *Cope* (S31: 182): Paddock, after restoration in 1990.
EXSICC.: **Kew Gardens.** Weed near Herbarium; 1 vii 1927; *Summerhayes* s.n. (K).
CULT.: 1768 (JH: 61); 1789, Nat. of Britain (HK1, 3: 142); 1812, Nat. of Britain (HK2, 4: 477); 1814, Nat. of Britain (HKE: 250); 1922, unspec. (K); 1924, unspec. (K); 1934, Arboretum (K). In cult.
CURRENT STATUS: It has not reappeared in the Paddock since 1990 but has been seen more recently by the Palm House Pond.

1329. Serratula tinctoria L.
saw-wort
CULT.: 1768 (JH: 68); 1789, Nat. of Britain (HK1, 3: 136); 1812, Nat. of Britain (HK2, 4: 472); 1814, Nat. of Britain (HKE: 250); 1972, plot 157-52 (K); 1974, plot 157-53 (K). In cult.

1330. Centaurea scabiosa L.
Sagmen scabiosa (L.) Hill
greater knapweed
FIRST RECORD: 1873/4 [1768].
NICHOLSON: **1873/4 survey** (JB: 46 and in SFS: 409): **Pal.** Here and there. **P.** Frequent about Winter Garden. **Strip.** A few plants. • **1906 revision** (AS: 82): **P.** Here and there. **Strip.** A few plants.
OTHER REFERENCES: **Kew Environs.** 1920, *Tremayne* (LN32: S166): By River Thames between Kew and Richmond.
CULT.: 1768 (JH: 66); 1789, Nat. of Britain (HK1, 3: 261); 1813, Nat. of Britain (HK2, 5: 150); 1814, Nat. of Britain (HKE: 276); 1903, unspec. (K); 1933, unspec. (K); 1934, unspec. (K); 1979, plot 157-53 (K). In cult.
CURRENT STATUS: Naturalised in the Robinsonian Meadow in front of 'Climbers & Creepers' where it still persists. Not recently seen anywhere as a wild plant.

1331. Centaurea cyanus L.
Cyanus cyanus (L.) Hill, nom. illegit.
cornflower
FIRST RECORD: 2000 [1768].
REFERENCES: **Kew Gardens.** 2000, *Cope* (S35: 647): Sown into the Millenium wheatfield display along the Broadwalk (124, 127, 163, 164) in 2000, and where it persisted for a year or two. Incorporated in the new wheatfield sown in 2002 at the south end of the Lake (261).
CULT.: 1768 (JH: 64); 1789, Nat. of Britain (HK1, 3: 258); 1813, Nat. of Britain (HK2, 5: 147); 1814, Nat. of Britain (HKE: 275); 1933, Arboretum (K); 1939, Herbarium Ground (K); 1970, Herbaceous Dept. (K); 1971, Herb. & Alpine Dept. (K). In cult.
CURRENT STATUS: Sown into the Wheatfields along the Broadwalk in 1999, but these have now been grassed over. It was later (2001) sown into the new Wheatfield in Syon Vista to the south of the Lake and appeared spontaneously in a bed near King William's Temple. It is occasionally found as an escape derived from these sources.

1332. Centaurea calcitrapa L.
Calcitrapa calcitrapa (L.) Hill, nom. illegit.
red star-thistle
FIRST RECORD: 1950 [1768].

Cornflower (*Centaurea cyanus*). A rare British species that has been sown on several occasions into 'wheatfield' displays and sporadically returns as a relict.

REFERENCES: **Kew Environs.** 1950, *Russell & Welch* (LFS: 324; LN30: 6; LN32: S167; S33: 749): Towpath by Kew Gardens, a single plant.
CULT.: 1768 (JH: 62); 1789, Nat. of England (HK1, 3: 264); 1813, Nat. of England (HK2, 5: 158); 1814, Nat. of England (HKE: 276); 1931, unspec. (K); 1970, Herbaceous Dept. (K). In cult.
CURRENT STATUS: No recent sightings. A rare and declining native species in Britain.

[1333. **Centaurea solstitialis** L.
yellow star-thistle
CULT.: 1789, Nat. of England (HK1, 3: 265); 1813, Nat. of England (HK2, 5: 155); 1814, Nat. of England (HKE: 275).
Note. Listed in error as native.]

1334. **Centaurea nigra** L.
common knapweed
FIRST RECORD: 1873/4 [1789].
NICHOLSON: **1873/4 survey** (JB: 46): In all the divisions. Most common in **P** and **Q.** • **1906 revision** (AS: 82): In all the divisions. Most common in **P** and **Q.**
CULT.: 1789, Nat. of Britain (HK1, 3: 258); 1813, Nat. of Britain (HK2, 5: 146); 1814, Nat. of Britain (HKE: 276); 1913, unspec. (K); 1918, unspec. (K); 1923, unspec. (K); 1933, unspec. (K). In cult.
CURRENT STATUS: Widespread in the less frequently cut grass and lightly wooded areas.

1335. ***Centaurea jacea** L.
C. angustifolia auct. non Mill., nec Schrank
Behen jacea (L.) Hill
brown knapweed
FIRST RECORD: 1870 [1768].
NICHOLSON: **1873/4 survey** (JB: 46): **P** and **Q.** A few plants near lake and an equal number in the [turf of the open "ride" nearest river] have held their own for several years. • **1906 revision** (AS: 82 and in SFS: 407): A few plants near lake and in Queen's Cottage Grounds.
OTHER REFERENCES: **Kew Gardens.** 1870, *Baker* (SFS: 408): Three plants, among *C. nigra*, Yarrow and Broom, in Kew pleasure-ground, near the lake (see also Exsicc.).
EXSICC.: **Kew Gardens.** Pleasure ground; 7 viii 1870; *Baker* s.n. (K). • [Unlocalised]; waste ground; 1927; *Knowles* s.n. (K).
CULT.: 1768 (JH: 68); 1789, Alien (HK1, 3: 261); 1813, Nat. of England (HK2, 5: 152); 1814, Nat. of England (HKE: 276); 1896, unspec. (K); 1931, unspec. (K). In cult.
CURRENT STATUS: No recent sightings.

1336. **Cichorium intybus** L.
chicory
FIRST RECORD: 1873/4 [1768].
NICHOLSON: **1873/4 survey** (JB: 47): **Strip.** Two or three plants between Isleworth Gate and the second seat from there. • **1906 revision** (AS: 83): **Strip.** A few plants along the gravelly sides of towing-path.
EXSICC.: **Kew Gardens.** Near Herbarium; amongst grasses; 8 viii 1933; *Hubbard* s.n. (K). • Near Tennis Courts; vii 1942; *Hutchinson* s.n. (K).
CULT.: 1768 (JH: 44); 1789, Nat. of Britain (HK1, 3: 135); 1812, Nat. of Britain (HK2, 4: 469); 1814, Nat. of Britain (HKE: 250); 1905, unspec. (K); 1933, unspec. (K); 1936, Herbaceous Ground (K); 1951, Herbarium Ground (K); 1972, plot 157-52 (K). In cult.
CURRENT STATUS: Rare; only two recent sightings within the Gardens. It has not been seen for many years along the towpath.

1337. **Arnoseris minima** (L.) Schweigg. & Körte
Hyoseris minima L.
Lapsana pusilla (Gaertn.) Willd.
lamb's succory
FIRST RECORD: 1985 [1768].
REFERENCES: **Kew Gardens.** 1985, *London Natural History Society* (LN65: 195; S33: 749): Around what is now the Princess of Wales Conservatory.

Chicory (*Cichorium intybus*). Only two recent sightings in the Gardens. It has disappeared from the towpath from where it was first recorded in 1950.

CULT.: 1768 (JH: 47); 1789, Nat. of Britain (HK1, 3: 131); 1812, Nat. of Britain (HK2, 4: 467); 1814, Nat. of Britain (HKE: 249). Not currently in cult.
CURRENT STATUS: No recent sightings. The species is now extinct in the wild in this country. The source of material cultivated between 1768 and 1815 is not known; if it had been sourced from within the British Isles its loss from cultivation is regrettable.

1338. Lapsana communis L.
nipplewort
FIRST RECORD: 1873/4 [1789].
NICHOLSON: **1873/4 survey** (JB: 47): **B.** A flower-bed weed. **Pal** and **P.** Here and there in shrubberies. • **1906 revision** (AS: 83): **B.** A flower-bed weed. **P, A, Q.** Here and there in shrubberies.
EXSICC.: **Kew Gardens.** [Unlocalised]; weed in cultivated ground; 2 x 1942; *Hutchinson* s.n. (K). • [Unlocalised]; common weed in cultivation; c. 1942; *Hutchinson* s.n. (K). • Naturalised between Herbarium and Experimental ground; by side of path; flowers lemon-yellow; note: this variant from Wiltshire was cult. in Expt. gr. in 1947; 16 viii 1948; *Hubbard* 13188 (K). • Mound of earth from excavation for foundations of new wing; 14 x 1965; *Verdcourt* 4241 (K). • Herbarium paddock; on earth tip; 21 vii 1966; *Jeffrey* s.n. (K). • Back of student plots; rough woodland; ix 1975; *Butler* 18 (K).
CULT.: 1789, Nat. of Britain (HK1, 3: 133); 1812, Nat. of Britain (HK2, 4: 467); 1814, Nat. of Britain (HKE: 249); 1933, unspec. (K); 1934, unspec. (K); 1947, Herbarium Ground (K). Not currently in cult.
CURRENT STATUS: Widespread and common throughout the Gardens.

1339. Hypochaeris radicata L.
cat's-ear
FIRST RECORD: 1873/4 [1768].
NICHOLSON: **1873/4 survey** (JB: 47): Everywhere. An extremely troublesome weed. • **1906 revision** (AS: 83): Everywhere. A very troublesome weed.
EXSICC.: **Kew Gardens.** Herb. Expt. Ground; weed; 27 vi 1927; *Turrill* s.n. (K). • Near Herbarium; among short grasses, in sandy soil; 20 vi 1928; *Hubbard* s.n. (K).
CULT.: 1768 (JH: 43); 1789, Nat. of Britain (HK1, 3: 133); 1812, Nat. of Britain (HK2, 4: 467); 1814, Nat. of Britain (HKE: 249); 1970, Herbaceous Dept. (K); 1972, plot 157-48 (K). Not currently in cult.
CURRENT STATUS: Abundant in grassland throughout the Gardens.

1340. Hypochaeris glabra L.
smooth cat's-ear
CULT.: 1789, Nat. of Britain (HK1, 3: 132); 1812, Nat. of Britain (HK2, 4: 467); 1814, Nat. of Britain (HKE: 249); 1929, unspec. (K). Not currently in cult.

1341. Hypochaeris maculata L.
spotted cat's-ear
CULT.: 1768 (JH: 43); 1789, Nat. of England (HK1, 3: 132); 1812, Nat. of England (HK2, 4: 466); 1814, Nat. of England (HKE: 249). Not currently in cult.

1342. Leontodon autumnalis L.
Apargia autumnalis (L.) Hoffm.
autumn hawkbit
FIRST RECORD: 1873/4 [1768].
NICHOLSON: **1873/4 survey** (JB: 47): **P** and **Q.** Very sparingly. • **1906 revision** (AS: 83): **P** and **Q.** Sparingly.
EXSICC.: **Kew Environs.** Thames banks, Kew; 31 viii 1884; *Fraser* s.n. (K).
CULT.: 1768 (JH: 45); 1789, Nat. of Britain (HK1, 3: 120); 1812, Nat. of Britain (HK2, 4: 446); 1814, Nat. of Britain (HKE: 246); 1895, unspec. (K); 1922, unspec. (K); 1934, unspec. (K); 1972, plot 157-48 (K). In cult.
CURRENT STATUS: Widespread but still rather scarce.

1343. Leontodon hispidus L.
Apargia hispida (L.) Willd.
rough hawkbit
FIRST RECORD: 1873/4 [1768].
NICHOLSON: **1873/4 survey** (JB: 47): **P.** Two plants in the turf near south end of Temperate House. • **1906 revision** (AS:83): **A.** In turf near south end of temperate house. **Q.**
EXSICC.: **Kew Environs.** Kew; 1787; *Goodenough* s.n. (K).
CULT.: 1768 (JH: 45); 1789, Nat. of Britain (HK1, 3: 120); 1812, Nat. of Britain (HK2, 4: 446); 1814, Nat. of Britain (HKE: 246); 1895, unspec. (K); 1898, unspec. (K); 1932, unspec. (K); 1933, unspec. (K); 1934, unspec. (K); 1937, Herbarium Ground (K); 1961, Order Beds (K); 1971, Herb. & Alpine Dept. (K); 1971, plot 157-47 (K). Not currently in cult.
CURRENT STATUS: Rare and sporadic over a wide area between Kew Green and the Temperate House.

1344. Leontodon saxatilis Lam.
L. hirtus auct. non L.
Thrincia hirta Roth
Apargia taraxaci (L.) Willd.
Leontodon taraxacoides (Vill.) Mérat, nom. illegit.
lesser hawkbit
FIRST RECORD: 1873/4 [1768].
NICHOLSON: **1873/4 survey** (JB: 47 and in SFS: 423): Common, particularly on the walks and in the dryer parts of districts **P** and **Q.** • **1906 revision** (AS: 83): Common, particularly in drier parts of divisions **P.** and **Q.**
EXSICC.: **Kew Gardens.** Near herbarium; in sandy soil among short grasses; 20 vi 1928; *Hubbard* s.n. (K).
CULT.: 1768 (JH: 45); 1812, Nat. of Britain (HK2, 4: 445 as *A. taraxaci* and 447 as *T. hirta*); 1814, Nat. of Britain (HKE: 246). Not currently in cult.
CURRENT STATUS: Fairly common in short grass in most parts of the Gardens.

1345. Picris echioides L.
Helmintha echioides (L.) Gaertn.
bristly oxtongue
FIRST RECORD: 1863 [1768].
REFERENCES: **Kew Gardens.** 1998, *Cope* (S33: 749): Herbaceous: Duke's Garden (135); North Arb.: near Princess Walk (161), behind Water Lily House (163); Ferry Lane, roadside verge between Banks Building and Kew Palace entrance.
EXSICC.: **Kew Environs.** Kew; waste ground near a brick kiln; viii 1863; *Brocas* s.n. (K).
CULT.: 1768 (JH: 46); 1789, Nat. of England (HK1, 3: 114); 1812, Nat. of Britain (HK2, 4: 461); 1814, Nat. of Britain (HKE: 248); 1970, Herbaceous Dept. (K). Not currently in cult.
CURRENT STATUS: Widely scattered in rough grassland and apparently increasing.

1346. Picris hieracioides L.
hawkweed oxtongue
FIRST RECORD: 1998 [1789].
REFERENCES: **Kew Gardens.** 1998, *Cope* (S33: 750): North Arb.: Paddock (104); Herbaceous: under trees by wall along Kew Road (158); West Arb.: Riverside Walk.
CULT.: 1789, Nat. of England (HK1, 3: 114); 1812, Nat. of England (HK2, 4: 447); 1814, Nat. of England (HKE: 246); 1935, Herbarium Ground (K); 1970, Herbaceous Dept. (K). In cult.
CURRENT STATUS: Scarce but apparently increasing.

1347. Scorzonera humilis L.
viper's-grass
CULT.: 1789, Nat. of Scotland (HK1, 3: 112); 1812, Alien (HK2, 4: 434); 1814, Alien (HKE: 244). Not currently in cult.
Note. This is a very rare native that has only ever been recorded from Dorset, Glamorgan and Warwickshire. The entries in *Hortus Kewensis* are therefore hard to understand.

1348. Tragopogon pratensis L.
goat's-beard
a. subsp. **pratensis**
CULT.: 1768 (JH: 41). Not currently in cult.

b. subsp. **minor** (Mill.) Wahlenb.
T. pratensis var. *minor* Billot
FIRST RECORD: 1873/4 [1789].
NICHOLSON: **1873/4 survey** (JB: 47): **Q** and **Strip.** Seems to be the var. *minor* and is not uncommon. • **1906 revision** (AS: 83): **Q** and **Strip.** Not uncommon.
EXSICC.: **Kew Gardens.** Queen's Cottage Grounds; amongst grasses; 5 vi 1933; *Hubbard* s.n. (K). **Kew Environs.** Thames bank, Kew; 10 vi 1888; *Fraser* s.n. (K).
CULT.: 1789, Nat. of Britain (HK1, 3: 110); 1812, Nat. of Britain (HK2, 4: 432); 1814, Nat. of Britain (HKE: 244). In cult. under glass.
CURRENT STATUS: Widely scattered and becoming common in places, especially around the Herbarium.

[**1349. Tragopogon porrifolius** L.
salsify
CULT.: 1789, Nat. of England (HK1, 3: 111); 1812, Nat. of England (HK2, 4: 432); 1814, Nat. of England (HKE: 244).
Note. Listed in error as native.]

1350. *Tragopogon hybridus L.
T. glaber Hill
slender salsify
FIRST RECORD: 1964 [1768].
REFERENCES: **Kew Gardens.** 1964, *Taylor* (S29: 408): In garden of 53 Kew Green. Probably from 'Swoop' seed, with *Phalaris, Linum*, etc.
CULT.: 1768 (JH: 41). Not currently in cult.
CURRENT STATUS: No recent sightings.

1351. Sonchus palustris L.
marsh sowthistle
FIRST RECORD: 1873/4 [1768].
NICHOLSON: **1873/4 survey** (JB: 47): {A few plants on the smallest island in lake. Planted 1873}. • **1906 revision**: Omitted.
EXSICC.: **Kew Gardens.** By Lake in Pleasure Grounds, nr. N. end; ix 1874; *anon.* s.n. (K).
CULT.: 1768 (JH: 42); 1789, Nat. of England (HK1, 3: 114); 1812, Nat. of England (HK2, 4: 437); 1814, Nat. of England (HKE: 245). Not currently in cult.
CURRENT STATUS: No recent sightings. One of the species mentioned by Nicholson as having been planted and occurring in a half-wild condition, but which he did not expect to survive.

1352. Sonchus arvensis L.
perennial sowthistle
FIRST RECORD: 1873/4 [1768].
NICHOLSON: **1873/4 survey** (JB: 47): **B.** Plentiful in shrubbery behind Museum No. 3. **P.** In beds near lake. A few plants grow out of wall facing river. • **1906 revision** (AS: 83): **A, B** and **P.** Here and there in shrubberies. A few plants grow out of wall facing river.
EXSICC.: **Kew Gardens.** Riverside Nursery; weed; 11 ix 1944; *Melville* s.n. (K).

Prickly sowthistle (*Sonchus asper*). Widespread and abundant in grassland, flower-beds and other disturbed places.

CULT.: 1768 (JH: 42); 1789, Nat. of Britain (HK1, 3: 115); 1812, Nat. of Britain (HK2, 4: 437); 1814, Nat. of Britain (HKE: 245); 1898, unspec. (K); 1932, unspec. (K). Not currently in cult.
CURRENT STATUS: Widely scattered but rather uncommon; it usually grows in neglected areas where it temporarily escapes weeding.

1353. Sonchus oleraceus L.
S. lacerus Willd.
smooth sowthistle
FIRST RECORD: 1873/4 [1789].
NICHOLSON: **1873/4 survey** (JB: 47): Fairly common everywhere, though not so frequent as [*Sonchus asper*]. • **1906 revision** (AS: 83): Fairly common everywhere, though not so frequent as *S. asper.*
OTHER REFERENCES: **Kew Environs.** 1877, *Baker* (BEC18: 8; BSEC9: 689): Kew, Surrey. An interesting form of *S. oleraceus.* – C.C.Babington.
EXSICC.: **Kew Gardens.** Riverside Nursery; weed; 12 ix 1944; *Hutchinson* s.n. (K) • Ground by Jodrell Laboratory; 20 vii 1948; *Burtt* 247 (K). **Kew Environs.** Kew; 1931; *Fraser* s.n. (K).
CULT.: 1789, Nat. of Britain (HK1, 3: 115); 1812, Nat. of Britain (HK2, 4: 437); 1814, Nat. of Britain (HKE: 245); 1949, Herbarium Ground (K); 1953, unspec. (K); 1955, unspec. (K). Not currently in cult.
CURRENT STATUS: Almost ubiquitous in the Gardens.

1354. Sonchus asper (L.) Hill
S. asper var. *integrifolius* Lej.
prickly sowthistle
FIRST RECORD: 1873/4 [1953].
NICHOLSON: **1873/4 survey** (JB: 47): Here and there on every piece of dug ground. • **1906 revision** (AS: 83): Here and there on every piece of dug ground.
OTHER REFERENCES: **Kew Gardens.** 1961, *J.L.Gilbert* (S29: 408): Near tennis courts [as var. *integrifolius*]. • 1999, *Stones* (EAK: 2): By Lake.
EXSICC.: **Kew Gardens.** [unlocalised]; weed; x 1942; *Hutchinson* s.n. (K). • Riverside Nursery; weed; 11 ix 1944; *Melville* s.n. (K). • Riverside Nursery; weed; 12 ix 1944; *Melville* s.n. (K). • Ground by Jodrell Laboratory; 20 vii 1948; *Burtt* 248 (K). • Ground by Jodrell Laboratory; 21 vii 1948, *Burtt* 249 (K). • Queen's Garden; waste ground; 5 vii 1978; *Campbell* 8 (K). **Kew Environs.** Kew; 10 vi 1931; *Fraser* s.n. (K).
CULT.: 1953, Herbarium Ground (K); 1963, unspec. (K). Not currently in cult.
CURRENT STATUS: Widespread and abundant throughout the Gardens.

1355. Lactuca serriola L.
L. scariola L.
prickly lettuce
FIRST RECORD: 1946 [1768].
REFERENCES: **Kew Gardens.** 1990, *Cope* (S31: 182): Paddock, after restoration in 1990. • 1999, *Stones* (EAK: 2): By Lake.

Exsicc.: **Kew Gardens.** South end of Bird Cage Walk; under white poplar tree; 8 viii 1946; *Turrill* s.n. (K). **Kew Environs.** Kew Green; 20 ix 1946; *Airy Shaw* s.n. (K).
Cult.: 1768 (JH: 43); 1789, Nat. of England (HK1, 3: 117); 1812, Nat. of England (HK2, 4: 441); 1814, Nat. of England (HKE: 246); 1933, unspec. (K); 1934, unspec. (K); 1951, Herbarium Ground (K). In cult.
Current status: Fairly common throughout the Gardens.

1356. Lactuca virosa L.
great lettuce
First record: 1988 [1768].
References: **Kew Gardens.** 1988, *Cope* (S31: 182): Behind Sir Joseph Banks Building before restoration of the 'Paddock.'
Cult.: 1768 (JH: 43); 1789, Nat. of Britain (HK1, 3: 117); 1812, Nat. of Britain (HK2, 4: 441); 1814, Nat. of Britain (HKE: 246); 1903, unspec. (K); 1933, Herb Garden (K); 1935, Herbarium Ground (K); 1936, Herbarium Ground (K); 1937, Herbaceous Ground (K). Not currently in cult.
Current status: Occasional, but seems to be spreading mostly in the southern part of the Gardens; much less common than the preceding species.

1357. Lactuca saligna L.
least lettuce
Cult.: 1768 (JH: 43); 1789, Nat. of England (HK1, 3: 117); 1812, Nat. of England (HK2, 4: 442); 1814, Nat. of England (HKE: 246); 1903, unspec. (K); 1981, unspec. (K). Not currently in cult.

1358. Cicerbita alpina (L.) Wallr.
Sonchus alpinus L.
alpine blue-sow-thistle
Cult.: 1789, Nat. of England (HK1, 3: 115); 1812, Nat. of Scotland (HK2, 4: 438); 1814, Nat. of Scotland (HKE: 245). Not currently in cult.

1359. Mycelis muralis (L.) Dumort.
Lactuca muralis (L.) Gaertn.
Prenanthes muralis L.
wall lettuce
First record: 1873/4 [1768].
Nicholson: **1873/4 survey** (JB: 47): **B.** A few plants in wall above "Icehouse door" (behind "Rockwork"). • **1906 revision** (AS: 83 and in SFS: 427): **B.** On hardy fernery wall, etc.
Cult.: 1768 (JH: 47); 1789, Nat. of Britain (HK1, 3: 119); 1812, Nat. of Britain (HK2, 4: 444); 1814, Nat. of Britain (HKE: 246); 1906, unspec. (K); 1933, unspec. (K); 1970, Herbaceous Dept. (K). Not currently in cult.
Current status: Rare; just a few recent sightings.

1360–1367. Taraxacum officinale sensu lato
dandelions
Nicholson: **1873/4 survey** (JB: 47): Plentiful on most of the lawns, also on walks in shrubberies, etc. • **1906 revision** (AS: 83): Common, everywhere.
Indet Exsicc.: **Kew Gardens.** Field near the Herbarium; [10 sheets] to show range from one locality; 1920; *Turrill* s.n. (K). • Herbarium Experimental Ground; in loose soil; 30 iv 1929; *Summerhayes* s.n. (K). • Herbarium field near gate; in short turf; 21 x 1931; *Turrill & A.K.Jackson* s.n. (K). • Near the tennis courts; waste ground; vi 1932; *Turrill* s.n. (K). **Kew Environs.** Kew; v 1912; *Divers* s.n. (K). • Kew Green; 1926; *Horwood* s.n. (K). • Towing path side betwn. Kew & Richmond; 14 v 1951; *Turrill* s.n. (K).
Indet Cult.: 1789, Nat. of Britain (HK1, 3: 120); 1812, Nat. of Britain (HK2, 4: 444); 1814, Nat. of Britain (HKE: 246); 1933, unspec. (K). In cult.
Notes: The Horwood specimen was originally determined as *T. obliquum* (q.v.) but later redetermined by Haworth as 'not *obliquum*.' Unidentified dandelions occur throughout the Gardens, often abundantly.
Current status: Dandelions are ubiquitous in the Gardens occurring in every piece of grass and rough ground, but none has yet been positively identified to microspecies.

1360. Taraxacum brachyglossum (Dahlst.) Dahlst.
T. erythrospermum Andrz. ex Besser
First record: 1921.
References: **Kew Environs.** 1931, *Fraser* (SFS: 425; S33: 750 and see Exsicc.): Kew Green.
Exsicc.: **Kew Environs.** Kew Green; 17 v 1921; *Fraser* s.n. (K).
Current status: Not known.

1361. Taraxacum glauciniforme Dahlst.
First record: 1911.
Exsicc.: **Kew Gardens.** [Unlocalised]; wild flora of Kew Gardens; 2 v 1911; *Turrill* s.n. (K).
Current status: Not known. The Turrill specimen was originally determined by Handel-Mazzetti in 1912 as *T. obliquum* (q.v.) but redetermined by van Soest in 1960 as *T. glauciniforme.*

1362. Taraxacum obliquum (Fries) Dahlst.
First record: 1913.
References: **Kew Gardens.** 1913, *Turrill* (S16: 216): Near the Herbarium.
Current status: Not known.

1363. Taraxacum palustre (Lyons) Symons
Leontodon palustre (Lyons) Sm.
marsh dandelion
Cult.: 1812, Nat. of Britain (HK2, 4: 444); 1814, Nat. of Britain (HKE: 246). Not currently in cult.

1364. Taraxacum gelertii Raunk.
T. adamii sensu Richards *non* Claire
T. kewense G.E.Haglund, *nom. nud.*
FIRST RECORD: 1936.
EXSICC.: **Kew Gardens.** Near Isleworth Gate; in grassland; 8 vi 1936; *Gilmour & Turrill* s.n. (K).
CURRENT STATUS: Not known.

1365. Taraxacum expallidiforme Dahlst.
FIRST RECORD: 1936.
REFERENCES: **Kew Gardens.** 1936, *Nannfeldt* (BSEC11: 265; S33: 750 and see Exsicc.): Lawn, Kew Gardens (seed taken to Lund and cult. by Haglund).
EXSICC.: **Kew Gardens.** [Unlocalised]; lawn; 17 v 1935; *Haglund* s.n. (K; cult in Lund).
CURRENT STATUS: Not known.

1366. Taraxacum stenacrum Dahlst.
FIRST RECORD: 1931.
EXSICC.: **Kew Environs.** Towing path side between Kew & Richmond; 14 v 1931; *Turrill* s.n. (K).
CURRENT STATUS: Not known.

1367. Taraxacum valdedentatum Dahst.
FIRST RECORD: 1975.
EXSICC.: **Kew Gardens.** Near Herbarium, by 'allotment' and near outside wall, GR 51/186776; 9 v 1975; *Pankhurst & Richards* 75/17 (K).
CURRENT STATUS: Not known. The species was included in Richards, *The Taraxacum flora of the British Isles* (1972), but was not mentioned, even in synonymy, in Dudman & Richards, *Dandelions of Great Britain and Ireland* (1997). Its taxonomic fate is unknown.

1368. Crepis paludosa (L.) Moench
Hieracium paludosum L.
marsh hawk's-beard
CULT.: 1768 (JH: 44); 1789, Nat. of Britain (HK1, 3: 123); 1812, Nat. of Britain (HK2, 4: 452); 1814, Nat. of Britain (HKE: 247). Not currently in cult.

1369. Crepis mollis (Jacq.) Asch.
Hieracium molle Jacq.
northern hawk's-beard
CULT.: 1789, Nat. of Scotland (HK1, 3: 124); 1812, Nat. of Scotland (HK2, 4: 451); 1814, Nat. of Scotland (HKE: 247). Not currently in cult.

1370. Crepis biennis L.
rough hawk's-beard
FIRST RECORD: 1931 [1768].
REFERENCES: **Kew Gardens.** 1931, *C.E.Salmon* (SFS: 416; S33: 750): Abundant among hay, Queen's Cottage Grounds, Kew. G.Nicholson. [Not in fact recorded in either of Nicholson's lists.]
CULT.: 1768 (JH: 46); 1789, Nat. of England (HK1, 3: 127); 1812, Nat. of England (HK2, 4: 460); 1814, Nat. of England (HKE: 248); 1922, unspec. (K); 1930, unspec. (K); 1971, Herb. & Alpine Dept. (K). Not currently in cult.
CURRENT STATUS: No recent sightings.

1371. Crepis capillaris (L.) Wallr.
C. tectorum L.
C. virens L.
smooth hawk's-beard
FIRST RECORD: 1873/4 [1768].
NICHOLSON: **1873/4 survey** (JB: 47): One of the commonest Composites in divisions **P** and **Q.** • **1906 revision** (AS: 83): One of the most common of Kew composites.
EXSICC.: **Kew Gardens.** Outside the Herbarium; among grass on waste land; 18 ix 1924; *Burtt-Davy* 19104 (K). • [Unlocalised]; on dry bank in full sun; 14 vii 1944; *Souster* 118 (K). **Kew Environs.** Kew Green; 18 viii 1922; *Hughs* s.n. (K).
CULT.: 1768 (JH: 46); 1789, Nat. of Britain (HK1, 3: 126); 1812, Nat. of Britain (HK2, 4: 459); 1814, Nat. of Britain (HKE: 248); 1986, Alpine (K). Not currently in cult.
CURRENT STATUS: Widespread and common in all types of grassland.

1372. *Crepis vesicaria subsp. **taraxacifolia** (Thuill.) Thell. ex Schinz & R.Keller
C. taraxacifolia Thuill.
beaked hawk's-beard
FIRST RECORD: 1906 [1934].
NICHOLSON: **1906 revision** (AS: 83): **Q.** In the open vista near river.
EXSICC.: **Kew Gardens.** Herbarium Experimental Ground; weed; 30 v 1946; *Sandwith* 3112 (K). • Herbarium Experimental Plot; as a weed; 21 vi 1946; *Sandwith* s.n. (K). **Kew Environs.** Kew, near River Thames; on waste ground; 28 v 1933; *Hubbard* s.n. (K). • Kew, towing path by Thames; 18 v 1939; *Wood* s.n. (K).
CULT.: 1934, unspec. (K). Not currently in cult.
CURRENT STATUS: Widespread; common in the longer grass.

1373. *Crepis setosa Haller f.
bristly hawk's-beard
FIRST RECORD: 1987 [1895].
REFERENCES: **Kew Gardens.** 1987, *Hastings* (LN67: 173): Below tit-feeders. • 1990, *Cope* (S31: 182): Paddock, after restoration in 1990.
CULT.: 1895, unspec. (K); 1986, plot 157-48 (K). Not currently in cult.
CURRENT STATUS: No recent sightings.

1374. Crepis foetida L.
stinking hawk's-beard
CULT.: 1768 (JH: 46); 1789, Nat. of England (HK1, 3: 126); 1812, Nat. of England (HK2, 4: 458); 1814, Nat. of England (HKE: 248); 1924, unspec. (K); 1933, Herbarium Ground (K); 1936, Herbarium Ground (K). Not currently in cult.

[1375. Crepis pulchra L.
small-flowered hawk's-beard
CULT.: 1789, Alien (HK1, 3: 128); 1812, Nat. of Scotland (HK2, 4: 457); 1814, Nat. of Scotland (HKE: 248).
Note. Listed in error as native.]

1376. Pilosella officinarum F.W.Schultz & Sch.Bip.
Hieracium alpinum L.
H. auricula L.
H. dubium L.
H. pilosella L.
mouse-ear-hawkweed
FIRST RECORD: 1782 [1768].

Mouse-ear-hawkweed (*Pilosella officinarum*). Widespread in grassland, and especially abundant in the Conservation Area.

NICHOLSON: **1873/4 survey** (JB: 47): Common on every dry slope within the present limits. • **1906 revision** (AS: 83): Common on every dry slope within our limits.
EXSICC.: **Kew Gardens.** Near Herbarium; among grasses, sandy soil; 16 vi 1928; *Hubbard* s.n. (K). • [Unlocalised]; in short grass on dry bank; 16 v 1944; *Souster* 63 (K). **Kew Environs.** Kew; 1782; *Goodenough* (K).
CULT.: 1768 (JH: 43, 44); 1789, Nat. of Britain (HK1, 3: 121 as *pilosella*), Nat. of England (ibid.: as *auricula* and *dubium*), Nat. of Scotland (ibid.: as *alpinum*); 1812, Nat. of Britain (HK2, 4: 448 as *pilosella, dubium* and *alpinum*), Nat. of England (ibid.: 449 as *auricula*); 1814, Nat. of Britain (HKE: 247 as *pilosella, dubium* and *alpinum*), Nat. of England (ibid.: as *auricula*); 1933, Herb. Ground (K); 1936, Herbaceous Ground (K). Not currently in cult.
CURRENT STATUS: Widespread in the Gardens and often abundant in short grass.

1377. *Pilosella aurantiaca (L.) F.W.Schultz & Sch.Bip.
Hieracium aurantiacum L.
fox-and-cubs
FIRST RECORD: 1889–1898 [1768].
NICHOLSON: **1906 revision:** Not recorded.
OTHER REFERENCES: **Kew Gardens.** 1889–1898, *G. Nicholson* in *Hanbury* (SFS: 420; S33: 750): Among grass, wild part of wood, Kew Gardens.
CULT.: 1768 (JH: 44); 1789, Alien (HK1, 3: 122); 1812, Nat. of Scotland (HK2, 4: 449); 1814, Nat. of Scotland (HKE: 247). Not currently in cult.
CURRENT STATUS: No recent sightings; it was last seen as a weed in 1982.

1378–1388. Hieracium vulgatum Fries sensu lato
hawkweeds
Most of the Hawkweeds in the Gardens have been named to microspecies; for exceptions see under *H. vulgatum* sensu stricto. Many of the names used in the older catalogues cannot be assigned to modern accepted taxa with any degree of certainty without voucher specimens. Hawkweeds are not common in the Gardens.

1378. Hieracium subaudum L.
H. perpropinquum (Zahn) Druce
FIRST RECORD: 1957 [1768].
REFERENCES: **Kew Gardens.** 1957, *Souster* (S24: 189): On bank of old filter-bed in Stable Yard.
CULT.: 1768 (JH: 44); 1789, Nat. of Britain (HK1, 3: 125); 1812, Nat. of Britain (HK2, 4: 456); 1814, Nat. of Britain (HKE: 247). Not currently in cult.
CURRENT STATUS: Rare; confined to the extreme southern end of the Gardens not far from its original station in the Stable Yard.

1379. Hieracium prenanthoides Vill.
H. spicatum All.
H. denticulatum Sm.
CULT.: 1789, Nat. of Scotland (HK1, 3: 125 as *spicatum*); 1812, Nat. of Scotland (HK2, 4: 456 as *denticulatum* and *prenanthoides*); 1814, Nat. of Scotland (HKE: 247 as *denticulatum* and *prenanthoides*). Not currently in cult.

1380. Hieracium umbellatum L.
FIRST RECORD: 1999 [1768].
REFERENCES: **Kew Gardens.** 1999, *Cope* (S35: 647): In rough grassland near the Azalea Garden (234).
CULT.: 1768 (JH: 44); 1789, Nat. of Britain (HK1, 3: 125); 1812, Nat. of Britain (HK2, 4: 456); 1814, Nat. of Britain (HKE: 247). Not currently in cult.
CURRENT STATUS: Seen for certain only twice so far, in widely separated localities. The identity of both populations may need checking.

1381. Hieracium latobrigorum (Zahn) Roffey
CULT.: 1880, unspec. (K). Not currently in cult.

1382. Hieracium calcaricola (F.Hanb.) Roffey
H. tridentatum auct. non Fries
FIRST RECORD: 1880.
NICHOLSON: **1906 revision:** Not recorded.
OTHER REFERENCES: **Kew Gardens.** 1880, *Baker* (BEC1: 33; S33: 750): Hort. Kew.
CURRENT STATUS: No recent sightings.

1383. Hieracium vulgatum Fries sensu stricto
H. murorum L.
H. sylvaticum Willd.
FIRST RECORD: 1873/4 [1768].
NICHOLSON: **1873/4 survey** (JB: 47): **Pal.** A few plants. **P.** A plot of about 50 good plants in wood south of "Engine House." • **1906 revision** (AS: 83): **P.** A few plants. **A.** A large plot in wood near pumping station.
EXSICC.: **Kew Gardens.** Wild flora near Arboretum Pits; 30 vi 1931; *Turrill* s.n. (K).
CULT.: 1768 (JH: 44); 1789, Nat. of Britain (HK1, 3: 122 as *murorum*); 1812, Nat. of Britain (HK2, 4: 451 as *murorum* and 452 as *sylvaticum*); 1814, Nat. of Britain (HKE: 247 as *murorum* and *sylvaticum*). Not currently in cult.
CURRENT STATUS: Only one population, near the Water Lily Pond, and the one herbarium specimen, have not yet been indentified to microspecies. The others, cited above under Nicholson, are *H. vulgatum* sensu stricto.

1384. Hieracium diaphanum Fries
FIRST RECORD: 1927.
EXSICC.: **Kew Gardens.** Herbarium grounds; vii 1927; *Horwood* s.n. (K).
CURRENT STATUS: No recent sightings.

1385. Hieracium maculatum Sm.
H. vulgatum var. *maculatum* (Sm.) F.Hanb.
FIRST RECORD: 1873/4 [1812].
NICHOLSON: **1873/4 survey** (JB: 47 and in SFS: 418): This is as plentiful as [*Hieracium vulgatum*] and grows not far from it. • **1906 revision** (AS: 83): **P, A.** With [*Hieracium vulgatum*]. **B.** As a weed in rockery.
EXSICC.: **Kew Gardens.** Garden, originally brought from Kew Gardens, so labelled [*H. maculatum*]; *Watson* in *Herb. Watson* (K).
CULT.: 1812, Nat. of England (HK2, 4: 451); 1814, Nat. of England (HKE: 247). Not currently in cult.
CURRENT STATUS: After a long absence it apppeared at the base of the wall around the grounds of St Anne's Church in 2008.

1386. Hieracium amplexicaule L.
CULT.: 1768 (JH: 44). Not currently in cult.

1387. Hieracium anglicum Fr.
H. cerinthoides auct. non L.
H. lawsonii auct. non Vill.
H. villosum auct. non L.
CULT.: 1768 (JH: 44); 1789, Alien (HK1, 3: 123 as *cerinthoides*), Nat. of Scotland (ibid.: 124 as *villosum*); 1812, Nat. of Scotland (HK2, 4: 453 as *cerinthoides*, 454 as *villosum*), Nat. of Britain (ibid.: 450 as *lawsonii*); 1814, Nat. of Scotland (HKE: 247 as *cerinthoides* and *villosum*), Nat. of Britain (ibid.: as *lawsonii*). Not currently in cult.

1388. Hieracium sinuans F.Hanb.
H. pulmonarium Sm.
CULT.: 1812, Nat. of Scotland (HK2, 4: 452); 1814, Nat. of Scotland (HKE: 247). Not currently in cult.

1389. Filago vulgaris Lam.
Filago germanica L. non Huds.
Gnaphalium germanicum L.
common cudweed
FIRST RECORD: 1873/4 [1768].
NICHOLSON: **1873/4 survey** (JB: 46): **P.** Several patches near river end of lake. • **1906 revision** (AS: 81): A weed in shrubberies and bare gravelly spots.
CULT.: 1768 (JH: 36); 1789, Nat. of Britain (HK1, 3: 279); 1813, Nat. of Britain (HK2, 5: 19); 1814, Nat. of Britain (HKE: 258). Not currently in cult.
CURRENT STATUS: Very rare; seen at the base of the wall of the Pavilion Restaurant and in recently turned soil in the Ancient Meadow.

1390. Filago pyramidata L.
broad-leaved cudweed
FIRST RECORD: 1998.
REFERENCES: **Kew Gardens.** 1998, *Cope* (S33: 750): West Arb.: edge of lawn near Riverside Walk (plot 212).
CULT.: In cult.
CURRENT STATUS: Very rare; only one recent sighting, above.

1391. Filago minima (Sm.) Pers.
Gnaphalium minimum Sm.
Filago montana L.
small cudweed
FIRST RECORD: 1873/4 [1789].
NICHOLSON: **1873/4 survey** (JB: 46; SFS: 377): Common in all the very dry places. Wherever the turf gets rather bare, a plentiful crop of this plant appears. • **1906 revision** (AS: 81): Common in all the very dry places. In some places where the turf gets badly worn, a plentiful crop of this species appears.
CULT.: 1789, Nat. of Britain (HK1, 3: 279); 1813, Nat. of Britain (HK2, 5: 20); 1814, Nat. of Britain (HKE: 258). Not currently in cult.
CURRENT STATUS: No recent sightings.

1392. Filago gallica L.
Gnaphalium gallicum L.
narrow-leaved cudweed
CULT.: 1768 (JH: 36); 1789, Nat. of England (HK1, 3: 279); 1813, Nat. of England (HK2, 5: 20); 1814, Nat. of England (HKE: 258). Not currently in cult.
Note. Originally grown as a native but now considered to be an archaeophyte.

1393. Antennaria dioica (L.) Gaertn.
Gnaphalium dioicum L.
mountain everlasting
CULT.: 1768 (JH: 35); 1789, Nat. of Britain (HK1, 3: 178); 1813, Nat. of Britain (HK2, 5: 17); 1814, Nat. of Britain (HKE: 258); 1940, Herb Garden (K). In cult.

[1394. **Anaphalis margaritacea** (L.) Benth.
Gnaphalium margaritaceum L.
pearly everlasting
CULT.: 1789, Nat. of England (HK1, 3: 177); 1813, Nat. of England (HK2, 5: 17); 1814, Nat. of England (HKE: 258).
Note. Listed in error as native.]

1395. Gnaphalium sylvaticum L.
G. rectum Sm.
heath cudweed
FIRST RECORD: 1873/4 [1768].
NICHOLSON: **1873/4 survey** (JB: 46): **P.** Frequent in the strip of turf facing Palace Grounds from Princess's Gate to within 100 yards of Brentford Gate. • **1906 revision** (AS: 81 and in SFS: 379): **A.** Frequent in turf in Birch collection.
CULT.: 1768 (JH: 35); 1789, Nat. of Britain (HK1, 3: 178); 1813, Nat. of Britain (HK2, 5: 18); 1814, Nat. of Britain (HKE: 258). Not currently in cult.
CURRENT STATUS: No recent sightings.

1396. Gnaphalium supinum L.
dwarf cudweed
CULT.: 1813, Nat. of Scotland (HK2, 5: 18); 1814, Nat. of Scotland (HKE: 258). Not currently in cult.

1397. *Gnaphalium purpureum L.
Gamochaeta purpurea (L.) Cabrera
Gnaphalium pennsylvanicum Willd., nom. illegit.
American cudweed
FIRST RECORD: 1944.
REFERENCES: **Kew Gardens.** 1987, *Leslie* (FSSC: 50; S33: 750): Weed in the gardens. **Kew Environs.** 1944, *anon.* (LN58: 65; S33: 750): Outside Cumberland Gate. • 1946, *Airy Shaw* (FSSC: 50; S33: 750): At foot of wall, Kew Gardens Road, opposite Cumberland Gate. • 1984, *Latham* (BSN38: 20; S33: 750): Now well established in the churchyard of St. Ann's, Kew. It originally escaped from the order beds in the Royal Botanic Gardens where it is uncritically labelled as *G. purpureum*. • 1978, *Wild Flower Society Excursion*; 1983, *Leslie*; 1985, *Latham* (all FSSC: 50; LN58: 65; S33: 750): St. Ann's churchyard. • 1993, *Sheahan* (LN73: 193; S33: 750): Growing out of a wall in Kew Road.
CURRENT STATUS: After a long absence it made a brief appearance in disturbed ground near the Jodrell Laboratory during building works for the new extension.

1398. Gnaphalium uliginosum L.
marsh cudweed
FIRST RECORD: 1873/4 [1768].
NICHOLSON: **1873/4 survey** (JB: 46): In all the divisions. Plentiful both on waste and cultivated ground. • **1906 revision** (AS: 81): In all the divisions. Plentiful both on waste and cultivated ground.
EXSICC.: **Kew Gardens.** Weed of cultivated land, on dry sandy soil; 29 vi 1950; *Souster* 1128 (K).
CULT.: 1768 (JH: 36); 1789, Nat. of Britain (HK1, 3: 178); 1813, Nat. of Britain (HK2, 5: 19); 1814, Nat. of Britain (HKE: 258); 1922, unspec. (K). Not currently in cult.
CURRENT STATUS: Widely scattered but scarce.

1399. Pseudognaphalium luteoalbum (L.) Hilliard & B.L.Burtt
Gnaphalium luteoalbum L.
Jersey cudweed
FIRST RECORD: 1998 [1768].
REFERENCES: **Kew Gardens.** 1998, *Cope* (S22: 750): North Arb.: under a weeping beech tree in the lawn near the weather station (124) (removed within three days during mowing and weeding operations). **Kew Environs.** 2006, *Hounsome* (LN86: 179): St Anne's churchyard, three plants.
CULT.: 1768 (JH: 35); 1789, Nat. of England (HK1, 3: 176); 1813, Nat. of England (HK2, 5: 14); 1814, Nat. of England (HKE: 258); 1911, unspec. (K). Not currently in cult.

CURRENT STATUS: The plant has been lost from its original locality, but was recently found near the Rockery, again near the Weather Station and more recently outside the Jodrell Laboratory. The species is on Schedule VIII of fully protected plants in the British Isles, but it is probably native only in the Channel Islands and its inclusion in the Red Data List for mainland Britain is surely an error. In the Gardens it is only ever found as a weed and does not persist.

1400. *Ammobium alatum E.A.Br.
winged everlasting
FIRST RECORD: 2004 [1884].
CULT.: 1884, unspec. (K); 1898, unspec. (K); 1960, unspec. (K). In cult.
CURRENT STATUS: Several plants grew from seed in topsoil stored by the Princess of Wales Conservatory. It did not return the following year and the area has now been levelled and grassed over.

1401. Inula helenium L.
elecampane
FIRST RECORD: 1953 [1768].
REFERENCES: **Kew Environs.** 1953, *Matthews* (LN33: 52; S33: 750): Near Kew Bridge.
CULT.: 1768 (JH: 14); 1789, Nat. of Britain (HK1, 3: 222); 1813, Nat. of Britain (HK2, 5: 76); 1814, Nat. of Britain (HKE: 266); 1922, unspec. (K); 1933, unspec. (K). In cult.
CURRENT STATUS: No recent sightings.

1402. Inula conyzae (Griess.) Meikle
Conyza squarrosa L.
ploughman's-spikenard
CULT.: 1768 (JH: 33); 1789, Nat. of Britain (HK1, 3: 181); 1813, Nat. of Britain (HK2, 5: 26); 1814, Nat. of Britain (HKE: 260); 1921, unspec. (K). Not currently in cult.

1403. Inula crithmoides L.
I. crithmifolia Willd.
golden-samphire
CULT.: 1768 (JH: 14); 1789, Nat. of England (HK1, 3: 225); 1813, Nat. of England (HK2, 5: 80); 1814, Nat. of England (HKE: 266); 1911, unspec. (K). In cult.

1404. Pulicaria dysenterica (L.) Bernh.
Inula dysenterica L.
common fleabane
FIRST RECORD: 1873/4 [1768].
NICHOLSON: **1873/4 survey** (JB: 47): **Strip.** Frequent about Isleworth Gate. • **1906 revision** (AS: 81): **Strip.** Frequent along ha-ha.
CULT.: 1768 (JH: 14); 1789, Nat. of England (HK1, 3: 223); 1813, Nat. of England (HK2, 5: 77); 1814, Nat. of England (HKE: 266); 1881, unspec. (K); 1936, Herbaceous Ground (K); 1967, Jodrell (K); 1948, Herbarium Ground (K). Not currently in cult.
CURRENT STATUS: The only recent record is a plant that grew from seed in topsoil stored in the Paddock.

1405. Pulicaria vulgaris Gaertn.
Inula pulicaria L.
small fleabane
CULT.: 1768 (JH: 14); 1789, Nat. of England (HK1, 3: 223); 1813, Nat. of England (HK2, 5: 77); 1814, Nat. of England (HKE: 266); 1949, Herbarium Ground (K). Not currently in cult.

1406. *Buphthalmum salicifolium L.
B. grandiflorum L.
willow-leaved yellow-oxeye
FIRST RECORD: 2001 [1768].
REFERENCES: **Kew Gardens.** 2001, *Cope* (S35: 647): Incorporated in the Robinsonian Meadow sown in front of the temporary Cycad House (224) and now naturalised.
CULT.: 1768 (JH: 13); 1789, Alien (HK1, 3: 247); 1813, Alien (HK2, 5: 124); 1814, Alien (HKE: 272); 1896, unspec. (K); 1985, plot 157.69 (K). Not currently in cult.
CURRENT STATUS: Incorporated in the Robinsonian Meadow in front of 'Climbers & Creepers' but unlikely to persist.

1407. Solidago virgaurea L.
S. cambrica Huds.
goldenrod
CULT.: 1768 (JH: 20); 1789, Nat. of Britain (HK1, 3: 218), Nat. of Wales (ibid.: as *cambrica*); 1813, Nat. of Britain (HK2, 5: 70), Nat. of Wales (ibid.: as *cambrica*); 1814, Nat. of Britain (HKE: 265), Nat. of Wales (ibid.: as *cambrica*); 1894, unspec. (K); 1896, unspec. (K); 1921, unspec. (K); 1922, unspec. (K); 1932, unspec. (K); 1934, unspec. (K); c. 1940?, Herb Garden (K); 1970, Herbaceous Dept. (K); 1971, Herb. & Alpine Dept. (K). In cult.

1408. *Solidago canadensis L.
Canadian goldenrod
FIRST RECORD: 1999 [1768].
REFERENCES: **Kew Gardens.** 1999, *Cope* (S35: 647): In wooded area beside Syon Vista (255).
CULT.: 1768 (JH: 20); 1789, Alien (HK1, 3: 210); 1813, Alien (HK2, 5: 64); 1814, Alien (HKE: 264); 1977, plot 157.72 (K). In cult.
CURRENT STATUS: A substantial specimen grows in the wall of the Palm House Pond. It was first noted on the towpath in 1976 and is widely scattered, but scarce, within the Gardens.

1409. *Solidago graminifolia (L.) Salisb.
grass-leaved goldenrod
FIRST RECORD: 19th cent. [1977].
NICHOLSON: **1873/4 survey:** Not recorded. • **1906 revision:** Not recorded.
EXSICC.: **Kew Gardens.** [Unlocalised]; [19th cent.]; *Leighton* s.n. (K).
CULT.: 1977, plot 157.72 (K). Not currently in cult.
CURRENT STATUS: No recent sightings.

1410. *Aster novi-belgii L.
confused michaelmas-daisy
FIRST RECORD: 1920 [1768].
EXSICC.: **Kew Environs.** By the Thames above Kew; 22 ix 1920; *Britton* 2316 (K). • By the Thames above Kew; 24 ix 1920; *Britton* 2326 (K).
CULT.: 1768 (JH: 16); 1789, Alien (HK1, 3: 206); 1813, Alien (HK2, 5: 62); 1814, Alien (HKE: 264); 1880s, unspec. (K); 1956, unspec. (K). In cult. under glass.
CURRENT STATUS: No recent sightings; it was last seen in 1980 on the towpath.

1410.1. *Aster × salignus Willd. (novi-belgii × lanceolatus)
common michaelmas-daisy
FIRST RECORD: 2008
CURRENT STATUS: A single plant in rough ground adjacent to the Stable Yard

1411. *Aster lanceolatus Willd.
narrow-leaved michaelmas-daisy
FIRST RECORD: 2003 [2006].
CULT.: In cult.
CURRENT STATUS: In patches along the towpath. The determination is not certain; the plant is either true *A. lanceolatus* or a hybrid of which it is a parent. It was seen again near Cumberland Gate in 2004.

1412. Aster tripolium L.
sea aster
CULT.: 1768 (JH: 15); 1789, Nat. of England (HK1, 3: 199); 1813, Nat. of Britain (HK2, 5: 58); 1814, Nat. of Britain (HKE: 263); 1921, unspec. (K); 1922, unspec. (K); 1930, unspec. (K). Not currently in cult.

1413. Aster linosyris (L.) Bernh.
Chrysocoma linosyris L.
Goldilocks aster
CULT.: 1768 (JH: 33); 1789, Alien (HK1, 3: 163); 1812, Alien (HK2, 4: 514); 1814, Nat. of England (HKE: 255); 1897, unspec. (K); 1914, unspec. (K); 1921, unspec. (K); 1958, Order Beds (K). In cult.
Note. Not discovered in Britain until 1813.

1414. Erigeron borealis (Vierh.) Simmons
E. alpinus auct. non L.
E. uniflorus auct. non L.
alpine fleabane
CULT.: 1789, Alien (HK1, 3: 187); 1813, Nat. of Scotland (HK2, 5: 33); 1814, Nat. of Scotland (HKE: 261); 1979, unspec. (K). In cult.
Note. Not known as a British plant until 1790.

1415. *Erigeron karvinskianus DC.
Mexican fleabane
FIRST RECORD: 1999 [1873]
CULT.: 1873, unspec. (K). In cult.
CURRENT STATUS: Found as an escape around the Princess of Wales Conservatory, in the Director's Garden and occasionally in the Paddock.

1416. Erigeron acer L.
blue fleabane
FIRST RECORD: 1942 [1768].
REFERENCES: **Kew Environs.** 1942, *Welch* (LN32: S148; S33: 750): Between Richmond and Kew.
CULT.: 1768 (JH: 23); 1789, Nat. of Britain (HK1, 3: 187); 1813, Nat. of Britain (HK2, 5: 33); 1814, Nat. of Britain (HKE: 260); 1912, unspec. (K); 1923, unspec. (K). Not currently in cult.
CURRENT STATUS: No recent sightings.

1417. *Conyza canadensis (L.) Cronquist
Erigeron canadensis L.
Canadian Fleabane
FIRST RECORD: 1873/4 [1768].
NICHOLSON: **1873/4 survey** (JB: 47): **B** and **P.** In nearly every shrubbery. • **1906 revision** (AS:81): **A, B, P, Q.** A common weed in shrubberies, etc.
OTHER REFERENCES: **Kew Gardens.** 1931, *Gough* (SFS: 374): Waste ground, Kew Gardens.
EXSICC.: **Kew Environs.** Naturalised in banks; 1871; *Lowne* s.n. (K). • Near Kew; 10 x 1891; *Britton* s.n. (K). • Kew; 12 viii 1900; *Clarke* 94 (K). • Thames embankment, Kew; x 1939; *Worsdell* s.n. (K). • Kew; garden weed; viii 1942; *Hutchinson* s.n. (K). • Weed at end of Bowling Green, Kew; 1942; *Hutchinson* s.n. (K).
CULT.: 1768 (JH: 23); 1789, Nat. of England (HK1, 3: 185); 1813, Nat. of England (HK2, 5: 32); 1814, Nat. of England (HKE: 260); 1899, unspec. (K); 1906, unspec. (K); 1934, Arboretum (K). Not currently in cult.
CURRENT STATUS: Widepsread and increasing as a weed in all types of ground.

1418. *Conyza sumatrensis (Retz.) E.Walker
Guernsey fleabane
FIRST RECORD: 1992.
REFERENCES: **Kew Gardens.** 1998, *Cope* (S33: 750): North Arb.: Paddock (104), roof of Banks Building (105), behind Banks Building (114); West Arb.:

Riverside Walk (211). **Kew Environs.** 1992, *Wurzell* (BSN62: 39; S33: 750): The most westerly occurrences of all were in Kew Road and again in Kew Churchyard where one large plant actually embraces the immortal *Sisymbrium strictissimum*. Now Sumatran Fleabane is literally poised within a few yards of Kew Gardens themselves, if it has not already entered their gates.
CURRENT STATUS: Widespread and increasing as a weed in all types of ground; more common at the moment than the preceding species.

1419. *Conyza bonariensis (L.) Cronquist
Erigeron bonariensis L.
Argentine fleabane
FIRST RECORD: 1843 [1768].
EXSICC.: **Kew Environs.** Kew Bridge; 1 v 1843; *Irvine* s.n. (K).
CULT.: 1768 (JH: 23); 1789, Alien (HK1, 3: 186); 1813, Alien (HK2, 5: 32); 1814, Alien (HKE: 260). Not currently in cult.
CURRENT STATUS: No recent sightings.

1420. *Conyza bilbaoana J.Rémy
FIRST RECORD: 2002.
REFERENCES: **Kew Gardens.** 2002, *Cope & Phillips* (S35: 647): In disturbed ground near the Rhododendron Dell (226). A relatively new invader in Southeast England, increasing in the London area.
CURRENT STATUS: A new arrival in the British Isles, spreading rapidly through parts of London, and recently arrived in the Gardens; only two plants have been seen thus far.

1421. Bellis perennis L.
daisy
FIRST RECORD: 1873/4 [1789].
NICHOLSON: **1873/4 survey** (JB: 47): Everywhere. • **1906 revision** (AS: 81): Everywhere.
EXSICC.: **Kew Gardens.** [Unlocalised]; in cultivated ground; xi 1922; *Turrill* s.n. (K). • [Unlocalised]; in lawns; 28 iv 1928; *Coates* s.n. (K). • [Unlocalised]; in lawns on sandy soil; 10 vi 1928; *Hubbard* s.n. (K). • [Unlocalised]; gravel path in partial shade; x 1939; *Ridley* s.n. (K). • [Unlocalised]; lawns, paths, beds, sunny open positions on sandy loam soils; 20 vi 1981; *Foster* 11 (K).
CULT.: 1789, Nat. of Britain (HK1, 3: 227); 1813, Nat. of Britain (HK2, 5: 86); 1814, Nat. of Britain (HKE: 267); 1934, unspec. (K). Not currently in cult.
CURRENT STATUS: Abundant in every piece of grass throughout the Gardens.

1422. Tanacetum parthenium (L.) Sch.Bip.
Chrysanthemum parthenium (L.) Bernh.
Matricaria parthenium L.
Pyrethrum parthenium (L.) Sm.
feverfew
FIRST RECORD: 1873/4 [1768].
NICHOLSON: **1873/4 survey** (JB: 46): On most of the soil heaps, and as a weed in many of the flower-beds and shrubberies. • **1906 revision** (AS: 82): A weed in flower beds and shrubberies.
EXSICC.: **Kew Gardens.** Herbaceous Ground; 1937; *Howes* s.n. (K).
CULT.: 1768 (JH: 19); 1789, Nat. of Britain (HK1, 3: 233); 1813, Nat. of Britain (HK2, 5: 99); 1814, Nat. of Britain (HKE: 269); 1933, Herb Garden (K); 1985, unspec. (K). Not currently in cult.
CURRENT STATUS: Scattered but not common. Double-flowered forms are clearly relicts of former cultivation and have not been recorded in the survey.

1423. Tanacetum vulgare L.
tansy
FIRST RECORD: 1873/4 [1768].
NICHOLSON: **1873/4 survey** (JB: 46): **Strip.** Here and there near river. • **1906 revision** (AS: 82): **Strip.** Here and there near ha-ha and river.

Daisy (*Bellis perennis*). Ubiquitous; in every patch of grass in the Gardens.

CULT.: 1768 (JH: 32); 1789, Nat. of Britain (HK1, 3: 168); 1813, Nat. of Britain (HK2, 5: 2); 1814, Nat. of Britain (HKE: 257). In cult.
CURRENT STATUS: Rare; near the Conservation Area and along the towpath.

1424. Seriphidium maritimum (L.) Poljakov
Artemisia maritima L.
sea wormwood
CULT.: 1768 (JH: 34); 1789, Nat. of Britain (HK1, 3: 171); 1813, Nat. of Britain (HK2, 5: 6); 1814, Nat. of Britain (HKE: 257); 1922, unspec. (K). In cult.

1425. Artemisia vulgaris L.
A. integrifolia L.
mugwort
FIRST RECORD: 1873/4 [1768].
NICHOLSON: **1873/4 survey** (JB: 46): **Strip.** A few plants by side of towing path. **Q.** Here and there. • **1906 revision** (AS: 82): **Strip.** By towing-path. **Q.** Here and there.
EXSICC.: **Kew Environs.** Thames bank, Kew; from crevices of new stone wall, near top; 2 viii 1928; *Summerhayes & Turrill* s.n. (K). • Bank of R. Thames near Kew Gardens; 23 viii 1944; *Blakelock* s.n. (K). • Thames bank, near Kew Gardens; 14 ix 1944; *Summerhayes & Blakelock* s.n. (K).
CULT.: 1768 (JH: 35); 1789, Nat. of Britain (HK1, 3: 172); 1813, Nat. of Britain (HK2, 5: 9); 1814, Nat. of Britain (HKE: 257); 1896, unspec. (K); 1918, unspec. (K); 1933, Arboretum (K); 1936, Herbaceous Ground (K). In cult.
CURRENT STATUS: Widespread in the rougher grassy areas but not common.

1426. *Artemisia verlotiorum Lamotte
Chinese mugwort
FIRST RECORD: 1942.
EXSICC.: **Kew Environs.** By Thames at Kew; waste ground; 3 xi 1942; *Sandwith* 3179 (K). • Towing path near Kew; 16 ii 1946; *Lousley* s.n. (K). • Riverbank, Kew; 15 i 1948; *Alston* s.n. (K).
CURRENT STATUS: Judging by the current sites in which this plant still grows it is likely that all records are from the wrong side of Kew Bridge and the species should perhaps not be included in this Catalogue.

1427. Artemisia absinthium L.
wormwood
CULT.: 1768 (JH: 35); 1789, Nat. of Britain (HK1, 3: 172); 1813, Nat. of Britain (HK2, 5: 8); 1814, Nat. of Britain (HKE: 257); 1918, unspec. (K); 1933, unspec. (K); 1934, Arboretum (K); 1936, Herbaceous Ground (K); 1955, Order Beds (K); 1971, Herb. & Alpine Dept. (K); 1971, plot 157.58 (K); 1980s, unspec. (K). In cult.

1428. Artemisia norvegica Fr.
Norwegian mugwort
CULT.: In cult. under glass.

1429. *Artemisia annua L.
annual mugwort
FIRST RECORD: 1931 [1768].
REFERENCES: **Kew Gardens.** 1965, *Wurzell* (FSSC: 14; S33: 750): Disturbed ground, Queen's Cottage grounds. • 1980, *Clement* (BSN26: 14; S33: 750): At Kew it regularly seeds itself out of its own Order Bed, but is not found outside the gardens. • 1983, *Burton* (FLA: 155; S33: 750): ... in 1965 several plants ... were on disturbed ground in the Queen's Cottage grounds. **Kew Environs.** 1956, *Lousley* (LN36: S351; S33: 750): Towing path between Kew and Richmond.
EXSICC.: **Kew Gardens.** [Unlocalised]; on rubbish heap; viii 1931; *Hubbard* s.n. (K).
CULT.: 1768 (JH: 35); 1789, Alien (HK1, 3: 172); 1813, Alien (HK2, 5: 8); 1814, Alien (HKE: 257); 1918, unspec. (K). Not currently in cult.
CURRENT STATUS: Appeared in the lawn behind the Herbarium after disturbance during building works in 1999 but failed to reappear the following year. Not recently seen elsewhere but may return to the Herbarium lawn when the current construction works have finished and the lawn is once again restored.

1430. Artemisia campestris L.
field wormwood
CULT.: 1768 (JH: 34); 1789, Nat. of England (HK1, 3: 170); 1813, Nat. of England (HK2, 5: 6); 1814, Nat. of England (HKE: 257); c. 1800, unspec. (K); 1897, unspec. (K); 1974, plot 157.59 (K). Not currently in cult.

[1431. Artemisia caerulescens L.
a. subsp. **caerulescens**
CULT.: 1789, Alien (HK1, 3: 172); 1813, Nat. of England (HK2, 5: 10); 1814, Nat. of England (HKE: 257).

b. subsp. **gallica** (Willd.) K.M.Perss.
A. gallica Willd.
CULT.: 1813, Nat. of Britain (HK2, 5: 7); 1814, Nat. of Britain (HKE: 257).
Note. Both subspecies listed in error as native.**]**

1432. Otanthus maritimus (L.) Hoffmans. & Link
Athanasia maritima L.
Santolina maritima (L.) Crantz
cottonweed
CULT.: 1789, Nat. of England (HK1, 3: 165); 1812, Nat. of England (HK2, 4: 518); 1814, Nat. of England (HKE: 256). Not currently in cult.

Chamomile (*Chamaemelum nobile*). In several places but the most extensive, in a good year, is in a lawn by the Palm House Pond.

1433. Achillea ptarmica L.
sneezewort
FIRST RECORD: 1966 [1768].
REFERENCES: **Kew Gardens.** 1966, *J.L.Gilbert* (S29: 408): On disturbed soil near Herbarium.
CULT.: 1768 (JH: 20); 1789, Nat. of Britain (HK1, 3: 240); 1813, Nat. of Britain (HK2, 5: 110); 1814, Nat. of Britain (HKE: 271). 1897, unspec. (K); 1974, plot 132.05 (K); 1974, plot 157.59 (K). In cult.
CURRENT STATUS: Cultivated in one or two places but plants in the Conservation Area, on the edge of the Brick Pit, appear to be spontaneous.

1434. Achillea millefolium L.
A. magna L.
yarrow
FIRST RECORD: 1873/4 [1768].
NICHOLSON: **1873/4 survey** (JB: 46): Everywhere. A common factor of the open turf. • **1906 revision** (AS: 82): Everywhere. A common factor of the open turf.
EXSICC.: **Kew Gardens.** Arboretum; among grass; 9 vi 1941; *Melville* s.n. (K). • [Unlocalised]; rough grass; 28 vi 1944; *Souster* 107 (K). • Temperate House surrounds; weed of lawns; 29 ix 1978; *A. Roberts* 9 (K). **Kew Environs.** Kew Bowling Green; 10 x 1942; *Hutchinson* s.n. (K).
CULT.: 1768 (JH: 20); 1789, Nat. of Britain (HK1, 3: 242); 1813, Nat. of Britain (HK2, 5: 116); 1814, Nat. of Britain (HKE: 271); 1887, unspec. (K); 1933, Herb Garden (K); 1935, unspec. (K); 1937, Herb Garden (K). In cult.
CURRENT STATUS: Widespread and abundant in all grassy areas. Plants in the Order Beds were once parasitised by *Orobanche purpurea* but the host was removed when some of the beds were converted to students' vegetable plots. Some decorative cultivars are still grown.

[**1435. Achillea decolorans** Schrad.
A. serrata Sm.
serrated yarrow
CULT.: 1789, Alien (HK1, 3: 241); 1813, Alien (HK2, 5: 111); 1814, Nat. of England (HKE: 271). Note. Listed in error as native.]

[**1436. Achillea tomentosa** L.
yellow milfoil
CULT.: 1789, Alien (HK1, 3: 239); 1813, Alien (HK2, 5: 116); 1814, Nat. of Britain (HKE: 271). Note. Listed in error as native.]

1437. Chamaemelum nobile (L.) All.
Anthemis nobilis L.
A. nobilis var. *discoidalis* Hesl.-Harr.
chamomile
FIRST RECORD: 1873/4 [1768].
NICHOLSON: **1873/4 survey** (JB: 46): **B.** Very common on the lawn behind Herb. Ground wall. • **1906 revision** (AS: 82): **B.** Common as a component of the turf in various lawns.
OTHER REFERENCES: **Kew Gardens.** 1931, *G.Nicholson* fide *C.E.Salmon* (SFS: 387): Dry grassy places, Kew [var. *discoidalis*]. • 1943/4, *Sandwith* (BSEC12: 730): In the lawns of Kew Gardens, the form recorded by G.Nicholson (see Salmon, *Flora of Surrey*, 387), and still there [var. *discoidalis*]. • 1949, *Sandwith* (S21: 236): still there, 1941–2. • 1952, *Cooke*; *D.H.Kent* (LN32: S153): Kew Gardens. • c. 1960, *Milne-Redhead* (S32: 656–7): Gardens on lawn by T-range, natural colony (destroyed when Gramineae beds were extended). • 1976, *Lousley* (LFS: 316): ... still plentiful on lawns in Kew Gardens. • 2000, *Wendt* (CHAM: 2, 3): A large, dense and prominent colony at TQ1875.7705 in the grassy area in the "L" between the Palm House pond and the mound

with the temple of Aeolus (plot 187); top of the bank, about halfway down the lake, on the Richmond side (plot 265); two clumps, either side of the main north-south park in grass strips, near the specimen grown in the N.O. beds (plot 157); between the waterlily ponds and the Jodrell laboratory, extending around both ends of the ponds at c. TQ1895.7730. This now a nursery bed, although it may survive nearby (plot 152); one patch exists in the northern side in the western area (in plots 141 and 142), and four in the eastern area. It also grows on the right hand side of the path from the Orangery, going towards Cambridge Cottage (plot 134); main ride, west of the Princess of Wales conservatory, and on Kew Green [Note: this is not Kew Green], on the north side of the Ice House (plot 143); near the *Escallonia* beds, by building 127, on the way from the Broad Walk to the Palm House (plot 164); a small to medium clump, in the grassy area on the side of the path from the entrance to the Princess of Wales conservatory on the Kew Green [Note: this is not Kew Green] (plot 127).
EXSICC.: **Kew Gardens.** [Unlocalised]; in lawns; 18 viii 1933; *Hubbard* s.n. (K); [Unlocalised]; in lawns; viii 1934; *Hubbard* s.n. (K).
CULT.: 1768 (JH: 9); 1789, Nat. of Britain (HK1, 3: 237); 1813, Nat. of Britain (HK2, 5: 106); 1814, Nat. of Britain (HKE: 270); 1927, unspec. (K); 1932, unspec. (K); 1933, unspec. (K); 1937, Herb Garden (K); 1945, Lower Nursery (K). In cult.
CURRENT STATUS: See references for the year 2000 above although there is some confusion over what constitutes Kew Green. It is also in one or two other places not mentioned in the Chamomile Survey. At least some of the populations may have originated as trials, during the Second World War, as a means of disguising airfields. The species was first recorded for the UK in 1548 when W.Turner found it on Richmond Green.

1438. Anthemis arvensis L.
corn chamomile
FIRST RECORD: 1873/4 [1789].
NICHOLSON: **1873/4 survey** (JB: 46 and in SFS: 386): **P.** A few plants about lake. • **1906 revision** (AS: 82): **A.** A few plants about lake.
EXSICC.: **Kew Gardens.** Cut grass lawn on E. side of Herbarium building; 1 x 1970; *Edwards* 402 (K).
CULT.: 1789, Nat. of Britain (HK1, 3: 237); 1813, Nat. of Britain (HK2, 5: 107); 1814, Nat. of Britain (HKE: 270); 1895, unspec. (K); 1930, unspec. (K). Not currently in cult.
CURRENT STATUS: Scattered, but not common; it occurred spontaneously in the Wheatfields sown along the Broad Walk in 1999 and flowered during 2000. It occasionally reappears in this and other areas.

1439. Anthemis cotula L.
stinking chamomile
CULT.: 1768 (JH: 9); 1789, Nat. of Britain (HK1, 3: 237); 1813, Nat. of Britain (HK2, 5: 107); 1814, Nat. of Britain (HKE: 270); 1934, Arboretum (K); 1974, plot 157.58 (K); 1987, plot 157.59 (K). Not currently in cult.

1440. *Anthemis tinctoria L.
yellow chamomile
FIRST RECORD: 1872 [1768]
EXSICC.: **Kew Environs.** Waste ground by Kew Bridge; ix 1872; *Watson* in *Herb. Watson* (K).
CULT.: 1768 (JH: 9); 1789, Nat. of England (HK1, 3: 238); 1813, Nat. of Britain (HK2, 5: 109); 1814, Nat. of Britain (HKE: 270); 1894, unspec. (K); 1923, unspec. (K); 1970, Herbaceous Dept. (K); c. 1986, unspec. (K). In cult.
CURRENT STATUS: No recent sightings.

[1441. Anthemis maritima L.
CULT.: 1789, Nat. of England (HK1, 3: 236); 1813, Nat. of England (HK2, 5: 105); 1814, Nat. of England (HKE: 270).
Note. Listed in error as native.**]**

1442. Chrysanthemum segetum L.
corn marigold
FIRST RECORD: 1990 [1768].
REFERENCES: **Kew Gardens.** 1990, *Cope* (S31: 182): Paddock, after restoration in 1990.
CULT.: 1768 (JH: 15); 1789, Nat. of Britain (HK1, 3: 232); 1813, Nat. of Britain (HK2, 5: 96); 1814, Nat. of Britain (HKE: 268); 1887, unspec. (K); 1936, Herbarium Ground (K); 1969, Herb. & Alpine Dept. (K); 1970, Herbaceous Dept. (K). Not currently in cult.
CURRENT STATUS: Has not reappeared in the Paddock since 1990; it was sown into the Wheatfields along the Broad Walk in 1999 and flowered profusely the following year; the area is now back under grass. It was sown in 2001 in the new Wheatfield to the south of the Lake on Syon Vista. It can occasionally be found as a relict from these sites.

1443. *Chrysanthemum coronarium L.
crown daisy
FIRST RECORD: 1934 [1768].
REFERENCES: **Kew Gardens:** 1990, *Cope* (S31: 182): Paddock, after restoration in 1990.
EXSICC.: **Kew Gardens.** ?Kew Gardens; 1 x 1934; *Rake* s.n. (K).
CULT.: 1768 (JH: 15); 1789, Alien (HK1, 3: 233); 1813, Alien (HK2, 5: 96); 1814, Alien (HKE: 268); 1883, unspec. (K); 1961, Order Beds (K); 1969, Herb. & Alpine Dept. (K); 1971, Herb. & Alpine Dept. (K). Not currently in cult.
CURRENT STATUS: No recent sightings; it has not reappeared in the Paddock since 1990.

Scented mayweed (*Matricaria recutita*). A rather common weed of flower-beds and other disturbed places and may persist for a while in new grassland.

1444. Leucanthemum vulgare Lam.
Chrysanthemum leucanthemum L.
oxeye daisy
FIRST RECORD: 1873/4 [1768].
NICHOLSON: **1873/4 survey** (JB: 46): Very frequent in the turf. Common also both in cultivated and waste ground. • **1906 revision** (AS: 82): Very common in the turf; common also in cultivated ground.
EXSICC.: **Kew Gardens.** Herbarium grounds; 21 vi 1892; *anon.* s.n. (K). • Near Herbarium; in long grass on sandy soil; 16 vi 1928; *Hubbard* s.n. (K). • Round the Herbarium; c. 20'; vi 1928; *C.A.Smith* 6058 (K). • Queen's Cottage Grounds; in grassy glades; 14 vi 1937; *Hubbard* s.n. (K). • [Unlocalised]; creeping perennial herb growing in rough grass; 3 vi 1944; *Souster* 78 (K). • Near Tennis Courts; 4 vi 1944; *Hutchinson* s.n. (K). **Kew Environs.** Kew; *Goodenough* s.n. (K).
CULT.: 1768 (JH: 14); 1789, Nat. of Britain (HK1, 3: 232); 1813, Nat. of Britain (HK2, 5: 93); 1814, Nat. of Britain (HKE: 268); 1880s, unspec. (K); 1919, unspec. (K); 1932, unspec. (K); 1933, Herb Garden (K); 1936, Herbaceous Ground (K); 1961, Herbarium Ground (K); 1970, Herbaceous Dept. (K); 1971, Herb. & Alpine Dept. (K); no date, Herbarium Ground (K). In cult.
CURRENT STATUS: Widely scattered in long grass throughout the Gardens but can no longer be described as common.

1445. Matricaria recutita L.
M. chamomilla auct. non L.
scented Mayweed
FIRST RECORD: 1873/4 [1768].
NICHOLSON: **1873/4 survey** (JB: 46): Much more common than [*Tripleurospermum inodorum*]. A discoid form of this grows abundantly by the side of the Kew Road. • **1906 revision** (AS: 82): A common weed.
EXSICC.: **Kew Gardens.** Student vegetable plots; cultivated ground & waste places; 20 viii 1978; *Roberts* 14 (K).
CULT.: 1768 (JH: 19); 1789, Nat. of Britain (HK1, 3: 234); 1813, Nat. of Britain (HK2, 5: 101); 1814, Nat. of Britain (HKE: 269); 1922, unspec. (K); 1934, unspec. (K). Not currently in cult.
CURRENT STATUS: Widespread in disturbed ground but not common.

1446. *Matricaria discoidea DC.
M. matricarioides (Less.) Porter, nom. illegit.
M. suaveolens (Pursh) Buchenau non L.
pineappleweed
FIRST RECORD: 1869 [1768].
NICHOLSON: **1873/4 survey**: Not recorded. • **1906 revision** (AS: 82): Not uncommon in some localities within our area.
OTHER REFERENCES: **Kew Environs.** 1871, reported 1921, *Druce* (BSEC6: 291): First evidence: a plant collected at Kew in 1871 by Mr J.Gilbert Baker in Herb. Druce, labelled by Baker *M. chamomilla* var. *discoidea*. • 1900, *Dunn* (WBEC17: 20; SFS: 390; LFS: 317): Kew Green.
EXSICC.: **Kew Gardens.** [Unlocalised]; on waste ground; vii 1928; *Hubbard* s.n. (K). • Office of Works Nursery; weed round rubbish heap; 7 vi 1930; *Turrill* s.n. (K). • [Unlocalised]; weed on rubbish heap; 1 vii 1932; *Hubbard* s.n. (K). • Lower Nursery; in cultivated soil; 3 x 1980; *Goodfellow* 13 (K). **Kew Environs.** Banks of the Thames, near Kew; 20 x 1869; *Thomson* s.n. (K). • Westerlay Ware; 11 vii 1886; *Fraser* s.n. (K). • Westerley Ware; 23 vii 1886; *Fraser* s.n. (K). • Kew; 21 vii 1895; *Worsdell* s.n. (K). • Westerley Ware; 22 vii 1914; *Fraser* s.n. (K). • Thames bank opposite Brentford Docks; on new sloping stone wall, submerged at high tide; 26 viii 1929; *Summerhayes* 438 (K). • Kew Green; in gravelly soil bordering path & much trodden; 20 vi 1944; *Souster* 103 (K).
CULT.: 1768 (JH: 19); 1789, Nat. of Britain (HK1, 3: 234); 1813, Alien (HK2, 5: 100); 1814, Alien (HKE: 269); 1931, unspec. (K). Not currently in cult.
CURRENT STATUS: Uncommon; rather scattered as a weed of rough and worn ground.

1447. Tripleurospermum maritimum (L.) W.D.J. Koch
Matricaria maritima L.
Pyrethrum maritimum (L.) Sm.
sea mayweed
CULT.: 1768 (JH: 19); 1789, Nat. of Britain (HK1, 3: 234); 1813, Nat. of Britain (HK2, 5: 99); 1814, Nat. of Britain (HKE: 269); 1904, unspec. (K). Not currently in cult.

1448. Tripleurospermum inodorum (L.) Sch.Bip.
Matricaria inodora L.
Chrysanthemum inodorum (L.) L.
Pyrethrum inodorum (L.) Moench
scentless mayweed
FIRST RECORD: 1873/4 [1768].
NICHOLSON: **1873/4 survey** (JB: 46): With [*Tanacetum parthenium* — on most of the soil heaps, and as a weed in many of the flower-beds and shrubberies], also in the open turf near lake. • **1906 revision** (AS: 82): A weed in flower borders and shrubberies.
EXSICC.: **Kew Gardens.** Near the Herbarium; waste heap; vi 1928; *C.A.Smith* 6069 (K).
CULT.: 1768 (JH: 9); 1813, Nat. of Britain (HK2, 5: 99); 1814, Nat. of Britain (HKE: 269); 1933, unspec. (K). Not currently in cult.
CURRENT STATUS: An occasional weed of disturbed ground.

1449. *Cotula coronopifolia L.
buttonweed
FIRST RECORD: 1963 [1768].
REFERENCES: **Kew Gardens.** 1963, *Airy Shaw* (S29: 408): At base of propagating houses in Melon Yard, with *Chenopodium glaucum* [q.v.] and *Rorippa* spp.
CULT.: 1768 (JH: 9, 31); 1789, Alien (HK1, 3: 235); 1813, Alien (HK2, 5: 102); 1814, Alien (HKE: 270); 1930, unspec. (K). Not currently in cult.
CURRENT STATUS: No recent sightings.

[1450. Senecio fluviatilis Wallr.
S. sarracenicus L. pro parte
broad-leaved ragwort
CULT.: 1789, Nat. of England (HK1, 3: 195); 1813, Nat. of Britain (HK2, 5: 45); 1814, Nat. of Britain (HKE: 262).
Note. Listed in error as native.]

1451. Senecio paludosus L.
fen ragwort
CULT.: 1768 (JH: 25); 1789, Nat. of England (HK1, 3: 194); 1813, Nat. of England (HK2, 5: 44); 1814, Nat. of England (HKE: 262). Not currently in cult.

1452. Senecio jacobaea L.
common ragwort
FIRST RECORD: 1873/4 [1768].
NICHOLSON: **1873/4 survey** (JB: 46): **P** and **Q.** Here and there in open places and less shady parts of woods. **Strip:** Much less frequent than *Senecio aquaticus*. • **1906 revision** (AS: 82): **A, P, Q.** Here and there in open places and less shady parts of woods. **Strip:** Much less frequent than *S. aquaticus*.
OTHER REFERENCES: **Kew Gardens.** 1999, *Stones* (EAK: 5): By Palm House Pond.
EXSICC.: **Kew Gardens.** [Unlocalised]; vii 1900; *G. Nicholson* s.n. (K).
CULT.: 1768 (JH: 25); 1789, Nat. of Britain (HK1, 3: 193); 1813, Nat. of Britain (HK2, 5: 44); 1814, Nat. of Britain (HKE: 262); 1933, Herb Garden (K); 1936, Herbaceous Ground (K) 1937, Herbarium Ground (K). Not currently in cult.
CURRENT STATUS: Widespread and abundant in grass; occasionally showing signs of introgression from *S. aquaticus*.

1453. Senecio aquaticus Hill
marsh ragwort
FIRST RECORD: 1873/4 [1813].
NICHOLSON: **1873/4 survey** (JB: 46): **P.** A plant or two near lake. **Strip.** Common. • **1906 revision** (AS: 82): **A.** Here and there near lake. **Strip.** Common.
OTHER REFERENCES: **Kew Environs.** 1933, *anon.* (LN12: S61): Between Kew and Richmond.
EXSICC.: **Kew Environs.** On Thames bank; 20 vi 1928; *Hubbard* s.n. (K). • Kew, Thames bank; in middle zone with *Phalaris* on mud; 13 vii 1928; *Turrill* s.n. (K). • Between Kew & Richmond bridges along bank of Thames; vii 1928; *C.A.Smith* 6082 (K). • River-bank, Kew; 14 viii 1936; *A.K.Jackson* 1, 3 & 4 (K).
CULT.: 1813, Nat. of Britain (HK2, 5: 44); 1814, Nat. of Britain (HKE: 262). Not currently in cult.
CURRENT STATUS: Occasional; by the Lake and along the towpath; scattered elsewhere but showing signs of introgression from *S. jacobaea*.

1454. Senecio erucifolius L.
S. tenuifolius Jacq.
hoary ragwort
CULT.: 1768 (JH: 25); 1789, Nat. of Britain (HK1, 3: 193); 1813, Nat. of Britain (HK2, 5: 43); 1814, Nat. of Britain (HKE: 261); 1880s, unspec. (K); 1976, plot 152.05 (K). In cult.

1455. *Senecio squalidus L.
Oxford ragwort
FIRST RECORD: 1940 [1813].
REFERENCES: **Kew Gardens.** 1964, *Brenan* (S29: 408): Neglected allotment behind Wing A of Herbarium [see under *Senecio* × *baxteri*: not otherwide recorded]. • 1983/4, *Cope* (S31: 181): New student vegetable plots behind Hanover House.

EXSICC.: **Kew Gardens.** Weed in beds by the Main Gate; in slight shade; 26 v 1940; *Cotton* s.n. (K). • On rubble & brickwork of old reservoir; 24 iv 1944; *Souster* 34 (K). • Near the Herbarium; 4 v 1944; *Hutchinson* s.n. (K). • Garden of Hanover House; 4 vi 1944; *Hutchinson* s.n. (K). • Mound of earth from excavation for new wing; 14 x 1965; *Verdcourt* 4237 (K).
CULT.: 1813, Nat. of England (HK2, 5: 42); 1814, Nat. of England (HKE: 261); 1938, Herbarium Ground (K); 1950, Herbarium Ground (K). Not currently in cult.
CURRENT STATUS: Widely scattered but uncommon; mostly in disturbed areas in the northern part of the Gardens.

1456. Senecio × baxteri Druce
(squalidus × vulgaris)
FIRST RECORD: 1964.
REFERENCES: **Kew Gardens.** 1964, *Brenan* (S29: 408): One plant with the parents (much *vulgaris*, several *squalidus*) on neglected allotment behind Wing A of Herbarium. The plant was apparently perennating, or at least behaving as a biennial. Habit branching like *squalidus*; capitula intermediate, but nearer *vulgaris* in size, with short rays. Achenes quite sterile.
EXSICC.: **Kew Gardens.** Herbarium Grounds; 1964; *Brenan* s.n. (K).
CURRENT STATUS: No recent sightings. It is certain that the original plant did not survive and there are no records of the hybrid having been remade within the Gardens.

1457. Senecio vulgaris L.
S. vulgaris subsp. *denticulatus* (O.F. Müll.) P.D.Sell
groundsel
FIRST RECORD: 1873/4 [1789].
NICHOLSON: **1873/4 survey** (JB: 46): Common. • **1906 revision** (AS: 82): Common.
OTHER REFERENCES: **Kew Gardens.** 1964, *Brenan* (S29: 408): Neglected allotment behind Wing A of Herbarium [see under *Senecio × baxteri*]. • 1990, *Cope* (S31: 182): [subsp. *denticulatus*] Paddock, after restoration in 1990.
EXSICC.: **Kew Gardens.** [Unlocalised]; in cultivated ground; v 1928; *Hubbard* s.n. (K). • Herb. Expt. Ground; in pot in greenhouse; 8 v 1946; *Blakelock* s.n. (K). • Student plots; 12 viii 1981; *Borg* 6 (K). **Kew Environs.** Kew; waste ground; 25 v 1928; *Fraser* s.n. (K). • Thames banks east end of Kew; 14 iv 1930; *Fraser* s.n. (K). • Kew; 30 v 1931; *Fraser* s.n. (K). • Kew; 18 x 1933; *Fraser* s.n. (K). • Towpath, Kew – Richmond; 13 vi 1948; *Parker* s.n. (K).
CULT.: 1789, Nat. of Britain (HK1, 3: 191); 1813, Nat. of Britain (HK2, 5: 39); 1814, Nat. of Britain (HKE: 261). Not currently in cult.
CURRENT STATUS: Widespread and abundant throughout the Gardens.

1458. Senecio sylvaticus L.
S. lividus auct. non L.
heath groundsel
FIRST RECORD: 1873/4 [1768].
NICHOLSON: **1873/4 survey** (JB: 46): Seems to be much more abundant about lake and elsewhere than [*Senecio vulgaris*]. • **1906 revision** (AS: 82 and in SFS: 394): More abundant about lake and along borders of woods than [*S. vulgaris*].
CULT.: 1768 (JH: 25); 1789, Nat. of Britain (HK1, 3: 192); 1813, Nat. of Britain (HK2, 5: 41), Alien (ibid.: 40 as *lividus*); 1814, Nat. of Britain (HKE: 261), Nat. of England (ibid.: as *lividus*). Not currently in cult.
CURRENT STATUS: Now very rare; it has recently been seen only near the Bamboo Garden and outside the Stable Yard. It is certainly no longer commoner than *S. vulgaris* as reported by Nicholson.

[**1459. Senecio viscosus** L.
sticky groundsel
CULT.: 1789, Nat. of Britain (HK1, 3: 192); 1813, Nat. of Britain (HK2, 5: 40); 1814, Nat. of Britain (HKE: 261).
Note. Originally thought to be native but now regarded as a neophyte.]

1460. Tephroseris integrifolia (L.) Holub
Othonna integrifolia L.
Cineraria campestris Retz.
C. alpina auct. non L.
field fleawort
CULT.: 1768 (JH: 4); 1789, Nat. of England (HK1, 3: 222); 1813, Nat. of England (HK2, 5: 74); 1814, Nat. of England (HKE: 265). Not currently in cult.

1461. Tephroseris palustris (L.) Fourr.
Cineraria palustris
marsh fleawort
CULT.: 1789, Nat. of England (HK1, 3: 222); 1813, Nat. of England (HK2, 5: 74); 1814, Nat. of England (HKE: 265). Not currently in cult.

[**1462. Doronicum pardalianches** L.
leopard's-bane
CULT.: 1789, Nat. of Scotland (HK1, 3: 226); 1813, Nat. of Britain (HK2, 5: 83); 1814, Nat. of Britain (HKE: 267).
Note. Listed in error as native.]

1463. *Doronicum austriacum Jacq.
Austrian leopard's-bane
FIRST RECORD: 2002 [1901].
REFERENCES: **Kew Gardens.** 2002, *Cope* (S35: 647): In wooded area between the Bamboo Garden and Mount Pleasant (252).
CULT.: 1901, unspec. (K). In cult.

CURRENT STATUS: Incorporated in the Robinsonian Meadow in front of 'Climbers & Creepers' but does not show signs of persisting.

1464. Tussilago farfara L.
colt's-foot
FIRST RECORD: 1873/4 [1768].
NICHOLSON: **1873/4 survey** (JB: 47): **P.** About lake and in "Engine House" yard. • **1906 revision** (AS: 82): **A.** About lake and near pumping station.
EXSICC.: **Kew Gardens.** Herbarium Exper. Ground; in loose soil of recently dug bed; 5 vii 1928; *Summerhayes* 339 (K). • Thames bank, Kew – Richmond; at top of river-wall, Richmond side of Brentford Ferry; 21 iii 1929; *Summerhayes & Turrill* 3329 (K). • Herbarium field; 25 iii 1936; *Sealy* 427 (K). • Herbarium field; 20 iv 1960; *anon.* s.n. (K). **Kew Environs.** River bank, Kew; 24 iv 1934; *Turner* s.n. (K). • Towpath, Kew; 11 iv 1944; *Souster* 13 (K).
CULT.: 1768 (JH: 6); 1789, Nat. of Britain (HK1, 3: 188); 1813, Nat. of Britain (HK2, 5: 35); 1814, Nat. of Britain (HKE: 261); 1915, unspec. (K); 1933, unspec. (K); 1933, Herb Garden (K). Not currently in cult.
CURRENT STATUS: Rare; a small population still occurs beside the Lake and a slightly larger one on the towpath outside Brentford Gate.

1465. Petasites hybridus (L.) G.Gaertn., B.Mey. & Scherb.
Tussilago hybrida L. (applies to the ♀ plant)
T. petasites L. (applies to the ♂ plant)
Petasites officinalis Moench
P. ovatus Hill
P. vulgaris Desf.
butterbur
FIRST RECORD: 1873/4 [1768].
NICHOLSON: **1873/4 survey** (JB: 47): **Strip.** Very common on the river side of towing path. • **1906 revision** (AS: 82): **Strip.** Very common on the river side of towing-path.
OTHER REFERENCES: **Kew Gardens.** 1999, *Stones* (EAK: 2 & 5): By Lake and Palm House Pond. **Kew Environs.** 1918, *Tremayne* (LN32: S157): Thames-side between Kew and Richmond. • 1933, *anon.* (LN12: S60): By the Thames between Kew and Richmond.
EXSICC.: **Kew Gardens.** Thames bank opp. Sion Ho.; iv 1888; *Gamble* 19850 (K). • Thames-bank, Kew – Richmond; Richmond side of Isleworth Gate, top zone; 16 viii 1928; *Summerhayes & Turrill* s.n. (K). • Kew Towpath opposite Sion House; found in a very sandy situation close to river side; 3 iv 1948; *O.J.Ward* 21 (K). **Kew Environs.** Banks of Thames, Kew; 2 v 1885; *Levinge* s.n. (K). • Thames side, Kew; 15 iv 1923; *Knight* s.n. (K). • On tow path just outside Kew Gardens; in eroding grass-turf; 19 iii 1928; *Summerhayes & Turrill* s.n. (K). • Tow Path Kew – Richmond; found on river's edge in very sandy soil in association with *Urtica dioica* & *Galium aparine*; 15 iv 1948; *Parker* s.n. (K).
CULT.: 1768 (JH: 6, 30); 1789, Nat. of Britain (HK1, 3: 189 as ♂ & ♀); 1813, Nat. of Britain (HK2, 5: 36 as ♂); 1814, Nat. of Britain (HKE: 261 as ♂); 1977, unspec. (K). In cult.
CURRENT STATUS: Rare or extinct. Records from around the Lake and Palm House Pond are erroneous and belong to cultivated plants of a different species. One patch in the Conservation Area, also planted, needs its identity checking. It has not recently been seen on the Thames bank where it was once a conspicuous component of the riverside flora.

1466. Homogyne alpina Cass.
Tussilago alpina L.
purple cat's-foot
CULT.: 1768 (JH: 6). Not currently in cult.
Note. Possibly not a native species

1467. *Calendula arvensis L.
field marigold
FIRST RECORD: 1990 [1768].
REFERENCES: **Kew Gardens.** 1990, *Cope* (S31: 182): Paddock, after restoration in 1990.
CULT.: 1768 (JH: 10); 1789, Alien (HK1, 3: 270); 1813, Alien (HK2, 5: 166); 1814, Alien (HKE: 278); 1895, unspec. (K); 1970, Alpine & Herbaceous Dept. (K); 1971, South Arboretum (K); 1980, unspec. (K). Not currently in cult.
CURRENT STATUS: No recent sightings; it has not reappeared in the Paddock since 1990.

1468. *Ambrosia artemisiifolia L.
ragweed
FIRST RECORD: 2004 [1768].
CULT.: 1768 (JH: 366); 1789, Alien (HK1, 3: 345); 1813, Alien (HK2, 5: 269); 1814, Alien (HKE: 292); 1901, unspec. (K); 1971, plot 157.64 (K). Not currently in cult.
CURRENT STATUS: A single plant was found by the Lake, most probably originating from birdseed.

1469. *Iva xanthiifolia Nutt.
marsh-elder
FIRST RECORD: 2007 [1856].
CULT.: 1856, unspec. (K); 1875, unspec. (K); 1883, unspec. (K). In cult.
CURRENT STATUS: Three plants grew from seed in topsoil stored in the Paddock but did not persist.

[**1470. Xanthium strumarium** L.
rough cocklebur
CULT.: 1789, Nat. of England (HK1, 3: 343); 1813, Nat. of England (HK2, 5: 268); 1814, Nat. of England (HKE: 292).
Note. Listed in error as native.]

1471. *Sigesbeckia serrata* DC.
Western St Paul's-wort

FIRST RECORD: 2007.
CURRENT STATUS: One or two plants grew from seed in topsoil stored in the Paddock but did not persist.

1472. *Rudbeckia hirta* L.
bristly coneflower

FIRST RECORD: 1999 [1768].
REFERENCES: **Kew Gardens.** 1999, *Cope* (S35: 647): A patch in the Paddock near the Herbarium (104).
CULT.: 1768 (JH: 10); 1789, Alien (HK1, 3: 250); 1813, Alien (HK2, 5: 131); 1814, Alien (HKE: 273). In cult.
CURRENT STATUS: Briefly appeared in the Paddock, but has not returned.

1473. *Helianthus annuus* L.
sunflower

FIRST RECORD: 2006 [1768].
CULT.: 1768 (JH: 12); 1789, Alien (HK1, 3: 248); 1813, Alien (HK2, 5:126); 1814, Alien (HKE: 273); 1895, unspec. (K); 1896, unspec. (K); 1967, Alpine Dept. (K); 1971, plot 157 (K); 1976, Herbaceous Dept. (K). In cult.
CURRENT STATUS: Appeared in quantity in a heap of topsoil removed from the staff allotment area behind the Mycology Building and stored on the Herbarium lawn. It was planted one year in quantity for display purposes and often reappears as a relict.

1474. *Helianthus tuberosus* L.
Jerusalem artichoke

FIRST RECORD: 2006 [1768]
CULT.: 1768 (JH: 12); 1789, Alien (HK1, 3: 249); 1813, Alien (HK2, 5:126); 1814, Alien (HKE: 273); 1908, unspec. (K); 1921, unspec. (K); 1936, Herbaceous Ground (K); 1955, Herbarium Expt. Ground (K); 1982, Vegetable Plot (K). Not currently in cult.
CURRENT STATUS: Appeared in quantity in a heap of soil removed from the staff allotment area behind the Mycology Building and stored on the Herbarium lawn. Doubtless a relict of cultivation from these allotments.

1475. *Galinsoga parviflora* Cav.
G. hirsuta Baker
G. hirsuta var. *adenophora* Thell.
gallant-soldier

FIRST RECORD: 1861 [1796].
NICHOLSON: **1873/4 survey** (JB: 47): Not common. A few plants only exist in each division. This does not hold its own at all in our Flora, whilst in the market gardens of Kew and Mortlake it is the most troublesome of all weeds. It seems to delight in soil which is being often turned. • **1906 revision** (AS: 81): A troublesome weed in well-cultivated ground.

Gallant soldier (*Galinsoga parviflora*). An occasional weed that despite its association with Kew Gardens has never been common; it is also called 'Joey Hooker' after Kew's second Director.

OTHER REFERENCES: **Kew Gardens.** 1976, *Lousley* (LFS: 306): This South American plant is known to have been grown in Kew Gardens in 1796 and escaped from there [1860–61] ... **Kew Environs.** 1861, *Irvine* (TBEC4: 13 and see Exsicc.): Kew Bridge, probably an escape from Kew Gardens. • 1904, *Thompson* (WBEC21: 17): Waste land, Kew. •1917, *Marshall* (BSEC5: 228): All my British specimens (Kew and Milford, Surrey; Oxford) have glands on the peduncles. • 1983, *Burton* (FLA: 145): [Arrived in 1860] as an escape from Kew Gardens where it had been grown for study.
EXSICC.: **Kew Gardens.** Allotments; 28 viii 1942; *Hutchinson* s.n. (K). • Allotments by the Herbarium; 10 ix 1942; *Melville* s.n. (K). • Herbarium field; cultivated ground (allotments), plentiful; 7 x 1944; *Ross-Craig & Sealy* 1036 (K). • Herbarium field; 12 x 1944; *Ross-Craig & Sealy* 1037 (K). • Lower Nursery; 3 x 1980; *Goodfellow* 12 in part (K). • Allotments; vii 1981; *Innes* 5 (K). **Kew Environs.** Alien; Kew Bridge, Surrey; 1861; *Irvine* s.n. (K). • Kew Bridge; ix 1862; *Mill* s.n. (K). • Kew; vii 1864; *Syme* in *Herb. Watson* (K). • Market Garden Ground, Kew; 13 viii 1875; *G. Nicholson* 727 (K). • Kew; 20 ix 1882; *Crosfield* s.n. (K). • Kew Road; 27 vi 1933; *Fraser* s.n. (K). • Thames embankment, Kew; x 1939(?); *Worsdell* s.n. (K).
CULT.: 1813, Alien (HK2, 5: 122); 1814, alien (HKE: 272); 1900, unspec. (K). Not currently in cult.

Current status: Rather scarce; despite its association with Kew Gardens (another English name for it is Joey Hooker after Kew's second director) it has never been particularly common.

1476. *Galinsoga quadriradiata Ruiz & Pav.
G. ciliata (Raf.) S.F.Blake
G. quadriradiata var. *hispida* (DC.) Thell.
shaggy-soldier
First record: 1943 [1881].
References: **Kew Gardens.** 1983/4, *Cope* (S31: 181): New student vegetable plots behind Hanover House. **Kew Environs.** 1945, *L.G.Payne* (BSEC13: 297; LN25: 14): Kew.
Exsicc.: **Kew Gardens.** In allotments behind the Herbarium; 9 vii 1943; *Summerhayes* 1285 (K). • In allotments behind Herbarium, very local; 24 vii 1943; *Summerhayes* 1286 (K). • Herbarium field; 13 x 1944; *Ross-Craig & Sealy* 1039 (K). • Weed in Herbaceous Ground; 4 ix 1949; *Melville* s.n. (K). • Lower Nursery; 3 x 1980; *Goodfellow* 12 in part (K).
Cult.: 1881, unspec. (K); 1932, unspec. (K). Not currently in cult.
Current status: Rather scarce; more widespread than the preceding species but still not common.

1477. Bidens cernua L.
B. minima Huds.
nodding bur-marigold
First record: 1873/4 [1768].
Nicholson: **1873/4 survey** (JB: 47): **P.** Common about edge of lake. • **1906 revision** (AS: 81): **A.** Common about lake. **Strip.**
Cult.: 1768 (JH: 5, 30); 1789, Nat. of Britain (HK1, 3: 153); 1812, Nat. of Britain (HK2, 4: 494); 1814, Nat. of Britain (HKE: 253). Not currently in cult.
Current status: No recent sightings.

1478. Bidens tripartita L.
trifid bur-marigold
First record: 1873/4 [1768].
Nicholson: **1873/4 survey** (JB: 47 and in SFS: 382): **P.** Growing with, and much more abundant than, [*Bidens cernua*]; also in the *Corylaceae* collections on the right hand side of Syon Vista going towards the river. • **1906 revision** (AS: 81): Growing with [*Bidens cernua*], but more abundant.
Other references: **Kew Gardens.** 1999, *Stones* (EAK: 1): By Lake. **Kew Environs.** 1931, *Livett* (SFS: 382): Thames between Kew and Richmond.
Exsicc.: **Kew Environs.** Thames banks, Kew; 12 vi 1931; *Fraser* s.n. (K). • Towpath, Kew; 23 viii 1953; *Souster* 1456 (K).
Cult.: 1768 (JH: 5, 30); 1789, Nat. of Britain (HK1, 3: 153); 1812, Nat. of Britain (HK2, 4: 494); 1814, Nat. of Britain (HKE: 253). In cult.
Current status: Still occurs around the Lake, on the towpath and in the Old Deer Park.

1479. *Coreopsis sp.
tickseed
First record: 1990.
References: **Kew Gardens.** 1990, *Cope* (S31: 182): Paddock, after restoration in 1990.
Cult.: Various species are in cult.
Current status: No recent sightings; it has not reappeared in the Paddock since 1990. The plant, when found, was too young for determination and was not seen again. Several species of *Coreopsis* are, or have been, cultivated so it is impossible to guess which it may have been.

1480. Eupatorium cannabinum L.
hemp-agrimony
First record: 1953 [1768].
References: **Kew Environs.** 1953, *Cooke* (LN32: S144; S33: 750): Kew.
Cult.: 1768 (JH: 34); 1789, Nat. of Britain (HK1, 3: 160); 1812, Nat. of Britain (HK2, 4: 507); 1814, Nat. of Britain (HKE: 255); 1881, unspec. (K); 1895, unspec. (K); 1900, unspec. (K); 1933, Arboretum (K); 1936, Herbaceous Ground (K). In cult.
Current status: Rare; a new find within the Gardens; probably planted but now naturalised in the bee garden behind Kew Palace. It can also be found along the towpath where it is undoubtedly native. It spreads easily by seed and will probably increase.

Trifid bur-marigold (*Bidens tripartita*). Once abundant but now restricted to the Lake margins, on the towpath and in the Old Deer Park.

MONOCOTYLEDONS

BUTOMACEAE

1481. Butomus umbellatus L.
flowering-rush
FIRST RECORD: 1853 [1789].
NICHOLSON: **1873/4 survey** (JB: 73): **Strip.** Frequent at edge of moat. • **1906 revision** (AS: 88): **Strip.** Common along edge of ha-ha.
EXSICC.: **Kew Environs.** Stagnant ditch, parallel with the Thames, Kew; vi 1853; *Brocas* 1114 (K).
CULT.: 1789, Nat. of Britain (HK1, 2: 42); 1811, Nat. of Britain (HK2, 2: 432); 1814, Nat. of Britain (HKE: 119); 1972, Aquatic Garden (K). In cult.
CURRENT STATUS: No recent sightings; the Moat is no longer suitable for it.

ALISMATACEAE

1482. Sagittaria sagittifolia L.
arrowhead
CULT.: 1768 (JH: 160); 1789, Nat. of England (HK1, 3: 352); 1813, Nat. of England (HK2, 5: 282); 1814, Nat. of England (HKE: 294). In cult. under glass.

1483. Baldellia ranunculoides (L.) Parl.
Alisma ranunculoides L.
lesser water-plantain
CULT.: 1768 (JH: 161); 1789, Nat. of Britain (HK1, 1: 492); 1811, Nat. of Britain (HK2, 2: 332); 1814, Nat. of Britain (HKE: 105). Not currently in cult.

Flowering-rush (*Butomus umbellatus*). At one time common in the moat between the Gardens and the river but not recorded for over a century.

1484. Luronium natans (L.) Raf.
Alisma natans L.
floating water-plantain
CULT.: 1811, Nat. of Wales (HK2, 2: 332); 1814, Nat. of Wales (HKE: 105). In cult.

1485. Alisma plantago-aquatica L.
A. plantago L.
water-plantain
FIRST RECORD: 1873/4 [1768].
NICHOLSON: **1873/4 survey** (JB: 73): Common about lake and near moat. • **1906 revision** (AS: 88): Common about lake and near ha-ha.
EXSICC.: **Kew Environs.** Thames bank, Kew; from crevices of new stone wall, 3/4 way down; 2 viii 1928; *Turrill* s.n. (K).
CULT.: 1768 (JH: 161); 1789, Nat. of Britain (HK1, 1: 492); 1811, Nat. of Britain (HK2, 2: 332); 1814, Nat. of Britain (HKE: 105). In cult.
CURRENT STATUS: In the ditch alongside the towpath, one spot beside the Lake and in the Larch Pond in the Conservation Area. It has been planted near the Banks Building but otherwise seems to be genuinely wild in the Gardens.

1486. Alisma lanceolatum With.
A. plantago var. *lanceolatum* (With.) auct. dub.
narrow-leaved water-plantain
FIRST RECORD: 1868 [1868].
NICHOLSON: **1873/4 survey**: Not recorded. • **1906 revision**: Not recorded.
OTHER REFERENCES: **Kew Gardens.** 1868, *Baker* (LBEC11: 15; S33: 750): Kew Gardens. The wild state of var. *lanceolatum* is usually smaller than that of var. *genuinum*, but the cultivated specimens sent by Mr Baker are of large size, showing that var. *lanceolatum* is not merely a stunted state of *A. plantago*.
CULT.: 1868, unspec. (*Herb. Watson*, K). Not currently in cult.
CURRENT STATUS: At the southern end of the Lake and extremely rare. This is one of the species mentioned by Nicholson as having persisted in a 'half-wild condition' from original plantings. It was planted around the Princess of Wales Conservatory a while ago but is no longer in the living collections database.

1487. Damasonium alisma Mill.
Alisma damasonium L.
starfruit
CULT.: 1768 (JH: 161); 1789, Nat. of England (HK1, 1: 492); 1811, Nat. of England (HK2, 2: 332); 1814, Nat. of England (HKE: 105). Not currently in cult.

Narrow-leaved water-plantain (*Alisma lanceolatum*). Extremely rare, at the southern end of the Lake, but originally planted in the mid-19th century.

HYDROCHARITACEAE

1488. Hydrocharis morsus-ranae L.
frogbit
FIRST RECORD: 1873/4 [1768].
NICHOLSON: **1873/4 survey** (JB: 73): **Strip.** In moat nearly the whole length of "Queen's Cottage Grounds." • **1906 revision** (AS: 87): **Strip.** In ha-ha nearly the whole length of Queen's Cottage Grounds.
CULT.: 1768 (JH: 418); 1789, Nat. of Britain (HK1, 3: 409); 1813, Nat. of Britain (HK2, 5: 399); 1814, Nat. of Britain (HKE: 311). Not currently in cult.
CURRENT STATUS: No recent sightings; the Moat is no longer suitable for it

1489. Stratiotes aloides L.
water-soldier
CULT.: 1768 (JH: 160); 1789, Nat. of England (HK1, 2: 250); 1813, Nat. of England (HK2, 5: 402); 1814, Nat. of England (HKE: 312). Not currently in cult.

1490. *Elodea canadensis Michx.
Hydrophyllum canadense L.
Canadian waterweed
FIRST RECORD: 1853 [1789].
NICHOLSON: **1873/4 survey** (JB: 73): Abundant in every piece of water within our limits, except the pond in front of Palm House. Mr Smith, the curator of the Royal Gardens tells me that some years ago this piece of water was entirely choked up with this plant, *Confervae*, etc., and that it was a serious task to keep it anything like clear. For the last two years I have never been able to find any floating plant in it at all. Mr Smith could not furnish me with any reason for the disappearance of the *Elodea*, etc. He says about the same number, and nearly all the species, of aquatic birds now on the pond were there when it was so bad, and he does not give them credit for having effected such a change. • **1906 revision** (AS: 87): Abundant in every piece of water within our limits, except the pond, near Palm House — about 40 years ago, however, this particular sheet of water was choked up with *Elodea*.
EXSICC.: **Kew Environs.** Stagnant ditch, Thames bank, Kew; vi 1853; *Brocas* 1107 (K).
CULT.: 1789, Alien (HK1, 1: 197); 1810, Alien (HK2, 1: 313); 1814, Alien (HKE: 43); 1940, Aquatic Garden (K). Not currently in cult.
CURRENT STATUS: No recent sightings; past records of it may all have been referable to *E. nuttallii* but without vouchers it is impossible to be sure.

1491. *Elodea nuttallii (Planch.) H.St.John
Nuttall's waterweed
FIRST RECORD: 2002.
REFERENCES: **Kew Gardens.** 2002, *Cope & Phillips* (S35: 649): In the Lake. The Lake previously held *E. canadensis*, but during the present survey this could not be found anywhere.
CURRENT STATUS: In small quantity in the Lake and it may possibly account for early records of *E. canadensis*. The population was probably destroyed when the Lake was recently drained and refurbished.

Frogbit (*Hydrocharis morsus-ranae*) with duckweeds (*Spirodela polyrhiza* & *Lemna minor*). At one time frogbit grew in the whole length of the Moat, but is now extinct. [Photo: B.R.Tebbs]

Scheuchzeriaceae

1492. Scheuchzeria palustris L.
Rannoch-rush
Cult.: 1811, Nat. of England (HK2, 2: 324); 1814, Nat. of England (HKE: 104). Not currently in cult.

Juncaginaceae

1493. Triglochin palustre L.
marsh arrowgrass
Cult.: 1768 (JH: 161); 1789, Nat. of Britain (HK1, 1: 487); 1811, Nat. of Britain (HK2, 2: 325); 1814, Nat. of Britain (HKE: 104). In cult.

1494. Triglochin maritimum L.
sea arrowgrass
Cult.: 1768 (JH: 161); 1789, Nat. of Britain (HK1, 1: 488); 1811, Nat. of Britain (HK2, 2: 325); 1814, Nat. of Britain (HKE: 104). Not currently in cult.

Potamogetonaceae

1495. Potamogeton natans L.
broad-leaved pondweed
Cult.: 1768 (JH: 339); 1789, Nat. of Britain (HK1, 1: 171); 1810, Nat. of Britain (HK2, 1: 279); 1814, Nat. of Britain (HKE: 38). Not currently in cult.

1496. Potamogeton × lanceolatus Sm.
(berchtoldii × coloratus)
Cult.: 1814, Nat. of England (HKE: Add.). Not currently in cult.

1497. Potamogeton lucens L.
shining pondweed
Cult.: 1768 (JH: 339); 1789, Nat. of Britain (HK1, 1: 171); 1810, Nat. of Britain (HK2, 1: 280); 1814, Nat. of Britain (HKE: 38). Not currently in cult.

1498. Potamogeton × fluitans Roth
(lucens × natans)
Cult.: 1810, Nat. of Britain (HK2, 1: 279); 1814, Nat. of Britain (HKE: 38). Not currently in cult.

1499. Potamogeton gramineus L.
P. heterophyllus Schreb.
various-leaved pondweed
Cult.: 1768 (JH: 339); 1789, Nat. of Britain (HK1, 1: 172); 1810, Nat. of Britain (HK2, 1: 281, and: 280 as *heterophyllus*); 1814, Nat. of Britain (HKE: 38). Not currently in cult.

1500. Potamogeton perfoliatus L.
P. perfoliatus var. *ovatifolius* Wallr.
perfoliate pondweed
First record: 1920 [1768].
References: **Kew Gardens.** 1920, *Druce* (BSEC6: 151; S33: 750): Stream, west side of Kew Gardens, Surrey.
Cult.: 1768 (JH: 339); 1789, Nat. of Britain (HK1, 1: 171); 1810, Nat. of Britain (HK2, 1: 280); 1814, Nat. of Britain (HKE: 38). Not currently in cult.
Current status: No recent sightings. The 'stream' referred to can only have been the Moat.

1501. Potamogeton pusillus L.
lesser pondweed
First record: 1873/4 [1768].
Nicholson: **1873/4 survey** (JB: 73): **P.** Plentiful in small pond (fed with condensed steam from Engine House) at the lake end of Cedar Avenue. • **1906 revision** (AS: 88): **A.** Lake, water-lily pond.
Exsicc.: **Kew Gardens.** Small pond, Pleasure Grounds; 30 v 1874; *Nicholson* s.n. (K).
Cult.: 1768 (JH: 339); 1789, Nat. of Britain (HK1, 1: 172); 1810, Nat. of Britain (HK2, 1: 281); 1814, Nat. of Britain (HKE: 38). Not currently in cult.
Current status: Known recently only from the Lake, but the population may have been destroyed when the Lake was drained and refurbished. It has not been refound in the Water Lily Pond.

1502. Potamogeton compressus L.
grass-wrack pondweed
Cult.: 1768 (JH: 339); 1789, Nat. of Britain (HK1, 1: 171); 1810, Nat. of Britain (HK2, 1: 280); 1814, Nat. of Britain (HKE: 38). Not currently in cult.

1503. Potamogeton crispus L.
curled pondweed
First record: 1873/4 [1768].
Nicholson: **1873/4 survey** (JB: 73): **P.** A tuft in lake midway between the two islands nearest Palm House. • **1906 revision** (AS: 88): Lake and ha-ha.
Exsicc.: **Kew Gardens.** Ha-ha, Kew; vi 1927; *Findlay* s.n. (K).
Cult.: 1768 (JH: 339); 1789, Nat. of Britain (HK1, 1: 171); 1810, Nat. of Britain (HK2, 1: 280); 1814, Nat. of Britain (HKE: 38). Not currently in cult.
Current status: No recent sightings. Much of what was in the Lake was lost when it was recently drained and reconstructed.

1504. Potamogeton filiformis Pers.
P. marinum L.
slender-leaved pondweed
Cult.: 1768 (JH: 339). Not currently in cult.

1505. Potamogeton pectinatus L.
fennel pondweed
FIRST RECORD: 1999 [1768].
REFERENCES: **Kew Gardens**. 1999, *Stones* (EAK: 3): In Lake.
CULT.: 1768 (JH: 339); 1789, Nat. of Britain (HK1, 1: 172); 1810, Nat. of Britain (HK2, 1: 280); 1814, Nat. of Britain (HKE: 38). In cult. under glass.
CURRENT STATUS: Known only from the Lake, but the population may have been destroyed when the Lake was drained and refurbished.

1506. Groenlandia densa (L.) Fourr.
Potamogeton densus L.
P. serratus L.
opposite-leaved pondweed
FIRST RECORD: 1873/4 [1768].
NICHOLSON: **1873/4 survey** (JB: 73): **Strip.** Rather common. • **1906 revision** (AS: 88): **Strip.** Common in ha-ha.
OTHER REFERENCES: **Kew Environs.** 1948, *Boniface* (LN35: S286): Ditch by Thames near Kew.
EXSICC.: **Kew Environs.** Ha-ha, Kew; vi 1927; *Findlay* s.n. (K).
CULT.: 1768 (JH: 339); 1789, Nat. of Britain (HK1, 1: 171); 1810, Nat. of Britain (HK2, 1: 280); 1814, Nat. of Britain (HKE: 38). Not currently in cult.
CURRENT STATUS: No recent sightings.

RUPPIACEAE

1507. Ruppia maritima L.
beaked tasselweed
CULT.: 1810, Nat. of Britain (HK2, 1: 281); 1814, Nat. of Britain (HKE: 39). Not currently in cult.

ZANNICHELLIACEAE

1508. Zannichellia palustris L.
horned pondweed
FIRST RECORD: 1930 [1768].
REFERENCES: **Kew Environs.** 1930, *Lousley* (LN35: S287; S33: 750): Ditch near Kew Gardens.
EXSICC.: **Kew Gardens.** Pond between Arboretum Pits & Queen's Cottage; 30 vi 1931; *Turrill* s.n. (K).
CULT.: 1768 (JH: 411); 1789, Nat. of Britain (HK1, 3: 321); 1813, Nat. of Britain (HK2, 5: 229); 1814, Nat. of Britain (HKE: 287). Not currently in cult.
CURRENT STATUS: This was still in the Lake, at least until its recent refurbishment, but it has apparently gone from the Water Lily Pond.

ACORACEAE

1509. *Acorus calamus L.
sweet-flag
FIRST RECORD: 1873/4 [1768].
NICHOLSON: **1873/4 survey** (JB: 73): **B.** All round pond, very plentiful near the steps. **P.** Lake. **Strip.** Frequent. • **1906 revision** (AS: 88): Plentiful about lake, pond and ha-ha.
OTHER REFERENCES: **Kew Gardens.** 1999, *Stones* (EAK: 2 & 5): By Lake and Palm House Pond. **Kew Environs.** 1955, *Robbins* (LN35: S281): Here and there by the Thames between Kew and Kingston.
EXSICC.: **Kew Gardens.** Naturalised & abundant on margins of lake; 7 vii 1948; *Souster* 852 (K).
CULT.: 1768 (JH: 387); 1789, Nat. of England (HK1, 1: 474); 1811, Nat. of England (HK2, 2: 305); 1814, Nat. of England (HKE: 101); 1965, unspec. (K). In cult.
CURRENT STATUS: Plants around the Lake and Palm House Pond margins were all planted long ago and are now thoroughly naturalised. It has been lost from the Strip and the Ha-ha.

Sweet-flag (*Acorus calamus*). Planted wherever it occurs; an important stand can be found at the extreme south end of the Lake.

ARACEAE

1510. *Calla palustris L.
bog arum
FIRST RECORD: 1867.
NICHOLSON: **1873/4 survey**: Not recorded. • **1906 revision**: Not recorded.
EXSICC.: **Kew Gardens.** [Unlocalised]; 26 ix 1867; *Duncan* s.n. (K).
CULT.: 1965, unspec. (K); 1994, Alpine & Herbaceous: Rockery, plot 154-0410 (K). Not currently in cult.
CURRENT STATUS: Not known. Since Nicholson did not record the species it is probable that at the time the specimen was collected it was only in cultivation within the Gardens.

1511. Arum maculatum L.
lords-and-ladies
FIRST RECORD: 1873/4 [1768].
NICHOLSON: **1873/4 survey** (JB: 73): **B.** On the mound where the *Scrophularia vernalis* grows. **P.** At foot of wall from "Unicorn Gate" to opposite "Douglas Spar." • **1906 revision** (AS: 88): **B.** Ice-house mound. **A.** Under trees along wall between Unicorn Gate and North Gallery. **Q.** Here and there.
OTHER REFERENCES: **Kew Gardens.** 1962, *Airy Shaw* (S29: 408): In copse between tennis courts and river, near Kew Palace.
CULT.: 1768 (JH: 367); 1789, Nat. of Britain (HK1, 3: 317); 1813, Nat. of Britain (HK2, 5: 308); 1814, Nat. of Britain (HKE: 298); 1883, unspec. (K). In cult.
CURRENT STATUS: Occasional in wooded areas, probably originating as bird-sown seed from cultivated specimens.

Lords-and-ladies (*Arum maculatum*). Occasional in wooded and shady areas but likely to have been bird-sown from cultivated specimens.

1512. Arum italicum Mill.
Italian lords-and-ladies
a. subsp. ***italicum**
A. maculatum var. *italicum* (Mill.) O.Targ.Tozz.
FIRST RECORD: 1950 [1789].
REFERENCES: **Kew Environs.** 1950, *Welch* (LN35: S280; S33: 751): By river Thames between Kew and Richmond, one plant.
CULT.: 1789, Alien (HK1, 3: 317); 1813, Alien (HK2, 5: 309); 1814, Alien (HKE: 298). In cult.
CURRENT STATUS: Cultivated in a number of places but widely naturalised from bird-sown seed.

b. subsp. **neglectum** (F.Towns.) Prime
CULT.: 1935, Herb. Dept. (K); 1936, unspec. (K). Not currently in cult.

LEMNACEAE

1513. Spirodela polyrhiza (L.) Schleid.
Lemna polyrhiza L.
greater duckweed
FIRST RECORD: 1855 [1768].
NICHOLSON: **1873/4 survey** (JB: 73): Lake and moat, very abundant. • **1906 revision** (AS: 88): Lake and ha-ha, abundant.
OTHER REFERENCES: **Kew Environs.** 1929, *Lousley* (LN35: S281): Ditch outside Kew Gardens.
EXSICC.: **Kew Environs.** Ditches, Kew; 1855; *anon.* s.n. (K). • A pond, Kew; 1 viii 1930; *Worsdell* s.n. (K). • Towpath between Kew & Richmond; floating with *Lemna minor* & *L. gibba* in ditch; 2 viii 1948; *Souster* 921 (K).
CULT.: 1768 (JH: 367); 1789, Nat. of Britain (HK1, 3: 322); 1813, Nat. of Britain (HK2, 5: 234); 1814, Nat. of Britain (HKE: 287). Not currently in cult.
CURRENT STATUS: This can still to be found in a pond in the Old Deer Park adjacent to the southern boundary of the Gardens, but it can no longer be found in ponds within the Gardens or in the Moat.

1514. Lemna gibba L.
fat duckweed
FIRST RECORD: 1873/4 [1768].
NICHOLSON: **1873/4 survey** (JB: 73): **P.** On the lake in company with [*Lemna minor*] and [*Spirodela polyrhiza*]. A few plants were brought me by Mr T.Entwhistle, and I have since collected it myself. It is very uncommon. • **1906 revision** (AS: 88): Grows with [*Lemna minor*] and [*Spirodela polyrhiza*], but uncommon.
OTHER REFERENCES: **Kew Environs.** 1929, *Lousley* (LN35: S282): Ditch outside Kew Gardens.
EXSICC.: **Kew Environs.** A ditch, Kew; 1 viii 1930; *Worsdell* s.n. pro parte [mixed with *L. minor*] (K). • Towpath between Kew & Richmond; floating on still water in ditch, with *L. minor* but less common; 18 vii 1948; *Souster* 881 (K).

CULT.: 1768 (JH: 367); 1813, Nat. of England (HK2, 5: 233); 1814, Nat. of England (HKE: 287). Not currently in cult.
CURRENT STATUS: This duckweed can still be found in the Moat south of Brentford Gate but it has been lost for many years from the Lake.

1515. Lemna minor L.
common duck weed
FIRST RECORD: 1873/4 [1768].
NICHOLSON: **1873/4 survey** (JB: 73): Common on lake and moat. • **1906 revision** (AS: 88): Common in lake and ha-ha.
OTHER REFERENCES: **Kew Gardens.** 1999, *Stones* (EAK: 2 & 5): In Lake and Palm House Pond [those in the lake are almost certainly *Lemna minuta*]
EXSICC.: **Kew Environs.** A ditch, Kew; 1 viii 1930; *Worsdell* s.n., pro parte [mixed with *L. gibba*] (K). • Towpath, between Kew & Richmond; floating on still water in ditch; 18 vii 1948; *Souster* 880 (K).
CULT.: 1768 (JH: 367); 1789, Nat. of Britain (HK1, 3: 322); 1813, Nat. of Britain (HK2, 5: 233); 1814, Nat. of Britain (HKE: 287). In cult. under glass.
CURRENT STATUS: This is now rather scarce and is probably being out-competed by *L. minuta*.

1516. Lemna trisulca L.
ivy-leaved duckweed
CULT.: 1768 (JH: 367); 1789, Nat. of Britain (HK1, 3: 322); 1813, Nat. of Britain (HK2, 5: 233); 1814, Nat. of Britain (HKE: 287). Not currently in cult.

1517. *Lemna minuta Kunth
Lemna minuscula auctt.
least duckweed
FIRST RECORD: 1995.
REFERENCES: **Kew Gardens.** 1995, *Lock & Verdcourt* (S33: 751): Banks Building, small water feature adjacent to staff room. • 1998, *Cope* (S33: 751): Secluded Garden (141).
CURRENT STATUS: In the last five years or so this species has increased from a curiosity to the level of a pest. In summer and winter it has covered almost the entire Lake, and the lower pond behind the Banks Building has been permanently green with it for several years. It is now beginning to occur in other water features in the Gardens. Attempts at removal have met with only temporary success.

ERIOCAULACEAE

1518. Eriocaulon aquaticum (Hill) Druce
E. septangulare With.
pipewort
CULT.: 1810, Nat. of Scotland (HK2, 1: 183); 1814, Nat. of Scotland (HKE: 26). Not currently in cult.

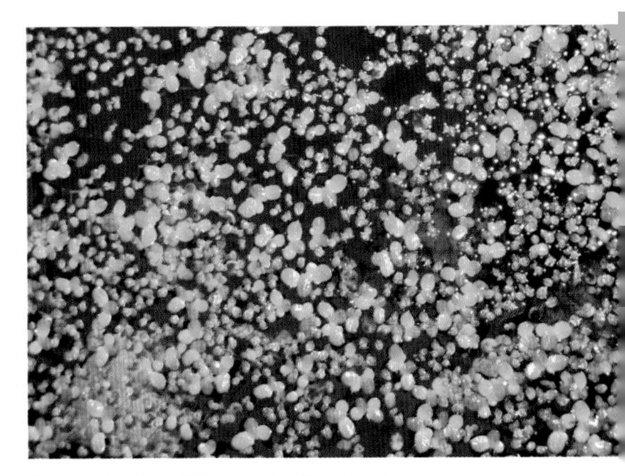

Common duckweed (*Lemna minor*) and least duckweed (*Lemna minuta*) growing together. The latter is beginning to replace the former in most of the ponds in Kew.
[Photo: P.J.Cribb]

JUNCACEAE

1519. Juncus squarrosus L.
heath rush
FIRST RECORD: 1873/4 [1789].
NICHOLSON: **1873/4 survey** (JB: 74): **P.** About a score plants in wood between Winter Garden and Engine House. • **1906 revision** (AS: 88): About a score plants in wood between temperate house and pumping station.
EXSICC.: **Kew Environs.** Riverside, Kew; 10 vi 1934; *Robinson* s.n. (K).
CULT.: 1789, Nat. of Britain (HK1, 1: 476); 1811, Nat. of Britain (HK2, 2: 308); 1814, Nat. of Britain (HKE: 101); 1954, Herbarium Ground (K); 1977, plot 152 (K). In cult.
CURRENT STATUS: No recent sightings.

[1520. Juncus tenuis Willd.
slender rush
CULT.: 1814, Nat. of Scotland (HKE: Add.).
Note. Listed in error as native.**]**

1521. Juncus compressus Jacq.
round-fruited rush
FIRST RECORD: 1929.
EXSICC.: **Kew Gardens.** Thames Bank opposite Brentford Docks; on lower part of new sloping stone wall; only 1 plant seen. Submerged at every high tide; 26 viii 1929; *Summerhayes* 3054 (K). **Kew Environs.** Kew; at top of river bank, in position subject to flooding at high tide (spring tides); 4 vii 1951; *Souster* 1208 (K).
CULT.: 1881, unspec. (K). Not currently in cult.
CURRENT STATUS: No recent sightings. It is probable that the records of Summerhayes and Souster refer to the same population although the latter's locality was even less precise than the former's.

1522. Juncus gerardii Loisel.
saltmarsh rush
CULT.: 1881, unspec. (K). Not currently in cult.

1523. Juncus trifidus L.
three-leaved rush
CULT.: 1789, Nat. of Scotland (HK1, 1: 475); 1811, Nat. of Scotland (HK2, 2: 308); 1814, Nat. of Scotland (HKE: 101). Not currently in cult.

1524. Juncus bufonius L.
toad rush
FIRST RECORD: 1833 [1789].
NICHOLSON: **1873/4 survey** (JB: 74): **P.** Forms quite a carpet on the bare wet places near edge of lake. • **1906 revision** (AS: 88): **A.** Abundant about lake.
EXSICC.: **Kew Gardens.** [Unlocalised]; weed in potting soil, probably introduced with loam; 10 vii 1945; *Souster* 265 (K). **Kew Environs.** Kew; 1833; *anon.* s.n. (K).
CULT.: 1789, Nat. of Britain (HK1, 1: 476); 1811, Nat. of Britain (HK2, 2: 309); 1814, Nat. of Britain (HKE: 101); 1977, plot 152-07 (K); 1978, plot 152-05 (K). Not currently in cult.
CURRENT STATUS: Widely scattered but surprisingly uncommon for such a successful weed. One of the earliest plants to have been recorded from Kew.

1525. Juncus subnodulosus Schrank
J. obtusiflorus Ehrh. ex Hoffm.
blunt-flowered rush
CULT.: 1811, Nat. of Britain (HK2, 2: 309); 1814, Nat. of Britain (HKE: 102); 1977, plot 159-02 (K). Not currently in cult.

1526. Juncus articulatus L.
J. lamprocarpus Ehrh.
jointed rush
FIRST RECORD: 1873/4 [1811].
NICHOLSON: **1873/4 survey** (JB: 74): **P.** Lake. Plentiful. • **1906 revision** (AS: 88): **A.** Plentiful about lake.
EXSICC.: **Kew Gardens.** Thames bank between Brentford Gate and Office of Works Gate; on high mud hump just below concrete wall, with *Eleocharis*, *Phalaris* etc; 26 viii 1929; *Summerhayes* 437 (K). **Kew Environs.** Thames bank, Kew; growing in crevices near top of river wall; 30 vii 1928; *Summerhayes & Turrill* 3006 (K). • Thames bank, Kew; crevices of new stone wall, near base; 2 viii 1928; *Summerhayes & Turrill* 3007 (K).
CULT.: 1811, Nat. of Britain (HK2, 2: 308); 1814, Nat. of Britain (HKE: 101); 1978, unspec. (K). In cult.
CURRENT STATUS: No recent sightings.

1527. Juncus acutiflorus Ehrh. ex Hoffm.
J. articulatus auct. non L.
sharp-flowered rush
CULT.: 1789, Nat. of Britain (HK1, 1: 476); 1811, Nat. of Britain (HK2, 2: 309); 1814, Nat. of Britain (HKE: 102); 1978, unspec. (K). Not currently in cult.

1528. Juncus bulbosus L.
J. uliginosus Kunth
bulbous rush
CULT.: 1789, Nat. of Britain (HK1, 1: 476); 1811, Nat. of Britain (HK2, 2: 309), Nat. of England (ibid.: as *uliginosus*); 1814, Nat. of Britain (HKE: 102), Nat. of England (ibid.: as *uliginosus*); 1977, plot 152-05 (K); 1977, plot 152-07 (K); 1977, plot 159-02 (K); 1978, unspec. (K). In cult.

1529. Juncus biglumis L.
two-flowered rush
CULT.: 1811, Nat. of Scotland (HK2, 2: 310); 1814, Nat. of Scotland (HKE: 102). Not currently in cult.

1530. Juncus triglumis L.
three-flowered rush
CULT.: 1811, Nat. of Britain (HK2, 2: 310); 1814, Nat. of Britain (HKE: 102). Not currently in cult.

1531. Juncus castaneus Sm.
chestnut rush
CULT.: 1811, Nat. of Scotland (HK2, 2: 310); 1814, Nat. of Scotland (HKE: 102). Not currently in cult. Note. Not known as a British plant until 1798.

1532. Juncus maritimus Lam.
sea rush
CULT.: 1811, Nat. of Britain (HK2, 2: 307); 1814, Nat. of Britain (HKE: 101). Not currently in cult.

1533. Juncus acutus L.
sharp rush
CULT.: 1789, Nat. of England (HK1, 1: 475); 1811, Nat. of England (HK2, 2: 307); 1814, Nat. of England (HKE: 102). Not currently in cult.

1534. Juncus balticus Willd.
baltic rush
CULT.: 1881, unspec. (K). Not currently in cult.

1535. Juncus filiformis L.
thread rush
CULT.: 1789, Nat. of England (HK1, 1: 475); 1811, Nat. of Britain (HK2, 2: 308); 1814, Nat. of Britain (HKE: 101); 1977, plot 159-02 (K). Not currently in cult.

1536. Juncus inflexus L.
J. glaucus Sibth.
hard rush
FIRST RECORD: 1873/4 [1811].
NICHOLSON: **1873/4 survey** (JB: 74): **P.** Lake. [Very stunted forms are not uncommon in some parts of the woods]. • **1906 revision** (AS: 88): **A.** Common about lake. [Stunted forms are not uncommon in some parts of the woods].
OTHER REFERENCES: **Kew Gardens.** 1999, *Stones* (EAK: 2): By Lake.
CULT.: 1811, Nat. of England (HK2, 2: 308); 1814, Nat. of England (HKE: 101); 1881, unspec. (K); 1977, plot 152-05 (K); 1977, plot 159-02 (K). In cult.
CURRENT STATUS: Known from the pond in the Conservation Area and the southwest corner of the Lake.

1537. Juncus effusus L.
soft-rush
FIRST RECORD: 1873/4 [1789].
NICHOLSON: **1873/4 survey** (JB: 74): **P.** Lake. Very stunted forms of both this and [*Juncus inflexus*] are not uncommon in some parts of the woods. • **1906 revision** (AS: 88): **A.** Lake. Stunted forms of this and [*Juncus inflexus*] are not uncommon in some parts of the woods.
CULT.: 1789, Nat. of Britain (HK1, 1: 475); 1811, Nat. of Britain (HK2, 2: 308); 1814, Nat. of Britain (HKE: 101); 1977, plot 152-03 (K); 1978, plot 152-03 (K). In cult.
CURRENT STATUS: Known only from the pond in the Conservation Area.

1538. Juncus conglomeratus L.
compact rush
FIRST RECORD: 1873/4 [1789].
NICHOLSON: **1873/4 survey** (JB: 74): **P.** This and [*Juncus effusus* and *Juncus inflexus*] are about equally common round lake. • **1906 revision** (AS: 88): **A.** Common about lake.
CULT.: 1789, Nat. of Britain (HK1, 1: 475); 1811, Nat. of Britain (HK2, 2: 307); 1814, Nat. of Britain (HKE: 101). In cult.
CURRENT STATUS: Known only from the pond in the Conservation Area.

1539. Luzula forsteri (Sm.) DC.
Juncus forsteri Sm.
southern wood-rush
CULT.: 1811, Nat. of England (HK2, 2: 311); 1814, Nat. of England (HKE: 102). Not currently in cult.

1540. Luzula pilosa (L.) Willd.
Juncus pilosus L.
hairy wood-rush
CULT.: 1789, Nat. of Britain (HK1, 1: 476); 1811, Nat. of Britain (HK2, 2: 310); 1814, Nat. of Britain (HKE: 102). Not currently in cult.

1541. Luzula sylvatica (Huds.) Gaudin
Juncus sylvaticus Huds.
J. maximus Reichard
Luzula maxima (Reichard) DC.
great wood-rush
FIRST RECORD: 1873/4 [1789].
NICHOLSON: **1873/4 survey** (JB: 74): **Q.** Two large tufts just within the fence opposite where *Potamogeton pusillus* grows. • **1906 revision** (AS: 88): Here and there in Queen's Cottage Grounds.
CULT.: 1789, Nat. of England (HK1, 1: 477); 1811, Nat. of Britain (HK2, 2: 311); 1814, Nat. of Britain (HKE: 102). Not currently in cult.
CURRENT STATUS: No recent sightings.

1542. Luzula campestris (L.) DC.
Juncus campestris L.
field wood-rush
FIRST RECORD: 1873/4 [1789].
NICHOLSON: **1873/4 survey** (JB: 74): Fairly common in every piece of turf. • **1906 revision** (AS: 88): Fairly common in every piece of turf.
EXSICC.: **Kew Gardens.** [Unlocalised]; among short grasses on sandy soil; 10 v 1928; *Hubbard* s.n. (K). • Meadow behind Herbarium; 27 iv 1930; *Bullock* 1 (K). • [Unlocalised]; in rough lawn; 14 iv 1944; *Souster* 17 (K).

Field wood-rush (*Luzula campestris*). Widespread in short grass, but its tufted habit makes it difficult to mow in ornamental lawns. [Photo: P.J.Cribb]

CULT.: 1789, Nat. of Britain (HK1, 1: 477); 1811, Nat. of Britain (HK2, 2: 311); 1814, Nat. of Britain (HKE: 102); 1953, Herb. Expt. Grnd. (K); 1977, Alpine Yard (K). Not currently in cult.
CURRENT STATUS: Widespread in short grass.

1543. Luzula multiflora (Ehrh.) Lej.
heath wood-rush

a. subsp. **multiflora**
L. multiflora subsp. *sudetica* (Willd.) Parl.
FIRST RECORD: 1873/4 [1978].
NICHOLSON: **1873/4 survey** (JB: 74): **P** and **Q.** Both [subspecies] occur, but [*congesta*] seems much the more frequent. • **1906 revision**: Not recorded.
CULT.: 1978, plot 159-02 (K). Not currently in cult.
CURRENT STATUS: No recent sightings. The lack of records in the 1906 report may be significant.

b. subsp. **congesta** (Thuill.) Archang.
FIRST RECORD: 1873/4.
NICHOLSON: **1873/4 survey** (JB: 74): **P** and **Q.** Both [subspecies] occur, but [*congesta*] seems much the more frequent. • **1906 revision**: Not recorded.
CURRENT STATUS: No recent sightings. The lack of records in the 1906 report may be significant.

1544. Luzula spicata (L.) DC.
Juncus spicatus L.
spiked wood-rush
CULT.: 1811, Nat. of Scotland (HK2, 2: 311); 1814, Nat. of Scotland (HKE: 102). In cult.

1545. *Luzula nivea (L.) DC.
Snow-white wood-rush
FIRST RECORD: 1873/4 [1978].
NICHOLSON: **1873/4** survey (JB: 9): {A good number of plants are growing in Q}. • **1906 revision** (AS: 88): {A considerable number of plants in Queen's Cottage Grounds.}
CULT.: 1978, plot 159-02 (K). Not currently in cult.
CURRENT STATUS: No recent sightings.

CYPERACEAE

1546. Eriophorum angustifolium Honck.
E. polystachion auct. non L.
common cottongrass
CULT.: 1768 (JH: 397); 1789, Nat. of Britain (HK1, 1: 83); 1810, Nat. of Britain (HK2, 1: 133); 1814, Nat. of Britain (HKE: 19). Not currently in cult.

1547. Eriophorum latifolium Hoppe
E. polystachion L. pro parte
broad-leaved cottongrass
CULT.: 1810, Nat. of Britain (HK2, 1: 133); 1814, Nat. of Britain (HKE: 19). Not currently in cult.

1548. Eriophorum gracile W.D.J.Koch ex Roth
slender cottongrass
CULT.: 1814, Nat. of Scotland (HKE: Add.). Not currently in cult.
Note. Said not to have discovered in Britain until 1835, but Aiton clearly knew it from Scotland much earlier.

1549. Eriophorum vaginatum L.
E. capitatum Host
hare's-tail cottongrass
CULT.: 1768 (JH: 397); 1789, Nat. of Britain (HK1, 1: 83); 1810, Nat. of Britain (HK2, 1: 133); 1814, Nat. of Britain (HKE: 19), Nat. of Scotland (ibid.: Add. as *capitatum*). In cult.

1550. Trichophorum alpinum (L.) Pers.
Eriophorum alpinum L.
cotton deergrass
CULT.: 1810, Nat. of Scotland (HK2, 1: 134); 1814, Nat. of Scotland (HKE: 19). Not currently in cult.
Note. Not found as a British plant until 1794.

1551. Trichophorum cespitosum (L.) Hartm.
Scirpus cespitosus L.
deergrass
CULT.: 1768 (JH: 397); 1789, Nat. of Britain (HK1, 1: 80); 1810, Nat. of Britain (HK2, 1: 130); 1814, Nat. of Britain (HKE: 19). Not currently in cult.

1552. Eleocharis palustris subsp. **vulgaris** Walters
Heleocharis palustris [mis-spelling of the above]
Scirpus palustris L., sens. lat.
common spike-rush
FIRST RECORD: 1873/4 [1768].
NICHOLSON: **1873/4 survey** (JB: 74): **P.** Very common on the Syon Vista side of lake. • **1906 revision** (AS: 89): Very common about lake and along ha-ha.
EXSICC.: **Kew Environs.** Kew, Thames shore; in lowest zone in pure society, being washed out from the mud; 13 vii 1928; *Turrill* s.n. (K).
CULT.: 1768 (JH: 397); 1789, Nat. of Britain (HK1, 1:80); 1810, Nat. of Britain (HK2, 1: 130); 1814, Nat. of Britain (HKE: 19). In cult.
CURRENT STATUS: No recent sightings.

1553. Eleocharis multicaulis (Sm.) Desv.
Scirpus multicaulis Sm.
many-stalked spike-rush
CULT.: 1810, Nat. of Britain (HK2, 1: 130); 1814, Nat. of Britain (HKE: 19). Not currently in cult.
Note. Not known as a British plant until 1800.

Sea club-rush (*Bolboschoenus maritimus*). Occurs in mud at low tide near the southern end of the Gardens, and in the Lake around the inlet from the river. Recorded outside before it was inside, but has been cultivated for a long time.

1554. Eleocharis quinqueflora (Hartmann) O.Schwarz
few-flowered spike-rush
Scirpus pauciflorus Lightf.
CULT.: 1810, Nat. of Britain (HK2, 1: 131); 1814, Nat. of Britain (HKE: 19). Not currently in cult.

1555. Eleocharis acicularis (L.) Roem. & Schult.
needle spike-rush
Scirpus acicularis L.
CULT.: 1789, Nat. of Britain (HK1, 1: 80); 1810, Nat. of Britain (HK2, 1: 131); 1814, Nat. of Britain (HKE: 19). In cult.

1556. Bolboschoenus maritimus (L.) Palla
Scirpus maritimus L.
sea club-rush
FIRST RECORD: 1908 [1768].
REFERENCES: **Kew Gardens.** 1950, *Welch*, and 1961, *Young* (LFS: 368): By the Thames ... opposite S boundary of Kew Gardens. • 1982, *Hastings* (FSSC: 99): Beside lake. • 1999, *Stones* (EAK: 2): By Lake. **Kew Environs.** 1908, *A.B.Jackson* (S7: 125): This species, which has been inadvertantly omitted from the Kew list, occurs within our area on mud bank by the river associated with *Eleocharis palustros*, [*Schoenoplectus triqueter*], [*S.* × *carinatus*] and other moisture-loving plants. • 1935, *anon.* (LN15: S100): Kew. • 1947, *Bangerter* (LN35: S288): By Thames, Kew to Richmond. • 1966, *Young* (LFS: 368): By the Thames ... Kew. • 1983, *Leslie* (FSSC: 99): Base of embankment, south-east of Kew Bridge.
EXSICC.: **Kew Gardens.** Towpath between Kew and Richmond; growing in mud by riverside, opposite Isleworth; 2 viii 1948; *Souster* s.n. (K). **Kew Environs.** By the Thames, above Kew; 1 vii 1920; *Britton* 2205 (K). • Kew, Thames bank; lower zone, ± submerged at high tide, on mud, pure society or with other *Scirpus* spp.; 13 vii 1928; *Turrill* s.n. (K).
CULT.: 1768 (JH: 397); 1789, Nat. of Britain (HK1, 1: 82); 1810, Nat. of Britain (HK2, 1: 132); 1814, Nat. of Britain (HKE: 19). In cult.
CURRENT STATUS: In mud below the towpath at the extreme southern end of the Gardens and by the Lake around the inlet from the river.

1557. Scirpus sylvaticus L.
wood club-rush
CULT.: 1768 (JH: 397); 1789, Nat. of Britain (HK1, 1: 82); 1810, Nat. of Britain (HK2, 1: 132); 1814, Nat. of Britain (HKE: 19); 1907, unspec. (K). In cult.

1558. Scirpoides holoschoenus (L.) Soják
round-headed club-rush
Scirpus holoschoenus L.
S. romanus L.
CULT.: 1768 (JH: 397); 1789, Native of England (HK1, 1: 81), Alien (ibid.: as *romanus*); 1810, Nat. of England (HK2, 1: 131, and: 132 as *romanus*); 1814, Nat. of England (HKE: 19, and as *romanus*); 1898, unspec. (WSY). Not currently in cult.

1559. Schoenoplectus lacustris (L.) Palla
common club-rush
Scirpus lacustris L.
FIRST RECORD: 1873/4 [1768].
NICHOLSON: **1873/4 survey** (JB: 74): **Strip.** Large masses in many places near river. • **1906 revision** (AS: 89): **Strip.** Large masses on banks of mud by river.
OTHER REFERENCES: **Kew Environs.** 1907, *C.S. Nicholson* (LN35: S289): Thames, Richmond to Kew. • 1948, *Bangerter* (LN35: S289): Thames, Richmond to Kew.
EXSICC.: **Kew Gardens.** Towpath between Kew and Richmond; in mud by riverside opposite Isleworth; 2 viii 1948; *Souster* 925 (K). **Kew Environs.** R. Thames, Kew; 1 viii 1930; *Worsdell* s.n. (K).
CULT.: 1768 (JH: 397); 1789, Nat. of Britain (HK1, 1: 81); 1810, Nat. of Britain (HK2, 1: 131); 1814, Nat. of Britain (HKE: 19); 1995, Aquatic Garden (K). In cult.
CURRENT STATUS: Rare; there is now only a small amount in the Conservation Area.

1560. Schoenoplectus × carinatus (Sm.) Palla (lacustris × triqueter)
Scirpus × carinatus Sm.
FIRST RECORD: 1873/4 [1814].
NICHOLSON: **1873/4 survey** (JB: 74): This grows on the banks of the Thames at Kew, but whether within our limits or just outside them I cannot at present recollect. • **1906 revision** (AS: 89): **Strip.** Side of river opposite Palace Grounds.
OTHER REFERENCES: **Kew Environs.** 1930, *Lousley* (LN35: S289): Kew Bridge. • 1931, *Lousley* (LN35: S289): Opposite Old Deer Park between Kew and Richmond. • 1976, *Lousley* (LFS: 369): Extinct in Surrey. This interesting hybrid grew on estuarine mud with [*S. triqueter*] nearly to Richmond and its habitats have ... been destroyed. It persisted at Old Deer Park until about 1936 and above Kew Bridge until about 1946.
EXSICC.: **Kew Environs.** Kew; viii 1877; *Baker* s.n. (K). • Thames bank, Kew; viii 1885; *Fraser* s.n. (K). • Thames banks above Mortlake; 10 ix 1921; *Fraser* s.n. (K). • Thames bank, Kew; on mud, stone below, mixed with *Scirpus triqueter* and *S. maritimus*; 13 vii 1928; *Turrill* s.n. (K). • River bed, Thames at Kew; 5 viii 1939; *Worsdell* s.n. (K).
CULT.: 1814, Nat. of England (HKE: Add.). Not currently in cult.
CURRENT STATUS: Long extinct in Surrey.

1561. Schoenoplectus tabernaemontani (C.C.Gmel.) Palla
Scirpus glaucus Sm.
grey club-rush
CULT.: 1814, Nat. of Britain (HKE: Add.). In cult.

1562. Schoenoplectus triqueter (L.) Palla
Scirpus triqueter L.
triangular club-rush
FIRST RECORD: 1873/4 [1810].
NICHOLSON: **1873/4 survey** (JB: 74): **Strip.** Some large tufts near edge of river opposite "Docks." • **1906 revision** (AS: 89): **Strip.** Side of river opposite Palace Grounds; also south of Brentford Ferry.
OTHER REFERENCES: **Kew Environs.** 1907, *C.S. Nicholson* (LN35: S289): Kew. • 1907, *A.B.Jackson* (BECS2: 314): Mud banks by the Thames, between Richmond and Hammersmith. • 1930–31, *Lousley* (LFS: 368–9): Extinct in Surrey ... grew formerly by the Thames at intervals ... to about 2 km above Kew Bridge. Just above Kew Bridge (survived until 1946 when a major reconstruction of the river-wall destroyed the last patch of estuarine mud on which the plant grew); opposite Old Deer Park. • 1931, *Lousley* (WBEC48: 144; LN35: S289): Just above Kew Bridge. The most luxuriant plants of this species which I have ever seen. • 1935, *anon.* (LN15: S100): River Thames at Mortlake and Kew.
EXSICC.: **Kew Environs.** Thames banks, Kew; 11 viii 1898; *Fraser* s.n. (K). • Kew, Thames bank; on mud mixed with *Scirpus carinatus* and *S. maritimus*; 13 vii 1928; *Turrill* s.n. (K).
CULT.: 1810, Nat. of England (HK2, 1: 132); 1814, Nat. of England (HKE: 19); 1893, unspec. (K); 1896, unspec. (K). In cult.
CURRENT STATUS: Long extinct in Surrey.

1563. Schoenoplectus pungens (Vahl) Palla
Scirpus mucronatus L.
sharp club-rush
CULT.: 1768 (JH: 397); 1789, Nat. of England (HK1, 1: 82). Not currently in cult.

1564. Isolepis setacea (L.) R.Br.
Scirpus setaceus L.
bristle cub-rush
CULT.: 1768 (JH: 397); 1789, Nat. of England (HK1, 1: 81); 1810, Nat. of Britain (HK2, 1: 132); 1814, Nat. of Britain (HKE: 19). Not currently in cult.

1565. Isolepis cernua (Vahl) Roem. & Schult.
slender club-rush
CULT.: 1967, Herbarium Ground (K). Not currently in cult.

1566. Eleogiton fluitans (L.) Link
Scirpus fluitans L.
floating club-rush
CULT.: 1768 (JH: 397); 1789, Nat. of Britain (HK1, 1: 81); 1810, Nat. of Britain (HK2, 1: 131); 1814, Nat. of Britain (HKE: 19). Not currently in cult.

1567. Blysmus compressus (L.) Panz. ex Link
Schoenus compressus L.
flat-sedge
CULT.: 1768 (JH: 396); 1789, Nat. of Britain (HK1, 1: 78); 1810, Nat. of Britain (HK2, 1: 126); 1814, Nat. of Britain (HKE: 18). In cult.

1568. Blysmus rufus (Huds.) Link
Schoenus rufus Huds.
saltmarsh flat-sedge
CULT.: 1810, Nat. of Scotland (HK2, 1: 127); 1814, Nat. of Scotland (HKE: 18). Not currently in cult.

1569. Cyperus longus L.
galingale
FIRST RECORD: 2001 [1768].
REFERENCES: **Kew Gardens.** 2001, *Cope* (S35: 649): Naturalised by the Lily Pond (333), but doubtless originally planted.
CULT.: 1768 (JH: 396); 1789, Nat. of England (HK1, 1: 79); 1810, Nat. of England (HK2, 1: 129); 1814, Nat. of England (HKE: 18); 1995, Aquatic Garden (K). In cult.
CURRENT STATUS: Originally planted, but now thoroughly naturalised by the Lily Pond.

1570. *Cyperus eragrostis Lam.
pale galingale
FIRST RECORD: 2000 [1929].
REFERENCES: **Kew Gardens.** 2000, *Cope* (S35: 649): Outside the Mycology Building, near the vegetable plots (113). Subsequently, in the Director's Garden (133) and near the Lily House (163).
CULT.: 1929, unspec. (K); 1957, Rock Garden (K); 1961, Temperate Dept. (K); 1962, Trop. Pits (K); 1962, Order Beds (K); 1965, unspec. (K). In cult.
CURRENT STATUS: An occasional escape from cultivation into rough ground.

1571. Cyperus fuscus L.
brown galingale
CULT.: 1789, Alien (HK1, 1: 80); 1810, Alien (HK2, 1: 128); 1814, Alien (HKE: 18). In cult.
Note. Not known as a British plant until 1821.

1572. Schoenus nigricans L.
black bog-rush
CULT.: 1768 (JH: 396); 1789, Nat. of Britain (HK1, 1: 78); 1810, Nat. of Britain (HK2, 1: 126); 1814, Nat. of Britain (HKE: 18). Not currently in cult.

1573. Schoenus ferrugineus L.
brown bog-rush
CULT.: 1768 (JH: 396); 1789, Nat. of Britain (HK1, 1: 78). Not currently in cult.

1574. Rhynchospora alba (L.) Vahl
Schoenus albus L.
white beak-sedge
CULT.: 1768 (JH: 396); 1789, Nat. of Britain (HK1, 1: 78); 1810, Nat. of Britain (HK2, 1: 127); 1814, Nat. of Britain (HKE: 18). Not currently in cult.

1575. Rhynchospora fusca (L.) W.T.Aiton
brown beak-sedge
CULT.: 1810, Nat. of Wales & Ireland (HK2, 1: 127); 1814, Nat. of Wales & Ireland (HKE: 18). Not currently in cult.

1576. Cladium mariscus (L.) Pohl
Schoenus mariscus L.
great fen-sedge
CULT.: 1768 (JH: 396); 1789, Nat. of England (HK1, 1: 77); 1810, Nat. of England (HK2, 1: 126); 1814, Nat. of England (HKE: 18). In cult.

1577. Kobresia simpliciuscula (Wahlenb.) Mack.
false sedge
Schoenus monoicus Sm.
CULT.: 1810, Nat. of England (HK2, 1: 127); 1814, Nat. of England (HKE: 18). Not currently in cult.
Note. Not known as a British plant until 1805.

1578. Carex paniculata L.
greater tussock-sedge
CULT.: 1789, Nat. of England (HK1, 3: 328); 1813, Nat. of England (HK2, 5: 244); 1814, Nat. of England (HKE: 289); 1964, Order Beds (K). Not currently in cult.

1579. Carex diandra Schrank
C. teretiuscula Gooden.
lesser tussock-sedge
CULT.: 1813, Nat. of Britain (HK2, 5: 243); 1814, Nat. of Britain (HKE: 289). Not currently in cult.
Note. Not known as a British plant until 1792.

1580. Carex vulpina L.
true fox-sedge
FIRST RECORD: 1873/4 [1789].
NICHOLSON: **1873/4 survey** (JB: 74): **P.** Here and there by lake. **Strip.** Not uncommon. • **1906 revision** (AS: 89): **A.** Here and there by lake. **Strip.** Not uncommon.
EXSICC.: **Kew Gardens.** Thames banks opposite Isleworth; 1883; *Fraser* s.n. (K).
CULT.: 1789, Nat. of Britain (HK1, 3: 327); 1813, Nat. of Britain (HK2, 5: 241); 1814, Nat. of Britain (HKE: 288). In cult.
CURRENT STATUS: No recent sightings.

1581. Carex otrubae Podp.
false fox-sedge
FIRST RECORD: 2004 [1964]
CULT.: 1964, Order Beds (K). Not currently in cult.
CURRENT STATUS: In a pond on the golf course in the Old Deer Park.

1582. Carex × pseudoaxillaris K.Richt.
(otrubae × remota)
C. axillaris Gooden.
CULT.: 1813, Nat. of England (HK2, 5: 242); 1814, Nat. of England (HKE: 289). Not currently in cult.

1583. *Carex vulpinoidea Michx.
American fox-sedge
FIRST RECORD: 1880 [1896].
REFERENCES: **Kew Environs.** 1880, G.*Nicholson* (BECB1 (2): 40; BSEC11: 459; BSEC9: 693; S33: 751, and see Exsicc.): Banks of the Thames at Kew.
EXSICC.: **Kew Environs.** Banks of the Thames, Kew; vi 1880; *G.Nicholson* s.n. (K).
CULT.: 1896, unspec. (K). Not currently in cult.
CURRENT STATUS: No recent sightings.

1584. Carex spicata Huds.
spiked sedge
FIRST RECORD: 1898 [1789].
REFERENCES: **Kew Environs.** 1925, *Lousley* (LN35: S299; S33: 751): Thames bank near Kew.

EXSICC.: **Kew Gardens.** Herbaceous Dept.; 14 v 1932; *Nelmes* 184 (K). **Kew Environs.** Kew; 1898; *Worsdell* s.n. (K). • Thames banks, Kew; 29 vi 1899; *Fraser* s.n. (K). • Bank of Thames betw. Richmond & Kew bridges; vii 1928; *C.A.Smith* 6074 (K). • Old Deer Park, near golf course; sandy soil; 12 vi 1947; *Meikle & Airy Shaw* s.n. (K).
CULT.: 1789, Nat. of Britain (HK1, 3: 327); 1967, unspec. (K). In cult.
CURRENT STATUS: No recent sightings; it was last seen on the towpath in 1979.

1585. Carex muricata subsp. **lamprocarpa** Čelak.
C. muricata auct. non L. sens. str.
C. pairii F.W.Schultz
prickly sedge
FIRST RECORD: 1873/4 [1789].
NICHOLSON: **1873/4 survey** (JB: 74): **B.** Several large plants near the wooden fence separating lawn in front of Palace from Botanic Garden. **P.** Here and there in Pagoda Avenue and in many other dry places. • **1906 revision** (AS: 89): **B, P, A.** Here and there in dry places.
OTHER REFERENCES: **Kew Gardens.** 1966, *J.L.Gilbert* (S29: 408): In turf close to S. side of Wing B of Herbarium.
EXSICC.: **Kew Gardens.** Near tennis courts and palace; in turf in light soil; 22 vii 1957; *Brenan* s.n. (K). **Kew Environs.** Old Deer Park, near golf course; sandy soil; 12 vi 1947; *Meikle & Airy Shaw* s.n. (K).
CULT.: 1789, Nat. of Britain (HK1, 3: 327); 1813, Nat. of Britain (HK2, 5: 241); 1814, Nat. of Britain (HKE: 289). Not currently in cult.
CURRENT STATUS: Rare; just two or three recent sightings.

1586. Carex divulsa Stokes
grey sedge
FIRST RECORD: 1981 [1813].
REFERENCES: **Kew Environs.** 1981, *Norman* (LN61: 103; S33: 751): Tow-path nearby [to Kew Green].
CULT.: 1813, Nat. of Britain (HK2, 5: 242); 1814, Nat. of Britain (HKE: 289); 1964, Order Beds (K). Not currently in cult.
CURRENT STATUS: Very rare; represented by subsp. *divulsa*. One of its recent localities has probably been destroyed by the current building works around the Herbarium but there are considerable quantities of it on the edge of the Brentford Gate car park.

1587. Carex arenaria L.
sand sedge
CULT.: 1789, Nat. of Britain (HK1, 3: 327); 1813, Nat. of Britain (HK2, 5: 240); 1814, Nat. of Britain (HKE: 288). Not currently in cult.

1588. Carex disticha Huds.
C. intermedia Gooden.
brown sedge
CULT.: 1789, Nat. of Britain (HK1, 3: 326); 1813, Nat. of Britain (HK2, 5: 240); 1814, Nat. of Britain (HKE: 288); 1964, Order Beds (K). Not currently in cult.

1589. Carex chordorrhiza L.f.
string sedge
CULT.: In cult.

1590. Carex divisa Huds.
divided sedge
CULT.: 1789, Nat. of Britain (HK1, 3: 327); 1813, Nat. of Britain (HK2, 5: 241); 1814, Nat. of Britain (HKE: 289). Not currently in cult.

1591. Carex maritima Gunnerus
C. incurva Lightf.
curved sedge
CULT.: 1813, Nat. of Scotland (HK2, 5: 239); 1814, Nat. of Scotland (HKE: 288). Not currently in cult.

1592. Carex remota L.
remote sedge
FIRST RECORD: 1873/4 [1789].
NICHOLSON: **1873/4 survey** (JB: 74): **Strip.** A few plants by moat from Isleworth Gate to "Old Deer Park." • **1906 revision** (AS: 89): **Strip.** By ha-ha and along towing-path near Isleworth Gate.
EXSICC.: **Kew Environs.** Tow-path betw. Kew & Richmond; 22 vi 1924; *Nelmes* 31 (K).
CULT.: 1789, Nat. of Britain (HK1, 3: 328); 1813, Nat. of Britain (HK2, 5: 242); 1814, Nat. of Britain (HKE: 289); 1868, unspec. (K). Not currently in cult.
CURRENT STATUS: Still on the towpath.

1593. Carex ovalis Gooden.
C. leporina auct. non L.
oval sedge
FIRST RECORD: 1873/4 [1789].
NICHOLSON: **1873/4 survey** (JB: 74): **P.** Lake. Several plants. **Q.** Not unfrequent. **Strip.** By moat. • **1906 revision** (AS: 89): **A.** Near lake. **Q.** Not unfrequent. **Strip.** By ha-ha.
EXSICC.: **Kew Environs.** River Bank, Kew – Richmond, S-W of Royal Gardens; 28 v 1932; *Nelmes & Bullock* s.n. (K).
CULT.: 1789, Nat. of Britain (HK1, 3: 327); 1813, Nat. of Britain (HK2, 5: 240); 1814, Nat. of Britain (HKE: 288); 1936, unspec. (K). In cult.
CURRENT STATUS: No recent sightings.

1594. Carex echinata Murray
C. stellulata Gooden.
star sedge
CULT.: 1813, Nat. of Britain (HK2, 5: 242); 1814, Nat. of Britain (HKE: 289). Not currently in cult.

1595. Carex dioica L.
dioecious sedge
CULT.: 1789, Nat. of Britain (HK1, 3: 326); 1813, Nat. of Britain (HK2, 5: 237); 1814, Nat. of Britain (HKE: 288). Not currently in cult.

1596. Carex davalliana Sm.
Davall's sedge
CULT.: 1813, Nat. of Britain (HK2, 5: 237); 1814, Nat. of Britain (HKE: 288). Not currently in cult.
Note. Species not described until 1800.

1597. Carex elongata L.
elongated sedge
CULT.: 1813, Nat. of England (HK2, 5: 242); 1814, Nat. of England (HKE: 289); 1967, unspec. (K). Not currently in cult.
Note. Not known as a British plant until 1808.

1598. Carex curta Gooden.
C. canescens auct. non L.
white sedge
CULT.: 1789, Nat. of Britain (HK1, 3: 328); 1813, Nat. of Britain (HK2, 5: 243); 1814, Nat. of Britain (HKE: 289). Not currently in cult.

1599. Carex hirta L.
hairy sedge
FIRST RECORD: 1873/4 [1789].
NICHOLSON: **1873/4 survey** (JB: 75): **Strip.** Very common. **Q.** Abundant within a line parallel to moat, and about 50 yards from it. • **1906 revision** (AS: 89): **Strip.** Common. **Q.** Abundant along a strip of about 50 yards broad from wall of ha-ha.
OTHER REFERENCES: **Kew Gardens.** 1999, *Stones* (EAK: 2): By Lake.
EXSICC.: **Kew Environs.** Kew, near bank of R. Thames; 16 vi 1928; *Hubbard* s.n. (K). • River Bank, Kew – Richmond, S-W of Royal Gardens; 28 v 1932; *Nelmes & Bullock* s.n. (K).
CULT.: 1789, Nat. of Britain (HK1, 3: 332); 1813, Nat. of Britain (HK2, 5: 253); 1814, Nat. of Britain (HKE: 290). In cult.
CURRENT STATUS: Rare; only a handful of recent sightings.

1600. Carex acutiformis Ehrh.
C. paludosa Gooden.
lesser pond-sedge
FIRST RECORD: 1873/4 [1813].
NICHOLSON: **1873/4 survey** (JB: 75): **Strip.** Here and there by moat. [This and *C. acuta* are the rarest species in our district]. • **1906 revision** (AS: 89): **Strip.** Here and there by ha-ha. [This and *C. acuta* are rather rare within our limits].
OTHER REFERENCES: **Kew Gardens.** 1999, *Stones* (EAK: 2): By Lake.
EXSICC.: **Kew Environs.** Tow-path betw. Kew & Richmond; 24 v 1924; *Nelmes* 20 (K). • Ha-ha, opposite nearby beech on golf links, SW of Kew Gardens; 25 v 1929; *Nelmes* 81 (K). • Same, or nearby plant, as 81 of 25 v 1929; 16 vi 1929; *Nelmes* 82 (K). • Ha-ha, S. of Kew Gardens; 16 vi 1929; *Nelmes* 83 (K). • Ha-ha edge, tow path, between Kew and Richmond; 2 v 1937; *Nelmes* 290 (K). • Ha-ha by tow path between Kew and Richmond; 12 vi 1937; *Nelmes* 335 (K).
CULT.: 1813, Nat. of Britain (HK2, 5: 252); 1814, Nat. of Britain (HKE: 290). Not currently in cult.
CURRENT STATUS: No recent sightings.

1601. Carex riparia Curtis
greater pond-sedge
FIRST RECORD: 1837 [1789].
NICHOLSON: **1873/4 survey** (JB: 75): Common near all pieces of water. • **1906 revision** (AS: 89): Common, nearly all pieces of water.
EXSICC.: **Kew Environs.** Kew; riverside; 1837; *Hooker* s.n. (K). • Ha-ha edge, tow path, between Kew and Richmond; 2 v 1937; *Nelmes* 291 (K). • Edge of Ha-ha by tow-path between Kew and Richmond; 12 vi 1937; *Nelmes* 340 (K). • Towpath, off R. Mid-Surrey Golf Course; by edge of fresh water ditch, with *Glyceria maxima* & *Iris pseudacorus*; 11 v 1950; *Souster* 1072 (K). • Towpath between Kew & Richmond (same station as 1072); in ditch, by edge of water, with *Glyceria maxima*, *Iris pseudacorus* etc.; 24 vi 1951; *Souster* 1205 (K).
CULT.: 1789, Nat. of Britain (HK1, 3: 331); 1813, Nat. of Britain (HK2, 5: 252); 1814, Nat. of Britain (HKE: 290). Not currently in cult.
CURRENT STATUS: By ponds in the Old Deer Park. It has not been seen recently on the towpath.

1602. Carex pseudocyperus L.
cyperus sedge
CULT.: 1789, Nat. of Britain (HK1, 3: 330); 1813, Nat. of Britain (HK2, 5: 251); 1814, Nat. of Britain (HKE: 289). In cult.

1603. Carex rostrata Stokes
C. ampullacea Gooden.
C. inflata auct. non Huds.
bottle sedge
CULT.: 1789, Nat. of England (HK1, 3: 330); 1813, Nat. of Britain (HK2, 5: 253); 1814, Nat. of Britain (HKE: 290); 1967, Herbarium Ground (K). Not currently in cult.

1604. Carex vesicaria L.
bladder-sedge
CULT.: 1789, Nat. of Britain (HK1, 3: 332); 1813, Nat. of Britain (HK2, 5: 253); 1814, Nat. of Britain (HKE: 290); 1980, Alpine 196-51 (K). Not currently in cult.

1605. Carex saxatilis L.
C. pulla Gooden.
russet sedge
CULT.: 1813, Nat. of Britain (HK2, 5: 247); 1814, Nat. of Britain (HKE: 289). In cult.

1606. Carex pendula Huds.
pendulous sedge
FIRST RECORD: 1977 [1789].
REFERENCES: **Kew Gardens.** 1977, *Cope* (unpubl.). • 2001, *Cope* (S35: 649): Neglected corner of the Lower Nursery (120). Subsequently in several widely scattered locations. Probably escaped from cultivation.
EXSICC.: **Kew Gardens.** Waste ground SE of Aiton House; [undated]; *Verdcourt* 5524 (K).
CULT.: 1789, Nat. of Britain (HK1, 3: 330); 1813, Nat. of Britain (HK2, 5: 249); 1814, Nat. of Britain (HKE: 289). In cult.
CURRENT STATUS: Probably not native in the Gardens, having escaped from cultivation, but now well established in several areas.

1607. Carex sylvatica Huds.
wood-sedge
CULT.: 1789, Nat. of Britain (HK1, 3: 330); 1813, Nat. of Britain (HK2, 5: 251); 1814, Nat. of Britain (HKE: 289). Not currently in cult.

1608. Carex capillaris L.
hair sedge
CULT.: 1813, Nat. of Britain (HK2, 5: 249); 1814, Nat. of Britain (HKE: 289). Not currently in cult.

1609. Carex strigosa Huds.
thin-spiked wood-sedge
CULT.: 1813, Nat. of England (HK2, 5: 249); 1814, Nat. of England (HKE: 289); 1964, Order Beds (K). Not currently in cult.

1610. Carex flacca Schreb.
C. recurva Huds.
glaucous sedge
CULT.: 1789, Nat. of England (HK1, 3: 331); 1813, Nat. of England (HK2, 5: 251); 1814, Nat. of England (HKE: 289); 1964, Order Beds (K). In cult.

1611. Carex panicea L.
carnation sedge
CULT.: 1789, Nat. of Britain (HK1, 3: 329); 1813, Nat. of Britain (HK2, 5: 248); 1814, Nat. of Britain (HKE: 289); 1964, Order Beds (K). In cult.

1612. Carex vaginata Tausch
C. mielichhoferi Sm.
sheathed sedge
CULT.: 1813, Nat. of Scotland (HK2, 5: 247); 1814, Nat. of Scotland (HKE: 289). Not currently in cult.
Note. Not known as a British plant until 1811.

1613. Carex depauperata Curtis ex With.
starved wood-sedge
CULT.: 1813, Nat. of England (HK2, 5: 247); 1814, Nat. of England (HKE: 289); 1967, Herbarium Ground (K). In cult.

1614. Carex laevigata Sm.
smooth-stalked sedge
CULT.: 1813, Nat. of Britain (HK2, 5: 251); 1814, Nat. of Britain (HKE: 289). Not currently in cult.
Note. Not known as a British plant until 1800.

1615. Carex binervis Sm.
green-ribbed sedge
CULT.: 1813, Nat. of Britain (HK2, 5: 246); 1814, Nat. of Britain (HKE: 289). In cult.
Note. Not known as a British plant until 1800.

1616. Carex distans L.
distant sedge
CULT.: 1789, Nat. of Britain (HK1, 3: 330); 1813, Nat. of Britain (HK2, 5: 246); 1814, Nat. of Britain (HKE: 289). Not currently in cult.

1617. Carex punctata Gaudin
dotted sedge
CULT.: 1963, Herbac. Dept. Order Bed (K); 1964, Order Beds (K). Not currently in cult.

1618. Carex extensa Gooden.
long-bracted sedge
CULT.: 1813, Nat. of Britain (HK2, 5: 245); 1814, Nat. of Britain (HKE: 289). Not currently in cult.
Note. Not known as a British plant until 1792.

1619. Carex × fulva Gooden.
(hostiana × viridula)
CULT.: 1813, Nat. of Britain (HK2, 5: 246); 1814, Nat. of Britain (HKE: 289). Not currently in cult.

1620. Carex flava L.
large yellow-sedge
CULT.: 1789, Nat. of Britain (HK1, 3: 328); 1813, Nat. of Britain (HK2, 5: 245); 1814, Nat. of Britain (HKE: 289). Not currently in cult.

1621. Carex viridula Michx.
yellow-sedge
a. subsp. **viridula**
C. oederi auct. non Retz.
CULT.: 1813, Nat. of England (HK2, 5: 246); 1814, Nat. of England (HKE: 289). Not currently in cult.
Note. Not known as a British plant until 1802.

b. subsp. **oedocarpa** (Andersson) B.Schmid
C. demissa Hornem.
CULT.: 1978, Alpine Dept. (K). Not currently in cult.

1622. Carex pallescens L.
pale sedge
CULT.: 1789, Nat. of Britain (HK1, 3: 329); 1813, Nat. of Britain (HK2, 5: 250); 1814, Nat. of Britain (HKE: 289); 1964, Order Beds (K); 1967, unspec. (K). In cult.

1623. Carex digitata L.
fingered sedge
CULT.: 1789, Nat. of England (HK1, 3: 328); 1813, Nat. of England (HK2, 5: 244); 1814, Nat. of England (HKE: 289). Not currently in cult.

1624. Carex humilis Leyss.
C. clandestina Gooden.
dwarf sedge
CULT.: 1813, Nat. of England (HK2, 5: 244); 1814, Nat. of England (HKE: 289). Not currently in cult.
Note. Not known as a British plant until 1792.

1625. Carex caryophyllea Latourr.
C. praecox auct. non Schreb.
spring-sedge
CULT.: 1813, Nat. of Britain (HK2, 5: 245); 1814, Nat. of Britain (HKE: 289); 1967, unspec. (K). Not currently in cult.

1626. Carex filiformis L.
C. tomentosa L.
downy-fruited dedge
CULT.: 1813, Nat. of Britain (HK2, 5: 252), Nat. of England (ibid.: 245 as *tomentosa*); 1814, Nat. of Britain (HKE: 290), Nat. of England (ibid.: 289 as *tomentosa*); 1894, unspec. (K); 1964, Order Beds (K). Not currently in cult.

1627. Carex montana L.
soft-leaved sedge
CULT.: 1789, Nat. of Britain (HK1, 3: 329); 1929, Natural Order Beds (K); 1964, Order Beds (K). Not currently in cult.

1628. Carex pilulifera L.
pill sedge
FIRST RECORD: 2003 [1789].
CULT.: 1789, Nat. of Britain (HK1, 3: 329); 1813, Nat. of Britain (HK2, 5: 244); 1814, Nat. of Britain (HKE: 289). Not currently in cult.
CURRENT STATUS: In patches in the Conservation Area. It is unlikely to be a recent arrival; it is more likely that Nicholson either failed to recognise it, although it is difficult to know with which species he may have confused it, or missed it altogether.

1629. Carex atrofusca Schkuhr
C. ustulata Wahlenb.
scorched alpine-sedge
CULT.: 1813, Nat. of Scotland (HK2, 5: 250); 1814, Nat. of Scotland (HKE: 289). Not currently in cult.

1630. Carex limosa L.
bog-sedge
CULT.: 1813, Nat. of Britain (HK2, 5: 250); 1814, Nat. of Britain (HKE: 289). Not currently in cult.

1631. Carex rariflora (Wahlenb.) Sm.
mountain bog-sedge
CULT.: 1813, Nat. of Scotland (HK2, 5: 250); 1814, Nat. of Scotland (HKE: 289). Not currently in cult.
Note. Not known as a British plant until 1813.

1632. Carex atrata L.
black alpine-sedge
CULT.: 1789, Nat. of Britain (HK1, 3: 329); 1813, Nat. of Britain (HK2, 5: 239); 1814, Nat. of Britain (HKE: 288). In cult.

1633. Carex buxbaumii Wahlenb.
club sedge
CULT.: 1964, Order Beds (K). Not currently in cult.

1634. Carex norvegica Retz.
close-headed alpine-sedge
CULT.: 1971, unspec. (K). Not currently in cult.

1635. Carex acuta L.
C. gracilis Curtis
slender tufted-sedge
FIRST RECORD: 1873/4 [1789].
NICHOLSON: **1873/4 survey** (JB: 75): **Strip.** This and [*C. acutiformis*] are the rarest species in our district. • **1906 revision** (AS: 89): **Strip.** This and [*C. acutiformis*] are rather rare within our limits.
OTHER REFERENCES: **Kew Environs.** 1932, *J.E.Lousley* (LN35: S298): By river Thames between Kew and Richmond.
EXSICC.: **Kew Gardens.** Ha-ha, north of Isleworth Ferry Gate; 25 v 1929; *Nelmes* 77/I (K). **Kew Environs.** Towing Path betw. Kew and Richmond; 22 vi 1924; *Nelmes* 39 (K). • Ha-ha, some 50 yds. beyond S.W. end of Kew Gardens; 25 v 1929; *Nelmes* 77/III (K). • Ha-ha, just beyond S.W. end of Kew Gardens; 25 v 1929; *Nelmes* 77/II (K). • Same plant, or near, as No. 77 of 25 v 1929; 16 vi 1929; *Nelmes* 78 (K). • Ha-ha, few yds. S. of No. 77 of 25 v 1929; 25 v 1929; *Nelmes* 79 (K). • Same plant as No. 79 of 25 v 1929; 16 vi 1929; *Nelmes* 80 (K). • Ha-ha, near S.W. end of Kew Gardens; 16 vi 1929; *Nelmes* 84 (K). • Ha-ha, S. of Kew Gardens; 16 vi 1929; *Nelmes* 85 (K). • By stream between Kew & Richmond; 25 vi 1932; *Lousley* s.n. (K). • Tow-path, betw. Kew & Richmond; vi 1933; *Nelmes* s.n. (K). • Ha-ha edge, tow path, between Kew and Richmond; 2 v 1937; *Nelmes* 289 (K). • Edge of ha-ha by tow-path between Kew and Richmond; 12 vi 1937; *Nelmes* 337, 338, 339 (K).
CULT.: 1789, Nat. of Britain (HK1, 3: 331), Nat. of England (ibid.: as *gracilis*); 1813, Nat. of Britain (HK2, 5: 252); 1814, Nat. of Britain (HKE: 290); 1929, Natural Order Beds (K); 1964, Order Beds (K). In cult.
CURRENT STATUS: No recent sightings.

1636. Carex trinervis Degl.
three-nerved sedge
CULT.: 1936, unspec. (K); 1937, Herbac. Dept. (K); 1939, unspec. (K); 1967, Herbarium Ground (K). Not currently in cult.

1637. Carex nigra (L.) Reichard
common sedge
CULT.: In cult.

1638. Carex elata All.
C. caespitosa auct. non L.
C. stricta Gooden.
tufted-sedge
CULT.: 1789, Nat. of Britain (HK1, 3: 331); 1813, Nat. of Britain (HK2, 5: 248 as *stricta*, and: 249 as *caespitosa*); 1814, Nat. of Britain (HKE: 289 as *stricta* and *caespitosa*); 1936, Herbac. Dept. (K); 1971, Bog Garden (K); 1980, Alpine Sect 156-51 (K). Not currently in cult.

1639. Carex bigelowii Torr. ex Schwein.
C. rigida Gooden.
stiff sedge
CULT.: 1813, Nat. of Britain (HK2, 5: 247); 1814, Nat. of Britain (HKE: 289). In cult.
Note. Not known as a British plant until 1792.

1640. Carex pauciflora Lightf.
few-flowered sedge
CULT.: 1813, Nat. of Britain (HK2, 5: 238); 1814, Nat. of Britain (HKE: 288). Not currently in cult.

1641. Carex pulicaris L.
flea sedge
CULT.: 1789, Nat. of Britain (HK1, 3: 326); 1813, Nat. of Britain (HK2, 5: 238); 1814, Nat. of Britain (HKE: 288). Not currently in cult.

[**1642. Carex capitata** L.
CULT.: 1789, Nat. of England (HK1, 3: 326).
Note. Listed in error as native, and known as an introduction only in the Outer Hebrides.]

POACEAE

1643. Leersia oryzoides (L.) Sw.
cut-grass
FIRST RECORD: Probably between 1779 and 1805 [1951].
EXSICC.: **Kew Environs.** Kew; September; *Goodenough* s.n. (K).
CULT.: 1951, Aquatic Garden (K). In cult.
CURRENT STATUS: This extremely rare British grass was recently found by a pond on the golf course in the Old Deer Park just south of the Gardens' boundary. Goodenough's record may have been the first for the country and the site in the Old Deer Park may well have been near where he gathered his specimen. The plants by the pond have only been recognised for what they are since 1999; they must have arrived either from seed in the soil or seed washed down the Thames from perhaps the Byfleet area. No mention has ever been made in the literature of Goodenough's specimen, and its significance for dating the first record for Great Britain has only recently been realised. It considerably predates the supposed first records for the country (1844) and for Surrey (1851). The species has been cultivated in the Gardens only since 1951 (according to available records) and it is therefore unlikely that Goodenough's plant either came from within the Gardens or had escaped from the Gardens. Sadly, the same cannot be said for certainty for the current population but there is every hope that it is indeed entirely spontaneous. The species is fully protected in the British Isles and it should be safe in its current station.

1644. *Ehrharta erecta Lam.
E. panicea Sm. ex Thunb.
Lamarck's ehrharta
FIRST RECORD: 1793.
EXSICC.: **Kew Environs.** Kew; May 1793; *Goodenough* s.n. (K).
CULT.: Late 18th century (possibly 1790), unspec. (LIV, photo K). In cult.
CURRENT STATUS: The species was introduced to the Botanic Garden in 1790 and it is possible that Goodenough's specimen came from this material; however, since Goodenough does not appear to have made any collections from within the Garden, the species may already have escaped by 1793. The LIV specimen, known to have come from the Garden, has no details beyond the words 'Hort. Kew'. Today, the species is a rare casual not deemed worthy of inclusion in any British plant list.

1645. Nardus stricta L.
mat-grass
FIRST RECORD: 1873/4 [1768].
NICHOLSON: **1873/4 survey** (JB: 77): **P.** On both sides of Syon Vista (clear of broad portion kept so closely cut). In wood near "Old Deer Park." **Q.** Here and there. • **1906 revision** (AS: 91): **A.** Both sides of Syon Vista; near south end of holly collection. **Q.** Here and there.
CULT.: 1768 (JH: 398); 1789, Nat. of Britain (HK1, 1: 83); 1810, Nat. of Britain (HK2, 1: 134); 1814, Nat. of Britain (HKE: 19). In cult.
CURRENT STATUS: No recent sightings; it may still be present but close mowing of the turf makes it difficult to recognise.

[1646. Stipa pennata L.
CULT.: 1789, Nat. of England (HK1, 1: 111); 1810, Nat. of England (HK2, 1: 170); 1814, Nat. of England (HKE: 24).
Note. Listed in error as native.**]**

1647. *Stipa tenuissima Trin.
FIRST RECORD: 2004 [1978].
CULT.: In cult.
CURRENT STATUS: Several plants grew from seed in topsoil stored in the Paddock.

1648. Milium effusum L.
wood millet
FIRST RECORD: 1873/4 [1768].
NICHOLSON: **1873/4 survey** (JB: 75): **Q.** Plentiful in wood skirting Pleasure Grounds. • **1906 revision** (AS: 89): **Q.** Plentiful in wood skirting pinetum.
EXSICC.: **Kew Gardens.** [Unlocalised]; 16 vi 1936; *Worsdell* s.n. (K). • Queen's Cottage Grounds; growing under trees in shade, moist position in association with various other grasses; 7 vi 1937; *Rawlings* 144 (K).
CULT.: 1768 (JH: 400); 1789, Nat. of Britain (HK1, 1: 93); 1810, Nat. of Britain (HK2,1: 147); 1814, Nat. of Britain (HKE: 21). In cult.
CURRENT STATUS: No recent sightings of wild material. Both the cultivated var. '*Aureum*' and the wild type can be found on Cumberland Mound but both have been planted.

1649. Festuca pratensis Huds.
meadow fescue
FIRST RECORD: 1873/4 [1810].
NICHOLSON: **1873/4 survey** (JB: 76): **P.** Here and there in turf and beds near river end of lake. • **1906 revision** (AS: 90): [**A:** Here and there. **Strip:** By towing-path. **Q:** Sparingly].
EXSICC.: **Kew Gardens.** [Unlocalised]; 16 vi 1936; *Worsdell* s.n. (K). • Cottage Grounds; open situation in rough grassland; 12 vi 1950; *Souster* 1103 (K).
CULT.: 1810, Nat. of Britain (HK2, 1: 165); 1814, Nat. of Britain (HKE: 23). Not currently in cult.
CURRENT STATUS: No recent sightings.

1650. Festuca arundinacea Schreb.
Festuca elatior L.
tall fescue
FIRST RECORD: 1873/4 [1768].
NICHOLSON: **1873/4 survey** (JB: 76): A single plant in wood behind Winter Garden. Mr J.M.Smith. Mr Hemsley has specimens collected some years ago between Winter Garden and Kew Road, where, he says, it was not uncommon. None exist there now. • **1906 revision** (AS: 90): **A.** Here and there. **Strip.** By towing-path. **Q.** Sparingly.
EXSICC.: **Kew Gardens.** River-bank Kew – Richmond; at top of river wall near Brentford Ferry; 25 vi 1929; *Summerhayes & Turrill* s.n. (K). • Thames Bank between Kew & Richmond near Isleworth Gate; only few plants seen; 9 vii 1929; *Hubbard* 1671 (K). **Kew Environs.** Thames bank, Kew; vii 1888; *Gamble* 20189 (K). • Thames banks, Kew; 9 vii 1930; *Fraser* s.n. (K).
CULT.: 1768 (JH: 403); 1789, Nat. of Britain (HK1, 1: 108); 1810, Nat. of Britain (HK2, 1: 165); 1814, Nat. of Britain (HKE: 23). In cult.
CURRENT STATUS: Widely scattered but uncommon.

1651. Festuca gigantea (L.) Vill.
Bromus giganteus L.
giant fescue
FIRST RECORD: 1873/4 [1768].
NICHOLSON: **1873/4 survey** (JB: 76): **Strip.** Uncommon. • **1906 revision** (AS: 90): **Strip.** Uncommon. **Q.** Here and there.
EXSICC.: **Kew Environs.** Banks of Thames, Old Deer Park, Richmond; 1884; *Fraser* s.n. (K). • River bank, Kew; vii 1888; *Gamble* 20215 (K). • Between Kew & Richmond, in shade near bank of R. Thames; abundant; 2 viii 1927; *Hubbard* G.69 (K).

CULT.: 1768 (JH: 404); 1789, Nat. of Britain (HK1, 1: 110); 1810, Nat. of Britain (HK2, 1: 164); 1814, Nat. of Britain (HKE: 23); 1902, Herbarium Grounds (K); 1973, Seed plot, S. Arboretum (K). Not currently in cult.
CURRENT STATUS: Still to be found on the towpath south of Brentford Gate.

1652. Festuca altissima All.
F. calamaria Sm.
F. decidua Bellardi ex Sm.
wood fescue
CULT.: 1810, Nat. of Scotland (HK2, 1: 165); 1814, Nat. of Scotland (HKE: 23 as *calamaria*), Nat. of England (ibid.: Add. as *decidua*). Not currently in cult.

1653. *Festuca heterophylla Lam.
various-leaved fescue
FIRST RECORD: 1932 [1926].
REFERENCES: **Kew Gardens.** 1936, *Hubbard* (LFS: 383; S32: 657; LN35: S316; see also Exsicc.): Queen's Cottage Grounds, in shade of *Fagus sylvatica*, locally common. • 1961, *Welch* (LFS: 383): Queen's Cottage Grounds. • 1983, *Burton* (FLA: 195): Well naturalised in Queen's Cottage grounds attached to Kew Gardens.
EXSICC.: **Kew Gardens.** Queen's Cottage Grounds, in shade of beech trees, common; vi 1932; *Hubbard* s.n. (K). • Queen's Cottage Grounds, in shade of *Fagus sylvatica*, locally common; 6 vi 1936; *Hubbard* 90 (K). • Queen's Cottage Grounds; 16 vi 1936; *Worsdell* s.n. (K). • Queen's Cottage Grounds; 11 vi 1948; *O.J.Ward* 48 (K). • Queen's Cottage Grounds; 14 vi 1949, *Souster* s.n. (K).
CULT.: 1926, Herb. Expt. Ground (K). In cult.
CURRENT STATUS: Still abundant in the Conservation Area and spreading to the adjacent golf course in the Old Deer Park.

1654. Festuca rubra L.
incl. *F. rubra* subsp. *commutata* Gaudin
red fescue
FIRST RECORD: 1888 [1768].
REFERENCES: **Kew Gardens.** 1990, *Cope* (S31: 182): Paddock, after restoration in 1990 [as subsp. *commutata*].
EXSICC.: **Kew Gardens.** Herbarium ground; 6 vi 1921; *anon.* s.n. (K). • [Unlocalised]; amongst long grasses on sandy soil; 31 v 1927; *Hubbard* G.72 (K). • Near tennis-courts; sandy soil; 10 vi 1927; *Hubbard* G.72C (K). • Herb. Expt. Ground; 27 vi 1927; *Hubbard* G73A (K). • Nr. Herbarium; waste ground; v 1928; *Ballard* s.n. (K). • Queen's Cottage Grounds; sheltered shady spot, soil – leafy sand; 11 vi 1948; *O.J.Ward* 49 (K). • Cottage Grounds; 14 vi 1949; *Souster* 1005 (K). • [Unlocalised; presumably Queen's Cottage Grounds]; shaded position on light sandy soil of untended land, ass. plants:- *Festuca heterophylla*, *Poa chaixii*; 12 vi 1950; *Naylor* 595 (K). • Herbarium field; on margin of recently sown grass, with indigenous grasses such as *Trisetum* & *Anthoxanthum*; sandy gravelly soil; 6 vi 1974; *Hubbard* F4/74FR (K). • Near Herbarium; margin of recently sown (1973) grass turf; 29 v 1975; *Hubbard* A29 (K). • Behind the Herbarium; cut lawn sown about 1968, few plants only; 16 vi 1976; *Hubbard* 16676FR2 (K). **Kew Environs.** Thames bank, Kew; vii 1888; *Gamble* 20188 (K).
CULT.: 1768 (JH: 403); 1789, Nat. of Britain (HK1, 1: 107); 1810, Nat. of Britain (HK2, 1: 164); 1814, Nat. of Britain (HKE: 23); 1935, unspec. (K). In cult.
CURRENT STATUS: Observations have indicated that a number of plants originally recorded as subsp. *rubra* appeared to be subsp. *commutata* in the particularly hot, dry summer of 2003. There are suggestions that habit in red fescue may be controlled more by environment than by genetics and that the two subspecies may not be so different after all. The species is abundant in turf throughout the Gardens.

1655. Festuca ovina L.
sheep's-fescue
FIRST RECORD: 1873/4 [1768].
NICHOLSON: **1873/4 survey** (JB: 76): **B, P** and **Q.** In many dry places where the gravel comes close to the surface, this grass is by far the principal factor in the turf. • **1906 revision** (AS: 91): In dry gravelly soils this is the principal factor in the turf.
CULT.: 1768 (JH: 403); 1789, Nat. of Britain (HK1, 1: 107); 1810, Nat. of Britain (HK2, 1: 163); 1814, Nat. of Britain (HKE: 23). In cult.
CURRENT STATUS: Rare; currently known only from the Conservation Area. Its decline since Nicholson's time is quite remarkable.

1656. Festuca vivipara (L.) Sm.
F. ovina var. *vivipara* L.
viviparous sheep's-fescue
CULT.: 1768 (JH: 403); 1789, Nat. of Britain (HK1, 1: 107); 1810, Nat. of Britain (HK2, 1: 163); 1814, Nat. of Britain (HKE: 23); 1937, Herbaceous Ground (K). Not currently in cult.

1657. Festuca filiformis Pourr.
F. capillata Lam.
F. ovina var. *tenuifolia* (Sibth.) Sm.
F. tenuifolia Sibth.
fine-leaved sheep's-fescue
FIRST RECORD: 1906.
NICHOLSON: **1906 revision** (AS: 91): Very common.
OTHER REFERENCES: **Kew Gardens.** 1925, *Druce* (BSEC7: 906): Queen's Cottage Grounds. • 1935, *Hubbard* (S32: 657; LN36: S317): Queen's Cottage Grounds, Kew Gardens.

EXSICC.: **Kew Gardens.** Queens Cottage Grounds; [c. 1914]; *Flippance* s.n. (K). • Near Arboretum; in shade of beech trees; 31 v 1927; *Hubbard* G.65 (K). • Queen's Cottage; vi 1929; *Pearce* s.n. (K). • Queen's Cottage Grounds; in shade of trees, sandy soil; vi 1932; *Hubbard* s.n. (K). • Queen's Cottage Ground; in shade of *Fagus sylvatica*, in sandy soil, abundant; 19 vi 1935; *Hubbard* 75 (K). • Queen's Cottage Grounds; 16 vi 1936; *Worsdell* s.n. (K). • Queens Cottage Ground; growing in clumps in shady position under Oak tree, soil sandy & moist; 11 vi 1948; *O.J. Ward* 61 (K).
CULT.: 1929, Grass garden (K); 1935, unspec. (K). In cult.
CURRENT STATUS: Two populations are known from within the Gardens. It occurs in a very characteristic assemblage of species found in a small patch of poor ground north of the Rhododendron Dell and in dry ground on Mount Pleasant; it also occurs, more widely, in adjacent parts of the golf course in the Old Deer Park.

1658. Festuca longifolia Thuill.
F. caesia Sm.
blue fescue
CULT.: 1814, Nat. of England (HKE: Add.); 1974, unspec. (K). In cult.

1659. *Festuca brevipila R.Tracey
F. rubra var. *duriuscula* Gaudin
F. duriuscula auct. non L.
F. lemanii auct. non Bastard?
hard fescue
FIRST RECORD: 1873/4 [1768].
NICHOLSON: **1873/4 survey** (JB: 76): **B, Pal, P** and **Q.** Frequent. • **1906 revision** (AS: 91): Common.
OTHER REFERENCES: **Kew Gardens.** 1998, *Cope* (S33: 751): North Arb.: on the RBG side of Kew Green (118) [as *F. lemanii*].
EXSICC.: **Kew Gardens.** Nr. Lake; in cult. ground; [no date]; *Hubbard* s.n. (K). • Came up in Herbarium Expt. Ground; vi 1932; *Hubbard* s.n. (K). • In turf in front of the Herbarium; dense tuft; 26 v 1940; *Hubbard* s.n. (K). • Herbarium grounds; 25 v 1940; *Marshall* 8 (K). **Kew Environs.** Kew: on green near Church; 29 v 1929, *Hubbard* 1523 (K).
CULT.: 1768 (JH: 403); 1789, Nat. of Britain (HK1, 1: 107); 1810, Nat. of Britain (HK2, 1: 164); 1814, Nat. of Britain (HKE: 23). In cult.
CURRENT STATUS: Rare; still to be found in front of the Herbarium, but frequently mown and difficult to spot. Because of unresolved taxonomic confusion the species has also been recorded as *F. lemanii*

1660. × Festulolium loliaceum (Huds.) P.Fourn. (Festuca pratensis × Lolium perenne)
Festuca loliacea Huds.
hybrid fescue
FIRST RECORD: 1936 [1810].
REFERENCES: **Kew Gardens.** 1955, *D.H. Kent* (S32: 657; LN35: S316): Kew Gardens.
EXSICC.: **Kew Gardens.** [Unlocalised]; 16 vi 1936; *Worsdell* s.n. (K).
CULT.: 1810, Nat. of England (HK2, 1: 165); 1814, Nat. of England (HKE: 23). Not currently in cult.
CURRENT STATUS: No recent sightings.

1661. × Festulolium brinkmanii (A.Braun) Asch. & Graebn. (Festuca gigantea × Lolium perenne)
FIRST RECORD: 1888.
EXSICC.: **Kew Environs.** Thames bank, Kew; vii 1888; *Gamble* 20207 (K).
CURRENT STATUS: No recent sightings.

1662. Lolium perenne L.
L. perenne var. *muticum* DC.
perennial rye-grass
FIRST RECORD: 1854 [1768].
NICHOLSON: **1873/4 survey** (JB: 76): Nearly everywhere both in long and short grass forming a large proportion of the turf. • **1906 revision** (AS: 91): Very common.
OTHER REFERENCES: **Kew Gardens.** 1907, *A.B.Jackson & Domin* (SFS: 650): Rough grassy places, Kew Gardens.
EXSICC.: **Kew Gardens.** *Berberis* Dell; 23 x 1915; *Turrill* s.n. (K). • [Unlocalised]; 14 vi 1925; *Hubbard* 17 (K). • [Unlocalised]; in shade, sandy soil; 1 vi 1927; *Hubbard* G.86 (K). • Lower nursery; amongst weeds on rubbish tip; 16 x 1929; *Hubbard* s.n. (K). • Tennis Courts; ix 1930; *Bullock* s.n. (K). • [Unlocalised]; on rubbish-tip; 10 vi 1938; *Hubbard* 127 (K). • Queen's Cottage Grounds; in shade, on leaf-soil by path; 14 vi 1938; *Hubbard* 128 (K). • Herbarium grounds; 5 vi 1940; *Marshall* 7 (K). **Kew Environs.** Kew; 1854; *anon.* s.n. (K). • Thames bank, Kew; vii 1888; *Gamble* 20194 (K). • Between Kew & Richmond, in shade on banks of R. Thames; 9 vii 1929; *Hubbard* 1694 (K). • Thames banks, Kew; 9 vii 1930; *Fraser* s.n. (K). • Thames bank, Kew; ix 1934; *Philipson* s.n. (K).
CULT.: 1768 (JH: 394); 1789, Nat. of Britain (HK1, 1: 116); 1810, Nat. of Britain (HK2, 1: 175); 1814, Nat. of Britain (HKE: 25); 1934, unspec. (K); 1935, unspec. (K). Not currently in cult.
CURRENT STATUS: Abundant throughout the Gardens in all types of grassland.

1663. *Lolium multiflorum Lam.
L. italicum A.Br.
L. italicum var. *muticum* Parl.
Italian rye-grass
FIRST RECORD: 1873/4 [1935].
NICHOLSON: **1873/4 survey** (JB: 77): **P.** Common about lake on newly-sown land. Frequent elsewhere in older turf. • **1906 revision** (AS: 91): In all the divisions.

OTHER REFERENCES: **Kew Gardens.** 1907, *A.B.Jackson & Domin* (BEC2: 323, and see Exsicc.): rough grassy places. The Italian rye grass was abundant in rough grassy places on the west side of Kew Gardens last summer, and assumed a great variety of forms, among them being [var. *muticum*], which is distinguished by its awnless glumes. • 1908, *A.B.Jackson* (S7: 125–6): Rough unmown places about the Arboretum, especially near the Palace. This awnless variety [var. *muticum*] of the Italian Rye-grass grows here with the type, and is liable to be passed over for *L. perenne*. A very variable species; when growing under luxuriant conditions as it does at Kew the spikes often become branched.
EXSICC.: **Kew Gardens.** Lower nursery; on rubbish heap; 1 vii 1932; *Hubbard* s.n. (K). • [Unlocalised]; on rubbish heap; 3 vii 1935; *Hubbard* s.n. (K).
CULT.: 1935, unspec. (K). In cult.
CURRENT STATUS: No recent sightings; it was last noted in 1977. It is possible that some records may refer to *L.* × *boucheanum*.

1664. Lolium × boucheanum Kunth (multiflorum × perenne)
L. italicum × *perenne*
FIRST RECORD: 1907 [1939].
REFERENCES: **Kew Gardens.** 1907, *A.B.Jackson & Domin* (BEC2: 324): Observed in Kew Gardens. • 1908, *A.B.Jackson* (S7: 126): A grass which K. Domin thought was this combination was not uncommon at Kew last August wherever the parents occurred.
EXSICC.: **Kew Gardens.** [Unlocalised]; rough grassy places; viii 1907; *A.B.Jackson & Domin* s.n. (K). • Arboretum; ix 1907; *A.B.Jackson & Domin* s.n. (K). • Queen's Cottage Grounds; in glade; 14 vi 1938; *Hubbard* 9346 (K).
CULT.: 1939, unspec., root originally from the Queen's Cottage Grounds (K). Not currently in cult.
CURRENT STATUS: Uncommon; there have been a few widely scattered recent sightings, some of which may account for records of *L. multiflorum*. Since *L. multiflorum* has not recently been confirmed within the Gardens it is likely that this hybrid has originated from commercial grass-seed mixes.

1665. Lolium temulentum L.
L. arvense With.
darnel
FIRST RECORD: 1935 [1768].
EXSICC.: **Kew Gardens.** [Unlocalised]; on rubbish heap; 3 vii 1935; *Hubbard* s.n. (K).
CULT.: 1768 (JH: 394); 1789, Nat. of Britain (HK1, 1: 116); 1810, Nat. of Britain (HK2, 1: 175), Nat. of England (ibid.: as *arvense*); 1814, Nat. of Britain (HKE: 25), Nat. of England (ibid.: as *arvense*); 1925, Herb. Expt. Ground (K); 1961, Jodrell, Order Beds (K). In cult.
CURRENT STATUS: No recent sightings.

1666. Vulpia fasciculata (Forssk.) Fritsch
Festuca uniglumis Aiton
dune fescue
CULT.: 1789, Nat. of England (HK1, 1: 108); 1810, Nat. of England (HK2, 1: 164); 1814, Nat. of England (HKE: 23). Not currently in cult.

1667. Vulpia bromoides (L.) Gray
Festuca bromoides L.
F. sciuroides Roth
squirreltail fescue
FIRST RECORD: 1873/4 [1768].
NICHOLSON: **1873/4 survey** (JB: 76): **P.** In dry beds and turf near Isleworth Gate. • **1906 revision** (AS: 91): **A.** In dry beds and turf near Isleworth Gate.
CULT.: 1768 (JH: 403); 1789, Nat. of Britain (HK1, 1: 107); 1810, Nat. of Britain (HK2, 1: 163); 1814, Nat. of Britain (HKE: 23). In cult.
CURRENT STATUS: No recent sightings.

1668. Vulpia myuros (L.) C.C.Gmel.
Festuca myuros L.
Vulpia megalura (Nutt.) Rydb.
rat's-tail fescue
FIRST RECORD: 1936 [1768].
REFERENCES: **Kew Gardens.** 1953, *Souster* (S24: 189, and see Exsicc.): From flax straw from W. Australia, used for mulch in private garden, and also in Yard.
EXSICC.: **Kew Gardens.** [Unlocalised]; on rubbish heap; 4 vi 1936; *Hubbard* s.n. (K) • [Unlocalised]; came up among remains of flax straw from Chelsea Show of 1953; 27 viii 1953; *Souster* s.n. (K).
CULT.: 1768 (JH: 403); 1789, Nat. of England (HK1, 1: 108); 1810, Nat. of England (HK2, 1: 164); 1814, Nat. of England (HKE: 23); 1925, unspec. (K); 1955, Herbaceous Ground (K). In cult.
CURRENT STATUS: Widespread but uncommon.

1669. Vulpia ciliata Dumort.
bearded fescue
Cult.: In cult.

1670. Cynosurus cristatus L.
crested dog's-tail
FIRST RECORD: 1873/4 [1768].
NICHOLSON: **1873/4 survey** (JB: 76): Common everywhere. Forms a fair share of most of the turf. • **1906 revision** (AS: 90): Common everywhere. Forms a fair share of most of the turf.
EXSICC.: **Kew Gardens.** Tennis Courts; ix 1930; *Bullock* s.n. (K). • [Unlocalised]; in rough grass; 12 vi 1945; *Souster* 262 (K). • Queen's Cottage Grounds; 4 vi 1951; *Kennedy-O'Byrne* 132 (K).
CULT.: 1768 (JH: 406); 1789, Nat. of Britain (HK1, 1: 104); 1810, Nat. of Britain (HK2, 1: 161); 1814, Nat. of Britain (HKE: 23); 1935, unspec. (K). In cult.
CURRENT STATUS: Still present, but hard to find as a consequence of frequent mowing. It is certainly not as common as it was in Nicholson's time and seems to have disappeared from most of the lawns.

1671. *Cynosurus echinatus L.
rough dog's-tail
FIRST RECORD: 1990 [1768].
REFERENCES: **Kew Gardens.** 1990, *Cope* (S31: 182): Paddock, after restoration in 1990.
CULT.: 1768 (JH: 406); 1789, Nat. of England (HK1, 1: 104); 1810, Nat. of England (HK2, 1: 161); 1814, Nat. of England (HKE: 23); 1929, Herbaceous Ground (K). In cult. (2006). In cult.
CURRENT STATUS: No recent sightings; it has not reappeared in the Paddock since 1990.

1672. Puccinellia maritima (Huds.) Parl.
Poa maritima Huds.
common saltmarsh-grass
FIRST RECORD: Probably between 1779 and 1805 [1789].
EXSICC.: **Kew Environs.** Kew; *Goodenough* (K).
CULT.: 1789, Nat. of Britain (HK1, 1: 100); 1810, Nat. of Britain (HK2, 1: 156); 1814, Nat. of Britain (HKE: 22). In cult. under glass.
CURRENT STATUS: No recent sightings.

1673. Puccinellia distans (Jacq.) Parl.
Poa distans Jacq.
reflexed saltmarsh-grass
CULT.: 1810, Nat. of England (HK2, 1: 158); 1814, Nat. of England (HKE: 22). In cult.

1674. Puccinellia × pannonica (Hack.) Holmb.
(distans × rupestris)
CULT.: 1948, Herbarium Ground (K). Not currently in cult.

1675. Puccinellia rupestris (With.) Fernald & Weath.
Poa procumbens Curtis
stiff saltmarsh-grass
CULT.: 1810, Nat. of Britain (HK2, 1: 154); 1814, Nat. of Britain (HKE: 22). Not currently in cult.
Note. Not known as a British plant until 1794.

1676. Briza media L.
quaking-grass
FIRST RECORD: 1873/4 [1789].
NICHOLSON: **1873/4 survey** (JB: 76): **Pal.** Frequent in turf before Palace. **P.** Common in the open part near locality given for *Stachys sylvatica* [in wood near large cedar (the one so conspicuous from behind Palm House)] and sparingly on the Richmond side of lake. **Q.** Here and there. • **1906 revision** (AS: 90): **P.** Frequent in turf near palace. **A.** Common at northern part of Ash collection. **Q.** Here and there.
CULT.: 1789, Nat. of Britain (HK1, 1: 103); 1810, Nat. of Britain (HK2, 1: 159); 1814, Nat. of Britain (HKE: 22); 1925, Herbaceous Ground (K); 1961, Order Beds (K); 1976, unspec. (K). In cult.
CURRENT STATUS: No recent sightings. It is possible that it is still present but the close mowing of grass makes it difficult to find.

1677. Briza minor L.
lesser quaking-grass
CULT.: 1768 (JH: 402); 1789, Nat. of England (HK1, 1: 102); 1810, Nat. of England (HK2, 1: 158); 1814, Nat. of England (HKE: 22); 1975, unspec. (K). In cult.
Note. Originally grown as a native but now considered to be an archaeophyte.

1678. *Briza maxima L.
greater quaking-grass
FIRST RECORD: 2006 [1768].
CULT.: 1768 (JH: 402); 1789, Nat. of the South of Europe (HK1, 1: 103); 1810, Nat. of the South of Europe (HK2, 1: 159); 1814, S. of Europe (HKE: 22). In cult.
CURRENT STATUS: A number of plants outside the Jodrell Laboratory in 2006, but the area is regularly tidied and plants do not survive for long. Doubtless they escaped from the nearby grass beds.

1679. Poa annua L.
Poa annua f. *purpurea* M.L.Grant
annual meadow-grass
FIRST RECORD: 1873/4 [1768].
NICHOLSON: **1873/4 survey** (JB: 76): A large proportion of nearly every piece of turf is made up of this plant. • **1906 revision** (AS: 90): A large proportion of nearly every piece of turf is made up of this plant.
EXSICC.: **Kew Gardens.** Herbarium Expt. Ground; in sandy soil; 26 ii 1928; *Hubbard* s.n. (K). • Thames bank, opposite Brentford Docks; on lower part of sloping wall with *Veronica anagallis* etc.; 26 viii 1929; *Summerhayes* s.n. (K). • Lower Nursery; on leaf-heap; 16 x 1929; *Hubbard* s.n. (K). • Meadow behind Herbarium; abundant on otherwise bare ground, under trees; 27 iv 1930; *Bullock* 18 (K). • Queen's Cottage Grounds; 4 vi 1951; *Kennedy-O'Byrne* 138 (K). • Near Herb. Expt. Gr.; weed on allotment ground; [?June 1951]; *Hubbard* 13353 (K). • On old allotment next Herb. Expt. Ground; 19 vi 1951; *Hubbard* 13354, 13355, 13356, 13357 (all K). • Herbarium Expt. Ground; on piece of old allotment, still alive in 1952; 20 vi 1951; *Hubbard* 13430e (K). • Expt. Ground, Herb.; in moist place near water tank; 22 ix 1952; *Hubbard* s.n. (K). • In Iris garden; 24 vi 1953; *Hubbard* 13627 (K). • In shrub bed near Unicorn Gate; 14 vii 1954; *White* s.n. (K). • Root from shrubbery near Unicorn Gate; cultivated in pot in Herbarium Expt. Ground; leaves striped cream-white and green; 30 xi 1954; *Souster* in *Hubbard* 13630 (K). • Behind Herbarium; weed in allotments; 14 ix 1956; *Hubbard* s.n. (K). • By Herbarium; between paving stones; 6 ix 1971; *Hubbard* s.n. (K). • Grass nursery; partial

shade/shelter of mature trees and shrubs to the south; 10 viii 1981; *Macdonald* 6 (K). • Student plots; well-drained, stony, sandy soil; x 1981; *Locke* 4 (K). • Plot 411 adjacent to King William's Temple; form with brownish-purple foliage; 12 viii 2002, *Cope* 703 (K). **Kew Environs.** Kew [unlocalised]; 3 vii. 1900; *Clarke* 49829 (K). • On Thames bank, between Kew & Richmond; submerged at high tide; 9 vii 1929; *Hubbard* 1693 (K). • Kew, banks of R. Thames; submerged at high tide; 28 v 1933; *Hubbard* s.n. (K). • Thames embankment, Ferry Lane, Kew; 9 x 1956; *Kennedy-O'Byrne* s.n. (K).
CULT.: 1768 (JH: 403); 1789, Nat. of Britain (HK1, 1: 100); 1810, Nat. of Britain (HK2, 1: 155); 1814, Nat. of Britain (HKE: 22). Not currently in cult.
CURRENT STATUS: Abundant throughout the Gardens in all habitats. The newly described f. *purpurea* occurs occasionally as a weed of flower-beds; its distinctive cryptic coloration makes it hard to see when weeding and it is thus likely to spread; it is behaving in exactly the same way at Wisley from where it was first described.

1680. Poa trivialis L.
rough meadow-grass
FIRST RECORD: 1854 [1768].
NICHOLSON: **1873/4 survey** (JB: 76): Not so frequent as [*P. pratensis*], if I have observed correctly. • **1906 revision** (AS: 90): Not so frequent as *P. pratensis*.
OTHER REFERENCES: **Kew Gardens.** 1990, *Cope* (S31: 182): Paddock, after restoration in 1990.
EXSICC.: **Kew Gardens.** Nr. Tennis ground; waste ground; vi 1913; *Turrill* s.n. (K). • Queen's Cottage Grounds; [c. 1914]; *Flippance* s.n. (K). • [Unlocalised]; 14 v 1925; *Hubbard* 16 (K). • [Unlocalised]; on rubbish heap; 31 v 1927; *Hubbard* G.102 (K). • Lower Nursery; on waste stony ground; 16 i 1928; *Hubbard* s.n. (K). • In field behind Herbarium; v 1928; *Ballard* s.n. (K). • Thames bank, in *Petasites* zone near Isleworth Gate; 25 vi 1929; *Summerhayes & Turrill* s.n. (K). • On humus in Nursery Yard; 1 vi 1933; *Hubbard* 159 (K). • Queen's Cottage Grounds; in shade of trees; 6 vi 1934; *Hubbard* s.n. (K). • Queens Cottage Grounds; in shade of Beech trees; 14 vi 1938; *Hubbard* 9526 (K). • [Unlocalised]; open position on untended land, sandy loam, ass. plants:- *Holcus lanatus*, *Festuca tenuifolia*; 12 vi 1950; *Naylor* 589 (K). • Herbarium Expt. Ground; 20 viii 1951; *Hubbard* s.n. (K). • [Unlocalised]; under trees; 6 vi 1960; *Evans & Evans* s.n. (K). **Kew Environs.** Kew; 1854; *anon.* s.n. (K). • On muddy bank of R. Thames; locally common on bare mud, submerged at high tide; 4 viii 1951; *Hubbard* 13325 (K).
CULT.: 1768 (JH: 403); 1789, Nat. of Britain (HK1, 1: 99); 1810, Nat. of Britain (HK2, 1: 155); 1814, Nat. of Britain (HKE: 22); 1929, Grass garden (K); 1939, Herbarium Expt. Ground (K). Not currently in cult.
CURRENT STATUS: Common in grassy areas throughout the Gardens.

1681. Poa pratensis L.
a. subsp. **pratensis**
smooth meadow-grass
FIRST RECORD: 1873/4 [1768].
NICHOLSON: **1873/4 survey** (JB: 76): Common in all the divisions. • **1906 revision** (AS: 90): Common in all the divisions.
EXSICC.: **Kew Gardens.** [Unlocalised]; among long grasses; 16 v 1927; *Hubbard* G.103 (K). • Near herbarium; in recently sown grass; vi 1930; *Hubbard* 9054 (K). • Herb. Exp. Ground; growing on path; 5 viii 1930; *Summerhayes* 543 (K). • Queen's Cottage Grounds; in shade, sandy soil; 5 vi 1933; *Hubbard* s.n. (K). • [Unlocalised]; on rubbish heap; 10 vi 1938; *Hubbard* 9433 (K). • Herbarium Grounds; 4 vii 1940; *Marshall* 4 (K). **Kew Environs.** Kew, on bank of Thames; in sandy soil near footpath; 9 vii 1929; *Hubbard* 1668 (K). • On bank of R. Thames between Kew & Richmond; amongst stones, just above high tide mark; 9 vii 1929; *Hubbard* 1695 (K).
CULT.: 1768 (JH: 403); 1789, Nat. of Britain (HK1, 1: 100); 1810, Nat. of Britain (HK2, 1: 155); 1814, Nat. of Britain (HKE: 22). Not currently in cult.
CURRENT STATUS: Widespread in grassland. It is not as common as it once was, but it is difficult to recognise in mown grass. It is probable that plants added to seed mixes for lawns are selected intermediates between this and subsp. *irrigata*. All records need to be checked in case of confusion between the various subspecies and their intermediates.

b. subsp. **angustifolia** (L.) Lej.
Poa angustifolia L.
narrow-leaved meadow-grass
FIRST RECORD: 1929 [1768].
REFERENCES: **Kew Gardens.** 1929 (S33: 751 and see Exsicc.).
EXSICC.: **Kew Gardens.** Against Herbarium wall; 16 x 1929; *Hubbard* s.n. (K). • [Unlocalised]; open position on untended land, sandy loam, ass. plants:- *Holcus lanatus*, *Festuca tenuifolia*; 12 vi 1950; *Naylor* 589 (K).
CULT.: 1768 (JH: 403); 1789, Nat. of Britain (HK1, 1: 99). In cult.
CURRENT STATUS: Several widely scattered sightings and apparently more common than expected. It is sometimes difficult to distinguish from subsp. *pratensis* though it usually occurs in drier ground.

c. subsp. **irrigata** (Lindm.) H.Lindb.
P. humilis Ehrh. ex Hoffm.
P. subcaerulea Sm.
spreading meadow-grass
FIRST RECORD: 1927 [1810].
REFERENCES: **Kew Gardens.** 1927 (S33: 751 and see Exsicc.).
EXSICC.: **Kew Gardens.** In short unmown grass in poplar collection; 31 v 1927; *Hubbard* G98 (K). • Queen's Cottage Grounds; in shade of trees,

sandy soil, in mass of *Deschampsia flexuosa*; very glaucous (whitish) leaves & inflorescence; 5 vi 1933; *Hubbard* s.n. (K). • Queen's Cottage Grounds; in partial shade, whole plant intensely glaucous, forming loose mat amongst *Festuca capillata* and *Deschampsia flexuosa*; 16 vi 1936; *Hubbard* s.n. (K). • [Unlocalised]; 16 vi 1936; *Worsdell* s.n. (K). • Queen's Cottage Grounds; 16 vi 1937; *Hubbard* s.n. (K). • Cottage grounds; 14 vi 1949; *Souster* 1004 (K).
CULT.: 1810, Nat. of Britain (HK2, 1: 155); 1814, Nat. of Britain (HKE: 22). Not currently in cult.
CURRENT STATUS: Rare; it can be difficult to distinguish from subsp. *pratensis* and all records of the latter need to be re-examined. There was, until recently, a fine stand of it on the edge of the Paddock but the site has been buried under hardcore to make a temporary car park.

1682. *Poa chaixii Vill.
P. sudetica Haenke
P. sylvatica Chaix
broad-leaved meadow-grass
FIRST RECORD: 1864.
NICHOLSON: **1873/4 survey** (JB: 76): In the shady parts of woods in **P** and **Q**, the long, dark-green leaves and dense tussocks of this grass render it very conspicuous. It seems to stand drought much better than many of our British grasses. • **1906 revision** (AS: 90): **A** and **Q:** In shady parts of woods the long dark green leaves and dense tussocks of this grass render it very conspicuous. It seems to stand drought much better than many of our British grasses.
OTHER REFERENCES: **Kew Gardens.** 1933, *Hubbard* (LFS: 389; LN35: S313; see also Exsicc.): Queen's Cottage Grounds, well known as established here. • 1983, *Burton* (FLA: 199): In the Queen's Cottage ground at Kew it has not survived since 1933 in the way that *Festuca heterophylla* has.
EXSICC.: **Kew Gardens.** Pleasure Grounds; 1864; *Hooker* s.n. (K) and in *Herb. Watson* (K). • Pleasure ground of the Royal Gardens, Kew — origin there not known — from Dr Hooker; 1865 or 6; *Herb. Watson* (K). • Queen's Cottage Grounds; vi 1877; *Nicholson* 1559 (K). • [Unlocalised]; 29 vi 1918; *Worsdell* s.n. (K). • Queens Cottage Grounds; 25 vi 1931; *Hubbard* s.n. (K). • Queen's Cottage Grounds; shade of trees, in grass turf, sandy soil, not common; vi 1932; *Hubbard* s.n. (K). • Queen's Cottage Grounds; in grassy glade between trees, in partial shade, sandy and gravelly soil, common; 4 vi 1933; *Hubbard* 149 (K). • Queen's Cottage Grounds; 6 vi 1934; *Hubbard* s.n. (K). • Queen's Cottage Ground; situated in damp shady position; 11 vi 1948; *O.J.Ward* 55 (K). • [Unlocalised; probably Queen's Cottage Grounds]; woodland grass, light sandy soil, ass. plants:- *Festuca rubra*, *Festuca heterophylla*; 15 vi 1950; *Naylor* 589 (K).
CULT.: In cult.
CURRENT STATUS: Not known; it has not recently been seen in the Conservation Area and may have been lost.

1683. Poa flexuosa Sm.
wavy meadow-grass
CULT.: 1810, Nat. of Scotland (HK2, 1: 154); 1814, Nat. of Scotland (HKE: 22). Not currently in cult.

1684. Poa compressa L.
flattened meadow-grass
FIRST RECORD: 1873/4 [1768].
NICHOLSON: **1873/4 survey** (JB: 76): **P.** Here and there on the dry slopes near Winter Garden, and in a dry spot or two in the wood. • **1906 revision** (AS: 90): **A.** Here and there on dry slopes near winter garden and in dry spots in wood. **P.** On wall of herbarium and palace grounds.
OTHER REFERENCES: **Kew Environs.** 1925, *Lousley* (LN35: S313): Towing path near Kew.
EXSICC.: **Kew Environs.** Thames banks, Kew; 20 vi 1922; *Fraser* s.n. (K).
CULT.: 1768 (JH: 403); 1789, Nat. of Britain (HK1, 1: 101); 1810, Nat. of Britain (HK2, 1: 157); 1814, Nat. of Britain (HKE: 22). Not currently in cult.
CURRENT STATUS: No recent sightings.

1685. *Poa palustris L.
P. palustris var. *effusa* (Kit.) Asch. & Graebn.
P. serotina Ehrt. ex Hoffm.
swamp meadow-grass
FIRST RECORD: 1879 [1768].
REFERENCES: **Kew Environs.** 1879, *Nicholson* (BECB1 (1): 24; LFS: 388; BSEC9: 693; S33: 751): Naturalised on the bank of the Thames, Kew, 30th June 1879. Sent by Mr George Nicholson with the following note:- "There was a considerable quantity of this species on banks of Thames at Kew and Mortlake last year, and there seems every probability of its retaining its hold." 1923, *Fraser* (SFS: 644; S33: 751; see also Exsicc.): Kew Green [var. *effusa*].
EXSICC.: **Kew Gardens.** Queen's Cottage Grounds; shade of trees, sandy soil, not common; vi 1932; *Hubbard* s.n. (K). **Kew Environs.** Kew Green; 24 vi 1920; *Fraser* s.n. (K).
CULT.: 1768 (JH: 403); 1929, Herbaceous Ground (K); 1934, unspec. (K); 1935, unspec. (K); 1936, unspec. (K). Not currently in cult.
CURRENT STATUS: No recent sightings.

1686. Poa glauca Vahl
P. caesia Sm.
glaucous meadow-grass
CULT.: 1810, Nat. of Britain (HK2, 1: 157), Nat. of Scotland (ibid.: 154 as *caesia*); 1814, Nat. of Britain (HKE: 22), Nat. of Scotland (ibid.: as *caesia*); 1929, Herbaceous Ground (K); 1936, Herbaceous Ground (K). Not currently in cult.

1687. Poa nemoralis L.

wood meadow-grass

FIRST RECORD: 1873/4 [1768].

NICHOLSON: **1873/4 survey** (JB: 76): **P.** Here and there about lake. In **Q** it is very abundant under trees, and in several places seems to form nearly the whole of the turf. • **1906 revision** (AS: 90): Abundant under trees in all the divisions, in some places forming nearly the whole of the turf.

EXSICC.: **Kew Gardens.** Queen's Cottage Grounds; beneath trees; vi 1932; *Hubbard* s.n. (K). • Queen's Cottage Grounds; 1932; *Hubbard* s.n. (K). • Queen's Cottage Grounds; dense masses beneath *Fagus sylvatica*, in sandy and gravelly soil with humus; 5 vi 1933; *Hubbard* 153 (K). • Queen's Cottage Grounds; in shade; 14 vi 1938; *Hubbard* 9348 (K). • Between Pagoda & Queen's Cottage Grounds; in shade; 14 vi 1938; *Hubbard* 9349 (K). **Kew Environs.** Thames bank & walls, Kew; 12 vii 1888; *Gamble* 20203 (K). • Towing path between Richmond & Kew; 25 vi 1905; *Sprague* s.n. (K).

CULT.: 1768 (JH: 403); 1789, Nat. of Britain (HK1, 1: 101); 1810, Nat. of Britain (HK2, 1: 157); 1814, Nat. of Britain (HKE: 22); 1929, Grass garden (K); 1934, unspec. (K); 1935, unspec. (K). In cult.

CURRENT STATUS: Certainly still in the Conservation Area and near the Lake but it can no longer be described as abundant.

1688. Poa bulbosa L.

bulbous meadow-grass

CULT.: 1768 (JH: 403); 1789, Nat. of England (HK1, 1: 102); 1810, Nat. of England (HK2, 1: 157); 1814, Nat. of England (HKE: 22). In cult.

1689. Poa alpina L.

alpine meadow-grass

CULT.: 1789, Nat. of Britain (HK1, 1: 99); 1810, Nat. of Scotland (HK2, 1: 154); 1814, Nat. of Scotland (HKE: 22). Not currently in cult.

1690. Dactylis glomerata L.

cock's-foot

FIRST RECORD: Probably between 1789 and 1805 [1768].

NICHOLSON: **1873/4 survey** (JB: 76): Very generally diffused over the whole of the ground included within the limits of our Flora. • **1906 revision** (AS: 90): Very generally diffused over the whole of the ground included within the limits of the flora.

EXSICC.: **Kew Gardens.** [Unlocalised]; sandy soil, in long grass; 16 v 1927; *Hubbard* G.39 (K). • Near the Tennis Courts; in uncut grass under trees; 15 vi 1927; *Hubbard & Turrill* s.n. (K). • [Unlocalised]; 3 vi 1944; *Souster* 84 (K). • Near Tennis Courts; in partial shade; 20 vii 1951; *Hubbard* 13372 (K). **Kew Environs.** Kew; *Goodenough* (K). • Thames banks, Kew; 10 vi 1888; *Fraser* s.n. (K).

CULT.: 1768 (JH: 400); 1789, Nat. of Britain (HK1, 1: 104); 1810, Nat. of Britain (HK2, 1: 160); 1814, Nat. of Britain (HKE: 23); 1929, Grass Garden (K); 1929, Grass garden (K); 1952, Herbarium Expt. Ground (K); 1953, Herb. Expt. Ground (K). In cult.

CURRENT STATUS: Throughout the Gardens in all types of grassland.

1691. Catabrosa aquatica (L.) P.Beauv.

Aira aquatica L.

whorl-grass

CULT.: 1768 (JH: 401); 1789, Nat. of Britain (HK1, 1: 96); 1810, Nat. of Britain (HK2, 1: 151); 1814, Nat. of Britain (HKE: 21); 1961, Jodrell (K). Not currently in cult.

1692. Catapodium rigidum (L.) C.E.Hubb.

Poa rigida L.

fern-grass

FIRST RECORD: 1999 [1768].

REFERENCES: **Kew Gardens.** 1999, *Cope* (S35: 649): In dry ground near the Princess of Wales Conservatory. This reasonably conspicuous grass seems to have been overlooked. It occurs in several other places in dry ground or on wall-tops.

CULT.: 1768 (JH: 403); 1789, Nat. of England (HK1, 1: 101); 1810, Nat. of England (HK2, 1: 157); 1814, Nat. of England (HKE: 22); 1925, unspec. (K); 1934, Herbaceous Ground (K); 1939, Herbaceous Ground (K). In cult.

CURRENT STATUS: Surely overlooked in the past; not uncommon on wall-tops and dry ground at the base of walls. It is abundant in gardens along Kew Road.

1693. Catapodium marinum (L.) C.E.Hubb.

Triticum loliaceum (Huds.) Sm.

T. unilaterale auct. non L.

sea fern-grass

CULT.: 1789, Nat. of Britain (HK1, 1: 122); 1810, Nat. of Britain (HK2, 1: 182); 1814, Nat. of Britain (HKE: 26); 1936, Herbaceous Ground (K); 1947, Herbarium Ground (K). Not currently in cult.

1694. Sesleria caerulea (L.) Ard.

Cynosurus caeruleus L.

blue moor-grass

CULT.: 1768 (JH: 406); 1789, Nat. of Britain (HK1, 1: 105); 1810, Nat. of Britain (HK2, 1: 153); 1814, Nat. of Britain (HKE: 22); 1929, Grass garden (K). In cult.

1695. Parapholis incurva (L.) C.E.Hubb.

Aegilops incurvata L.

Rottboellia incurvata (L.) L.f.

curved hard-grass

CULT.: 1768 (JH: 409); 1789, Nat. of England (HK1, 1: 116); 1810, Nat. of England (HK2, 1: 175); 1814, Nat. of England (HKE: 25). In cult.

1696. Parapholis strigosa (Dumort.) C.E.Hubb.
hard-grass
Cult.: 1924, unspec. (K). Not currently in cult.

1697. Glyceria maxima (Hartm.) Holmb.
Poa aquatica L.
Glyceria aquatica (L.) Wahlb.
reed sweet-grass
First record: 1873/4 [1768].
Nicholson: **1873/4 survey** (JB: 76): **Strip.** Abundant by moat. • **1906 revision** (AS: 90): **Strip.** Abundant by ha-ha.
Exsicc.: **Kew Environs.** Thames banks, Kew; viii 1883; *Fraser* s.n. (K). • Thames bank, Kew; 12 vii 1888; *Gamble* 20198 (K). • Between Richmond & Kew; in ditch by side of Thames; 2 viii 1927; *Hubbard* G.93 (K). • Between Kew & Richmond; in ditch by side of R. Thames; 9 vii 1929; *Hubbard* 1670 (K).
Cult.: 1768 (JH: 403); 1789, Nat. of England (HK1, 1: 99); 1810, Nat. of Britain (HK2, 1: 153); 1814, Nat. of Britain (HKE: 22). In cult.
Current status: Recently seen only in the Conservation Area, by the Water Lily Pond and on the golf course in the Old Deer Park. It no longer forms stands beside the river as it did in the nineteenth century.

1698. Glyceria fluitans (L.) R.Br.
Festuca fluitans L.
Poa fluitans (L.) Scop.
floating sweet-grass
First record: 1873/4 [1768].
Nicholson: **1873/4 survey** (JB: 76): Lake and **Strip.** Common. • **1906 revision** (AS: 90): Lake and **Strip.** Common.
Exsicc.: **Kew Environs.** Banks of Thames, Kew; 1883; *Fraser* s.n. (K).
Cult.: 1768 (JH: 403); 1789, Nat. of Britain (HK1, 1: 108); 1810, Nat. of Britain (HK2, 1: 154); 1814, Nat. of Britain (HKE: 22). In cult.
Current status: No recent sightings. There is no longer any suitable habitat for this species either within the Gardens or on the towpath.

1699. Glyceria declinata Bréb.
small sweet-grass
Cult.: 1936, 1937, 1938 & 1940, Aquatic Garden (all K). Not currently in cult.

1700. Glyceria notata Chevall.
plicate sweet-grass
First record: 1936 [1936].
Exsicc.: **Kew Environs.** Along banks of Thames River, Kew; 1 ix 1936; *E.T. & H.N.Moldenke* 9779 (K).
Cult.: 1936 & 1940, Aquatic Garden (both K). In cult.
Current status: No recent sightings.

1701. Melica nutans L.
mountain melick
Cult.: 1768 (JH: 402); 1789, Nat. of Britain (HK1, 1: 98); 1810, Nat. of Britain (HK2, 1: 152); 1814, Nat. of Britain (HKE: 22); 1929, Grass Garden (K). In cult.

1702. Melica uniflora Retz.
wood melick
Cult.: 1789, Nat. of Britain (HK1, 1: 98); 1810, Nat. of Britain (HK2, 1: 152); 1814, Nat. of Britain (HKE: 22); 1929, Grass Garden (K); 1940, Herbaceous Ground (K). In cult.

1703. Helictotrichon pubescens (Huds.) Pilg.
Avena pubescens Huds.
downy oat-grass
First record: 1873/4 [1789].
Nicholson: **1873/4 survey** (JB: 75): **P.** Frequent in turf among young trees from "Douglas Spar" to near Pagoda. **Q.** Here and there towards river. • **1906 revision** (AS: 90): Same distribution as [*H. pratense* – **A:** Frequent in turf from flagstaff to pagoda. **P:** Not uncommon. **Q:** In turf on side bordering river], but much commoner. [see comment under *H. pratense*!]
Exsicc.: **Kew Gardens.** In field near herbarium; sandy & gravelly soil; 23 v 1927; *Hubbard* G.30 (K). • Near Herbarium; waste ground; v 1928; *Ballard* s.n. (K). • Edge of herbarium field; 28 v 1929; *Hubbard* 1528 (K).
Cult.: 1789, Nat. of Britain (HK1, 1: 113); 1810, Nat. of Britain (HK2, 1: 172); 1814, Nat. of Britain (HKE: 24). Not currently in cult.
Current status: No recent sightings; possibly lost through close mowing of the turf.

1704. Helictotrichon pratense (L.) Besser
Avena pratensis L.
A. planiculmis auct. non Schrad.
meadow oat-grass
First record: 1873/4 [1768].
Nicholson: **1873/4 survey** (JB: 75): Same distribution as [*H. pubescens* – **P.** Frequent in turf among young trees from "Douglas Spar" to near Pagoda. **Q.** Here and there towards river], though much the commoner species. [see comment under *H. pubescens*!]. • **1906 revision** (AS: 90): **A.** Frequent in turf from flafstaff to pagoda. **P.** Not uncommon. **Q.** In turf on side bordering river.
Cult.: 1768 (JH: 405); 1789, Nat. of Britain (HK1, 1: 114); 1810, Nat. of Britain (HK2, 1: 172); 1814, Nat. of Britain (HKE: 24), Nat. of Scotland (ibid.: Add. as *planiculmis*). Not currently in cult.
Current status: No recent sightings; possibly lost through close mowing of the turf.

1705. Arrhenatherum elatius (L.) P.Beauv. ex J.&C.Presl
a. subsp. **elatius**
Avena elatior L.
Holcus avenaceus Scop.
Arrhenatherum avenaceum (Scop.) P.Beauv.
false oat-grass
FIRST RECORD: 1873/4 [1768].
NICHOLSON: **1873/4 survey** (JB: 75): Everywhere. Forms a large proportion of the rougher turf. Is a most troublesome weed on every piece of dug ground. • **1906 revision** (AS: 90): Everywhere forms a large proportion of the rougher turf. A troublesome weed in dug ground.
EXSICC.: **Kew Gardens.** Near tennis courts; in partial shade on edge of shrubbery; 16 x 1926; *Hubbard* s.n. (K). • [Unlocalised]; in shrubbery; 31 v 1927; *Hubbard* G.28.B (K). • Lower Nursery; grassy patch; 10 vi 1927; *Hubbard* s.n. (K). • [Unlocalised]; viii 1930; *Bullock* 126 (K). • Herbarium field; 13 x 1939; *Hubbard* s.n. (K). • Herbarium grounds; 4 vii 1940; *Marshall* 3 (K). • Queen's Cottage Ground; growing in open rather dry position; 11 vi 1948; *O.J.Ward* 57 (K). • In Herbarium field; 19 vi 1951; *Hubbard* 13359 (K). • In Herbarium field; [specimen flood-damaged; ?26 vi 1951]; *Kennedy-O'Byrne* [?144] (K). **Kew Environs.** Thames side, Kew; 1854; *anon.* s.n. (K). • Thames Banks, Kew; 1 vii 1921; *Fraser* s.n. (K). • Kew, on bank of Thames, near footpath; 9 vii 1929; *Hubbard* 1669 (K). • Towpath, Kew; 20 vi 1944; *Souster* 101 (K). • River bank, Kew; 14 vi 1964; *S.T. Blake* 22186 (K, BR).
CULT.: 1768 (JH: 404); 1789, Nat. of Britain (HK1, 1: 112); 1813, Nat. of Britain (HK2, 5: 431); 1814, Nat. of Britain (HKE: 316) 1939, unspec. (K). In cult.
CURRENT STATUS: Abundant throughout the Gardens in areas of long grass.

b. subsp. **bulbosum** (Willd.) St-Amans
onion couch
CULT.: 1953, Herbaceous Ground (K). Not currently in cult.

1706. *Avena strigosa Schreb.
bristle oat
FIRST RECORD: 1878 [1810].
REFERENCES: **Kew Environs.** 1878, *Baker* (S33: 751 and see Exsicc.).
EXSICC.: **Kew Environs.** Thames side, Kew; vii 1878; *Baker* 1529 (K). • Kew; undated, presented 1880; *Rizzi* s.n. (K).
CULT.: 1810, Nat. of Britain (HK2, 1: 171); 1814, Nat. of Britain (HKE: 24). In cult.
CURRENT STATUS: No recent sightings.

1707. Avena fatua L.
wild-oat
FIRST RECORD: 1990 [1768].
REFERENCES: **Kew Gardens.** 1990, *Cope* (S31: 182): Paddock, after restoration in 1990.
CULT.: 1768 (JH: 404); 1789, Nat. of Britain (HK1, 1: 113); 1810, Nat. of Britain (HK2, 1: 171); 1814, Nat. of Britain (HKE: 24). Not currently in cult.
CURRENT STATUS: It has not reappeared in the Paddock since 1990 but can occasionally be found in neglected corners.

1708. *Avena sativa L.
oat
FIRST RECORD: 1931 [1768].
REFERENCES: **Kew Gardens.** 1990, *Cope* (S31: 182): Paddock, after restoration in 1990.
EXSICC.: **Kew Gardens.** [Unlocalised]; on rubbish heap; viii 1931; *Hubbard* s.n. (K).
CULT.: 1768 (JH: 404); 1789, Cult. (HK1, 1: 113); 1810, Cult. (HK2, 1: 171); 1814, Cult. (HKE: 24). In cult.
CURRENT STATUS: Occasional and probably resulting from spillage from bird feeders.

1709. Gaudinia fragilis (L.) P.Beauv.
Avena fragilis L.
french oat-grass
CULT.: 1768 (JH: 405); 1925, unspec. (K). In cult.

1710. Trisetum flavescens (L.) P.Beauv.
Avena flavescens L.
yellow oat-grass
FIRST RECORD: 1873/4 [1768].
NICHOLSON: **1873/4 survey** (JB: 75): Common in every open piece of turf within our limits. • **1906 revision** (AS: 90): Common in every piece of turf within our limits.
EXSICC.: **Kew Environs.** Kew; vii 1888; *Gamble* 20186 (K). • Thames bank, Kew; 12 vii 1888; *Gamble* 20200 (K).
CULT.: 1768 (JH: 404); 1789, Nat. of Britain (HK1, 1: 114); 1810, Nat. of Britain (HK2, 1: 172); 1814, Nat. of Britain (HKE: 24). In cult.
CURRENT STATUS: Occasional in the less frequently cut grass. Its distribution is hard to establish since much of the grassland is regularly cut, making it difficult to spot. It is probably not as abundant or widespread now as it was at the time of Nicholson's survey.

1711. Koeleria vallesiana (Honck.) Gaudin
Somerset hair-grass
CULT.: 1939, unspec. (K); 1961, Jodrell, Order Beds (K). In cult.

1712. Koeleria macrantha (Ledeb.) Schult.
Poa cristata (L.) L.
Aira cristata L.
Koeleria cristata auct. non (L.) Pers.
crested hair-grass
FIRST RECORD: 1873/4 [1768].
NICHOLSON: **1873/4 survey** (JB: 75): Very frequent in open turf in **P** and **Q.** In **B** this forms the greater part of the turf in places on both sides of the wooden fence in front of the Palace. • **1906 revision** (AS: 90): Very frequent in open turf.
EXSICC.: **Kew Gardens.** Ha Ha near Kew Palace; 1883; *Fraser* s.n. (K). • [Unlocalised]; 9 vii 1925; *Hubbard* 88 (K). • On edge of tennis courts; sandy soil; 31 v 1927; *Hubbard* G.47 (K). • Edges of tennis courts; v 1928; *Ballard* s.n. (K). • Near Tennis courts; on sandy soil; 10 vi 1928; *Hubbard* s.n. (K). • Near Office of Works yard; in patch of grass (original), on sandy and gravelly soil, frequent; 18 vi 1936; *Hubbard* 121 (K). • Queen's Cottage Grounds; 4 vi 1951; *Kennedy-O'Byrne* 134 (K).
CULT.: 1768 (JH: 401); 1789, Nat. of Britain (HK1, 1: 102); 1810, Nat. of Britain (HK2, 1: 151); 1814, Nat. of Britain (HKE: 21). In cult.
CURRENT STATUS: No recent sightings. As with several other grasses, it is distinctly possible that this species still exists in the Gardens. However, if it does not grow in one of the patches that are left uncut for the benefit of invertebrates it is unlikely ever to be spotted.

1713. Deschampsia cespitosa (L.) P.Beauv.
tufted hair-grass

a. subsp. **cespitosa**
Aira cespitosa L.
FIRST RECORD: 1869 [1768].
NICHOLSON: **1873/4 survey** (JB: 75): **P, Q** and **Strip.** Very common about lake and on the borders of plantations nearest it, also near moat. • **1906 revision** (AS: 89): **A, Q** and **Strip.** Very common about lake and on the borders of plantations nearest it; also near ha-ha.
EXSICC.: **Kew Gardens.** The Lake, Pleasure Grounds; 20 vii 1869; *Hooker* s.n. (K).
CULT.: 1768 (JH: 401); 1789, Nat. of Britain (HK1, 1: 96); 1810, Nat. of Britain (HK2, 1: 151); 1814, Nat. of Britain (HKE: 21); 1929, unspec. (K); 1953, Herbaceous Ground (K); 1970, Grass Beds (K). In cult.
CURRENT STATUS: Very rare; only one plant has recently been seen in the Gardens although it is common on the golf course in the Old Deer Park.

b. subsp. **parviflora** (Thuill.) Dumort.
CULT.: In cult.

c. subsp. **alpina** (L.) Hook.f.
Aira laevigata Sm.
CULT.: 1814, Nat. of Scotland (HKE: Add.). Not currently in cult.
Note. Not known as a British plant until 1810.

1714. Deschampsia flexuosa (L.) Trin.
Aira flexuosa L.
A. montana L.
wavy hair-grass
FIRST RECORD: 1873/4 [1768].
NICHOLSON: **1873/4 survey** (JB: 75): This seems to affect shade much more than [*Deschampsia cespitosa*]. It is common under trees in divisions **P** and **Q.** • **1906 revision** (AS: 89): **A, Q.** Affects shade more than [*Deschampsia cespitosa*]. Common under trees.
EXSICC.: **Kew Gardens.** [Unlocalised]; vi 1924; *Hubbard* s.n. (K). • On edge of tennis courts; among short grass; 31 v 1927; *Hubbard* s.n. (K). • Queen's Cottage Grounds; in shade; vi 1932; *Hubbard* s.n. (K). • [Unlocalised]; under trees on acid sandy soil; 14 vii 1945; *Souster* 268 (K). • Queens Cottage Ground; in dryish soil shaded by trees; 11 vi 1948; *O.J.Ward* 58 (K). • Queen's Cottage Grounds; 4 vi 1951; *Kennedy-O'Byrne* 137 (K). **Kew Environs.** Kew; moist position on sandy loam, open situation; 12 vi 1950; *Naylor* 596 (K).
CULT.: 1768 (JH: 401); 1789, Nat. of Britain (HK1, 1: 97); 1810, Nat. of Britain (HK2, 1: 151); 1814, Nat. of Britain (HKE: 21). In cult.
CURRENT STATUS: Rare; just a few widely scattered localities in the Gardens; more common on the golf course in the Old Deer Park.

1715. Holcus lanatus L.
Yorkshire-fog
FIRST RECORD: 1873/4 [1768].
NICHOLSON: **1873/4 survey** (JB: 75): Plentiful in all the divisions. • **1906 revision** (AS: 89): Plentiful in all the divisions.
EXSICC.: **Kew Gardens.** Queen's Cottage Grounds; 15 vi 1914; *Flippance* s.n. (K). • Herbarium field; 24 vii 1925; *Hubbard* 137 (K). • Lower nursery; on rubbish heap; 16 x 1929; *Hubbard* s.n. (K). • Queen's Cottage Grounds; growing amongst *Milium effusum*, *Bromus sterilis*, *Anthoxanthum odoratum*, *Arrhenatherum elatius* & various other grasses; 7 vi 1937; *Rawlings* s.n. (K). • Herbarium grounds; 4 vii 1940; *Marshall* 2 (K). • [Unlocalised]; common grass in rough places among trees; 3 vi 1944; *Souster* 83 (K). • By tennis courts; rough grass area; 30 v 1952; *Kennedy-O'Byrne* 125 (K). **Kew Environs.** Kew; open position on untended land, sandy loam, dry, ass. plants:- *Festuca tenuifolia*, *Poa pratensis*; 12 vi 1950; *Naylor* 573 (K).
CULT.: 1768 (JH: 408); 1789, Nat. of Britain (HK1, 3: 425); 1813, Nat. of Britain (HK2, 5: 431); 1814, Nat.

of Britain (HKE: 316); 1929, Herbarium Expmtl. Ground (K). In cult.
CURRENT STATUS: Abundant in all types of grassland throughout the Gardens.

1716. Holcus mollis L.
creeping soft-grass
FIRST RECORD: 1873/4 [1768].
NICHOLSON: **1873/4 survey** (JB: 75): **P.** Common in turf and shrubberies round King William's Temple. Abundant in old Broom beds. **B.** In hedge facing Museum No. 2. • **1906 revision** (AS: 90): Common both in turf and shrubberies.
EXSICC.: **Kew Gardens.** Pleasure Ground; [undated]; *Prior* s.n. (K). • [Unlocalised]; in shade on sandy soil; 9 vii 1929; *Hubbard* 1667 (K). • [Unlocalised]; 29 vii 1932; *Worsdell* s.n. (K). • [Unlocalised]; 3 viii 1949; *Souster* 1037 (K). • Herbarium field; x 1950; *Hubbard* s.n. (K).
CULT.: 1768 (JH: 408); 1789, Nat. of Britain (HK1, 3: 425); 1813, Nat. of Britain (HK2, 5: 431); 1814, Nat. of Britain (HKE: 316). In cult.
CURRENT STATUS: Common in all types of grassland throughout the Gardens.

1717. Corynephorus canescens (L.) P.Beauv.
Aira canescens L.
grey hair-grass
CULT.: 1768 (JH: 401); 1789, Nat. of England (HK1, 1: 97); 1810, Nat. of England (HK2, 1: 151); 1814, Nat. of England (HKE: 21). In cult.

1718. Aira caryophyllea L.
silver hair-grass
FIRST RECORD: 1873/4 [1768].
NICHOLSON: **1873/4 survey** (JB: 75): **P.** Plentiful on waste ground near Winter Garden. Very abundant in old Broom beds near Pagoda, where it grows very luxuriantly. • **1906 revision** (AS: 89): Plentiful in dry spots — along hedge-rows, etc.
CULT.: 1768 (JH: 402); 1789, Nat. of Britain (HK1, 1: 97); 1810, Nat. of Britain (HK2, 1: 152); 1814, Nat. of Britain (HKE: 21). In cult.
CURRENT STATUS: No recent sightings.

1719. Aira praecox L.
early hair-grass
FIRST RECORD: 1873/4 [1768].
NICHOLSON: **1873/4 survey** (JB: 75): In every dry place both in turf and elsewhere. • **1906 revision** (AS: 89): In every dry place both in turf and elsewhere.
EXSICC.: **Kew Gardens.** Queen's Cottage Grounds; on cart track ; 16 vi 1936; *Hubbard* s.n. (K). • Queen's Cottage Grounds; growing in acid sandy soil with *Deschampsia flexuosa*; 11 vi 1948; *Senogles* 75 (K). • [Unlocalised]; sandy dry position, shaded, ass. plants:- *Festuca rubra*; 12 vi 1950; *Naylor* 570 (K).
CULT.: 1768 (JH: 402); 1789, Nat. of Britain (HK1, 1: 97); 1810, Nat. of Britain (HK2, 1: 151); 1814, Nat. of Britain (HKE: 21). Not currently in cult.
CURRENT STATUS: Occasional in short dry grassland, especially with *Festuca filiformis* and abundant on Mount Pleasant.

1720. Anthoxanthum nitens (Weber) Y.Schouten & Veldk.
Hierochloe odorata (L.) P.Beauv.
holy-grass
CULT.: 1788, unspec. (K); 1840-1850, unspec. (K); 1952, Herbarium Expt. Ground (K). In cult.

1721. Anthoxanthum odoratum L.
sweet vernal-grass
FIRST RECORD: 1854 [1789].
NICHOLSON: **1873/4 survey** (JB: 75): Common, particularly in district **Q.** • **1906 revision** (AS: 89): Common in all the divisions.
EXSICC.: **Kew Gardens.** Pleasure Grounds; *Prior* s.n. (K). • Herbarium field; 5 v 1925; *Hubbard* 4 (K). • [Unlocalised]; in shade of beech trees on sandy soil; 16 v 1927; *Hubbard* G.4 (K). • Queen's Cottage Grounds; in shade; 14 vi 1938; *Hubbard* 9347 (K). • Queen's Cottage Grounds; growing in acid soil in association with *Festuca tenuifolia*; 11 vi 1948; *Senogles* 73 K). • Queen's Cottage; 16 vi 1949; *Cook* s.n. (K). **Kew Environs.** Kew; vi 1854; *anon.* s.n. (K).
CULT.: 1789, Nat. of Britain (HK1, 1: 48); 1810, Nat. of Britain (HK2, 1: 68); 1814, Nat. of Britain (HKE: 10); 1939, unspec. (K). In cult.
CURRENT STATUS: Common in grassland throughout the Gardens.

1722. Phalaris arundinacea L.
Digraphis arundinacea (L.) Trin.
Arundo colorata Aiton
reed canary-grass
FIRST RECORD: 1873/4 [1768].
NICHOLSON: **1873/4 survey** (JB: 75): **P.** Sparingly near Lake. **Q.** A large plot in wood. **Strip.** Abundant. • **1906 revision** (AS: 89): **A.** Near lake. **Strip.** Abundant along ha-ha.
EXSICC.: **Kew Gardens.** River-bank Kew – Richmond, Richmond side of Isleworth Gate; middle zone with *Oenanthe* & *Caltha*; 16 viii 1928; *Summerhayes & Turrill* s.n. (K). • Queen's Cottage Ground; situation shaded & very moist at the roots; 11 vi 1948; *O.J. Ward* 51 (K). **Kew Environs.** Thames banks, Kew; 1884; *Fraser* s.n. (K). • Thames bank, Kew; vii 1888; *Gamble* 20195 (K). • River bank between Kew & Richmond; 25 vi 1905; *Sprague* s.n. (K). • Kew, Thames bank; middle zone on mud with *Oenanthe*, *Senecio* etc.; 13 vii 1928; *Turrill* s.n. (K).
CULT.: 1768 (JH: 399); 1789, Nat. of Britain (HK1, 1: 116); 1810, Nat. of England (HK2, 1: 174); 1814, Nat. of England (HKE: 25). In cult.

CURRENT STATUS: Uncommon; near the Stable Yard and along the towpath. Var. *picta* (Gardener's Garters) is in a pond on the adjacent golf course.

1723. *Phalaris canariensis L.

canary-grass

FIRST RECORD: 1925 [1768].
REFERENCES: **Kew Gardens.** 1925 (S33: 751 and see Exsicc.).
EXSICC.: **Kew Gardens.** Lower Nursery; on rubbish heap; 22 vi 1925; *Hubbard* 45 (K). • Near Works Department; on rubbish heap; vi 1928; *Ballard* s.n. (K). • Herbarium, in pavement by wing A; 27 vi 1967; *anon.* s.n. (K). • By the succulent house; 'weed' or escape from cultivated section, growing at edge of path; 22 vi 1976; *Lewis* s.n. (K).
CULT.: 1768 (JH: 399); 1789, Nat. of Britain (HK1, 1: 85); 1810, Nat. of Britain (HK2, 1: 137); 1814, Nat. of Britain (HKE: 20). Not currently in cult.
CURRENT STATUS: Occasional in disturbed ground and doubtless resulting from spillage from bird feeders.

1724. *Phalaris minor Retz.

lesser canary-grass

FIRST RECORD: 1877.
REFERENCES: **Kew Gardens.** 1990, *Cope* (S31: 182): Paddock, after restoration in 1990.
EXSICC.: **Kew Environs.** Kew; viii 1877; *Baker* s.n. (K).
CURRENT STATUS: Rare and doubtless it originally arose as spillage from bird feeders.

1725. *Phalaris paradoxa L.

awned canary-grass

FIRST RECORD: 1967.
REFERENCES: **Kew Gardens.** 1967 (S33: 751 and see Exsicc.).
EXSICC.: **Kew Gardens.** Herbarium, pavement by Wing A; 27 vi 1967; *anon.* s.n. (K). • Waste ground in Nursery; 16 vii 1975; *Hubbard* s.n. (K).
CURRENT STATUS: No recent sightings.

1726. *Phalaris coerulescens Desf.

FIRST RECORD: 1975 [1933]
EXSICC.: **Kew Gardens.** Weed in waste ground in Nursery; 16 vii 1975; *Hubbard* s.n. (K).
CULT.: 1933, unspec. (K); 1961, Jodrell, Order Beds (K). In cult.
CURRENT STATUS: No recent sightings.

1727. Agrostis capillaris L.

A. vulgaris With.
A. castellana auct. non Boiss. & Reut.

common bent

FIRST RECORD: 1873/4 [1768].
NICHOLSON: **1873/4 survey** (JB: 75): Abundant in all the divisions. • **1906 revision** (AS: 89): Abundant in all divisions.
OTHER REFERENCES: **Kew Gardens.** 1990, *Cope* (S31: 182): Paddock, after restoration in 1990 [as *A. castellana*]
EXSICC.: **Kew Gardens.** Near Brentford Gate, inside the Gardens; 23 vi 1925; *Turrill* s.n. (K). • Edge of Herbarium expt. ground; 12 vii 1927; *Turrill* G.60A (K). • [Unlocalised]; in rough grass under trees near river; 13 vii 1945; *Souster* 267 (K) • [unlocalised]; with other grasses, used as a lawn grass; 1 vii 1964, *Blake* 22280 (K, MEL).
CULT.: 1768 (JH: 400); 1789, Nat. of Britain (HK1, 1: 95); 1810, Nat. of Britain (HK2, 1: 149); 1814, Nat. of Britain (HKE: 21); 1934, unspec. (K). In cult.
CURRENT STATUS: Abundant in all types of grassland throughout the Gardens. When sown in the Paddock in 1990 it was recorded as Highland Bent (*A. castellana*) because the boundary between these two species had not been properly defined. It seems that the name 'Highland Bent', currently given to *A. castellana*, properly belongs to cultivars of *A. capillaris* (*fide* E.J.Clement, pers. comm.). True *A. castellana* has not been recorded in the Gardens.

1728. Agrostis gigantea Roth

black bent

FIRST RECORD: 1934 [1934].
REFERENCES: **Kew Gardens.** 1934 (S33: 751 and see Exsicc.).
EXSICC.: **Kew Gardens.** Weed in experimental plots; 18 vii 1934; *Philipson* 228 (K). • [Unlocalised]; on waste ground; 22 vii 1949; *Hubbard* 13253 (K).
CULT.: 1934, unspec. (K). In cult.
CURRENT STATUS: No recent sightings. It is possible that the specimen collected from the experimental plots and designated 'cultivated' was the same plant that Philipson collected as a weed.

1729. Agrostis stolonifera L.

A. alba auct. non L.

creeping bent

FIRST RECORD: 1873/4 [1768]
NICHOLSON: **1873/4 survey** (JB: 75): **P.** About lake and on borders of plantations. • **1906 revision** (AS: 89): **A.** About lake and on borders of plantations.
OTHER REFERENCES: **Kew Gardens.** 1999, *Stones* (EAK: 2): By Lake.
EXSICC.: **Kew Gardens.** Lower Nursery; creeping over ashes; 16 x 1929; *Hubbard* s.n. (K). **Kew Environs.** Thames bank, Kew; vii 1888; *Gamble* 20187 (K). • On bank of R. Thames between Kew & Richmond; edge of path; 9 vii 1929; *Hubbard* 1672 (K). • S. bank of Thames at Kew; in open forming matted turf; viii 1933; *Philipson* 311 (K). • S. bank of Thames, Kew; dry top of wall; viii 1933; *Philipson* 373 (K). • S banks of Thames at Kew; dry top of wall, next Old Deer Park; viii 1933; *Philipson* 374 (both K). • S. bank of Thames at Kew; in damp and shaded places; viii 1933; *Philipson* 313, 314, 315, 316, 317 (all K). • Kew, bank of river Thames; in wet soil, creeping round edge of *Salix* bushes; 17 vii 1938; *Hubbard* 24 (K).

CULT.: 1768 (JH: 400); 1789, Nat. of Britain (HK1, 1: 95); 1810, Nat. of Britain (HK2, 1: 149); 1814, Nat. of Britain (HKE: 21); 1934, unspec. (K). Not currently in cult.
CURRENT STATUS: Abundant throughout the Gardens.

1730. Agrostis curtisii Kerguélen
A. setacea Curtis
bristle bent
CULT.: 1810, Nat. of Britain (HK2, 1: 149); 1814, Nat. of Britain (HKE: 21). In cult.

1731. Agrostis canina L.
velvet bent
FIRST RECORD: Probably between 1779 and 1805 [1768].
EXSICC: **Kew Environs.** Kew; *Goodenough* s.n. (K).
CULT.: 1768 (JH: 400); 1789, Nat. of Britain (HK1, 1: 95); 1810, Nat. of Britain (HK2, 1: 148); 1814, Nat. of Britain (HKE: 21). In cult.
CURRENT STATUS: No recent sightings.

1732. × Agropogon lutosus (Poir.) P.Fourn.
(Agrostis stolonifera × Polypogon monspeliensis)
Agrostis littoralis sensu Sm. *non* With.
× *Agropogon littoralis* (Sm.) C.E.Hubb.
perennial beard-grass
FIRST RECORD: 1931 [1810]
EXSICC.: **Kew Gardens.** Nursery; waste stony ground; ix 1931; *Bullock* s.n. (K).
CULT.: 1810, Nat. of England (HK2, 1: 148); 1814, Nat. of England (HKE: 21). Not currently in cult.
CURRENT STATUS: No recent sightings. It was still in cultivation until the late 1970s or 80s but was finally lost. It has not been seen again as an escape.

1733. Calamagrostis epigeios (L.) Roth
Arundo epigeios L.
wood small-reed
CULT.: 1768 (JH: 405); 1789, Nat. of England (HK1, 1: 115); 1810, Nat. of Britain (HK2, 1: 174); 1814, Nat. of Britain (HKE: 25). Not currently in cult.

1734. Calamagrostis canescens (F.H.Wigg.) Roth
Arundo calamagrostis L.
purple small-reed
CULT.: 1768 (JH: 405); 1789, Nat. of Britain (HK1, 1: 115); 1810, Nat. of England (HK2, 1: 174); 1814, Nat. of England (HKE: 25). Not currently in cult.

1735. Calamagrostis purpurea (Trin.) Trin.
Scandinavian small-reed
CULT.: In cult.

1736. Calamagrostis stricta (Timm) Koeler
Arundo stricta Timm
narrow small-reed
CULT.: 1814, Nat. of Scotland (HKE: Add.). Not currently in cult.

1737. Ammophila arenaria (L.) Link
Arundo arenaria L.
marram
CULT.: 1768 (JH: 405); 1789, Nat. of Britain (HK1, 1: 116); 1810, Nat. of Britain (HK2, 1: 174); 1814, Nat. of Britain (HKE: 25). In cult.

1738. Gastridium ventricosum (Gouan) Schinz & Thell.
Milium lendigerum L.
Agrostis rubra L.
nit-grass
CULT.: 1768 (JH: 400); 1789, Nat. of England (HK1, 1: 93); 1810, Nat. of England (HK2, 1: 147); 1814, Nat. of England (HKE: 21). Not currently in cult.

[**1739. Lagurus ovatus** L.
hare's-tail
CULT.: 1789, Alien (HK1, 1: 114); 1810, Nat. of Guernsey (HK2, 1: 173); 1814, Nat. of Guernsey (HKE: 24).
Note. Listed in error as native.]

1740. Apera spica-venti (L.) P.Beauv.
Agrostis spica-venti L.
loose silky-bent
FIRST RECORD: 1873/4 [1768].
NICHOLSON: **1873/4 survey** (JB: 75): **B.** A few plants in private ground near Museum No. 2. **P.** Two or three on waste ground near Winter Garden. • **1906 revision** (AS: 89): **B.** A weed in cultivated ground.
CULT.: 1768 (JH: 400); 1789, Nat. of England (HK1, 1: 94); 1810, Nat. of England (HK2, 1: 148); 1814, Nat. of England (HKE: 21). In cult.
CURRENT STATUS: No recent sightings. It appeared in the Paddock after disturbance in 1987 but has never returned.

1741. Mibora minima (L.) Desv.
Agrostis minima L.
early sand-grass
CULT.: 1768 (JH: 400); 1810, Nat. of Wales (HK2, 1: 149); 1814, Nat. of Wales (HKE: 21); 1925, unspec. (K). Not currently in cult.

1742. Polypogon monspeliensis (L.) Desf.
Alopecurus paniceus L.
Agrostis panicea (L.) Aiton
annual beard-grass
FIRST RECORD: 2008 [1768].
CULT.: 1768 (JH: 399); 1789, Nat. of England (HK1, 1: 94); 1810, Nat. of England (HK2, 1: 148); 1814, Nat. of England (HKE: 21). In cult.
CURRENT STATUS: Found in considerable quantity in rough ground in the Lower Nursery in 2008.

1742.1. Polypogon viridis (Gouan) Breistr.
water bent
FIRST RECORD: 2008.
CULT.: 1982 (Order Beds). In cult.
CURRENT STATUS: Found in considerable quantity in rough ground in the Lower Nursery in 2008.

1743. Alopecurus pratensis L.
meadow foxtail
FIRST RECORD: 1873/4 [1768].
NICHOLSON: **1873/4 survey** (JB: 75): Common. In some places forming the principal factor in the turf. • **1906 revision** (AS: 89): Common. In some places forming the principal factor in the turf.
EXSICC.: **Kew Gardens.** [Unlocalised]; on sandy soil, in partial shade; 31 v 1927; *Hubbard* G.12 (K). • Meadow behind Herbarium; 3 v 1930; *Bullock* 20 (K). • Field near herbarium; *Hubbard* 1 (K). • Herbarium field; damp loamy soil over gravel, common; 13 v 1933; *Hubbard* 7 (K). • Herbarium field; abundant; 15 v 1933; *Hubbard* s.n. (K). • Herbarium grounds; 5 vi 1940; *Marshall* 6 (K). • [Unlocalised]; a common grass on rough land in open & partial shade; 6 v 1944; *Souster* 56 (K). • By Willows colln., Queen's Cottage Grounds; 24 v 1944; *Melville* s.n. (K). • Queen's Cottage Ground; somewhat of an open dry sandy position; 11 vi 1948; *O.J.Ward* 50 (K). • In field by tennis courts; in partial shade; 30 v 1952; *Kennedy-O'Byrne* 127 (K). • [Unlocalised, presumably near the tennis courts]; in rough grass; 30 v 1952; *Kennedy-O'Byrne* 128 (K).• By tennis courts; in rough grass; 30 v 1952; *Kennedy-O'Byrne* 129 (K).
CULT.: 1768 (JH: 399); 1789, Nat. of Britain (HK1, 1: 92); 1810, Nat. of Britain (HK2, 1: 146); 1814, Nat. of Britain (HKE: 21). In cult.
CURRENT STATUS: Common in all grassland areas and especially prominent is those that are less frequently cut.

1744. Alopecurus geniculatus L.
marsh foxtail
FIRST RECORD: 2004 [1768].
CULT.: 1768 (JH: 399); 1789, Nat. of Britain (HK1, 1: 93); 1810, Nat. of Britain (HK2, 1: 146); 1814, Nat. of Britain (HKE: 21). In cult.
CURRENT STATUS: By a pond in the Old Deer Park, on the edge of the golf course.

1745. Alopecurus bulbosus Gouan
bulbous foxtail
CULT.: 1768 (JH: 399); 1789, Nat. of England (HK1, 1: 92); 1810, Nat. of England (HK2, 1: 146); 1814, Nat. of England (HKE: 21). Not currently in cult.

1746. Alopecurus aequalis Sobol.
A. fulvus Sm.
orange foxtail
CULT.: 1810, Nat. of England (HK2, 1: 147); 1814, Nat. of England (HKE: 21). Not currently in cult.
Note. Not known as a British plant until 1796.

1747. Alopecurus ovatus Knapp
Alopecurus borealis Trin.
A. alpinus Sm. non Vis.
alpine foxtail
CULT.: 1810, Nat. of Scotland (HK2, 1: 146); 1814, Nat. of Scotland (HKE: 21). Not currently in cult.
Note. Not known as a British plant until 1803. May possibly be conspecific with *A. magellanicus* Lam., an older name that would take priority.

1748. Alopecurus myosuroides Huds.
A. agrestis L.
black-grass
FIRST RECORD: 1873/4 [1768].
NICHOLSON: **1873/4 survey** (JB: 75): **P.** A couple of plants near Winter Garden. Mr J.M.Smith. **Pal.** Several in kitchen garden ground. • **1906 revision** (AS: 89): Here and there in cultivated ground.
CULT.: 1768 (JH: 399); 1789, Nat. of Britain (HK1, 1: 92); 1810, Nat. of Britain (HK2, 1: 146); 1814, Nat. of Britain (HKE: 21). In cult.
CURRENT STATUS: Reputed to be one of the world's worst weeds, but rare in the Gardens and only occasionally found in disturbed ground.

1749. Phleum pratense L.
Timothy
FIRST RECORD: 1928 [1768].
EXSICC.: **Kew Gardens.** Near Works Department; on rubbish heap; vi 1928; *Ballard* s.n. (K).
CULT.: 1768 (JH: 399); 1789, Nat. of Britain (HK1, 1: 92); 1810, Nat. of Britain (HK2, 1: 145); 1814, Nat. of Britain (HKE: 21). In cult.
CURRENT STATUS: Occasional, in the less frequently cut grass. Most early records almost certainly refer to *P. bertolonii* (q.v.).

1750. Phleum bertolonii DC.
P. pratense subsp. *bertolonii* (DC.) Bornm.
P. nodosum L.
P. pratense var. *nodosum* (L.) Schrad.
P. pratense auct. non L. sens. str.
smaller cat's-tail
FIRST RECORD: 1873/4 [1768].
NICHOLSON: **1873/4 survey** (JB: 75): Generally distributed over the whole of the open turf. • **1906 revision** (AS: 89): Generally distributed over the whole of the open turf.
OTHER REFERENCES: **Kew Gardens.** 1990, *Cope* (S31: 182): Paddock, after restoration in 1990.
EXSICC.: **Kew Gardens.** [Unlocalised]; in rough grassland; 13 vii 1945; *Souster* 266 (K). • Bird Cage Walk, Kew Green; 21 vii 1946; *Hutchinson* s.n. (K). • Near Rhodo. Dell; in rough grass; 9 vii 1951; *Souster* 1223 (K). **Kew Environs.** Thames bank, Kew; 12 vii 1888; *Gamble* 20197 (K).
CULT.: 1768 (JH: 399); 1789, Nat. of Britain (HK1, 1: 92). Not currently in cult.
CURRENT STATUS: Abundant throughout the Gardens in all kinds of grassland. There is little doubt that most of the early records of *P. pratense* should be referred to this species since the two have not always been considered distinct.

1751. Phleum alpinum L.
alpine cat's-tail
CULT.: 1810, Nat. of Scotland (HK2, 1: 145); 1814, Nat. of Scotland (HKE: 21). In cult.

1752. Phleum phleoides (L.) H.Karst.
Phalaris phleoides L.
purple-stem cat's-tail
CULT.: 1789, Nat. of England (HK1, 1: 86); 1810, Nat. of England (HK2, 1: 137); 1814, Nat. of England (HKE: 20). In cult.

1753. Phleum arenarium L.
Phalaris arenaria (L.) Huds.
sand cat's-tail
CULT.: 1789, Nat. of England (HK1, 1: 86); 1810, Nat. of England (HK2, 1: 138); 1814, Nat. of England (HKE: 20). In cult.

[1754. Phleum hirsutum Honck.
P. michelii All.
CULT.: 1814, Nat. of Scotland (HKE: Add.).
Note. Listed in error as native.**]**

[1755. Phleum paniculatum Huds.
Phalaris paniculata (Huds.) Aiton
CULT.: 1789, Nat. of England (HK1, 1: 87); 1810, Nat. of England (HK2, 1: 145); 1814, Nat. of England (HKE: 20).
Note. Listed in error as native.**]**

1756. *Bromus arvensis L.
field brome
FIRST RECORD: 1905 [1768].
REFERENCES: **Kew Environs.** 1905 (S33: 751 and see Exsicc.).
EXSICC.: **Kew Environs.** Border of Kew Green; 19 vii 1905; *Thompson* s.n. (K).
CULT.: 1768 (JH: 404); 1789, Nat. of Britain (HK1, 1: 110); 1810, Nat. of Britain (HK2, 1: 168); 1814, Nat. of Britain (HKE: 24). In cult.
CURRENT STATUS: Undoubtedly extinct. It has not been seen for over a century.

1757. Bromus racemosus L.
B. commutatus Schrad.
B. multiflorus Host
B. pratensis Hoffm.
smooth brome
FIRST RECORD: 1873/4 [1768].
NICHOLSON: **1873/4 survey** (JB: 76): **Pal.** Hay-grass between Palace and Brentford Ferry. • **1906 revision** (AS: 91): **P.** Hay-grass near palace.
CULT.: 1768 (JH: 404, as *racemosus*); 1810, Nat. of Britain (HK2, 1: 166, as *commutatus*), 1810, Nat. of England (HK2, 1: 169, as *racemosus*) ; 1814, Nat. of Britain (HKE: 23, as *commutatus*), 1814, Nat. of England (HKE: 24. and: Add., as *pratensis*). Not currently in cult.
CURRENT STATUS: Rare; it has persisted in the region of the Palace though as a weed of flower-beds adjacent to the boundary wall rather than as a component of grassland.

1758. Bromus hordeaceus L.
B. mollis L.
B. pseudothominei P.M.Sm.
B. thominei auct. non Hardouin
B. lepidus auct. non Holmb.
soft-brome
FIRST RECORD: 1873/4 [1768].
NICHOLSON: **1873/4 survey** (JB: 76): Here and there over the whole turf. • **1906 revision** (AS: 91): Everywhere.
OTHER REFERENCES: **Kew Gardens.** 1938 (S33: 751 and see Exsicc.).
EXSICC.: **Kew Gardens.** Field near Herbarium; 29 v 1922; *Turrill* s.n. (K). • Herbarium field; 5 v 1925; *Hubbard* 6 (K). • Herbarium field, in rich less sandy soil; 12 v 1927; *Hubbard* 122F (K). • Herbarium fields; in patches, on moist soil, among *Alopecurus pratensis*; 12 v 1927; *Hubbard* G.122H (K). • Herbarium field; on dry soil among short grasses; 12 v 1927; *Hubbard* 122N (K). • Herbarium field; among short perennial grasses in dry sandy soil; 12 v 1927; *Hubbard* G.123 (K). • [Unlocalised]; among long grasses, sandy soil; 16 v 1927; *Hubbard* G.121 (K). • Lower Nursery; on rubbish heap; 31 v 1927; *Hubbard* G.122E (K). • Waste ground, nr. Herbarium; v 1928; *Ballard* s.n.

Smooth brome (*Bromus racemosus*). Rare but persistent near Kew Palace, but in flower-beds rather than grassland.

(K). • Thames-bank, Kew - Richmond, on river wall near Brentford Ferry; 25 vi 1929; *Summerhayes & Turrill* s.n. (K). • Herbarium field; common; 16 x 1929; *Hubbard* s.n. (K). • Herbarium field; abundant, on gravelly soil; 29 v 1933; *Hubbard* 57 (K). • Herbarium grounds; 4 vi 1935; *Ross-Craig* 413 (K). • Between the Pagoda and Queen's Cottage; in partial shade, one small patch; 14 vi 1938; *Hubbard* 9350 (K). • Queen's Cottage Grounds; growing in acid sandy soil with *Anthoxanthum odoratum*; 11 vi 1948; *Senogles* 79 (K). **Kew Environs.** Kew Road, Richmond; 1883; *Fraser* s.n. (K).
CULT.: 1768 (JH: 404); 1789, Nat. of Britain (HK1, 1: 109); 1810, Nat. of Britain (HK2, 1: 166); 1814, Nat. of Britain (HKE: 24); 1929, Herbaceous Ground (K); 1933, Herbarium Expt. Ground (K); 1936, Herbaceous Ground (K); 1952, Herbarium Ground (K). In cult.
CURRENT STATUS: Fairly common in grassland. The distinction between this and the supposed hybrid with *B. lepidus* (*B.* × *pseudothominei*) cannot be supported. The record for *B. lepidus* also appears to be *B. hordeaceus*.

1759. Bromus interruptus (Hack.) Druce
interrupted brome
CULT.: 1932, Herb. Expt. Ground (K). In cult.

1760. Bromus secalinus L.
rye brome
CULT.: 1768 (JH: 404); 1789, Nat. of England (HK1, 1: 109); 1810, Nat. of England (HK2, 1: 166); 1814, Nat. of England (HKE: 23). Not currently in cult.

1761. *Bromus lanceolatus Roth
large-headed brome
FIRST RECORD: 1878.
NICHOLSON: **1906 revision:** Not recorded.
OTHER REFERENCES: **Kew Gardens.** 1990, *Cope* (S31: 182): Paddock, after restoration in 1990.
EXSICC.: **Kew Gardens.** [Unlocalised]; vii 1878, *Nicholson* s.n. (K). • [Unlocalised]; on rubbish heap; 10 vi 1938; *Hubbard* 9435 (K).
CULT.: 1925, unspec. (K); 1936, Herbaceous Ground (K). In cult.
CURRENT STATUS: No recent sightings; it has not reappeared in the Paddock since 1990.

[1762. Bromus squarrosus L.
CULT.: 1789, Nat. of England (HK1, 1: 109); 1810, Nat. of England (HK2, 1: 167); 1814, Nat. of Britain (HKE: 24).
Note. Listed in error as native.**]**

1763. Bromus ramosus Huds.
B. asper Murray
hairy-brome
FIRST RECORD: 2008 [1789].
CULT.: 1789, Nat. of England (HK1, 1: 109); 1810, Nat. of England (HK2, 1: 167); 1814, Nat. of England (HKE: 24). Not currently in cult.
CURRENT STATUS: Found in the Conservation Area after clearance of scrub and brambles. Likely to be native although it was in cultivation a long time ago.

1764. Bromus erectus Huds.
upright brome
FIRST RECORD: 1873/4 [1810].
NICHOLSON: **1873/4 survey** (JB: 76): **Pal.** In hay-grass. **Strip.** About 150 yards north of Isleworth Gate. A good tuft grows out of wall near this place. • **1906 revision** (AS: 91): **P.** In hay grass. **Strip** and **Q.** Not uncommon.
EXSICC.: **Kew Gardens.** Near the tennis courts; 27 v 1921; *Turrill* s.n. (K). **Kew Environs.** Kew (probably between 1779 and 1805); *Goodenough* s.n. (K).
CULT.: 1810, Nat. of England (HK2, 1: 168); 1814, Nat. of England (HKE: 24). In cult.
CURRENT STATUS: No recent sightings.

1765. *Bromus diandrus Roth
B. rigidus auct. non Roth
B. madritensis auct. non L.
great brome
FIRST RECORD: 1902 [1789].
REFERENCES: **Kew Gardens.** 1990, *Cope* (S31: 182): Paddock, after restoration in 1990.
EXSICC.: **Kew Gardens.** Lower Nursery; on rubbish heap; 22 vi 1925; *Hubbard* 48 (K). • Near Arboretum; under shade of Beech trees; 19 v 1927; *Hubbard* G.109E (K). • Lower Nursery; on rubbish heap; 31 v 1927; *Hubbard* G.108 (K). • [Unlocalised]; on rubbish heap; 22 vii 1927; *Hubbard* G.106 (K). • [Unlocalised]; on decaying heap of leaves; 6 vi 1934; *Hubbard* 68 (K). • [Unlocalised]; grassy area; 4 vi 1936; *Worsdell* s.n. (K). • Herbarium grounds; 4 vii 1940; *Marshall* 5 (K). **Kew Environs.** Kew; 7 vii 1902; *M.Ward* 69 (K). • Kew; *M.Ward* 155 (K).
CULT.: 1789, Nat. of England (HK1, 1: 111); 1810, Nat. of England (HK2, 1: 169); 1814, Nat. of England (HKE: 24); 1902, unspec. (K). In cult.
CURRENT STATUS: No recent sightings.

1766. Bromus sterilis L.
barren brome
FIRST RECORD: 1873/4 [1768].
NICHOLSON: **1873/4 survey** (JB: 76): Common, particularly about Temperate House and near Brentford Ferry. • **1906 revision** (AS: 91): Common.
EXSICC.: **Kew Gardens.** Herbarium field; 5 v 1925; *Hubbard* 5 (K). • [Unlocalised]; area of grass; 4 vi 1936; *Worsdell* s.n. (K). • Queen's Cottage Grounds; moist loamy soil in association with *Festuca pratensis*, *Holcus lanatus* and various other grasses; 7 vi 1937; *Rawlings* 146 (K). • Field behind Herbarium; 15 vi 1955; *Kennedy-O'Byrne* s.n. (K). **Kew Environs.** Thames bank, Kew; vi 1888; *Gamble* 20199 (K). • Kew, on bank of R. Thames; roots flooded at high tide; 30 v 1933; *Hubbard* s.n. (K).
CULT.: 1768 (JH: 404); 1789, Nat. of Britain (HK1, 1: 110); 1810, Nat. of Britain (HK2, 1: 168); 1814, Nat. of Britain (HKE: 24); 1902, unspec. (K); 1948, Herbaceous Ground (K). In cult.
CURRENT STATUS: Common throughout the Gardens.

1767. *Bromus tectorum L.
drooping brome
FIRST RECORD: 1927 [1789].
REFERENCES: **Kew Gardens.** 1927 (S33: 751 and see Exsicc.).
EXSICC.: **Kew Gardens.** Lower Nursery; on rubbish heap; 22 vii 1927; *Hubbard* G106A (K). • Near Herbarium; amongst grasses; 25 v 1933; *Hubbard* s.n. (K).
CULT.: 1789, Alien (HK1, 1: 110); 1810, Alien (HK2, 1: 168); 1814, Alien (HKE: 24); 1925, unspec. (K). Not currently in cult.
CURRENT STATUS: A single plant was recently found on the golf course, in the rough in an area in which herbicides had been used to kill off broad-leaved herbs. There are no recent records from within the Gardens.

1768. *Bromus madritensis L.
compact brome
FIRST RECORD: 1925 [1925].
REFERENCES: **Kew Gardens.** 1990, *Cope* (S31: 182): Paddock, after restoration in 1990.
EXSICC.: **Kew Gardens.** Field near the Herbarium; mixed with and crossing with *B. sterilis*; 12 vi 1925; *Turrill* s.n. (K). • [Unlocalised]; on rubbish tip; 10 vi 1938; *Hubbard* 9434 (K). • Herbarium field; an escape from the gardens; 6 vi 1951; *Kennedy-O'Byrne* 141 (K).
CULT.: 1925, unspec. (K); 1936, Herbaceous Ground (K). In cult.
CURRENT STATUS: After a long absence it has reappeared in rough ground in the Lower Nursery.

1769. *Bromus carinatus Hook. & Arn.
Ceratochloa carinata (Hook. & Arn.) Tutin
California brome
FIRST RECORD: 1919 [1929].
REFERENCES: **Kew Gardens.** 1961, *Kennedy-O'Byrne* (S24: 189): Now a firmly established alien at Kew. From about [1841], the plant has been grown in the Herbaceous Ground at Kew for over a hundred years, but it was not until comparatively recent times that it made its escape to become a naturalised alien. The Herbaceous Ground was extended in the early 1920's and much discarded material was removed to the rubbish tip in the Lower Nursery, where the plant became well established by 1925. It was first noticed outside the Gardens on Queen Elizabeth's Lawn in 1927, and since then it has spread in both directions along the Surrey side of the Thames embankment for a distance of about six miles between Richmond and Hammersmith. **Kew Environs.** 1919, *Clement* (BSN28: 12): Now dominates several miles of tow-path and road verges centred about Kew ... It was first found at Kew as an escape from the Royal Bot. Gdns by CEH[ubbard] about 1919. • 1954, *D.H.Kent* (BSP1: 160): Abundant by the Thames in the vicinity of the Royal Botanic Gardens, Kew, from whence it originally escaped. It was at Kew that the plant was shown to members of the Society at the start of the London Area excursion in September 1952. During the course of that excursion the rubbish-tip at Hanwell was visited, and it is possible that the seeds were accidentally introduced from Kew via the trouser turn-ups or shoes of a member, or members, of the society. • 1956, *Welch* (LN36: 12; LN36.S: 319): Greatly extended its range in recent years, and now extends along the Thames from Richmond to ... Hammersmith Bridge, having originally 'escaped'

from Kew Gardens. • 1983, *Burton* (FLA: 202): ... first found near the Thames at Kew in 1938 by C.E. Hubbard. By 1945 it was abundant there.
EXSICC.: **Kew Gardens.** Herbarium field; naturalised; 18 vi 1925; *Hubbard* 37 (K). • Lower Nursery; on rubbish heap; 22 vi 1925; *Hubbard* 47 (K). • Lower Nursery & Herbarium field; naturalised on rubbish heaps; 22 vii 1927; *Hubbard* G.107 (K). • Kew, near bank of Thames on Queen Elizabeth's Lawn; 23 vi 1936; *Hubbard* s.n. (K). • Lower Nursery; weed on hard ground by side of rubbish heap; 23 vi 1936; *Hubbard* s.n. (K). • Naturalised on Queen Elizabeth's lawn, near R. Thames; 29 vi 1936; *Hubbard* s.n. (K). • Herbarium field; abundant; 19 vi 1951; *Hubbard* 13358A, 13358B (K). • Near Herbarium; by side of hard ash path; 19 vi 1951; *Hubbard* 13360 (K). • Near Tennis Courts; on very hard dry ground; 20 vii 1951; *Hubbard* 13371 (K). • Herbarium field; abundant; 1952; *Hubbard* s.n. (K). • Near Experimental Plots; waste ground; 20 vi 1955; *Kennedy-O'Byrne* s.n. (K). • By Experimental Plots; waste ground; 1 vii 1955; *Kennedy-O'Byrne* s.n. (K). • By Experimental Plots; on waste ground; 10 vii 1955; *Kennedy-O'Byrne* s.n. (K). • By Expt. Plots; on waste ground; 30 vii 1955; *Kennedy-O'Byrne* s.n. (K). **Kew Environs.** Kew Bridge; naturalised on bank of R. Thames; 15 vii 1936; *Hubbard* 9297 (K). • Kew, tow-path of R. Thames; 7 viii 1944; *R.M.Payne* s.n. (K). • Kew, Thames embankment at end of Ferry Lane; vi 1955; *Kennedy-O'Byrne* s.n. (K).
CULT.: 1929, Herbaceous Ground (K). In cult.
CURRENT STATUS: A persistent weed rapidly spreading throughout the Gardens. It was certainly in cultivation before the voucher dated 1929 was collected because all other records would have originated from escaped cultivated material and the earliest of these was 1919.

1770. Brachypodium pinnatum (L.) P.Beauv.
Bromus pinnatus L.
tor-grass
CULT.: 1768 (JH: 404); 1789, Nat. of England (HK1, 1: 111); 1810, Nat. of England (HK2, 1: 169); 1814, Nat. of England (HKE: 24); 1929, Herbaceous Ground (K). In cult.

1771. Brachypodium sylvaticum (Huds.) P.Beauv.
Bromus sylvaticus (Huds.) Pollich
false brome
FIRST RECORD: 1873/4 [1810].
NICHOLSON: **1873/4 survey** (JB: 76): **P.** In the turf skirting moat. **Strip.** Frequent. • **1906 revision** (AS: 91): **A.** In turf by the river. **Strip.** Frequent.
EXSICC.: **Kew Environs.** Bank of Thames, Kew; 1883; *Fraser* s.n. (K). • Between Kew and Richmond; in dense shade near bank of Thames; 2 viii 1927; *Hubbard* G.120 (K).
CULT.: 1810, Nat. of Britain (HK2, 1: 169); 1814, Nat. of Britain (HKE: 24). In cult.
CURRENT STATUS: Widepsread in shady areas.

1772. *Brachypodium distachyum (L.) P.Beauv.
stiff brome
FIRST RECORD: 1927 [1925]
EXSICC.: **Kew Gardens.** Lower Nursery; on rubbish heap; 31 v 1927; *Hubbard* G.117A (K).
CULT.: 1925, unspec. (K). In cult.
CURRENT STATUS: No recent sightings.

1773. Elymus caninus (L.) L.
Triticum caninum L.
bearded couch
FIRST RECORD: 1974 [1810].
EXSICC.: **Kew Gardens.** Weed in Nursery; viii 1974; *Hubbard* s.n. (K).
CULT.: 1810, Nat. of Britain (HK2, 1: 182); 1814, Nat. of Britain (HKE: 26); 1855, unspec. (K); 1926, Herbarium Expt. Ground (K); 1929, Herbaceous Ground (K); 1951, Herbaceous Ground (K); 1952, Herb. Expt. Ground (K). In cult.
CURRENT STATUS: No recent sightings.

1774. Elymus repens (L.) Gould
Triticum repens L.
Agropyron repens (L.) P.Beauv.
common couch
FIRST RECORD: 1873/4 [1768].
NICHOLSON: **1873/4 survey** (JB: 76): In every division. This seems to be the principal ingredient in the rough parts of turf in some places in **P.** • **1906 revision** (AS: 91): The principal ingredient in the turf in some spots in arboretum, etc.
EXSICC.: **Kew Gardens.** Herbarium Experimental Ground; on loose bare loam; 10 iv 1928; *Summerhayes* s.n. (K). **Kew Environs.** Kew; vii 1888; *Gamble* 20190 (K). • Thames bank, Kew; vii 1888; *Gamble* 20190, 20191, 20192 (all K). • Thames bank, Kew; 12 vii 1888; *Gamble* 20205 (K). • Towing path, Kew; vii 1902; *Fraser* s.n. (K). • Between Kew and Richmond; common on bank of R. Thames; 2 viii 1927; *Hubbard* G.114C (K). • Bank of River Thames; frequent in gravelly soil; 23 vi 1936; *Hubbard* 42, 43 (both K).
CULT.: 1768 (JH: 405); 1789, Nat. of Britain (HK1, 1: 122); 1810, Nat. of Britian (HK2, 1: 181); 1814, Nat. of Britain (HKE: 26); 1933, unspec. (K). Not currently in cult.
CURRENT STATUS: A common weed throughout the Gardens.

1775. Elymus athericus (Link) Kerguélen
Agropyron pungens auct. non (Pers.) Roem. & Schult.
Elymus pycnanthus (Link) Godr.
sea couch
FIRST RECORD: 1915.
REFERENCES: **Kew Environs.** 1915, *Britton* (LFS: 394; S33: 751): Formerly at Kew [along river].
CURRENT STATUS: No recent sightings; it no longer appears to be along the river in the immediate area.

1776. Elymus farctus (Viv.) Runemark ex Melderis
Triticum junceum L.
sand couch
CULT.: 1768 (JH: 405); 1789, Nat. of Britain (HK1, 1: 121); 1810, Nat. of Britain (HK2, 1: 181); 1814, Nat. of Britain (HKE: 26). In cult.

1777. Leymus arenarius (L.) Hochst.
Elymus arenarius L.
E. geniculatus Curtis, nom. nud.
lyme-grass
CULT.: 1768 (JH: 395); 1789, Nat. of Britain (HK1, 1: 117); 1810, Nat. of Britain (HK2, 1: 175), Nat. of England (ibid.: 176 as *geniculatus*); 1814, Nat. of Britain (HKE: 25), Nat. of England (ibid.: as *geniculatus*). In cult.

1778. Hordelymus europaeus (L.) Jess. ex Harz
Elymus europaeus L.
wood barley
CULT.: 1810, Nat. of England (HK2, 1: 177); 1814, Nat. of England (HKE: 25). In cult.

1779. *Hordeum vulgare L.
barley
FIRST RECORD: 1990 [1768].
REFERENCES: **Kew Gardens.** 1990, *Cope* (S31: 182): Paddock, after restoration in 1990.
CULT.: 1768 (JH: 395); 1789, Cult. (HK1, 1: 118); 1810, Cult. (HK2, 1: 178); 1814, Cult. (HKE: 25); 1939, Herbaceous Ground (K). In cult.
CURRENT STATUS: Very rare and not recently found in the Paddock; it probably only occurs through spillage from bird feeders.

1780. Hordeum murinum L.
wall barley
FIRST RECORD: 1873/4 [1768].
NICHOLSON: **1873/4 survey** (JB: 77): **P.** Common about Winter Garden. **Strip.** Abundant, in company with *Bromus sterilis*. • **1906 revision** (AS: 91): **Strip.** Abundant. **P.** About palace.
EXSICC.: **Kew Gardens.** [Unlocalised]; waste land; 3 vi 1944; *Souster* 82 (K). **Kew Environs.** Kew Road, Richmond; wayside; 1883; *Fraser* s.n. (K). • Thames bank, Kew; vii 1888; *Gamble* 20193 (K). • Kew; roadside weed; 1 vii 1964; *Blake* 22281 (BR, K).
CULT.: 1768 (JH: 395); 1789, Nat. of Britain (HK1, 1: 119); 1810, Nat. of Britain (HK2, 1: 179); 1814, Nat. of Britain (HKE: 25). Not currently in cult.
CURRENT STATUS: Abundant throughout the Gardens.

1781. *Hordeum jubatum L.
foxtail barley
FIRST RECORD: 1990 [1789].
REFERENCES: **Kew Gardens.** 1990, *Cope* (S31: 182): Paddock, after restoration in 1990.
EXSICC.: **Kew Gardens.** on rubbish heap; viii 1931; *Hubbard* s.n. (K).
CULT.: 1789, Alien (HK1, 1: 120); 1810, Alien (HK2, 1: 180); 1814, Alien (HKE: 25); 1926, Hb. Expt. Ground (K). Not currently in cult.
CURRENT STATUS: No recent sightings; it has not reappeared in the Paddock since 1990 but did make a subsequent brief appearance outside the Stable Yard.

1782. Hordeum secalinum Schreb.
meadow barley
FIRST RECORD: 1998.
REFERENCES: **Kew Gardens.** 1998, *Cope* (S33: 751): North Arb.: lawn in Rose Garden, behind Palm House (170) (single flowering stem overlooked during regular mowing).
CURRENT STATUS: Probably very rare, and noticed only by chance in a strip of grass that had escaped mowing. It had not previously been reported from the Gardens and needed checking but the area in which it grew — the Rose Garden — has since been stripped of its turf ready for reconstruction. The species is, however, common on parts of the golf course in the Old Deer Park.

1783. Hordeum marinum Huds.
H. maritimum Stokes
sea barley
CULT.: 1810, Nat. of Britain (HK2, 1: 179); 1814, Nat. of Britain (HKE: 25). Not currently in cult.

[1784. Agropyron cristatum (L.) Gaertn.
bromus cristatus L.
CULT.: 1814, Nat. of Scotland (HKE: Add.).
Note. Listed in error as native.**]**

1785. *Eremopyrum bonaepartis (Spreng.) Nevski
Agropyron patulum (Willd.) Trin.
FIRST RECORD: 1878.
REFERENCES: **Kew Environs.** 1878, *Nicholson* (BSEC5: 136; S33: 751): Kew, Surrey.
CULT.: 1970, Grass Beds (K). In cult.
CURRENT STATUS: No recent sightings.

1786. *Secale cereale L.
rye
FIRST RECORD: 1928 [1768].
EXSICC.: **Kew Gardens.** On rubbish heap in Lower Nursery; 10 vi 1928; *Hubbard* s.n. (K).
CULT.: 1768 (JH: 395). In cult.
CURRENT STATUS: No recent sightings.

1787. *Triticum aestivum L.
bread wheat
FIRST RECORD: 1990 [1789].
REFERENCES: **Kew Gardens.** 1990, *Cope* (S31: 182): Paddock, after restoration in 1990.
CULT.: 1789, Cult. (HK1, 1: 120); 1810, Cult. (HK2, 1: 180); 1814, Cult. (HKE: 25). In cult.

Current status: Occasional as spillage from bird feeders; it was sown into the Wheatfields along the Broad Walk in 1999 and flowered profusely in 2000. More recently (2001) it was sown into the Wheatfield at the south end of the Lake. It now occurs additionally as a relict from these sowings.

1788. Danthonia decumbens (L.) DC.
Festuca decumbens L.
Poa decumbens (L.) Scop.
Triodia decumbens (L.) P. Beauv.
heath-grass
First record: 1873/4 [1768].
Nicholson: **1873/4 survey** (JB: 75): **P.** Common on both sides of Syon Vista towards river. Very common in turf between Unicorn Gate and Pagoda Avenue. • **1906 revision** (AS: 90): Common on both sides of Syon Vista towards river. Very common in turf between Unicorn Gate and Pagoda Vista.
Cult.: 1768 (JH: 403); 1789, Nat. of Britain (HK1, 1: 98); 1810, Nat. of Britain (HK2, 1: 153); 1814, Nat. of Britain (HKE: 22); 1925, Herbaceous Ground (K). In cult.
Current status: Difficult to find because of close mowing of most of the turf, but there is a substantial population N of the Rhododendron Dell in a patch of acidic grassland it shares with *Festuca filiformis*. As with several other species of turf grass it needed the right combination of conditions to be refound. So far just the one population is known.

1789. Molinia caerulea (L.) Moench
Aira caerulea L.
Melica caerulea (L.) L.
purple moor-grass
First record: 1873/4 [1768].
Nicholson: **1873/4 survey** (JB: 75–6): **P.** A strong plant with several panicles, opposite end of smallest island on the Brentford side of lake. Another in wood in company with *Juncus squarrosus*. • **1906 revision** (AS: 90): **A.** Here and there by lake; also in wood near pumping station.
Cult.: 1768 (JH: 401); 1789, Nat. of Britain (HK1, 1: 98); 1810, Nat. of Britain (HK2, 1: 153); 1814, Nat. of Britain (HKE: 22). In cult.
Current status: No recent sightings.

1790. Phragmites australis (Cav.) Trin. ex Steud.
Arundo phragmites L.
common reed
First record: 1999 [1768].
References: **Kew Gardens.** 1999, *Stones* (EAK: 2): By Lake.
Cult.: 1768 (JH: 405); 1789, Nat. of Britain (HK1, 1: 115); 1810, Nat. of Britain (HK2, 1: 174); 1814, Nat. of Britain (HKE: 25); 1937, Herbaceous Ground (K); 1939, Aquatic Garden (K). In cult.
Current status: By the Banks Building pond where it has recently been planted. The population by the Lake is not in the living collections database but must have been planted long ago.

1791. *Eragrostis pilosa (L.) P.Beauv.
Jersey love-grass
First record: 1999.
References: **Kew Gardens.** 1999, *Cope* (S35: 649): In soil heaps outside the Stable Yard (353).
Current status: Rare; found on a single occasion outside the Stable Yard.

1792. *Eleusine indica (L.) Gaertn.
yard-grass
Cynosurus indicus L.
First record: 1960 [1768].
References: **Kew Gardens.** 1960, *Simmonds* (specimen 220-1960 *per* G.E.Nicholson) (S29: 408): In pot of *Zeuxine* sp. (*Orchidac.*) from Fiji. • 1985, *London Natural History Society* (LN65: 196): [Around what is now the Princess of Wales Conservatory].
Cult.: 1768 (JH: 406). In cult.
Current status: No recent sightings. What may have been this species appeared in bare ground under a tree near the Brentford Gate, but it was removed before it had flowered.

1793. Cynodon dactylon (L.) Pers.
Panicum dactylon L.
Bermuda-grass
First record: 1844 [1789].
Nicholson: **1873/4 survey**: Not recorded • **1906 revision**: Not recorded.
Other references: **Kew Gardens.** 1911, *anon.* (S12: 375): Occurs abundantly in the turf to the west of the new range. • 1920, *Thiselton-Dyer* (S16: 216): Abundant in the turf to the west of the new Range. • 1955, *Hubbard* (LN35: S310): Kew Gardens in several places. • 1976, *Lousley* (LFS: 402): The 1844 record was from Kew Green, where it persisted for a time; it still occurs on several lawns in the Royal Botanic Gardens. **Kew Environs.** 1844, *Hill* (BFS: 272; SFS: 641): On Kew Green, in some abundance, in the month of August ... Doubtless an escape from Kew Gardens.
Exsicc.: **Kew Gardens.** Abundant in lawn by Herbaceous Border, between Succulent House and T-Range; known in this lawn for very long time; 7 viii 1947; *Hubbard* 13109 (K). • T-Range lawn; 4 ix 1959; *Hubbard* s.n. (K). **Kew Environs.** Kew Green; naturalised!; ix 1846 [or 1896?]; ?*W.S.* s.n. (K).
Cult.: 1789, Nat. of England (HK1, 1: 90); 1810, Nat. of England (HK2, 1: 142); 1814, Nat. of England (HKE: 20); 1936, Herbaceous Ground (K). In cult.

Cockspur (*Echinochloa crusgalli*). An exotic-looking grass that is on the increase, not only in Kew Gardens but in the south of England generally.

CURRENT STATUS: A substantial patch, in the grass around the old Grass Beds (T-Range lawn), was destroyed when the Princess of Wales Conservatory was built, but the plant has re-colonised the lawn around the new Grass Beds from the collection on display. It also occurs in dry ground at the south end of the Order Beds.

1794. Spartina maritima (Curtis) Fernald
Dactylis stricta Aiton
small cord-grass
CULT.: 1789, Nat. of England (HK1, 1: 104); 1810, Nat. of England (HK2, 1: 160); 1814, Nat. of England (HKE: 23); 1961, unspec. (K). Not currently in cult.

1795. Spartina × townsendii H. & J.Groves (alterniflora × maritima)
Townsend's cord-grass
CULT.: 1961, Jodrell (K). Not currently in cult.

1796. Spartina anglica C.E.Hubb.
common cord-grass
CULT.: 1961, Jodrell (K). Not currently in cult.

1797. *Panicum capillare L.
witch-grass
FIRST RECORD: 1877 [1768].
REFERENCES: **Kew Gardens.** 1985, *London Natural History Society* (LN65: 196; S33: 751): [Around what is now the Princess of Wales Conservatory].
EXSICC.: **Kew Environs.** Kew; waste places; viii 1877; *Baker* s.n. (K).
CULT.: 1768 (JH: 407); 1789, Alien (HK1, 1: 91); 1810, Alien (HK2, 1: 143); 1814, Alien (HKE: 20); 1967, Grass Beds (K). In cult.
CURRENT STATUS: Recently appeared in heavily disturbed soil outside the Stable Yard and as a weed in the Lavender display at the end of the Grass Beds.

1798. *Panicum miliaceum L.
common millet
FIRST RECORD: 1999 [1789].
REFERENCES: **Kew Gardens.** 1999, *Cope* (S35: 649): In soil heaps outside the Stable Yard (353).
CULT.: 1789, Alien (HK1, 1: 90); 1810, Alien (HK2, 1: 143); 1814, Alien (HKE: 20); 1935, Herb. Gd. (K). In cult.
CURRENT STATUS: Casual; in rough ground near the Stable Yard.

1799. *Echinochloa crusgalli (L.) P.Beauv.
Panicum crusgalli L.
cockspur
FIRST RECORD: 1936 [1768].
REFERENCES: **Kew Gardens.** 1985, *London Natural History Society* (LN65: 196): [Around what is now the Princess of Wales Conservatory]. • 1985, *Cope* (S31: 183): Alpine Yard.
EXSICC.: **Kew Gardens.** [Unlocalised]; weed in nursery; 24 vii 1936; *Hubbard* s.n. (K). **Kew Environs.** Kew; in garden from bird seed; ix 1956; *Bullock* s.n. (K).
CULT.: 1768 (JH: 407); 1789, Nat. of England (HK1, 1: 89); 1810, Nat. of England (HK2, 1: 141); 1814, Nat. of England (HKE: 20); 1929, Herbaceous Ground (K); 1935, Herb. Gd. (K). In cult.
CURRENT STATUS: Recently appeared in the grass around the student vegetable plots behind Hanover House, outside the Stable Yard and near the Brentford Gate. It is an increasingly common alien that naturalises easily in our progressively milder climate.

1800. *Echinochloa colona (L.) Link
Panicum colonum L.
shama millet
FIRST RECORD: 1967 [1768]
REFERENCES: **Kew Gardens.** 1967 (S33: 751 and see Exsicc.).

EXSICC.: **Kew Gardens.** Outside Herbarium; 12 x 1967; *Milne-Redhead* s.n. (K).
CULT.: 1768 (JH: 407); 1789, Alien (HK1, 1: 89); 1810, Alien (HK2, 1: 141); 1814, Alien (HKE: 20); 1951, warm greenhouse (K). In cult.
CURRENT STATUS: No recent sightings.

1801. *Paspalum dilatatum Poir.
dallis-grass
FIRST RECORD: 2003 [1847].
CULT.: 1847, unspec. (K); 1936, Herbaceous Ground (still there 1956) (K); 1953, Herbaceous Ground (K). In cult.
CURRENT STATUS: Persistent in the lawn around the plot in the Grass Beds from which it escaped.

1802. *Axonopus sp.
carpet-grass
FIRST RECORD: 1960.
REFERENCES: **Kew Gardens.** 1960, *G.E.Nicholson* (S29: 408): In pot of *Cymbidium* sp. (*Orchidac.*) (entry no. 582-1953) from Georgetown Botanic Garden, Guyana.
CURRENT STATUS: No recent sightings. Lack of a voucher specimen means the species must remain undetermined.

1803. *Setaria pumila (Poir.) Roem. & Schult.
yellow bristle-grass
FIRST RECORD: Probably between 1779 and 1805 [1925].
EXSICC.: **Kew Environs.** Kew; [in sunny places]; *Goodenough* (K). • Kew; 10 ix 1873; *Clarke* 47588 (K). • Kew; viii 1877; *Baker* 1486 (K).
CULT.: 1925, Herb. Expt. Ground (K); 1929, Herbaceous Ground (K). In cult.
CURRENT STATUS: Recently seen in heavily disturbed ground by the Stable Yard and in the Dell Nursery. These are the only certain records of the plant growing wild within the Gardens.

1804. *Setaria verticillata (L.) P.Beauv.
Panicum verticillatum L.
rough bristle-grass
FIRST RECORD: 1877 [1789].
REFERENCES: **Kew Gardens.** 1945, *Welch* (S32: 657; LN36: S359): Kew Gardens Allotments. • 1985, *London Natural History Society* (LN65: 196): [Around what is now the Princess of Wales Conservatory].
EXSICC.: **Kew Environs.** Kew; viii 1877; *Baker* s.n. (K).
CULT.: 1789, Nat. of England (HK1, 1: 88); 1810, Nat. of England (HK2, 1: 139); 1814, Nat. of England (HKE: 20); 1929, Herbaceous Ground (K); 1936, Herbaceous Ground (K). Not currently in cult.
CURRENT STATUS: In cracks between paving slabs around the Princess of Wales Conservatory where it has been seen sporadically for more than 20 years.

1805. *Setaria viridis (L.) P.Beauv.
Panicum viride L.
green bristle-grass
FIRST RECORD: 1928 [1768].
EXSICC.: **Kew Gardens.** Comes up as a weed in Herb. Expt. ground; 30 viii 1928; *Hubbard* s.n. (K).
CULT.: 1768 (JH: 407); 1789, Nat. of England (HK1, 1: 88); 1810, Nat. of England (HK2, 1: 140); 1814, Nat. of England (HKE: 20); 1935, Herb. Gd. (K). In cult.
CURRENT STATUS: A weed in places in the Order Beds, and appeared for a short time in grass behind Hanover House before the allotments were removed and building works begun.

1806. *Setaria italica (L.) P.Beauv.
Panicum italicum L.
foxtail bristle-grass
FIRST RECORD: 1985 [1768].
REFERENCES: **Kew Gardens.** 1985, *London Natural History Society* (LN65: 196; S33: 752): [Around what is now the Princess of Wales Conservatory].
CULT.: 1768 (JH: 407); 1789, Alien (HK1, 1: 89); 1810, Alien (HK2, 1: 140); 1814, Alien (HKE: 20); 1936, Herb. Gd. (K). In cult.
CURRENT STATUS: No recent sightings.

1807. *Digitaria ischaemum (Schreb. ex Schweigg.) Muhl.
smooth finger-grass
FIRST RECORD: 1938 [1936].
REFERENCES: **Kew Gardens.** 1938 (S33: 752 and see Exsicc.).
EXSICC.: **Kew Gardens.** Grass beds; weed; ix 1938; *Hubbard* s.n. (K).
CULT.: 1936, Herbaceous Ground (K). Not currently in cult.
CURRENT STATUS: No recent sightings.

1808. *Digitaria sanguinalis (L.) Scop.
Panicum sanguinale L.
hairy finger-grass
FIRST RECORD: 1872 [1768].
REFERENCES: **Kew Gardens.** 1955 (S33: 752 and see Exsicc.).
EXSICC.: **Kew Gardens.** Herbaceous ground; weed; 23 ix 1955; *Hubbard* s.n. (K). **Kew Environs.** Waste ground, by Kew Bridge; ix 1872; *Watson* in *Herb. Watson* (K).
CULT.: 1768 (JH: 407); 1789, Nat. of England (HK1, 1: 89); 1810, Nat. of England (HK2, 1: 142); 1814, Nat. of England (HKE: 20). Not currently in cult.
CURRENT STATUS: Appeared on bare ground around the new Alpine House during building operations but has not returned.

Branched bur-reed (*Sparganium erectum*). Once grew wild in the Moat but can now only be found by the Lake where it was planted in the mid-19th century

1809. *Pennisetum villosum R.Br. ex Fresen.
feathertop
FIRST RECORD: 2004 [1969].
CULT.: In cult.
CURRENT STATUS: Grew from seed in topsoil stored in the Paddock.

TYPHACEAE

1810. Sparganium erectum L.
S. ramosum Huds.
branched bur-reed
FIRST RECORD: 1873/4 [1768].
NICHOLSON: **1873/4 survey** (JB: 73): **Strip.** Really wild about moat. Planted near lake and elsewhere. • **1906 revision** (AS: 88): **Strip.** Really wild about ha-ha; planted elsewhere.
CULT.: 1768 (JH: 337, 376); 1789, Nat. of Britain (HK1, 3: 323); 1813, Nat. of Britain (HK2, 5: 235); 1814, Nat. of Britain (HKE: 288). In cult.
CURRENT STATUS: Two clumps can be found on the margin of the Lake where they were likely to have been planted, perhaps as long ago as the mid-nineteenth century. Certainly by the time Nicholson did his survey they were well established. None have been refound in or near the Moat.

1811. Sparganium emersum Rehmann
S. simplex Huds.
unbranched bur-reed
CULT.: 1789, Nat. of Britain (HK1, 3: 324); 1813, Nat. of Britain (HK2, 5: 235); 1814, Nat. of Britain (HKE: 288). Not currently in cult.

1812. Sparganium natans L.
least bur-reed
CULT.: 1768 (JH: 337, 376); 1813, Nat. of England (HK2, 5: 235); 1814, Nat. of England (HKE: 288). Not currently in cult.

1813. Typha latifolia L.
reedmace
FIRST RECORD: 1873/4 [1768].
NICHOLSON: **1873/4 survey** (JB: 73): **P.** Lake. **Strip.** Side of moat. • **1906 revision** (AS: 88): **A.** Lake. **Strip.** Side of ha-ha.
OTHER REFERENCES: **Kew Gardens.** 1999, *Stones* (EAK: 2): By Lake.
CULT.: 1768 (JH: 412); 1789, Nat. of Britain (HK1, 3: 323); 1813, Nat. of Britain (HK2, 5: 234); 1814, Nat. of Britain (HKE: 288). In cult.
CURRENT STATUS: Possibly native by the pond in the Conservation Area, and by the Lake; it was planted by the Banks Building Pond.

Reedmace (*Typha latifolia*). Planted near the Banks Building but likely to be native by the Lake and in the Conservation Area.

1814. Typha angustifolia L.
T. minor Curtis
lesser reedmace
FIRST RECORD: 1873/4 [1768].
NICHOLSON: **1873/4 survey** (JB: 73): {Not a truly wild Kew plant. Wherever it occurs in our Flora it has been planted}. • **1906 revision** (AS: 88): {This has been planted wherever it occurs in our flora}.
CULT.: 1768 (JH: 412); 1789, Nat. of England (HK1, 3: 323); 1813, Nat. of Britain (HK2, 5: 234), Nat. of England (ibid.: as *minor*); 1814, Nat. of Britain (HKE: 288), Nat. of England (ibid. as *minor*). In cult.
CURRENT STATUS: By the pond in the Conservation Area.

PONTEDERIACEAE

1815. *Pontederia cordata L.
pickerelweed
FIRST RECORD: 2001 [1789].
CULT.: 1789, Alien (HK1, 1: 403); 1811, Alien (HK2, 2: 206); 1814, Alien (HKE: 87). In cult.
CURRENT STATUS: Naturalised in a pond on the golf course in the Old Deer Park adjacent to the southern boundary of the Gardens.

MELANTHIACEAE

1816. Tofieldia pusilla (Michx.) Pers.
T. palustris auct. non Huds.
Scottish asphodel
CULT.: 1811, Nat. of Britain (HK2, 2: 234); 1814, Nat. of Britain (HKE: 104). In cult.

1817. Narthecium ossifragum (L.) Huds.
Anthericum ossifragum L.
bog asphodel
CULT.: 1768 (JH: 357); 1789, Nat. of Britain (HK1, 1: 450); 1811, Nat. of Britain (HK2, 2: 270); 1814, Nat. of Britain (HKE: 95). In cult.

HEMEROCALLIDACEAE

1818. *Hemerocallis fulva (L.) L.
orange day-lily
FIRST RECORD: 2001 [1768].
REFERENCES: **Kew Gardens.** 2001, *Cope* (S35: 649): Incorporated in the Robinsonian Meadow sown in front of the temporary Cycad House (224), and now naturalised.
CULT.: 1768 (JH: 334); 1789, Alien (HK1, 1: 474); 1811, Alien (HK2, 2: 304); 1814, Alien (HKE: 101); 1956, Herbaceous Dept. (K). In cult.
CURRENT STATUS: Incorporated in the Robinsonian Meadow in front of 'Climbers & Creepers' but declining and unlikely to persist.

COLCHICACEAE

1819. Colchicum autumnale L.
meadow saffron
CULT.: 1768 (JH: 334); 1789, Nat. of Britain (HK1, 1: 490); 1811, Nat. of Britain (HK2, 2: 329); 1814, Nat. of Britain (HKE: 105); 1856, unspec. (K); 1920, unspec. (K); 1933, unspec. (K); 1941, unspec. (K); 1996, Alpine & Herbaceous (K); no date, plot 155-06 (K). In cult.

LILIACEAE

1820. Lloydia serotina (L.) Rchb.
Anthericum serotinum L.
Snowdon lily
CULT.: 1811, Nat. of Wales (HK2, 2: 266); 1814, Nat. of Wales (HKE: 95). Not currently in cult.

1821. Gagea lutea (L.) Ker Gawl.
Ornithogalum luteum L.
yellow star-of-Bethlehem
CULT.: 1768 (JH: 356); 1789, Nat. of Britain (HK1, 1: 440); 1811, Nat. of Britain (HK2, 2: 257); 1814, Nat. of Britain (HKE: 94). Not currently in cult.

[**1822. Tulipa sylvestris** L.
wild tulip
CULT.: 1789, Alien (HK1, 1: 435); 1811, Nat. of England (HK2, 2: 248); 1814, Nat. of England (HKE: 93).
Note. Listed in error as native. It was found in the wild in 1790 but is not considered to be native.]

1823. Fritillaria meleagris L.
fritillary
FIRST RECORD: 2004 [1768]
CULT.: 1768 (JH: 354); 1789, Nat. of England (HK1, 1: 432); 1811, Nat. of England (HK2, 2: 244); 1814, Nat. of England (HKE: 92); 1887, unspec. (K); 1934, Herbaceous Dept. (K). In cult.
CURRENT STATUS: Planted in drifts in two places in the Gardens in 2003. Not known as a wild plant in Kew although it has been recorded nearby in Mortlake.

1824. *Lilium martagon L.
martagon lily
FIRST RECORD: 1992 [1768].
REFERENCES: **Kew Gardens.** 1992, *Cope* (unpubl.). • 1999, *Cope* (S35: 649): Known for many years in the Conservation Area (310) where originally planted but now thoroughly naturalised and increasing. Subsequently also found under trees at the south end of the Lake (264).
CULT.: 1768 (JH: 354); 1789, Alien (HK1, 1: 431); 1811, Alien (HK2, 2: 242); 1814, Alien (HKE: 92); 1909, unspec. (K); 1967, unspec. (K). In cult.

Martagon lily (*Lilium martagon*). Not native in the Gardens but a spectacular feature in the Conservation Area where it has been naturalised for many years.

CURRENT STATUS: Established in two places in the Conservation Area, and formerly in one place by the Lake. It was known in the Conservation Area for many years without ever being recorded. All populations would have originally been planted although it is now increasing naturally.

CONVALLARIACEAE

1825. Convallaria majalis L.
lily-of-the-valley
CULT.: 1768 (JH: 331); 1789, Nat. of Britain (HK1, 1: 455); 1811, Nat. of Britain (HK2, 2: 279); 1814, Nat. of Britain (HKE: 97). In cult.

1826. Polygonatum multiflorum (L.) All.
Convallaria multiflora L.
Solomon's-seal
CULT.: 1768 (JH: 331); 1789, Nat. of England (HK1, 1: 456); 1811, Nat. of Britain (HK2, 2: 280); 1814, Nat. of Britain (HKE: 97). In cult.

1827. Polygonatum odoratum (Mill.) Druce
Convallaria polygonatum L.
angular Solomon's-seal
CULT.: 1768 (JH: 331); 1789, Nat. of England (HK1, 1: 455); 1811, Nat. of England (HK2, 2: 279); 1814, Nat. of England (HKE: 97); 1915, unspec. (K). In cult.

1828. Polygonatum verticillatum (L.) All.
Convallaria verticillata L.
whorled Solomon's-seal
CULT.: 1768 (JH: 331); 1789, Alien (HK1, 1: 455); 1811, Nat. of Scotland (HK2, 2: 279); 1814, Nat. of Scotland (HKE: 97); 1971, back of Rock Garden near the Mound (K). In cult.
Note. Not known as a British plant until 1793.

1829. Maianthemum bifolium (L.) F.W.Schmidt
Convallaria bifolia L.
May lily
CULT.: 1768 (JH: 331). Not currently in cult.

1830. *Maianthemum racemosum (L.) Link
Convallaria racemosa L.
Smilacina racemosa (L.) Desf.
treacleberry
FIRST RECORD: 1969 [1768].
REFERENCES: **Kew Gardens.** 1969, *Palmer* (BSN79: 64; S33: 752): Patch naturalised in natural vegetation near Queen's Cottage, TQ/177.764.
CULT.: 1768 (JH: 331); 1789, Alien (HK1, 1: 456); 1811, Alien (HK2, 2: 280); 1814, Alien (HKE: 97); 1896, unspec. (K); 1901, unspec. (K); 1920, unspec. (K); 1935, unspec. (K). In cult.
CURRENT STATUS: Naturalised in a few places in rough grass under trees.

TRILLIACEAE

1831. Paris quadrifolia L.
herb-Paris
CULT.: 1768 (JH: 172/9); 1789, Nat. of Britain (HK1, 2: 37); 1811, Nat. of Britain (HK2, 2: 425); 1814, Nat. of Britain (HKE: 118); 1904, unspec. (K). In cult.

HYACINTHACEAE

1832. Ornithogalum pyrenaicum L.
spiked star-of-Bethlehem
CULT.: 1768 (JH: 356); 1789, Nat. of England (HK1, 1: 441); 1811, Nat. of England (HK2, 2: 258); 1814, Nat. of England (HKE: 94); 1888, unspec. (K); 1893, unspec. (K). In cult.

1833. *Ornithogalum angustifolium* Boreau
O. umbellatum auct. non L.
star-of-Bethlehem
FIRST RECORD: 1873/4 [1768].
NICHOLSON: **1873/4 survey** (JB: 74): **Strip.** A fine tuft or two near Isleworth Gate. **P.** A plant near pond containing *Potamogeton pusillus.* • **1906 revision** (AS: 88): **Strip.** By side of towing-path near Isleworth Gate. **P.** Near Palace in hay grass.
OTHER REFERENCES: **Kew Environs.** 2004, *O'Reilly & Coleman* (LN84: 221): Old Deer Park.
CULT.: 1768 (JH: 356); 1789, Nat. of England (HK1, 1: 440); 1811, Nat. of England (HK2, 2: 257); 1814, Nat. of England (HKE: 94). In cult. under glass.
CURRENT STATUS: Recently seen in the Herbarium lawn where it survives despite being mown; also in one or two other patches of long grass. It was probably planted or inadvertantly introduced wherever it occurs. It was long thought to be a British native but is now considered to have been introduced, probably in the seventeenth century. Populations of it occur not too far away upriver in Walton-on-Thames etc.

1834. *Ornithogalum nutans* L.
drooping star-of-Bethlehem
FIRST RECORD: 1906 [1789].
NICHOLSON: **1906 revision** (AS: 88): **P.** In quantity in hay grass near Old Palace.
CULT.: 1789, alien (HK1, 1: 443); 1811, Nat. of England (HK2, 2: 262); 1814, Nat. of England (HKE: 94); 1878, unspec. (K); 1951, Herbarium Ground (K); 1954, Herbarium Ground (K); 1972, Alpine Dept. (K). In cult.
CURRENT STATUS: Several patches in long grass, possibly all planted, but some may be self-sown. It still occurs in the vicinity of Kew Palace.

1835. *Scilla bifolia* L.
alpine squill
FIRST RECORD: 2001 [1768].
REFERENCES: **Kew Gardens.** 2001, *Cope* (S35: 649): Naturalised in flowerbeds near King William's Temple (412).
CULT.: 1768 (JH: 357); 1789, Alien (HK1, 1: 444); 1811, Nat. of England (HK2, 2: 264); 1814, Nat. of England (HKE: 95); 1886, unspec. (K); 1920, unspec. (K); 1970, unspec. (K); 1994, Jodrell Glass (K). In cult.
CURRENT STATUS: Originally cultivated but has escaped into flower-beds in many parts of the Gardens.

1836. *Scilla bithynica* Boiss.
Turkish squill
FIRST RECORD: 1991.
REFERENCES: **Kew Gardens.** 1991, *Cope* (S31: 183): Restored 'Paddock.'
CULT.: In cult.
CURRENT STATUS: A significant population comes up every year in the Paddock and may be increasing.

1837. Scilla verna Huds.
spring squill
CULT.: 1789, Nat. of England (HK1, 1: 445); 1811, Nat. of Britain (HK2, 2: 264); 1814, Nat. of Britain (HKE: 95); 1955, Herbarium Expt. Ground (K). In cult.

1838. Scilla autumnalis L.
autumn squill
CULT.: 1768 (JH: 357); 1789, Nat. of England (HK1, 1: 445); 1811, Nat. of England (HK2, 2: 264); 1814, Nat. of England (HKE: 95); 1895, unspec. (K). In cult.

1839. Hyacinthoides non-scripta (L.) Chouard ex Rothm.
Scilla nutans Sm.
Hyacinthus non-scriptus L.
bluebell
FIRST RECORD: 1873/4 [1768].
NICHOLSON: **1873/4 survey** (JB: 74): **P, Pal** and **Q.** Very abundant in all the woods. A large tuft with pure white flowers grew in Palace Grounds. • **1906 revision** (AS: 88): **A, P, Q.** Very abundant in all the woods.

Bluebell (*Hyacinthoides non-scripta*). Justifiably celebrated with its own festival every year in Kew at the beginning of May.

EXSICC.: **Kew Gardens.** From the mound; 5 v 1971; *Ross-Craig* s.n. (K).
CULT.: 1768 (JH: 332); 1789, Nat. of Britain (HK1, 1: 457); 1811, Nat. of Britain (HK2, 2: 282); 1814, Nat. of Britain (HKE: 97). In cult.
CURRENT STATUS: Throughout the Gardens where it easily naturalises in suitable places. It produces spectacular shows in the Conservation Area and around the Queen's Cottage. Its native status within the Gardens is sometimes considered doubtful though Nicholson did not hesitate to record it within the Queen's Cottage Grounds. It is undoubtedly widely planted and equally widely self-sown. The main threat to the species in the Gardens is hybridisation with *H. hispanica*.

1840. *Hyacinthoides hispanica (Miller) Rothm.
Spanish bluebell
FIRST RECORD: 1977.
REFERENCES: **Kew Gardens.** 1977, *Cope* (unpubl.). • 2002, *Cope* (S35: 649): Around the Orangery (126). Subsequently throughout the Gardens where it easily naturalises from cultivated specimens.
CULT.: 1872, unspec. (K); 1912, unspec. (K); 1926, unspec. (K). In cult.
CURRENT STATUS: Originally cultivated, but now found as a weed throughout the Gardens; hybridisation with *H. non-scripta* is potentially a serious problem.

1841. *Chionodoxa forbesii Baker
C. luciliae auct. non Boiss.
glory-of-the-snow
FIRST RECORD: 1969 [1884].
REFERENCES: **Kew Gardens.** 1983, *Burton* (FLA: 171; S33: 752): ... in Kew Gardens in 1969 it was said to be appearing everywhere, even in newly dug ground.
CULT.: 1884, unspec. (K); 1886, unspec. (K); 1894, unspec. (K); 1896, unspec. (K); 1897, unspec. (K); 1908, unspec. (K); 1947, Rock Garden (K). In cult. under glass.
CURRENT STATUS: It is probable that all plants that have been seen recently were presumed, by earlier recorders, to have been planted and therefore not noted by them. It was planted in a huge drift near White Peaks in 2001 and has naturalised in a number of flower-beds.

1842. Muscari neglectum Guss. ex Ten.
Hyacinthus racemosus L.
grape-hyacinth
CULT.: 1789, Alien (HK1, 1: 459); 1811, Nat. of Britain (HK2, 2: 284); 1814, Nat. of Britain (HKE: 97); 1869, unspec. (K); 1896, unspec. (K); 1931, unspec. (K); 1941, unspec. (K). In cult.
Note. Not known as a British plant until 1805.

1843. *Muscari armeniacum Leichtlin ex Baker
garden grape-hyacinth
FIRST RECORD: 1999 [1901].
REFERENCES: **Kew Gardens.** 1999, *Cope* (S35: 649): Edge of the Herbarium car park between the Paddock and the boundary wall (104).
CULT.: 1901, unspec. (K); 1930, Herbaceous Dept. (K); 1931, unspec. (K); 1941, unspec. (K); 1989, unspec. (K). In cult.
CURRENT STATUS: An escape from cultivation and found on the edge of the Herbarium car park near the boundary wall.

1844. *Muscari latifolium Kirk
FIRST RECORD: 2004 [1901]
CULT.: 1901, unspec. (K). Not currently in cult.
CURRENT STATUS: A contaminant of bulb stock and accidentally introduced into the two Fritillary meadows.

1845. *Camassia cusickii S.Watson
quamash
FIRST RECORD: 2003 [1941].
CULT.: 1941, Rock Garden (K); 1955, Herbaceous Dept. (K). In cult.
CURRENT STATUS: Planted in quantity for naturalisation in grass along the Riverside Walk in 2003 though it is not clear why this species should have been chosen for this purpose.

ALLIACEAE

1846. Allium schoenoprasum L.
chives
CULT.: 1768 (JH: 353); 1789, Alien (HK1, 1: 428); 1811, Nat. of Britain (HK2, 2: 239); 1814, Nat. of Britain (HKE: 91); 1934, unspec. (K); 1940, Herb Garden (K). In cult.

1847. *Allium roseum L.
A. ambiguum Sm.
rosy garlic
FIRST RECORD: 1906 [1789].
NICHOLSON: **1906 revision** (AS: 88): Abundant in turf near Old Palace.
OTHER REFERENCES: **Kew Gardens.** 1949, *Jerrard* (LFS: 352): Towpath nr Isleworth Ferry Gate. • 1958, *Welch* (LFS: 352): Shrubbery by towpath by gate to Kew Palace. • 1987, *Hastings* (LN67: 173): Well naturalised in grassland. **Kew Environs.** 1943, *Cooke* (LN34: S274): By the Thames between Kew and Richmond. No doubt originally thrown out from Kew Gardens.
EXSICC.: **Kew Gardens.** Root from Kew Gardens into which it was introduced from the locality in Kent — specimen in my own garden; 1847; *Watson* in *Herb. Watson* (K). • Near the Office of Works Yard; among grasses; 8 vi 1933; *Hubbard* s.n. (K). • Naturalised

among grasses, by Office of Works Yard; 26 v 1943; *Melville* s.n. (K). **Kew Environs.** Tow path, Kew Gardens; 8 vi 1938; *Perkins* s.n. (K). • Kew; growing on river bank in association with *Juncus* and various grasses; 9 vi 1938; *Rawlings* s.n. (K).
CULT.: 1789, Alien (HK1, 1: 423); 1811, Alien (HK2, 2: 234); 1814, Alien (HKE: 91); 1900, unspec. (K); 1934, unspec. (K); 1936, by Nursery (K); 1950, Herbarium Ground (K). In cult. under glass.
CURRENT STATUS: A common flower-bed weed throughout the Gardens and found in shades of pink and white.

1848. *Allium subhirsutum L.
hairy garlic
FIRST RECORD: 1990 [1768].
REFERENCES: **Kew Gardens.** 1990, *Cope* (S31: 182): Paddock, after restoration in 1990.
CULT.: 1768 (JH: 353); 1789, Alien (HK1, 1: 422); 1811, Alien (HK2, 2: 233); 1814, Alien (HKE: 90). In cult.
CURRENT STATUS: Can still be found in the Paddock but is very scarce and unpredictable; it may go several years without reappearing.

1849. *Allium triquetrum L.
three-cornered garlic
FIRST RECORD: 2002 [1789].
REFERENCES: **Kew Gardens.** 2002, *Cope* (S35: 649): In rough grass between the Bamboo Garden and Mount Pleasant (252).
CULT.: 1789, Alien (HK1, 1: 427); 1811, Alien (HK2, 2: 238); 1814, Alien (HKE: 91). Not currently in cult.
CURRENT STATUS: In flower-beds between the Rhododendron Dell and the Lake; not yet a serious weed but has the potential to become one.

Ramsons (*Allium ursinum*). Widespread and sometimes abundant in open woodland and shaded grassland.

Few-flowered garlic (*Allium paradoxum*). An occasional weed that spreads largely by the bulbils that replace some of the flowers. Sometimes seen en masse in shady places.

1850. *Allium paradoxum (M.Bieb.) G.Don
few-flowered garlic
FIRST RECORD: 1983 [1930].
REFERENCES: **Kew Gardens.** 1983, *Cope* (S31: 183): Near Herbarium.
CULT.: 1930, unspec. (K). Not currently in cult.
CURRENT STATUS: At one time abundant between the Herbarium and Kew Palace but the site was destroyed during construction of an electrical substation. It is otherwise widespread as a weed of flower-beds and is steadily increasing. The earliest observations, unpublished, were in 1975 on the towpath and 1977 in the Gardens.

1851. Allium ursinum L.
ramsons
FIRST RECORD: 1971 [1768].
REFERENCES: **Kew Gardens.** 1983, *Burton* (FLA: 173; S33: 752): ... safely assumed wild ... in the Spinney and Queen's Cottage Grounds.
EXSICC.: **Kew Gardens.** The mound; v 1971; [?*Ross-Craig*] s.n. (K).
CULT.: 1768 (JH: 353); 1789, Nat. of Britain (HK1, 1: 427); 1811, Nat. of Britain (HK2, 2: 238); 1814, Nat. of Britain (HKE: 91). In cult. under glass.
CURRENT STATUS: Widespread and sometimes abundant in open woodland and shaded grassland.

1852. Allium oleraceum L.
field garlic
Cult.: 1768 (JH: 353); 1789, Nat. of England (HK1, 1: 426); 1811, Nat. of England (HK2, 2: 236); 1814, Nat.of England (HKE: 91). In cult. under glass (2006).

[1853. Allium carinatum L.
keeled garlic
Cult.: 1789, Nat. of England (HK1, 1: 424); 1811, Nat. of England (HK2, 2: 234); 1814, Nat. of England (HKE: 91).
Note. Listed in error as native.]

1854. Allium ampeloprasum L.
wild leek
Cult.: 1768 (JH: 353); 1789, Nat. of England (HK1, 1: 421); 1811, Nat. of England (HK2, 2: 232); 1814, Nat. of England (HKE: 90); 1947, Herbarium Ground (K). In cult.
Note. Originally grown as a native but now considered to be an archaeophyte. The specimen from 1947 represents var. *babingtonii* (Borrer) Syme.

1855. Allium scorodoprasum L.
A. arenarium L.
sand leek
Cult.: 1768 (JH: 353); 1789, Nat. of Britain (HK1, 1: 424); 1811, Nat. of Britain (HK2, 2: 234); 1814, Nat. of Britain (HKE: 91). Not currently in cult.

1856. Allium sphaerocephalon L.
round-headed leek
Cult.: 1768 (JH: 353); 1789, Alien (HK1, 1: 424); 1811, Alien (HK2, 2: 235); 1814, Alien (HKE: 91); 1933, unspec. (K); 1934, unspec. (K). In cult.
Note. Not known from mainland Britain until 1847.

1857. Allium vineale L.
wild onion
First record: 1873/4 [1768].
Nicholson: **1873/4 survey** (JB: 74): **Strip.** Common between third and fourth seats from Brentford Ferry. • **1906 revision** (AS: 88): **Strip.** Common in turf by towing-path.
Exsicc.: **Kew Gardens.** [Unlocalised]; on rubbish tip; 6 vi 1939; *Hubbard* s.n. (K). • From site of chalk-garden near the Economic glass-houses; 21 vi 1971; [?*Ross-Craig*] s.n. (K).
Cult.: 1768 (JH: 353); 1789, Nat. of Britain (HK1, 1: 425); 1811, Nat. of Britain (HK2, 2: 236); 1814, Nat. of Britain (HKE: 91); 1946, Herbarium Ground (K). In cult. under glass.
Current status: In many areas of grassland and especially noticeable in those that are less frequently cut.

1858. *Nectaroscordum siculum Lindl.
honey garlic
First record: 2001.
References: **Kew Gardens.** 2001, *Cope* (S35: 649): Naturalised in the Conservation Area (310).
Cult.: In cult. under glass.
Current status: Naturalised in open places in the Conservation Area and sporadic in other parts of the Gardens.

1859. *Tristagma uniflorum (Lindl.) Traub
Ipheion uniflorum (Graham) Raf.
spring starflower
First record: 1991.
References: **Kew Gardens.** 1991, *Cope* (S31: 183): Restored 'Paddock.'
Cult.: In cult.
Current status: Returns every year in the Paddock, but is apparently not increasing; it also occurs near the Palm House at the base of a holly hedge.

Amaryllidaceae

1860. Leucojum aestivum L.
summer snowflake
First record: 2009 [1789].
Cult.: 1789, Nat. of England (HK1, 1: 406); 1811, Nat. of England (HK2, 2: 212); 1814, Nat. of England (HKE: 88). In cult.
Current status: Widely planted, mostly as the cultivar 'Gravetye Giant'. Well naturalised in several places, most notably in the Conservation Area.

1861. *Galanthus nivalis L.
snowdrop
First record: 1936 [1768].
Exsicc.: **Kew Gardens.** From the Wild Garden, Temple Mound; 19 ii 1936; *Sealy* 419 (K). • Wild Garden, Temple Mound; 13 iv 1936; *Sealy* 441 (K).
Cult.: 1768 (JH: 159); 1789, Alien (HK1, 1: 406); 1811, Nat. of England (HK2, 2: 211); 1814, Nat. of England (HKE: 88); 1911, unspec. (K); 1913, unspec. (K); 1932, unspec. (K); 1997, Jodrell Glass (K). In cult.
Current status: Doubtless originally planted but now thoroughly naturalised in many places. Still abundant on the Temple Mound and large drifts can be seen in the Conservation Area.

1862. *Galanthus latifolius Rupr.
G. ikariae Baker
green snowdrop
First record: 2004 [1933].
Cult.: 1933, 1934, 1935, Herbaceous Ground (K); 1934, Rock Garden (K). In cult.
Current status: Naturalised in a few widely scattered places.

Snowdrop (*Galanthus nivalis*). Abundant in the Conservation Area and on Temple Mound, but doubtless originally planted as there are no early records of it.

[1863. **Narcissus × medioluteus** Mill. (poeticus × tazetta)
N. × *biflorus* Curtis
primrose-peerless
CULT.: 1811, Nat. of England (HK2, 2: 214); 1814, Nat. of England (HKE: 88).
Note. Listed in error as native.]

[1864. **Narcissus poeticus** L.
pheasant's-eye daffodil
CULT.: 1789, Alien (HK1, 1: 408); 1811, Nat. of England (HK2, 2: 214); 1814, Nat. of England (HKE: 88).
Note. Listed in error as native.]

1865. *Narcissus × incomparabilis Mill. (poeticus × pseudonarcissus)
nonesuch daffodil
FIRST RECORD: 2004 [1856].
CULT.: 1856, unspec. (K). In cult. under glass.
CURRENT STATUS: Cv. 'Verger' has been planted for naturalisation in drifts along Birdcage Walk and escapes occur in various other parts of the Gardens.

1866. Narcissus pseudonarcissus L.
a. subsp. **pseudonarcissus**
daffodil
FIRST RECORD: 2002 [1768].
REFERENCES: **Kew Gardens.** 2002, *Cope* (S35: 649): Cumberland Mound (166). Originally planted but now thoroughly naturalised and increasing.
CULT.: 1768 (JH: 351); 1789, Nat. of England (HK1, 1: 408); 1811, Nat. of England (HK2, 2: 215); 1814, Nat. of England (HKE: 88); 1950, Herbm. Expt. Ground (K); 1952, Herb. Expt. Grnd. (K). In cult.
CURRENT STATUS: Planted on Cumberland Mound and now thoroughly naturalised. Recently planted in large numbers elsewhere, particularly near the Azalea Garden. Occasional plants that seem to correspond to this subspecies sporadically occur in many parts of the Gardens.

b. subsp. ***obvallaris** (Salisb.) A.Fern.
Tenby daffodil
FIRST RECORD: 2004.
CURRENT STATUS: Planted in large quantities for naturalisation in various parts of the Gardens and sometimes spreading spontaneously.

c. subsp. ***major** (Curtis) Baker
FIRST RECORD: 2004 [1983].
CULT.: In cult.

Daffodil (*Narcissus pseudonarcissus* subsp. *pseudonarcissus*). The native daffodil, but sadly not native in the Gardens. Well established on Temple Mound.

Butcher's-broom (*Ruscus aculeatus*). Widely cultivated in shrubberies, where it often self-seeds, and established as an escape on the towpath.

CURRENT STATUS: Numerous unidentified cultivated daffodils have been planted in, or have escaped to, various parts of the Gardens. No attempt has been made to resolve their identities but most of them are likely to be this subspecies or hybrids thereof. They seem to crop up wherever soil is deposited in any quantity.

1867. *Narcissus cyclamineus DC.

cyclamen-flowered daffodil

FIRST RECORD: 2004.
CURRENT STATUS: Naturalised in a small patch near the Riverside Walk.

1868. *Narcissus cyclamineus DC.
× **pseudonarcissus** L.

FIRST RECORD: 2004.
CURRENT STATUS: Cv. 'February Gold' was planted for naturalisation in drifts along Birdcage Walk and the Broad Walk, and more plantings near the Azalea Garden took place late in 2003. Odd escapes can be seen in various other parts of the Gardens.

ASPARAGACEAE

[1869. Asparagus officinalis L.
subsp. **officinalis**

garden asparagus

CULT.: 1789, Nat. of England (HK1, 1: 451); 1811, Nat. of England (HK2, 2: 273); 1814, Nat. of England (HKE: 96).
Note. Listed in error as native; subsp. *prostratus* (Dumort.) Corb., with its procumbent stems, is native, but plants grown at Kew were described as erect.]

RUSCACEAE

1870. Ruscus aculeatus L.

butcher's-broom

FIRST RECORD: 2001 [1768].
REFERENCES: **Kew Gardens.** 2001, *Cope* (S35: 649): Self-sown from cultivated plants along the Riverside Walk (211). Self-sown in several other places where it has been planted and recently found on the towpath outside the Gardens.
CULT.: 1768 (JH: 424); 1789, Nat. of England (HK1, 3: 418); 1813, Nat. of England (HK2, 5: 420); 1814, Nat. of England (HKE: 314); 1880, Arboretum (K); 1884, unspec. (K); 1934, Arboretum (K); 1971, unspec. (K). In cult.
CURRENT STATUS: This is very widely cultivated in shrubberies but self-seeding near cultivated plants has been observed and one clump has become established on the towpath.

IRIDACEAE

1871. Iris pseudacorus L.

yellow iris

FIRST RECORD: 1873/4 [1768].
NICHOLSON: **1873/4 survey** (JB: 74): Lake and moat. • **1906 revision** (AS:87): Lake and ha-ha.
OTHER REFERENCES: **Kew Gardens.** 1999, *Stones* (EAK: 1 & 5): By Lake and Palm House Pond.
EXSICC.: **Kew Environs.** Ha-ha, Old Deer Park, Richmond; 10 vi 1909; *Fraser* s.n. (K). • Tow Path between Kew and Richmond; 8 vi 1938; *Perkins* s.n. (K).
CULT.: 1768 (JH: 325); 1789, Nat. of Britain (HK1, 1: 71); 1810, Nat. of Britain (HK2, 1: 115); 1814, Nat. of Britain (HKE: 17); 1874, unspec. (K); 1940, Aquatic Garden (K); 1954, Herbaceous Dept. (K); 1964, Jodrell (K); 1965, unspec. (K); 1971, Water Garden (K). In cult.
CURRENT STATUS: In quantity around the Lake but less common in other wet areas. In one or two places it is cultivated but its occurrence around the Lake and on the edge of the golf course in the Old Deer Park are probably natural.

Yellow iris (*Iris pseudacorus*). Although sometimes cultivated, this is likely to be native around the Lake and in the Old Deer Park.

1872. Iris foetidissima L.

stinking iris

CULT.: 1768 (JH: 326); 1789, Nat. of Britain (HK1, 1: 71); 1810, Nat. of Britain (HK2, 1: 116); 1814, Nat. of Britain (HKE: 17). In cult.

1873. Romulea columnae Sebast. & Mauri

sand crocus

CULT.: In cult. under glass.

1874. *Crocus vernus (L.) Hill

spring crocus

FIRST RECORD: 2004 [1789].
CULT.: 1789, Nat. of England (HK1, 1: 56); 1810, Nat. of England (HK2, 1: 80); 1814, Nat. of England (HKE: 12); 1929, nr. Succ. House (K); 1978, unspec. (K); 1979, unspec. (K). In cult.
CURRENT STATUS: Widely scattered and naturalising well. Planted in huge drifts near the Victoria Gate but the population declines after a few years and has to be replenished.

1875. *Crocus tommasinianus Herb.

early crocus

FIRST RECORD: 2004 [1889].
CULT.: 1889, unspec. (K); 1897, unspec. (K); 1937, *Crocus* Wall (K); 1979, unspec. (K). In cult.
CURRENT STATUS: Planted long ago in the grass behind the Herbarium; it is now thoroughly naturalised and has increased markedly over the last two or three years. Recently planted in quantity with other bulbs near the Azalea Garden. Otherwise naturalised in many parts of the Gardens, often in quantity.

1876. *Crocus chrysanthus (Herb.) Herb.

Golden Crocus

FIRST RECORD: 2004 [1895].
CULT.: 1895, unspec. (K); 1897, unspec. (K); 1968, unspec. (K); 1971, unspec. (K); 1976, unspec. (K); 1978, unspec. (K); 1979, unspec. (K). In cult.
CURRENT STATUS: Naturalised in one or two places. It seems to have been planted accidentally among *Chionodoxa* near White Peaks.

Early crocus (*Crocus tommasinianus*). Formerly less common, but recently planted in large drifts for spring colour. Easily naturalises in grass and a population behind the Herbarium has steadily grown over the years.

1877. *Crocus × stellaris Haw. (angustifolius × flavus)
yellow crocus
FIRST RECORD: 2004.
CULT.: In cult.
CURRENT STATUS: Widely naturalised in grass.

1878. *Crocus nudiflorus Sm.
autumn crocus
FIRST RECORD: 2003 [1810].
CULT.: 1810, Nat. of England (HK2, 1: 82); 1814, Nat. of England (HKE: 12); 1947, Rock Garden (K); 1948, Rock Garden (K); 1947, Rock Garden (K). In cult.
CURRENT STATUS: Rarely naturalised in grass.

1879. *Crocus speciosus M.Bieb.
Bieberstein's crocus
FIRST RECORD: 2003 [1968].
CULT.: In cult.
CURRENT STATUS: Commonly naturalised in grass.

[**1880. Crocus sativus** L.
saffron crocus
CULT.: 1789, Nat. of England (HK1, 1: 56); 1810, Nat. of England (HK2, 1: 81); 1814, Nat. of England (HKE: 12).
Note. Listed in error as native.

1881. Gladiolus illyricus W.D.J.Koch
wild gladiolus
CULT.: In cult. under glass but a few have been planted outside.

1882. *Gladiolus communis L.
G. byzantinus Mill.
G. communis subsp. *byzantinus* (Mill.) A.P.Ham.
eastern gladiolus
FIRST RECORD: 2002 [1768].
REFERENCES: **Kew Gardens.** A small clump naturalised in long grass near the Azalea Garden (234).
CULT.: 1768 (JH: 325); 1789, Alien (HK1, 1: 62); 1810, Alien (HK2, 1: 102); 1814, Alien (HKE: 14); 1943, Herbarium Ground (K); 1974, unspec. (K). In cult.
CURRENT STATUS: A small group naturalised in long grass near the Azalea Garden, and recently planted in quantity alongside Princess's Walk.

DIOSCOREACEAE

1883. Tamus communis L.
black bryony
FIRST RECORD: 1873/4 [1768].
NICHOLSON: **1873/4 survey** (JB: 74): **P.** A plant or two at "Old Deer Park" end of Holly Walk. **Q.** Near Cottage. **B:** In a rhododendron bed near Palm House. • **1906 revision** (AS: 88): **A.** A few plants near Pagoda. **Q.** Several plants.
OTHER REFERENCES: **Kew Gardens.** 1993, *Verdcourt* (S31: 183): Var. with ellipsoid instead of spherical fruits. Gardens, hedges between Melon Yard and Jodrell Laboratory, shown to E.Milne-Redhead by I.H.Burkill many years ago and long since destroyed. No other record of such a variety is known.

Black bryony (*Tamus communis*). Three plants are known near Museum No. 1; all are male and are likely to have been planted. It used to be native along Holly Walk near the Pagoda but has been lost.

CULT.: 1768 (JH: 423); 1789, Nat. of England (HK1, 3: 400); 1813, Nat. of England (HK2, 5: 386); 1814, Nat. of Britain (HKE: 309). In cult.
CURRENT STATUS: Three plants occur near Museum No. 1, all of them male and they may have been originally planted. No others have been found recently.

ORCHIDACEAE

1884. Cypripedium calceolus L.
lady's-slipper
CULT.: 1768 (JH: 278); 1789, Nat. of England (HK1, 3: 302); 1813, Nat. of England (HK2, 5: 220); 1814, Nat. of England (HKE: 286). In cult. under glass.

1885. Cephalanthera damasonium (Mill.) Druce
Serapias grandiflora L.
Epipactis pallens Sw.
white helleborine
FIRST RECORD: 1937 [1789].
EXSICC.: **Kew Environs.** Kew Road, in railed grass enclosure opposite Cumberland Gate entrance to Kew Gardens; 20 v 1937; *Oram* s.n. (K).
CULT.: 1789, Nat. of Britain (HK1, 3: 301); 1813, Nat. of Britain (HK2, 5: 202); 1814, Nat. of Britain (HKE: 283). Not currently in cult.
CURRENT STATUS: Long extinct.

1886. Cephalanthera longifolia (L.) Fritsch
Epipactis ensifolia F.W. Schmidt
narrow-leaved helleborine
CULT.: 1813, Nat. of Britain (HK2, 5: 202); 1814, Nat. of Britain (HKE: 283). Not currently in cult.

1887. Cephalanthera rubra (L.) Rich.
Epipactis rubra (L.) F.W.Schmidt
red helleborine
CULT.: 1813, Nat. of Britain (HK2, 5: 202); 1814, Nat. of Britain (HKE: 283). Not currently in cult. Note. Not known as a British plant until 1797.

1888. Epipactis palustris (L.) Crantz
Serapias longifolia Huds.
marsh helleborine
CULT.: 1768 (JH: 348); 1789, Nat. of Britain (HK1, 3: 300); 1813, Nat. of Britain (HK2, 5: 202); 1814, Nat. of Britain (HKE: 283). In cult.

1889. Epipactis purpurata Sm.
violet helleborine
CULT.: In cult. under glass.

Broad-leaved helleborine (*Epipactis helleborine*). Extinct. First recorded as a wild plant in 1887, but the last plant near Victoria Gate was gone by 1906 when the tree under which it grew had to be felled.

1890. Epipactis helleborine (L.) Crantz
Serapias helleborine L.
S. latifolia (L.) Huds.
Epipactis latifolia (L.) All.
broad-leaved helleborine
FIRST RECORD: 1887 [1768].
NICHOLSON: **1906 revision** (AS: 87): Until within the last year or so this species throve under a large beech tree near Victoria Gate; the death of the tree and its consequent removal have resulted in the destruction of the *Epipactis*.
EXSICC.: **Kew Gardens.** Growing wild!; 8 viii 1887; *Nicholson* s.n. (K).
CULT.: 1768 (JH: 348); 1789, Nat. of Britain (HK1, 3: 300); 1813, Nat. of Britain (HK2, 5: 201); 1814, Nat. of Britain (HKE: 283). Not currently in cult.
CURRENT STATUS: Extinct.

1891. Neottia nidus-avis (L.) Rich.
Ophrys nidus-avis L.
bird's-nest orchid
CULT.: 1768 (JH: 347); 1789, Nat. of Britain (HK1, 3: 299). Not currently in cult.

1892. Listera ovata (L.) R.Br.
Ophrys ovata L.
common twayblade
CULT.: 1768 (JH: 347); 1789, Nat. of Britain (HK1, 3: 299); 1813, Nat. of Britain (HK2, 5: 201); 1814, Nat. of Britain (HKE: 283); 1896, unspec. (K). Not currently in cult.

1893. Listera cordata (L.) R.Br.
Ophrys cordata L.
lesser twayblade
CULT.: 1768 (JH: 347); 1813, Nat. of Britain (HK2, 5: 201); 1814, Nat. of Britain (HKE: 283). Not currently in cult.

1894. Spiranthes spiralis (L.) Chevall.
Ophrys spiralis L.
Neottia spiralis (L.) Sw.
autumn lady's-tresses
CULT.: 1768 (JH: 347); 1789, Nat. of Britain (HK1, 3: 299); 1813, Nat. of Britain (HK2, 5: 199); 1814, Nat. of Britain (HKE: 283). Not currently in cult.

1895. Spiranthes aestivalis (Poir.) Rich.
summer lady's-tresses
CULT.: In cult. under glass.
Note. Extinct in the British Isles since 1959. The specimen in cultivation was received in 2001.

1896. Goodyera repens (L.) R.Br.
Satyrium repens L.
creeping lady's-tresses
CULT.: 1789, Nat. of Scotland (HK1, 3: 298); 1813, Nat. of Scotland (HK2, 5: 198); 1814, Nat. of Scotland (HKE: 282). Not currently in cult.

1897. Liparis loeselii (L.) Rich.
Malaxis loeselii (L.) Sw.
fen orchid
CULT.: 1813, Nat. of England (HK2, 5: 208); 1814, Nat. of England (HKE: 284). Not currently in cult.

1898. Hammarbya paludosa (L.) Kuntze
Malaxis paludosa (L.) Sw.
bog orchid
CULT.: 1813, Nat. of England (HK2, 5: 208); 1814, Nat. of England (HKE: 284). Not currently in cult.

1899. Corallorhiza trifida Châtel.
C. innata R.Br.
coralroot orchid
CULT.: 1813, Nat. of Scotland (HK2, 5: 209); 1814, Nat. of Scotland (HKE: 284). Not currently in cult.

1900. Herminium monorchis (L.) R.Br.
Ophrys monorchis L.
musk orchid
CULT.: 1768 (JH: 347); 1789, Nat. of England (HK1, 3: 300); 1813, Nat. of England (HK2, 5: 191); 1814, Nat. of England (HKE: 281). Not currently in cult.

1901. Platanthera chlorantha (Custer) Rchb.
greater butterfly-orchid
CULT.: In cult. under glass.

1902. Platanthera bifolia (L.) Rich.
Orchis bifolia L.
Habenaria bifolia (L.) R.Br.
lesser butterfly-orchid
CULT.: 1768 (JH: 346); 1789, Nat. of Britain (HK1, 3: 295); 1813, Nat. of Britain (HK2, 5: 193); 1814, Nat. of Britain (HKE: 282). Not currently in cult.

1903. Anacamptis pyramidalis (L.) Rich.
Orchis pyramidalis L.
pyramidal orchid
CULT.: 1768 (JH: 346); 1789, Nat. of Britain (HK1, 3: 295); 1813, Nat. of Beitain (HK2, 5: 189); 1814, Nat. of Britain (HKE: 281); 1924, unspec. (K). In cult. under glass.

1904. Pseudorchis albida (L.) Á.&D.Löve
Satyrium albidum L.
Habenaria albida (L.) R.Br.
small-white orchid
CULT.: 1768 (JH: 347); 1813, Nat. of Britain (HK2, 5: 193); 1814, Nat. of Britain (HKE: 282). Not currently in cult.

1905. Gymnadenia conopsea (L.) R.Br.
Orchis conopsea L.
fragrant orchid
CULT.: 1768 (JH: 347); 1789, Nat. of Britain (HK1, 3: 297); 1813, Nat. of Britain (HK2, 5: 191); 1814, Nat. of Britain (HKE: 281). Not currently in cult.

1906. Coeloglossum viride (L.) Hartm.
Satyrium viride L.
Habenaria viridis (L.) R.Br.
frog orchid
CULT.: 1768 (JH: 347); 1789, Nat. of Britain (HK1, 3: 298); 1813, Nat. of Britain (HK2, 5: 192); 1814, Nat. of Britain (HKE: 281). Not currently in cult.

1907. Dactylorhiza fuchsii (Druce) Soó
common spotted-orchid
FIRST RECORD: 2000.
REFERENCES: **Kew Gardens.** 2000, *Cope* (S35: 648): Ancient Meadow (321). This species has made sporadic appearances in the Ancient Meadow over many years but seems to be declining. Micropropagated material has recently been planted in the Conservation Area (310).
CULT.: In cult.
CURRENT STATUS: Once known from the Ancient Meadow, but apparently now extinct. The micropropagated material referred to above is not this species, but the hybrid with *D. praetermissa* (*D.* × *grandis* (Druce) P.F.Hunt). It was recently moved from its original site to a better location near the Brick Pit. Curiously, in 2008, a number of flower-spikes were displaying apical branching, a mild form of fasciation.

1908. Dactylorhiza maculata (L.) Soó
Orchis maculata L.
heath spotted-orchid
CULT.: 1768 (JH: 347); 1789, Nat. of Britain (HK1, 3: 297); 1813, Nat. of Britain (HK2, 5: 190); 1814, Nat. of Britain (HKE: 281). In cult. under glass.

Common spotted-orchid (*Dactylorhiza fuchsii*). Extinct. Last seen in the 'Ancient Meadow' in the mid- or late 1970s.

Green-winged orchid (*Orchis morio*). Extinct. Last recorded, in several localities, over a century ago.

1909. Dactylorhiza incarnata (L.) Soó
Orchis latifolia L.
early marsh-orchid
CULT.: 1768 (JH: 347); 1789, Nat. of Britain (HK1, 3: 296); 1813, Nat. of Britain (HK2, 5: 190); 1814, Nat. of Britain (HKE: 281). Not currently in cult.

1910. Dactylorhiza praetermissa (Druce) Soó
southern marsh-orchid
CULT.: In cult.

1911. Orchis mascula (L.) L.
early-purple orchid
CULT.: 1768 (JH: 346); 1789, Nat. of Britain (HK1, 3: 296); 1813, Nat. of Britain (HK2, 5: 188); 1814, Nat. of Britain (HKE: 281); 1896, unspec. (K). Not currently in cult.

1912. Orchis morio L.
green-winged orchid
FIRST RECORD: 1873/4 [1768].
NICHOLSON: **1873/4 survey** (JB: 74): **B.** A plant cut down by the scythe in the American Garden behind Palm House, 1873. Mr A.Choules. **P.** One from near Palm House end of Pagoda Avenue, also mown down was brought me by p.-c. (police constable) Austin. One midway between "Railway Gate" and end of lake, Mr H.Murton. **Q.** A single plant about 100 yards from Isleworth Gate. • **1906 revision** (AS: 87): **A, B, Q.** A plant or two in each of the divisions named.

CULT.: 1768 (JH: 346); 1789, Nat. of Britain (HK1, 3: 295); 1813, Nat. of Britain (HK2, 5: 188); 1814, Nat. of Britain (HKE: 281). Not currently in cult.
CURRENT STATUS: Extinct.

1913. Orchis ustulata L.
burnt orchid
CULT.: 1768 (JH: 347); 1789, Nat. of England (HK1, 3: 296); 1813, Nat. of England (HK2, 5: 189); 1814, Nat. of England (HKE: 281). Not currently in cult.

1914. Orchis purpurea Huds.
O. fusca Jacq.
lady orchid
CULT.: 1813, Nat. of England (HK2, 5: 189); 1814, Nat. of England (HKE: 281). In cult. under glass.

1915. Orchis militaris L.
military orchid
CULT.: 1768 (JH: 347); 1789, Nat. of England (HK1, 3: 296); 1813, Nat. of England (HK2, 5: 189); 1814, Nat. of England (HKE: 281). Not currently in cult.

1916. Orchis simia Lam.
monkey orchid
CULT.: In cult.

1917. Aceras anthropophorum (L.) W.T.Aiton
Ophrys anthropophora L.
man orchid
CULT.: 1789, Nat. of England (HK1, 3: 300); 1813, Nat. of England (HK2, 5: 191); 1814, Nat. of England (HKE: 281). Not currently in cult.

1918. Himantoglossum hircinum (L.) Spreng.
Satyrium hircinum L.
Orchis hircina (L.) Crantz
Orchis coriophora auct. non L.
lizard orchid
CULT.: 1768 (JH: 347); 1789, Nat. of England (HK1, 3: 298), Nat. of Britain (ibid.: 295 as *coriophora*); 1813, Nat. of England (HK2, 5: 190); 1814, Nat. of England (HKE: 281). In cult. under glass.

1919. Ophrys insectifera L.
O. muscifera Huds.
fly orchid
CULT.: 1768 (JH: 347); 1789, Nat. of England (HK1, 3: 300); 1813, Nat. of England (HK2, 5: 196); 1814, Nat. of England (HKE: 282). Not currently in cult.

1920. Ophrys sphegodes Mill.
O. aranifera Huds.
early spider-orchid
CULT.: 1789, Nat. of England (HK1, 3: 300); 1813, Nat. of England (HK2, 5: 195); 1814, Nat. of England (HKE: 282). In cult. under glass.

1921. Ophrys apifera Huds.
bee orchid
CULT.: 1789, Nat. of England (HK1, 3: 300); 1813, Nat. of England (HK2, 5: 195); 1814, Nat. of England (HKE: 282). In cult. under glass.

Appendix

Notes on citation of the Botanical Exchange Club reports

Citation of the numerous manifestations of the Botanical Exchange Club Reports have proved to be particularly difficult, partly because of frequent changes of title, and partly because of occasional lack of volume or part numbers and errors in pagination. In most cases the fascicles issued were for the previous calendar year, but this was not always so. A means, therefore, had to be devised of citing these reports in the briefest possible way and the method finally chosen is outlined below. Where single quotation marks have been used below, these designate volumes or parts whose numbers have been deduced; these numbers were never officially assigned to the fascicles. These quotation marks have not been used in the body of the Checklist.

Between 1858 and 1865 the Club was entitled 'Thirsk Botanical Exchange Club' and eight unnumbered parts were issued. From 1866 until 1868, it was entitled 'London Botanical Exchange Club' and three parts were issued under this name. From 1869 until 1878 it was simply 'Botanical Exchange Club' and a further seven parts were issued. For convenience, the eighteen parts issued have been informally numbered sequentially and are cited in the Checklist thus: TBEC3:16; LBEC10:5; BEC12:20.

Thirsk Botanical Echange Club (TBEC)

Part		*Pages*		*For year*		*Published*	
	'1'		1–14		1858		1859
	'2'		1–20		1859		1860
	'3'		1–20		1860		1861
	'4'		1–22		1861		1862
	'5'		1–18		1862		1863
	'6'		1–16		1863		1864
	'7'		1–22		1864		1865
	'8'		1–20		1865		1866

London Botanical Exchange Club (LBEC)

Part	Pages	For year	Published
'9'	1–22	1866	1867
'10'	1–18	1867	1868
'11'	1–28	1868	1869

Botanical Exchange Club (BEC)

Part	Pages	For year	Published
'12'	1–24	1869	1870
'13'	1–26	1870	1871
'14'	1–30	1871	1872
'15'	1–46	1872–4	1875
'16'	1–32	1875	1876
'17'	1–40	1876	1878
'18'	1–20	1877–8	1879

The reports of these three forms of the Exchange Club are bound into a single volume in the Kew library and designated 'Thirsk Botanical Exchange Club' on the spine.

The 'Botanical Exchange Club' became the 'Botanical Exchange Club of the British Isles' (BECB) in 1879; a single volume with twenty-one parts was published between 1880 and 1901. This volume was clearly intended to have continuous pagination, but part '18' (for 1897) was numbered from page 533 through to page 580, while the following part (for 1898) was numbered from page 563 through to page 594; subsequent parts continued with the altered numbering. It is therefore partially misleading to cite just the page for this volume, so volume number and informal part number are cited in the Checklist thus: BECB1 (4): 95.

Botanical Exchange Club of the British Isles (BECB)

Volume	Part	Pages	For year	Published
1	'1'	1–24	1879	1880
	'2'	25–42	1880	1881
	'3'	43–82	1881	1882
	'4'	83–100	1883	1885
	'5'	101–120	1884	1885
	'6'	121–142	1885	1886
	'7'	143–166	1886	1887
	'8'	167–196	1887	1888
	'9'	197–242	1888	1889
	'10'	243–280	1889	1890
	'11'	281–322	1890	1891
	'12'	323–350	1891	1892
	'13'	351–396	1892	1893
	'14'	397–430	1893	1894
	'15'	431–464	1894	1895
	'16'	465–506	1895	1897
	'17'	507–532	1896	1898
	'18'	533–580	1897	1898
	'19'	563–594 (sic)	1898	1900
	'20'	595–616	1899	1901
	'21'	617–652	1900	1901

In 1901 the society was renamed the 'Botanical Exchange Club and Society of the British Isles' (BECS) and two volumes, numbered 2 & 3, were published under this name. Volume 2 was clearly meant to have continuous pagination but page numbers inadvertently restarted at 1 in parts 3, 4 and 5. Part 6 resumed numbering where part 5 would have finished had the numbering not restarted. So again it is impossible to cite page numbers in the early parts without indicating the part number. For volume 2 the 10 parts are informally numbered as follows and the reports are cited in the Checklist by volume number and informal part number thus: BECS2(8): 401.

Botanical Exchange Club and Society of the British Isles (BECS)

Volume	Part	Pages		
2	'1'	1–32	1901	1901
	'2'	33–66	1902	1903
	'3'	1–34	1903	1904
	'4'	1–40	1904	1905
	'5'	1–52	1905	1906
	'6'	193–252	1906	1907
	'7'	253–326	1907	1908
	'8'	327–408	1908	1909
	'9'	409–488	1909	1910
	'10'	489–610	1910	1911

Volume 3 was published in six numbered parts with continuous pagination and is not therefore a problem to cite. For the sake of uniformity, it is cited in the Checklist by both volume and part number, even though the latter is redundant.

In the 1915 volume the society had once again been renamed, this time as the 'Botanical Society and Exchange Club of the British Isles' (BSEC). Under this title it published volumes 4–13, each with six numbered parts (except volume 13 which had only four). Each volume has continuous pagination and is therefore straightforward to cite; I have not used part numbers for these volumes in the Checklist.

Other botanical exchange clubs

Two other exchange clubs have also existed alongside the Botanical Exchange Club; these were the 'Botanical Locality Record Club' (later the 'Botanical Record Club') and the 'Watson Botanical Exchange Club.'

The 'Botanical Locality Record Club' reports have only minor problems. They were issued in eleven parts between 1874 and 1887 and were obviously intended to be published in quinquennial volumes. The parts were not formally numbered, but within each volume pagination is continuous. The copy in the Kew library is bound in two volumes, the second comprising the 2nd and 3rd quinequennial volumes, the latter of which was never completed because of the winding up of the club. The word 'Locality' was dropped from the title of the club in volume 2 but the reports are all designated BLRC in the Checklist. In the Checklist the reports are cited by volume number and page thus: 1: 156, 2: 178, 3: 87.

Botanical Locality Record Club/Botanical Record Club (BLRC)

Volume		*Part*		*Pages*		*For year*		*Published*	
	1		'1'		1–32		1873		1874
			'2'		33–88		1874		1875
			'3'		89–152		1875		1876
			'4'		153–192		1876		1877
			'5'		193–306		1877		1878
	2		'1'		1–42		1878		1879
			'2'		43–118		1879		1880
			'3'		119–176		1880		1882
			'4'		177–256		1881–2		1883
	3		'1'		1–78		1883		1884
			'2'		79–156		1884–6		1887

The 'Watson Botanical Exchange Club' issued reports between 1885 and 1934, after which, failing to get a mandate to unite with the Botanical Exchange Club to form the Botanical Society of the British Isles, it closed down. It issued fifty annual reports, designated as such, but no. 22 was also designated 'Vol. 2 part 2.' Thereafter, the reports are not only individually numbered, but also assigned to a volume. In volume 1 the parts each have their own pagination, but from volume 2 onwards the pagination is continuous within the volume. In the Checklist I have used one form of citation, the annual report number, since this is printed unequivocally on the cover of each fascicle. All parts and volumes are as follows:

Ann. Rep. No.		*Volume (part)*		*Pages*		*For years*		*Publ.*	
	1		'1(1)'		1–6		1884–85		1885
	2		'1(2)'		1–16		1885–86		1886
	3		'1(3)'		1–16		1886–87		1887
	4		'1(4)'		1–16		1887–88		1888
	5		'1(5)'		1–8		1888–89		1889
	6		'1(6)'		1–12		1889–90		1890
	7		'1(7)'		1–12		1890–91		1891
	8		'1(2)'		1–16		1891–92		1892
	9		'1(9)'		1–20		1892–93		1893
	10		'1(10)'		1–16		1893–94		1894
	11		'1(11)'		1–16		1894–95		1895
	12		'1(12)'		1–18		1895–96		1896
	13		'1(13)'		1–18		1896–97		1898
	14		'1(14)'		1–24		1897–98		1899
	15		'1(15)'		1–28		1898–99		1899
	16		'1(16)'		1–30		1899–00		1900
	17		'1(17)'		1–40		1900–01		1901
	18		'1(18)'		1–32		1901–02		1902
	19		'1(19)'		1–30		1902–03		1903
	20		'1(20)'		1–20		1903–04		1904

Ann. Rep. No.	*Volume (part)*	*Pages*	*For years*	*Publ.*
21	'2(1)'	1–34	1904–05	1905
22	2(2)	35–70	1905–06	1906
23	2(3)	71–124	1906–07	1907
24	2(4)	125–168	1907–08	1908
25	2(5)	169–212	1908–09	1909
26	2(6)	213–270	1909–10	1910
27	2(7)	271–324	1910–11	1911
28	2(8)	325–372	1911–12	1913
29	2(9)	373–420	1912–13	1914
30	2(10)	421–472	1913–14	1915
31	2(11)	473–520	1914–15	1915
32	2(12)	521–562	1915–16	1916
33	3(1)	1–44	1916–17	1917
34	3(2)	45–84	1917–18	1918
35/36	3(3)	85–130	1918–19 & 19–20	1920
37	3(4)	131–156	1920–21	1921
38	3(5)	157–196	1921–22	1922
39	3(6)	197–236	1922–23	1923
40	3(7)	237–276	1923–24	1924
41	3(8)	277–326	1924–25	1925
42	3(9)	327–362	1925–26	1926
43	3(10)	363–408	1926–27	1927
44	3(11)	409–458	1927–28	1928
45	3(12)	459–504	1928–29	1929
46	4(1)	1–48	1929–30	1930
47	4(2)	49–96	1930–31	1931
48	4(3)	97–156	1931–32	1932
49	4(4)	157–200	1932–33	?1933
50	4(5)	201–248	1933–34	?1934

Index to English names

N.B. Numbers refer to species' numbers in the account, not page numbers.

INDEX TO BOTANICAL NAMES

Synonyms in italic, numbers refer to species' numbers in the account, not page numbers